ENVIRONMENTAL SCIENCE, ENGINEERING AND TECHNOLOGY

ENVIRONMENTAL CHEMISTRY OF ANIMAL MANURE

Environmental Science, Engineering and Technology

Additional books in this series can be found on Nova's website
under the Series tab.

Additional e-books in this series can be found on Nova's website
under the e-books tab.

Animal Science, Issues and Professions

Additional books in this series can be found on Nova's website
under the Series tab.

Additional e-books in this series can be found on Nova's website
under the e-books tab.

ENVIRONMENTAL SCIENCE, ENGINEERING AND TECHNOLOGY

ENVIRONMENTAL CHEMISTRY OF ANIMAL MANURE

ZHONGQI HE
EDITOR

New York

NOTICE TO THE READER

LIBRARY OF CONGRESS CATALOGING-IN-PUBLICATION DATA
Environmental chemistry of animal manure / editor, Zhongqi He.
    p. cm.
  Includes bibliographical references and index.
  ISBN: 978-1-62808-641-6 (softcover)
  1. Agricultural chemistry. 2. Chemistry, Analytic. 3. Farm manure. 4. Environmental chemistry. I. He, Zhongqi.
  S587.E58 2011
  631.8'61--dc22

                        2010051543

Published by Nova Science Publishers, Inc. † New York

CONTENTS

PREFACE

Animal manure is traditionally regarded as a valuable resource of plant nutrients. However, there is an increasing environmental concern associated with animal manure utilization due to high and locally concentrated volumes of manure produced in modern intensified animal production. Although considerable research has been conducted on environmental impacts and best management practices, the environmental chemistry of animal manure has not developed accordingly. Accurate and insightful knowledge of the environmental chemistry of animal manure is needed to effectively utilize animal manure while reducing its adverse environmental impacts. The primary goals of this book are to (1) synthesize and analyze the basic knowledge and latest research on the environmental chemistry of animal manure, (2) stimulate new research ideas and directions in this area, and (3) promote applications of the knowledge derived from basic research in the development and improvement of applied, sustainable manure management strategies in the field. This book will serve as a valuable reference source for university faculty, graduate students, extension specialists, animal and soil scientists, agricultural engineers, and government regulators who work and deal with various aspects of animal manure.

This book consists of four parts. Part I is manure organic matter characterization. Five chapters in this part examine the chemical composition and structural environments of organic matter in animal manure and relevant compost, using pyrolysis-mass spectrometry, infrared spectroscopy, solid state ^{13}C nuclear magnetic resonance spectroscopy, ultraviolet-visible spectroscopy, and fluorescence spectroscopy. Part II is focused on nitrogen and volatile compounds in animal manure. Four chapters in Part II examine ammonia emission from animal manure, key manure odor components, manure amino compounds, and manure nitrogen availability. Part III is manure phosphorus forms and lability. The first four chapters in Part III examine solubility, enzymatic hydrolysis, forms, and metal speciation of manure phosphorus using various wet and instrumental analysis. The last two chapters in Part III then examine the models used in predicting phosphorus transformations and runoff loss for surface-applied manure and reduction of runoff potential of manure phosphorus by alum amendment. Beyond the phosphorus concern, the alum chapter also comprehensively examines the sustainability of animal agriculture by treating manure with alum. Part IV covers heavy elements and environmental concerns. The first chapter in Part IV examines sources and contents of heavy metals and other trace elements in animal manures. Although not heavy metals in strict terms, arsenic and mercury in animal and soil have been frequently investigated with other toxic heavy metals. Thus, the last two chapters in Part IV examine fate and transport of arsenic from organoarsenicals fed to poultry and mercury in animal manure and impacts on environmental health, respectively.

Chapter contribution is by invitation only. Each chapter is designed to cover a specific topic. For each chapter to stand alone, there is occasionally some overlap in literature review, and some experiments have been used as examples in more than one chapter. All 18 chapters in the four parts were written by accomplished experts in the relevant fields, and were subject to the peer reviewing and revision processes. Positive comments from at least two reviewers were required to warrant the acceptance of a manuscript. I would like to thank all reviewers for their many helpful comments and suggestions which certainly improved the quality of this book.

ABOUT THE EDITOR

ZHONGQI HE is Research Chemist of Environmental Chemistry and Biochemistry of Plant Nutrients at the United States Department of Agriculture-Agricultural Research Service, New England Plant, Soil and Water Laboratory, Orono, Maine. He was a recipient of the National Research Council postdoctoral fellowship with the host of the United States Air Force Research Laboratory, Tyndall Air Force Base, Florida. The author or co-author of over 100 research articles, patents, proceedings, and book chapters, he has actively pursued basic and applied research in phosphorus, nitrogen, metals, and natural organic matter. He received the B.S degree (1982) in applied chemistry from Chongqing University, China, the M.S. degrees (1985 and 1992) in applied chemistry from South China University of Technology, Guangzhou, and in chemistry from the University of Georgia, Athens, and the Ph.D. degree (1996) in biochemistry from the University of Georgia, Athens, USA.

PART I. ORGANIC MATTER CHARACTERIZATION

In: Environmental Chemistry of Animal Manure
Editor: Zhongqi He

ISBN: 978-1-62808-641-6
© 2013 Nova Science Publishers, Inc.

Chapter 1

APPLICATION OF ANALYTICAL PYROLYSIS-MASS SPECTROMETRY IN CHARACTERIZATION OF ANIMAL MANURES

Jim J. Wang[1,], Syam K. Dodla[1] and Zhongqi He[2]*

1.1. INTRODUCTION

Analytical pyrolysis-mass spectrometry (Py-MS), principally in the format of pyrolysis-field ionization mass spectrometry (Py-FIMS) or pyrolysis-gas chromatography/mass spectrometry (Py-GC/MS), is a technique capable of providing information on complex organic matter at the molecular level. Unlike C-13 nuclear magnetic resonance (NMR) spectroscopy which provides an average structure of the whole organic material, analytical pyrolysis with mass spectrometry characterizes individual molecular composition through thermal "extraction" (pyrolysis) of the complex organic matter followed by either direct detection by MS or separation through GC then detection by MS. The technique provides a "fingerprint" that can be used to characterize a sample and statistically compare it to others. Besides the use mostly as a qualitative tool, its ability to quantitatively compare samples with similar organic and inorganic matrices makes analytical pyrolysis a powerful tool. Both Py-FIMS and Py-GC/MS have been widely used for the characterization of organic matter of various environmental matrices including aquatic and terrestrial natural organic matter (NOM), microorganisms, soils, and municipal wastes (Meuzelaar et al., 1974; Bracewell and Robertson, 1976; Saiz-Jimenez et al., 1979; Schnitzer and Schulten, 1995; Gonzalez-Vila et al., 1999; White et al., 2004; Leinweber et al., 2009). The major advantages of this technique in organic matter characterization as compared to other traditional techniques are (1) relatively small sample size (usually in the sub milligram range), (2) virtually negligible

---

[*] Corresponding Author: jjwang@agcenter.lsu.edu
[1]School of Plant, Environmental and Soil Sciences, Louisiana State University Agricultural Center, Baton Rouge, LA 70803, USA
[2]USDA-ARS, New England Plant, Soil and Water Laboratory, Orono, ME 04469, USA

sample preparation except for grinding and (3) short analysis time (typically one hour or less). Also, Py-GC/MS is much more affordable as compared to solid state NMR spectroscopy. Though used widely, there have been only limited studies investigating the chemistry of animal manures using Py-FIMS or Py-GC/MS. In this chapter, we review the current literature on the use of analytical pyrolysis in organic manure characterization and present molecular composition data of cattle manure and poultry litter as characterized by Py-GC/MS.

1.2. THE PRINCIPLE OF ANALYTICAL PYROLYSIS

Analytical pyrolysis involves the chemical analysis where non-volatile organic compounds are thermally broken down at high temperature and anoxic conditions for a very short period of time. Following this process, newly formed volatile compounds are either directly detected or separated using gas chromatography followed by detection via flame ionization detector (FID), Fourier transform infrared (FTIR) spectroscopy, or MS. Among all, pyrolysis coupled with FIMS or GC/MS especially the later has been the most popular (White et al., 2004). This is attributable to the fact that MS detection is highly sensitive, specific, and reliable for many organic compounds (Schnitzer and Schulten, 1995). When a mass spectrometer shatters compounds using electron impact, the compound is fragmented in a reproducible way, the ions are separated based on mass/charge ratios, and the result is a spectrum which is both qualitative and quantitative.

The breakdown mechanism of compounds in pyrolysis is a characteristic of initial compounds and resultant low molecular weight chemical moieties compositions are indicative of specific types of macromolecule in the sample analyzed (e.g. lignin, cellulose, chitin etc.) (White et al., 2004). According to Wampler (2007), the breakdown of the compounds that occur during pyrolysis is analogous to the processes that occur during the production of mass spectrum. By applying heat to a sample that is greater than the energy of specific bonds, the molecule will fragment in a reproducible way. The fragments are then separated by the analytical column to produce the chromatogram (pyrogram) which contains both qualitative and quantitative information. The number of peaks, the resolution by capillary GC, and the relative intensities of the peaks permit discrimination among many similar formulations, making Py-GC/MS a powerful tool in the identification of unknown samples (Wampler, 2007). The heating of the sample is often carried out through flash pyrolysis, which employs rapid heating of the samples normally in an inert atmosphere. Two modes of heating, inductive (Curie-point) and resistive (filament), are commonly used in flash pyrolysis. Research has shown little difference between the results of organic material characterization using Curie-point Py-GC/MS and resistive filament Py-GC/MS (Stankiewicz et al., 1998). Besides GC separation, the sample can be pyrolyzed under vacuum directly in the ion source of the mass spectrometer, and the volatile components are identified by soft ionization (field ionization or field desorption) mass spectrometry (Py-FIMS or Py-FDMS). While Py-GC/MS is able to take the advantage of GC separation of various pyrolysis fragments for mass spectrometry, Py-FIMS emphasizes on reduced mass fragments with a wide range of mass coverage.

Analytical pyrolysis has advanced characterization of complex organic matter in many ways. Most conventional methods in identifying or quantifying individual organic compounds require the target chemical be extracted from a solid or liquid matrix. This is often done using a liquid or supercritical fluid extraction. Solvents, particularly basic solutions, can partially oxidize, or otherwise modify the organic matter being studied. In addition, organic molecules can only be identified by conventional GC/MS if they remain volatile in an inert gas stream at $300^{\circ}C$ or less. Most organic matrices in the environment are composed of materials too large to volatilize at $300^{\circ}C$ and cannot be analyzed by traditional GC/MS. However, pyrolysis will thermally extract intact molecules or crack large molecules into fragments that can then be separated and/or directly identified by GC/MS. As such, pyrolysis is an alternative way to "extract" organic matter from complex matrices. The major advantages of Py-GC/MS are requirement of very small sample sizes lower than few milligrams, no requirement of initial processing, reproducible results, faster analysis times, and the ability to provide information about most potential soil organic matter (SOM) precursors such as carbohydrates, lignin, amino acids and lipids (Lehtonen, 2005). Nevertheless, analytical pyrolysis has some limitations from the use of instrumentation to its interpretation (Saiz-Jimenez, C. 1994; Wampler, 2007). In particular, pyrolysis is a destructive technique that fragments organic molecules and, at the same time, can result in side reactions that form new compounds such as ring structures (White et al., 2004). Overall, analytical pyrolysis, especially Py-GC/MS and Py-FIMS, has been considered as one of premiere tools for characterizing complex organic matter (White et al., 2004; Wampler, 2007; Leinweber et al., 2009).

1.3. APPLICATION OF ANALYTICAL PYROLYSIS IN CHARACTERIZING NATURAL ORGANIC MATTER

As early as 60 years ago, Zemany (1952) proposed an approach of using of Py-MS for the analysis of complex organic materials including proteins. Later, Nagar (1963) used Py-GC technique to examine the structure of soil humic acids and emphasized the importance of GC separation. Since then, there has been a great deal of work using analytical pyrolysis to investigate humic substances in soils and sediments and other natural biopolymers (Bracewell and Robertson, 1976; Saiz-Jimenez and De Leeuw, 1986; Hatcher et al., 1988; Abbt-Braun et al., 1989; Hempfling and Schulten, 1990; Fabbri et al., 1996; Stuczynski et al., 1997; Nierop et al., 2001; Chefetz et al., 2002; Buurman et al., 2007). Dignac et al. (2006) suggested that a polar (wax) column was better suited to characterize pyrolysis products originating from less humified OM, such as polysaccharides, proteins; alkanoic acids, and lignin-derived products. By contrast, the use of a non-polar column was more satisfactory to characterize the distribution of aliphatic structures producing alkanes and alkenes upon pyrolysis. Several excellent reviews on the use of analytical pyrolysis for studying organic matter can be found elsewhere (Saiz-Jimenez, 1994; Schnitzer and Schulten, 1995; Leinweber and Schulten, 1999; White et al., 2004; Leinweber et al., 2009). Analytical pyrolysis contributed significantly to the discovery of relationships between organic precursors and soil organic composition as well as between geographic origin and specific SOM constituents/soil functions (Leinweber and Schulten, 1999). In a very recent study of the SOM composition in natural ecosystems under different climatic regions using Py-GC/MS, Vancampenhout et al. (2009) found that

SOM in cold climates still resembled the composition of plant litter as evidenced by high quantities of levosugars and long alkanes relative to N-compounds and there was a clear odd-over-even dominance of the longer alkanes. On the other hand, SOM formed under temperate coniferous forests exhibits accumulation of aromatic and aliphatic moieties, whereas SOM under tropic region is generally characterized by a composition rich in N-compounds and low in lignin without any accumulation of recalcitrant fractions such as aliphatic and aromatic compounds (Vancampenhout et al., 2009). In another study that compared whole soil OM and different humic fractions in soils with contrasting land use based on pyrolysis molecular beam mass spectrometry (Py-MBMS), it was shown that agricultural cultivation generally increases the composition heterogeneity of SOM as compared to native vegetation (Plante et al., 2009). Also recently, a series of chemical parameters based on Py-GC/MS analysis were developed to better describe relations between vegetation shifts and aerobic/anaerobic decomposition of organic matter in peatlands (Schellekens et al., 2009). In a study of humic acids from different coastal wetlands, we also observed an increasing trend in the condensed domain of alkyl C, relatively more stable G-type structural unit of lignin residue, and more contribution of sulfur as a structural component in humic acids along an increasing salinity gradient (Dodla, 2009). Clearly analytical pyrolysis continues to be an important tool for researching soil and biogeochemical processes.

1.4. ANIMAL MANURE CHEMISTRY BY ANALYTICAL PYROLYSIS

There has been a long history of land application of animal manures to agricultural fields as a means of waste disposal and as a soil amendment in many parts of the world. The beneficial use of animal manures has been shown to maintain the SOM status, to increase the levels of plant-available nutrients, and to improve the physical, chemical, and biological soil properties that directly or indirectly affect soil fertility (Eck and Stewart, 1995; Briceño et al., 2007). On the other hand, various studies have demonstrated that animal manure application to agricultural lands may contribute to soil, water and air contamination by emitting and releasing ammonia, greenhouse gases, excess nutrients, pathogens, and odors as well as other substances such as antibiotics (Gerba and Smith, 2005; Kumar et al., 2005, Briceño et al., 2008; Paramasivam et al., 2009; Wang et al., 2010). Chemical composition of animal manure is found to be particularly important in influencing the sorption, mobility and transport of nutrients and contaminants (McGechan and Lewis, 2002; Jorgensen and Jensen, 2009). Recently, research has also focused on possibility of using animal manure as an alternative energy source (Cantrell, 2008; Zhang et al., 2009). All these studies have generated tremendous interest in understanding the organic matter composition and structure of various animal manures (Schnitzer et al., 2007, 2008; Aust et al., 2009). A summary of the various usage of analytical pyrolysis in animal manure characterization is given in Table 1.1.

Table 1.1. Studies of animal manure organic matter (OM) using analytical pyrolysis.

References	Samples	Goals	Technique[a]
Hervas et al., 1989	Humic acids of cow manure	Vermicompost OM Characterization	Py-GC/MS
Saiz-Jimenez et al., 1989	Humic acids of cow manure	Vermicompost OM Characterization	Py-GC/MS
Schnitzer, et al., 1993	Water extracts of four manures and composts	Compost biomaturity	Py-FIMS
Ayuso et al., 1996	Sheep manure	OM characterization	Py-GC/FID
van Bochove et al., 1996	Cow manure	Composting characterization	Py-FIMS
Liang et al., 1996	Water extracts of dairy manure	Dissolved OM characterization	Py-FIMS
Dinel et al., 1998	Pig slurry colloidal fractions	OM characterization	Py-FIMS
Dinel et al., 2001	Organic Extracts of duck manure/wood shaving	Lipids/sterols in composting	Py-GC/MS
Veeken et al., 2001	Pig manure/straw	Composting characterization	Py-GC/MS
Genevini et al., 2002	Humic fractions of pig manure/wheat straw	Composting characterization	Py-GC/MS
Genevini et al., 2003	Humic fractions of pig manure/wheat straw	Composting characterization	Py-GC/MS
Calderon et al., 2006	Dairy and beef manure	Decomposition characterization	Py-GC/MS
Schnitzer et al., 2007	Chicken manure	Biooils production	Py-GC/MS
Schnitzer et al., 2008	Chicken manure	Biooils production	Py-FIMS, Py-FDMS
Aust et al., 2009	Particle fractions of pig slurry	OM characterization	Py-FIMS

[a]Py-GC/MS; pyrolysis-gas chromatography/mass spectrometry; Py-GC/FID, pyrolysis-gas chromatography/field ionization detector; Py-FIMS, pyrolysis-field ionization mass spectrometry; Py-FDMS, pyrolysis-field desorption mass spectrometry.

Previous research on animal manure using analytical pyrolysis focused primarily on exploring OM changes in characteristics during composting animal wastes (Saiz-Jimenez et al., 1989; Van Bochove et al., 1996; Veeken et al., 2001; Calderon et al., 2006). Saiz-Jimenez and coworkers studied the process of vermicomposting cow manures using Py-GC/MS and showed that humic acids extracted from cow manures consisted of lignin and/or lignin residues similar to those grasses; the lignin components of humic acid fractions changed little during vermicomposting (Saiz-Jimenez et al., 1989; Hervas et al., 1989). Genevini et al. (2002, 2003) investigated humification during high rate composting of swine manures amended with wheat straw using Py-GC/MS and reported that alkali-insoluble humin-like substances played an important role by its solubilization in converting to humic acid-like matter. On the other hand, Py-FIMS data showed that dissolved organic matter (DOM) in water extracts from stockpiled and composted cow manures was quite different with phenols and lignin monomers dominating in the composted manure as compared to more N-containing compounds in the stockpiled manure (Liang et al., 1996). Significant changes in lipid composition were also observed during composting, based on Py-GC/MS characterization of chloroform extracts of duck excreta enriched with wood shavings (Dinel et al., 2001). These changes were likely related to the total N content in the system (Dinel et al., 2001).

Besides various extracts, Ayuso et al. (1996) investigated bulk samples of sheep manures during compositing using Py-GC/FID and indicated that although composting stabilized the organic matter, the structure-chemical composition of the compost was more similar to that of the fresh materials than to that of the more evolved materials. On the other hand, different rates of degradation of biomolecules were commonly observed in bulk samples of manure composts. For example, Veeken et al. (2001) showed high initial rates of degradation for aliphatics, hemicelluloses, and proteins but slow degradation rates of lignin during the composting of swine manures based on Py-GC/MS analysis along with solid state ^{13}C NMR characterization. Van Bochove et al. (1996) also examined organic matter changes during four phases (mesophilic, thermophilic, cooling, and maturation) of cow manure composting. Using Py-FIMS, they found that proportion of carbohydrates increased in thermophilic and cooling phases but all identifiable molecules decreased during the maturation phase. In a study of manure decomposition in soil, Py-GC/MS results of four dairy or beef manures in mesh bags buried in soil also showed changes in lignin-derived pyrolyzates but the changes were not consistent across manures, which could be due to the lignin composition of different manures (Calderon et al., 2006). These results suggested a significant influence of manure composition on composting products.

Dinel et al. (1998) characterized the OM distribution in colloidal fractions of pig slurry using Py-FIMS and found that sterols concentrations were relatively high, accounting for 10.1-12.7% of total ion current. The result indicated high propensity of their contribution to the contamination of soils and surface and subsurface waters if these pig manures are applied to agricultural land (Dinel et al., 1998). In a very recent study, Aust et al. (2009) also investigated the relationship between particle size and OM composition in pig slurry using Py-FIMS and showed that sterols were abundant primarily in large-sized fractions (10-2000 μm) but generally less abundant in <10 μm fractions especially < 0.45 μm. On the other hand, steroid profiles of pig slurry were found to be more unique than dairy and poultry manures and could be used as a sterol "fingerprint" to differentiate if a soil sample was once contaminated by pig slurry (Jardé et al. 2007).

Recent characterization of animal manures using analytical pyrolysis has concentrated on exploration of the relationship between manure and its conversion to biofuels and bioproducts such as bio-oil and biochar as chemical compositions of these products are closely related to the nature of biomass wastes including animal manure (Schnitzer et al., 2007; Das et al., 2009). Schnitzer et al. (2007) characterized chicken manure and converted bio-oil fractions (light and heavy) and char through fast pyrolysis using Py-GC/MS. They found that 42% of all compounds identified in the initial chicken manure and 50% of those in the heavy fraction oil were N-heterocycles while aliphatics made up 38% and 44%, respectively. In addition, carbocyclics were also prominent in the initial chicken manure and heavy bio-oil fraction but not in the light bio-oil fraction and char. Using Py-FIMS and FDMS, they showed that sterols were rich in the chicken manure and char followed by light and heavy bio-oil fractions (Schnitzer et al., 2008).

1.5. CASE STUDY I: COMPOUNDS IDENTIFIED IN SELECTED ANIMAL MANURES FROM CONVENTIONAL AND ORGANIC DAIRY FARMS BY PY-GC/MS

Figure 1.1 shows pyrograms of animal manures collected from a conventional dairy farm and an organic dairy farm in Maine, USA. The efficiency of animal and crop production in the conventional and organic dairy farms has been evaluated in past studies (Sundrum, 2001). In general, an organic dairy farm uses feeds produced with little or no inorganic fertilizers, pesticides, and antibiotics/growth promoters as compared to a conventional dairy farm (Sundrum, 2001). However, there has been little research on manure characteristics of these systems even though the feed inputs are usually different. The identified compounds with peaks > 0.1% of total ion intensity are classified into 8 categories: aliphatics, benzenes, carbocylics, carbohydrates, lignin monomers, N-containing compounds, phenols, and sterols (Table 1.2).

Major classes of identified compounds for manures of both conventional and organic dairy farms were lignin monomers (38.2% vs. 35.6%) followed by N-containing compounds (19.9% vs. 16.5%), aliphatics (7.3% vs. 13.3%), carbohydrates (10.1% vs. 5.6%), phenols (4.8% vs. 8.0%), carbocyclics (3.5% vs. 6.9%), benzenes (2.4% vs.1.8%) and sterols (0.4% vs. 0.3%). The overall identified compounds accounted for approximately 86% and 88% of the total ion current (TIC), respectively for the manures of conventional and organic farms. The close percentages in overall identified compounds suggest similar matrix compositions of the two types of manures. The high percentage of identified lignin monomers, an indication of plant source, in both manure samples suggest large quantity of bedding materials such as sawdust shavings being mixed with these manures as well as the presence of undigested forage feeds. Lignin content in cow manure has been shown to range from 12% to 19% depending on diets (Amon, et al. 2007), whereas sawdust typically contains approximately 25% lignin (Stiller et al. 1996).

Major identified lignin monomers included phenol, 2,6-dimethoxy- (L8); 4-methyl-2,5-dimethoxybenzaldehyde (L18); phenol, 4-methoxy- (L1); phenol, 2-methoxy-4-(1-propenyl)-,

Table 1.2. Compounds identified in dairy manure by Pyrolysis GC/MS analysis (Wang et al. unpublished data).

	Compound	Code	RT (min)	Major ions (m/z)
Aliphatics				
1	Acetic acid, heptyl ester	A1	2.23	43, 70
2	4-Octanol, 7-methyl-, acetate	A2	3.43	43, 55, 71
3	1-Octene, 4-methyl-	A3	6.39	43, 55
4	3-Octyne, 2-methyl-	A4	11.83	67, 82
5	2-Hexenoic acid, 3,4,4-trimethyl-5-oxo-,	A5	13.99	110, 95, 67
6	Hexane, 2-chloro-2,5-dimethyl-	A6	14.05	69, 57, 41
7	1,5-Heptadiene-3,4-diol	A7	14.28	71, 43
8	2,6-Dimethyl-1,3,6-heptatriene	A8	14.49	91, 79, 107
9	2,6-Octadien-1-ol, 3,7-dimethyl-, (Z)-	A9	15.09	69, 93, 41
10	Hexadecenoic acid, Z-11-	A10	25.76	55, 41, 69, 84
11	Tetradecene	A11	33.69	41, 57, 69
12	3,7,11,15-Tetramethyl-2-hexadecen-1-ol	A12	35.99	67, 81, 95
13	1-Dodecanol, 3,7,11-trimethyl-	A13	36.11	57, 41, 70
14	Pentadecanoic acid	A14	37.79	74, 43, 87
15	n-Hexadecanoic acid	A15	38.59	43, 41, 57, 73, 129
16	Oleic Acid	A16	41.89	55, 41, 69, 97
17	2-Methyl-Z,Z-3,13-octadecadienol	A17	42.03	55, 67, 41, 81
18	Octadecanoic acid	A18	42.23	43, 60, 73, 129
19	9-Hexacosene	A19	49.17	55, 97, 83, 69, 111
20	Squalene	A20	52.29	69, 81, 41, 95, 121
Benzenes				
21	Toluene	B1	5.51	91, 92
22	Ethylbenzene	B2	8.91	91, 43, 106
23	p-Xylene	B3	9.24	91, 106
24	Styrene	B4	10.09	104, 78
25	1,2-Benzenediol	B5	20.89	110, 64
26	1-Phenanthrenecarboxylic acid, 7-ethenyl	B6	44.73	241, 91, 256
Carbocyclics				
27	Cyclopentene	Cy1	1.91	67
28	1,3-Cyclohexadiene	Cy2	2.77	79
29	Cyclopentene,3-(2-propenyl)-	Cy3	5.05	67, 41
30	1,5-Hexadiene, 2-methyl-	Cy4	5.09	67, 81
31	Cyclohexanol, 2,3-dimethyl-	Cy5	6.61	95, 41
32	2-Cyclopenten-1-one, 2-methyl-	Cy6	10.62	67, 96
33	1,3-Cyclopentanedione	Cy7	11.53	99, 55
34	Cyclohexene, 1-methyl-4-(1-methylethenyl)	Cy8	12.52	67, 79, 93
35	2-Cyclopenten-1-one, 3-methyl-	Cy9	12.83	96, 53, 67
36	1,2-Cyclopentanedione, 3-methyl-	Cy10	14.04	112, 69, 41

	Compound	Code	RT (min)	Major ions (m/z)
37	Cyclopentene, 1-(1-methylethyl)-	Cy11	14.87	67, 95, 41, 118
38	2-Cyclopenten-1-one, 2-hydroxy-3-methyl-	Cy12	15.14	112, 55
39	2-Cyclopenten-1-one, 2,3-dimethyl-	Cy13	15.34	67, 110
40	2-Cyclopenten-1-one, 3-ethyl-2-hydroxy-	Cy14	18.05	126, 55, 83
41	Bicyclo[2.2.1]heptane-1,2-dicarboxylic acid	Cy15	18.91	112, 94, 66
Carbohydrates				
42	4-Penten-1-yl acetate	C1	1.89	43, 68
43	Acetic acid	C2	2.67	43, 60
44	Glyceric acid	C3	3.19	75, 43
45	2-Butanone, 1-(acetyloxy)-	C4	3.63	43, 57
46	Furan, 2,5-dimethyl-	C5	3.78	96, 53, 43
47	3-Furanmethanol	C6	8.81	98, 81, 41
48	Ethanone, 1-(2-furanyl)-	C7	10.81	95, 110
49	Propanoic acid, 2-methyl-, anhydride	C8	14.29	71, 41, 43
50	Maltol	C9	17.87	126, 71
51	Benzofuran, 2,3-dihydro-	C10	21.18	120, 91
52	Levoglucosan	C11	29.38	60, 42
Lignin monomers				
53	Phenol, 4-methoxy	L1	17.09	109, 124, 81
54	2-Methoxy-6-methylphenol	L2	19.86	123, 138
55	Phenol, 2-methoxy-4-methyl-	L3	20.27	138, 123
56	3,4-Dimethoxytoluene	L4	21.63	152, 137, 121
57	O-Methoxy-α methylbenzyl alcohol	L5	22.25	107, 137, 152
58	1,4-Benzenediol, 2-methoxy-	L6	22.36	140, 125, 97
59	Phenol, 4-ethyl-2-methoxy-	L7	22.73	137, 152
60	Phenol, 2,6-dimethoxy-	L8	24.78	154, 139
61	4-Allyl-2-methoxy phenol	L9	24.87	104, 149
62	Phenol, 2-methoxy-4-propyl-	L10	25.13	137, 166
63	Benzaldehyde, 3-hydroxy-4-methoxy-	L11	26.13	151
64	Phenol, 2-methoxy-4-(1-propenyl)-	L12	26.25	164
65	1,2,4-Trimethoxybenzene	L13	27.22	168, 153
66	Phenol, 2-methoxy-4-(1-propenyl)-, (E)-	L14	27.39	164
67	6-Methoxy-3-methylbenzofuran	L15	28.13	147, 162, 91
68	Ethanone, 1-(4-hydroxy-3-methoxyphenyl)-	L16	28.29	151, 166
69	2-Propanone, 1-(4-hydroxy-3-methoxyphenyl)-	L17	29.30	137, 180
70	4-Methyl-2,5-dimethoxybenzaldehyde	L18	30.12	180, 165
71	Phenol, 2,6-dimethoxy-4-(2-propenyl)-	L19	30.98	194, 91
72	1,2-Dimethoxy-4-(2-methoxyethenyl)benzene	L20	31.68	194, 151, 179
73	Phenol, 2,6-dimethoxy-4-(1-propenyl)-	L21	32.05	194, 91
74	Methyl-(2-hydoxy-3-ethoxy-benzyl)ether	L22	32.15	137, 182
75	Benzaldehyde, 4-hydroxy-3,5-dimethoxy-	L23	32.32	182
76	2-Propenoic acid,3-(4-hydroxy-3-methoxyphenyl)	L24	33.24	194, 179
77	Ethanone, 1-(4-hydroxy-3,5-dimethoxyphenyl	L25	33.91	181, 196

Table 1.2. (Continued).

	Compound	Code	RT (min)	Major ions (m/z)
78	4-Hydroxy-2-methoxycinnamaldehyde	L26	34.03	178, 135, 77
N containing compounds				
79	Ethylenediamine	N1	1.57	44, 43, 57
80	Guanidine	N2	1.86	43, 59
81	Pentane, 2-nitro-	N3	2.30	43, 55, 71
82	Acetamidoacetaldehyde	N4	2.36	43, 71
83	Cyanamide, dimethyl-	N5	2.92	41, 71
84	N-tert-Butylethylamine	N6	3.84	86, 58
85	1H-Pyrrole, 1-methyl-	N7	4.62	81, 69
86	Pyridine	N8	4.96	79, 52
87	2-Pentenenitrile, 5-hydroxy-, (E)-	N9	5.04	67, 41
88	4,4-Ethylenedioxy-1-pentylamine	N10	6.22	87, 57
89	1H-Imidazole-4-ethanamine, β-methyl	N11	7.28	95
90	Pyridine, 3-methyl-	N12	7.40	93, 66
91	1H-Tetrazole, 1-methyl-	N13	10.89	55, 84
92	Cyclobutanecarboxylic acid, 1-amino-	N14	10.95	42, 87
93	2-Amino-4-methyl-oxazole	N15	11.41	42, 70
94	Oxazole, 2-ethyl-4,5-dihydro-	N16	11.57	99, 56
95	Oxazolidine, 2,2-diethyl-3-methyl-	N17	14.20	114, 58
96	1-Benzoyl-3-amino-4-cyano-3-pyrroline	N18	16.40	105, 77, 51
97	Indole	N19	23.36	117, 90
98	Phenyl-1,2-diamine, N,4,5-trimethyl-	N20	23.82	150
99	1H-Indole, 4-methyl-	N21	25.86	130
100	α -Amino-3'-hydroxy-4'-methoxyacetophenone	N22	30.61	151
101	10-Formamido-10,11-dihydro-2,3-dimethoxydibenz(b,f) oxepin	N23	40.78	254, 239, 183
102	(6-Isopropyl-3,4-bis(methylamino)-2,4,6-(cycloheptatrienylidine) malanon	N24	42.94	254, 239
103	4-(4-Oxo-1,2,3,4,6,7,12,12b-octahydropyrido [2,1-a]- β carbolin -12b-yl) butanoic acid	N25	45.34	239
Phenols				
104	Phenol	P1	13.54	94, 66
105	Phenol, 2-methyl-	P2	16.02	108, 79
106	Phenol, 4-methyl-	P3	16.76	107, 71
107	Phenol, 2,6-dimethyl-	P4	17.75	122, 107
108	Phenol, 3-ethyl-	P5	18.67	107, 122
109	Phenol, 3,5-dimethyl-	P6	19.00	107, 121
110	Phenol, 2,3-dimethyl-	P7	19.09	107, 121
111	Phenol, 2-ethyl-	P8	19.59	107, 122
112	Phenol, 2,5-dimethyl-	P9	19.70	107, 122
113	2,3-Dimethylhydroquinone	P10	20.13	123, 138
114	1H-Inden-1-one, 2,3-dihydro-	P11	22.98	132, 104, 78
115	Hydroquinone mono-trimethylsilyl ether	P12	29.11	167, 182

	Compound	Code	RT (min)	Major ions (m/z)
116	4-Propyl-1,1'-diphenyl	P13	31.08	167, 196
117	1-Butanone, 1-(2,4,6-trihydroxy-3-methyl phenyl)	P14	34.62	167, 210
Sterols				
118	5α-Cholest-8-en-3-one, 14-methyl-	S1	54.74	57, 43, 215
119	β- Sitosterol acetate	S2	56.08	43, 147, 396

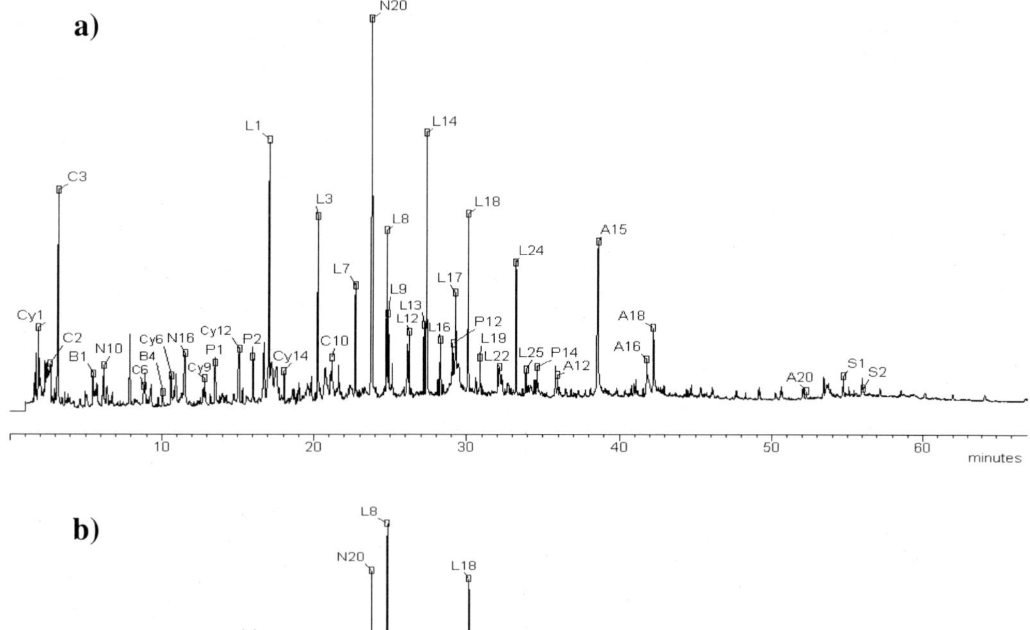

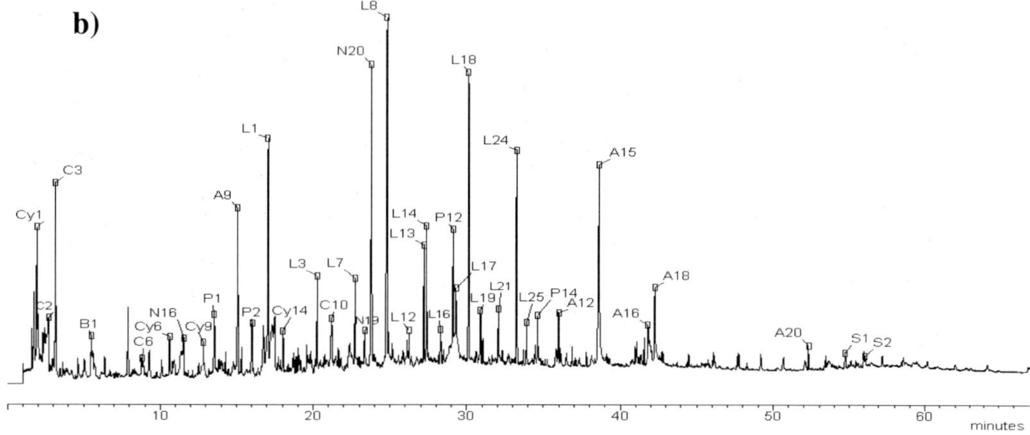

Figure 1.1. Total ion chromatogram obtained from pyrolysis-GC/MS of dairy manure collected from a conventional farm (a) and an organic farm (b) (Wang et al. unpublished data).

(E)-(L14); phenol, 2-methoxy-4-methyl-(L3); 2-propenoic acid, 3-(4-hydroxy-3-methoxyphenyl) (L24); 1,2,4-trimethoxybenzene (L13); phenol, 4-ethyl-2-methoxy- (L7); 2-propanone, 1-(4-hydroxy-3-methoxyphenyl)- (L17); 4-allyl-2-methoxy phenol (L9); phenol, 2,6-dimethoxy-4-(2-propenyl)- (L19); 1,2-dimethoxy-4-(2-methoxyethenyl)benzene (L20); phenol, 2-methoxy-4-(1-propenyl)- (L12); and ethanone, 1-(4-hydroxy-3,5-dimethoxyphenyl (L25) (Table 1.2). Of the two specific manures, organic dairy manure was dominated with lignin monomers derived more from syringyl (L8, L18, L19, L21, L25) structures whereas conventional dairy manure was dominated with those derived more from guaiacyl structures

(L14, L3, L9, L12). The dominance of guaiacyl and syringyl structures indicates that these dairy farm manures contains lignin monomers derived more from woody materials than from grasses as these structures are basic units of woody plant lignin (Hedges and Mann, 1979).

The major identified N-containing compounds were phenyl-1,2-diamine, N,4,5-trimethyl- (N20); indole (N19); oxazole, 2-ethyl-4,5-dihydro- (N16); and 4,4-ethylenedioxy-1-pentylamine (N10). Among these, indole was found in the manure of organic dairy farm but was absent in the manure of conventional dairy farm. This could be an indication of different crude proteins used between the farms since indoles are metabolites of tryptophan amino acid in crude proteins used for feeds (Mackie et al., 1998). On the other hand, some of N-heterocyclics such as pyrroles and pyridines listed in Table 1.2 could be produced by secondary reactions during pyrolysis. Recent studies showed that while the majority of N-heterocyclic's are likely the breakdown units from proteins, it is possible that some could be generated by the Maillard reaction during the pyrolysis (Schnitzer et al., 2007).

The major identified aliphatics included n-hexadecanoic acid (A15); 2,6-octadien-1-ol, 3,7-dimethyl-, (Z)- (A9); octadecanoic acid (A18); oleic Acid (A16); 3,7,11,15-tetramethyl-2-hexadecen-1-ol (A12); and squalene (A20). However, 2,6-octadien-1-ol, 3,7-dimethyl-, (Z)- (or nerol), a monoterpene, was only found in the manure sample from the organic dairy farm.

The major identified carbohydrates were glyceric acid (C3); acetic acid (C2); benzofuran, 2,3-dihydro- (C10); and 3-furanmethanol (C6). The major identified phenols were phenol (P1); phenol, 2-methyl- (P2); hydroquinone mono-trimethylsilyl ether (P12); and 1-butanone, 1-(2,4,6-trihydroxy-3-methyl phenyl) (P14). There was generally little difference in the relative distribution of the major compounds identified in these categories between the two dairy manure samples.

In addition, the major identified carbocyclics included cyclopentene (Cy1); 2-cyclopenten-1-one, 2-hydroxy-3-methyl- (Cy12); 2-cyclopenten-1-one, 3-ethyl-2-hydroxy- (Cy14); 2-cyclopenten-1-one, 2-methyl- (Cy6); and 2-cyclopenten-1-one, 3-methyl- (Cy9). The major identified benzenes were toluene (B1) and styrene (B4), and the major identified sterols were 5α-Cholest-8-en-3-one, 14-methyl- (S1) and β- sitosterol acetate (S2), respectively. There was also little difference in these categories with the exception that the organic dairy manure was higher in cyclopentene than the conventional dairy manure.

Previously, He et al., (2009) comparatively characterized P in organic and conventional dairy manure using solution and solid state ^{31}P NMR spectroscopic techniques. They found that the two types of manure had the same types of P compounds, but the concentrations varied. This Py-GC/MS work analyzed the whole chemical composition of the two types of manure. The observation on the whole chemical composition identified by Py-GC/MS is similar to that of P composition. That is, the chemical composition of the two types of manure is basically identical; however, the relative abundance of individual compounds is affected by the type of manures. For example, the top eight abundant compounds were in the order of N20 > L1 ≈ L14 > C3 > L3 ≈ L18 > A15 > L24 in the conventional dairy manure, but in the order of L8 > N20 > L18 > L1 ≈ L14 > C3 ≈ A15 > A9 in the organic dairy manure (Figure 1.1). Whereas this observation is based on one sample for each type of manure, Py-GC/MS characterization of more dairy manure samples from farms under different management practices is under way. Results from the on-going research should provide more insights on how organic farming impacts the chemical composition of dairy manure.

1.6. Case Study II: Impact of Tetramethylammonium Hydroxide Pretreatment on Pyrolysis-GC/MS Characterization of Chicken Litter

Figure 1.2 shows the Py-GC/MS pyrogram of a chicken litter sample from northern Louisiana that was collected using the same procedure as those dairy manure samples. Clearly, the chicken litter exhibits a rather different pyrogram. Most compounds identified in chicken litter pyrogram were listed in Table 1.2. However, some additional compounds such as acetohydroxamic acid (N33) and cholesta-3,5-diene (S31) were also identified. The pyrogram of the chicken litter sample is dominated by n-hexadecanoic acid (A15) followed by phenyl-1,2-diamine, N,4,5-trimethyl- (N20); phenol, 4-methoxy- (L1); acetohydroxamic acid (N33); guanidine (N2); phenol, 2,6-dimethoxy- (L8); indole (N19); oleic acid (A16); and phenol (P1).

Overall, the chicken litter sample contained less identified lignin monomers, accounting for 18% of TIC as compared to 36-38% for the dairy manures. The identified lignin monomers were dominated with more guaiacyl structures (L1, L7, L3, L14, L24). The chicken manure sample also showed less N-compounds (10.5% of TIC) but slightly more aliphatics (16.2% TIC) as compared to those of dairy manures. On the other hand, about 31% of the total peak areas of the pyrogram were not identified, much higher than the 12-14% unidentified for the dairy manures. This suggests different matrix chemical compositions between chicken litter and dairy manures. Using a Currie-point Py-GC/MS, Schnitzer et al. (2007) reported 43% of the peak area identification for a chicken manure sample that was used for biooil conversion. The difference in identification could be due to variations in chicken manure samples as well as heating modes of pyrolyzers used (resistive filament in this study vs. Curie point) and pyrolysis temperature and duration (620°C for 20 sec vs. 500°C for 10 sec) although previous research had showed no significant differences between Curie point Py-GC/MS and resistive filament Py-GC/MS in characterizing organic materials (Stankiewicz et al., 1998).

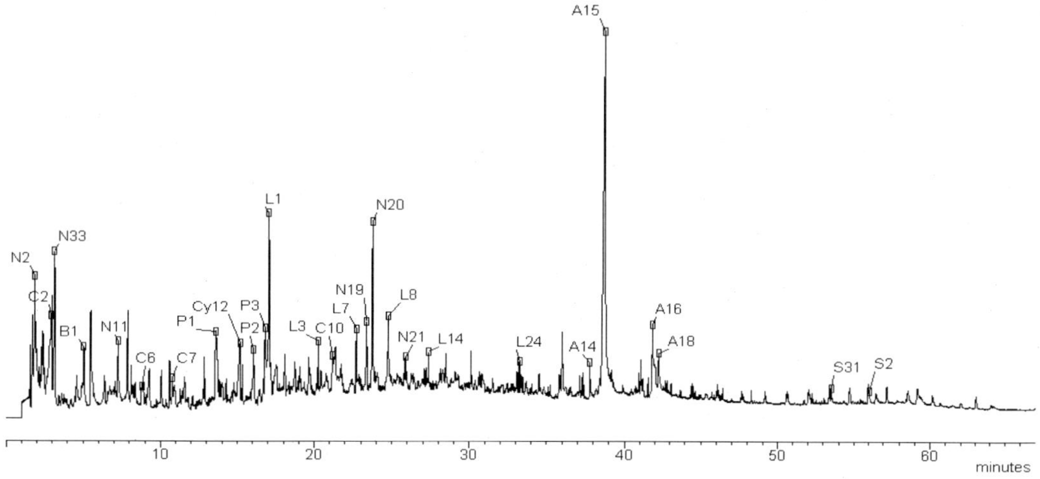

Figure 1.2. Total ion chromatogram of a chicken manure sample obtained from Pyrolysis-GC/MS (Wang et al. unpublished data).

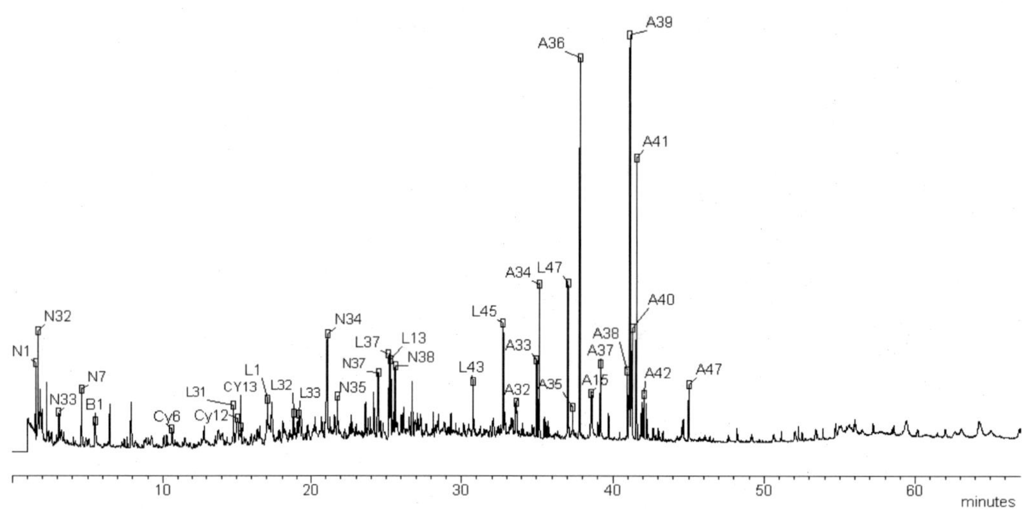

Figure 1.3. Total ion chromatogram of a chicken manure sample treated with TMAH obtained from Pyrolysis-GC/MS (Wang et al. unpublished data).

Figure 1.3 shows Py-GC/MS pyrogram of the same chicken litter sample treated with tetramethylammonium hydroxide (TMAH). Hatcher and Clifford (1994) indicated that highly polar oxygenated compounds such as polyhydric phenols, benzenecarboxylic acids, and hydroxybenzenecarboxylic acids require methylation of hydroxyl and carboxylic functional groups for effective gas chromatographic analysis. Various researchers have also demonstrated that the TMAH as well as trimethylsufonium hydroxide (TMSH) thermochemolysis technique hydrolyzes and methylates esters and ether linkages, assisting depolymerization and methylation, which makes most polar products volatile enough for gas chromatographic analysis (Chafetz et al., 2000; 2002; Kuroda, 2000; Nakanishi et al., 2003). The pyrogram with the TMAH treatment (Figure 1.3) is quite different from that without TMAH treatment (Figure 1.2). Many additional compounds identified are listed in Table 1.3. These additional compounds were dominated with fatty acids as methyl esters. The TMAH treatment greatly improved the identification of aliphatics, accounting for 44% of TIC as compared to only 16% for the chicken litter sample without TMAH treatment. The major identified aliphatic compounds were 9-octadecenoic acid methyl ester (A39); pentadecanoic acid, 14-methyl-, methyl ester (A36); heptadecanoic acid, 16-methyl-, methyl ester (A41); tetradecanoic acid, 12-methyl-, methyl ester (A34); 9-octadecenoic acid (Z)-, methyl ester (A40); pentadecanoic acid, methyl ester (A33); hexadecanoic acid, 14-methyl-, methyl ester (A37); 9,12-octadecadienoic acid (Z,Z)-, methyl ester (A38), and eicosanoic acid, methyl ester (A47). The methylation clearly enhanced signal intensity of oleic acid (A16) and octadecanoic acid (A18) as shown in Figure 1.2 by converting them to 9-octadecenoic acid, methyl ester (A39) and octadecanoic acid, methyl ester (A41), respectively.

Major identified N-containing compounds included quinoline, 6-methyl (N34); propanamide, 2-hydroxy-N-methyl- (N32); ethylenediamine (N1); carbamodithioic acid, diethyl-, methyl ester (N38); L-proline, 1-methyl-5-oxo-, methyl ester(N37); 1H-pyrrole, 1-methyl- (N7); and acetohydroxamic Acid (N33). The overall identified N-containing compounds changed very little with the TMAH treatment, accounting for11.1% of the TIC as compared to approximately 10.5% for untreated chicken litter sample.

Table 1.3. Compounds identified in chicken manure treated with TMAH by Pyrolysis GC/MS analysis (Wang et al. unpublished data).

	Compound	Code	RT (min)	Major ions (m/z)
Aliphatics				
1	Nonanedioic acid, dimethyl ester	A31	29.64	55, 152, 83, 185
2	Tridecanoic acid, 12-methyl-, methyl ester	A32	33.66	74, 87, 143, 199
3	Pentadecanoic acid, methyl ester	A33	35.00	74, 87, 213
4	Tetradecanoic acid, 12-methyl-, methyl ester	A34	35.18	74, 87, 199
5	9-Hexadecenoic acid, methyl ester, (Z)-	A35	37.37	55, 69, 41, 83
6	Pentadecanoic acid, 14-methyl-, methyl ester	A36	37.85	271
7	Hexadecanoic acid, 14-methyl-, methyl ester	A37	39.17	74, 87, 143
8	9,12-Octadecadienoic acid (Z,Z)-, methyl ester	A38	40.97	67, 81, 95
9	9-Octadecenoic acid methyl ester	A39	41.13	264, 74, 81
10	9-Octadecenoic acid (Z)-, methyl ester	A40	41.28	55, 69, 41, 264
11	Octadecanoic acid, methyl ester	A41	41.58	298, 74, 255
12	9,12-Octadecadienoic acid, methyl ester	A42	42.05	67, 81, 95
13	Octadecanoic acid, 10-methyl-, methyl ester	A43	42.23	74, 87, 143, 199
14	Cyclopropaneoctanoic acid, 2-octyl-, methyl ester	A44	43.01	55, 69, 97, 278
15	11-Eicosenoic acid, methyl ester	A45	44.57	55, 81, 292
16	Octadecanoic acid, 10-oxo-, methyl ester	A46	44.64	57, 43, 81, 125
17	Eicosanoic acid, methyl ester	A47	45.00	74, 43, 87, 143
18	Docosanoic acid, methyl ester	A48	48.19	74, 87, 43, 143,354
19	Tetracosanoic acid, methyl ester	A49	51.13	74, 87, 382, 143
20	2,6,10,14,18,22-Tetracosahexaene, 2,6,10,15,19,23 hexamethyl	A50	52.27	69, 81, 41, 121
21	Hexacosanoic acid, methyl ester	51	53.88	74, 87, 410, 143
Benzenes				
22	Benzeneacetic acid, methyl ester	B31	19.76	91, 150
23	Benzenepropanoic acid, methyl ester	B32	22.69	104, 91
Carbocyclics				
24	1-Buten-3-one, 1-(1-acetyl-5,5-dimethylcyclopentyl)-	Cy31	23.95	151, 166
Carbohydrates				
25	Furan, 2-methyl-	C31	2.39	82,53
Lignin monomers/dimmers				
26	Benzene, 1-methoxy-4-methyl-	L31	14.81	122
27	Benzene, 1,2-dimethoxy-	L32	18.78	138, 95, 123
28	Benzene, 1-ethenyl-4-methoxy-	L33	19.17	134, 91, 119
29	Benzene, 1,4-dimethoxy-	L34	19.37	123, 138, 95

Table 1.3. (Continued).

	Compound	Code	RT (min)	Major ions (m/z)
30	1,2,3-Trimethoxybenzene	L35	23.63	168, 153
31	2-Methoxy-4-vinylphenol	L36	23.78	135, 150
32	Benzene, 4-ethenyl-1,2-dimethoxy-	L37	25.19	164, 149, 91
33	Benzoic acid, 4-methoxy-, methyl ester	L38	25.47	135, 166
34	3-(4-Methoxyphenyl)propionic acid	L39	26.93	121, 180
35	Phenol, 3,4-dimethoxy-	L40	26.99	154, 139, 111
36	Benzaldehyde, 3,4-dimethoxy-	L41	28.15	166, 165, 95
37	Benzenepropanoic acid, 4-methoxy-, methyl	L42	29.31	121, 194, 134
38	Benzoic acid, 3,4-dimethoxy-, methyl ester	L43	30.76	165, 196
39	Benzeneacetic acid, 3,4-dimethoxy-, methyl ester	L44	31.49	151, 210
40	2-Propenoic acid, 3-(4-methoxyphenyl)-, methyl ester	L45	32.75	161, 192, 133
41	Benzoic acid, 3,4,5-trimethoxy-, methyl ester	L46	34.71	226, 211, 195
42	2-Propenoic acid, 3-(3,4-dimethoxyphenyl) methyl ester	L47	37.05	222, 191
N containing compounds				
43	Trimethylamine	N31	1.058	58, 59, 42
44	Propanamide, 2-hydroxy-N-methyl-	N32	1.72	58, 60, 45
45	Acetohydroxamic Acid	N33	1.97	43, 75
46	Quinoline, 6-methyl	N34	21.11	143
47	5-Ethyl-5-(1-methyl-3-butenyl)-hexahydropyrimidine-2,4,6-trione	N35	21.76	154
48	1H-Indole, 1-methyl-	N36	22.58	131, 130
49	L-Proline, 1-methyl-5-oxo-, methyl ester	N37	24.47	98
50	Carbamodithioic acid, diethyl-, methyl ester	N38	25.59	116, 163
51	1-[2-Pyridyl]-2,2-dimethyl-2-morpholino	N39	26.54	128
52	2,4(1H,3H)-Pyrimidinedione, 1,3,5-trimethyl	N40	27.34	154, 68
Phenols				
53	Phenol, 2-ethyl-5-methyl-	P31	17.89	121, 136
Sterols				
54	Cholesta-3,5-diene	S31	53.53	368, 147

The major identified lignin monomers were 2-propenoic acid, 3-(3,4-dimethoxyphenyl) methyl ester (L47); 2-propenoic acid, 3-(4-methoxyphenyl)-, methyl ester (L45); benzene, 4-ethenyl-1,2-dimethoxy- (L37); 1,2,4-trimethoxybenzene (L13); benzoic acid, 3,4-dimethoxy-, methyl ester (L43); and phenol, 4-methoxy (L1). There was a significant increase of 1,2,4-trimethoxybenzene (L13), which was absent in the pyrogram of untreated chick litter (Figure 1.2). Previous research has suggested that this compound could be induced from carbohydrates during thermochemolysis with TMAH (Fabbri and Helleur, 1999). On the other hand, the use of TMAH has been shown to improve identification of lignin compounds (Hardell and Nilvebrant 1996; Kuroda, 2000; Chefetz et al., 2002). The use of TMAH could prevent cyclization and aromatization of compounds especially in presence of soil clays (Faure et al., 2006). In this study, some compounds such as 2-propenoic acid,3-(4-hydroxy-3-methoxyphenyl) (L24) identified in the untreated chicken litter displayed an improved signal peak as 2-propenoic acid, 3-(3,4-dimethoxyphenyl) methyl ester (L47) as identified in TMAH-treated chicken litter sample (Figure 1. 3). However, the overall percentage of lignin monomers identified (18.4% TIC) was very similar to that for untreated chicken litter (18.2% of TIC). Past studies have shown the importance of TMAH:sample ratio to improve lignin identification (Hardell and Nilvebrant, 1999; Kuroda et al., 2001; Joll et al., 2003). Nonetheless, the effect of TMAH use on identification of different sample matrices could vary. While a high ratio of TMAH to sample can improve the yield to a certain extent, negative effects of the alkaline TMAH on the GC column and the pyrolysis system should be also taken into account (Joll et al., 2003). Increasing the time of incubation or using sonication has been also shown to compensate for losses of signal yield if less TMAH is applied in sample treatment (Kuroda et al., 2001).

The treatment of TMAH significantly suppressed carbocyclic compounds as well as those derived from carbohydrates. For instance, 3-furanmethanol (C6) and ethanone, 1-(2-furanyl)-(C7) were present in the pyrogram of untreated chicken litter (Figure 1.2) but were absent in the pyrogram of TMAH-treated chicken litter (Figure 1.3). With TMAH, overall carbocyclics was decreased from 6.8% to 0.8% and carbohydrates from 7.6% to 1.6% based on TIC. The underestimation of carbohydrates compounds has been noted by several workers (Clifford et al., 1995; Nierop and Verstraten, 2003). Possible formation of aromatic compounds was also suggested (Fabbri and Helleur, 1999). In addition, the TMAH technique has been shown to be unable to differentiate naturally occurring methyl esters and those formed during the thermochemolysis (Gonzalez-Vila et al., 2001). Clearly, the use of TMAH and similar reagents must be with care and specific purpose in order to benefit the molecular carbon characterization of complex matrix materials including animal manures.

1.7. CONCLUSION

Different analytical pyrolysis techniques have been used to characterize natural organic matter and synthesized organic polymers. The most common ones are pyrolysis followed by direct detection using MS such as in the Py-FIMS technique or pyrolysis followed by GC separation of pyrosates then detected by MS such as in Py-GC/MS.

Major advantages of analytical pyrolysis especially Py-GC/MS include small sample size requirement, little or no sample preparation, faster analysis time, reproducible results, and the ability to provide information about most organic matter precursors. Major limitations of analytical pyrolysis are its destructive nature of fragmenting organic molecules and at same time likely causing side reactions. The latter could be, however, reduced by TMAH thermochemolysis technique. Pyrolysis GC/MS analyses of two dairy manures showed slightly different molecular compositions. Although both were dominated in lignin monomers accounting for approximately 36-38% of the TIC, the manure from an organic dairy farm had more syringyl structures whereas that from a conventional dairy farm had guaiacyl structures, suggesting different origins of materials in the feeds and /or bedding materials mixed with manures. On the other hand, Py-GC/MS of a chicken manure sample showed very different molecular composition from dairy manures. The chicken manure sample contained a greater percentage of aliphatics but had less lignin monomers and N-containing compounds than dairy manure. In addition, the TMAH treatment greatly enhanced the identification of aliphatic compounds of the chicken manure but significantly reduced the signals from carbocyclics and carbohydrate-derived compounds. Nonetheless, these analyses demonstrated that analytical pyrolysis can provide unique molecular composition of organic matter in animal manure.

REFERENCES

Abbt-Braun, G., F.H. Frimmel, and H.R. Schulten. 1989. Structural investigations of aquatic humic substances by pyrolysis-field ionization mass spectrometry and pyrolysis-gas chromatography/mass spectrometry. *Water Res.* 23:1579–1591.

Amon, T., B. Amon, V. Kryvoruchko, W. Zollitsch, K. Mayer, and L. Gruber. 2007. Biogas production from maize and dairy cattle manure—Influence of biomass composition on the methane yield. *Agric. Ecosyst. Environ.* 118:173–182.

Aust, M.O., S.Thiele-Bruhn, K.U. Eckhardt, and P. Leinweber. 2009. Composition of organic matter in particle size fractionated pig slurry. *Bioresour. Technol.* 100:5736–5743.

Ayuso,M., T. Hernández, C. García and J. A. Pascual.1996. Biochemical and chemical-structural characterization of different organic materials used as manures. *Bioresour. Technol.* 57:201–207.

Bracewell, J.M. and G.W. Robertson. 1976. A pyrolysis gas chromatography method for discrimination of soil humus types. *J. Soil Sci.* 27:196–205.

Briceño, G., G. Palma, and N. Duran. 2007. Influence of organic amendment on the biodegradation and movement of pesticides. *Crit. Rev. Environ. Sci. Technol.* 37:233–271.

Briceño, G., R. Demanet, M. de la Luz Mora, and G. Palma. 2008. Effect of liquid cow manure on andisol properties and atrazine adsorption. *J. Environ. Qual.* 37:1519–1526.

Buurman, P., F. Peterse, and G. A. Martin. 2007. Soil organic matter chemistry in allophanic soils: a pyrolysis-GC/MS study of a Costa Rican Andosol catena. *Eur. J. Soil Sci.* 58:1330–1347.

Calderon, F.J., G.W. McCarty, and J.B. Reeves III. 2006. Pyrolysis-MS and FT-IR analysis of fresh and decomposed dairy manure. *J. Anal. Appl. Pyrolysis.* 76:14–23.

Cantrell, K.B., T. Ducey, K.S. Ro and P.G. Hunt. 2008. Livestock waste-to-bioenergy generation opportunities. *Bioresour. Technol.* 99:7941–7953.

Chefetz, B., M.J. Salloum, A.P. Deshmukh, P.G. Hatcher. 2002. Structural components of humic acids as determined by chemical modifications and Carbon-13 NMR, Pyrolysis-, and Thermochemolysis-Gas Chromatography/Mass Spectrometry. *Soil Sci. Am. J.* 66:1159–1171.

Chefetz, B., Y. Chen, C.E. Clapp, and P.G. Hatcher. 2000. Characterizing organic matter in soils by thermocheolysis using tetramethyl ammonium hydroxide (TMAH). *Soil Sci. Soc. Am. J.* 64:583–589.

Clifford, D.J., D.M. Carson, D.E. McKinney, J.M. Bortiatynski, P.G. Hatcher. 1995. A new rapid technique for the characterization of lignin in vascular plants: thermochemolysis with tetramethylammonium hydroxide (TMAH). *Org. Geochem.* 23:169–175.

Das, D.D., M.I. Schnitzer, C.M. Monreal, and P. Mayer. 2009. Chemical composition of acid–base fractions separated from biooil derived by fast pyrolysis of chicken manure. *Bioresour. Technol.* 100:6524–6532.

Dignac, M.F., S. Houot, and S. Derenne. 2006. How the polarity of the separation column may influence the characterization of compost organic matter by pyrolysis-GC/MS. *J. Anal. Appl. Pyrolysis.* 75:128–139.

Dinel, H., M. Schnitzer, H.R. Schulten.1998. Chemical and spectroscopic characterization of colloidal fractions separated from liquid hog manures. *Soil Sci.* 163:665–673.

Dinel, H., M. Schnitzer, T. Paré, L. Lemee, A. Ambles, and S. Lafond. 2001. Changes in lipids and sterols during composing. J. Environ. Sci. Health, Part B. 36:651–665.

Dodla, S.K. 2009. P*hysical and chemical factors controlling carbon gas emissions and organic matter transformation in coastal wetlands.* Ph.D. Dissertation, Louisiana State University, Baton Rouge, USA.

Eck, H.V. and B.A. Stewart. 1995. Manure. *In*: Rechcigl, J.E. (Ed.). *Soil Amendments and Environmental Quality.* pp. 169–198. CRC Press, Inc., Boca Raton, FL.

Fabbri, D., and R. Helleur. 1999. Characterization of the tetramethylammonium hydroxide thermochemolysis products of carbohydrates. *J. Anal. Appl. Pyrolysis* 49:277–293.

Fabbri, D., G. Chiavari, and G.C. Galletti. 1996. Characterization of soil humin by pyrolysis(/methylation)–gas chromatography/mass spectrometry; structural relationships with humic acids. *J. Anal. Appl. Pyrolysis* 37:161–172.

Faure, P., L. Schlepp, L. Mansuy-Huault, M. Elie, E. Jarde and M. Pelletier. 2006. Aromatization of organic matter induced by the presense of clays during flash pyrolysis-gas chromatography-mass spectrometry (Py-GC-MS): A major analytical artifact. *J. Anal. Appl. Pyrolysis* 75:1–10.

Genevini, P.L., F. Adani, A. Veeken, G.J. Nierop, B. Scaglia, and C. Dijkema, 2002. Qualitative modifications of humic acid-like and core-humic acid-like during high-rate composting of pig faeces amended with wheat straw. *Soil Sci. Plant Nutr.* 48:143–150.

Genevini, P.L., F. Tambone, F. Adani, A.H.M. Veeken, K.G.J. Nierop, and E. Montoneri. 2003. Evolution and qualitative modifications of humic-like matter during high rate composting of pig faeces amended with wheat straw. *Soil Sci. Plant Nutr.* 49:785–792

Gerba, C.P, and Smith, J.E. 2005. Sources of pathogenic microorganisms and their fate during land application of wastes. *J. Environ. Qual.* 34:42–48.

Gonzalez-Vila, F.J., G. Almendros, and F. Madrid. 1999; Molecular alterations of organic fractions from urban waste in the course of composting and their further transformation in amended soil. *Sci. Total Environ.* 236:215–229.

Gonzalez-Vila, F.J., U. Lankes, and H.-D. Ludemann. 2001. Comparison of the information gained by pyrolytic techniques and NMR spectroscopy on the structural features of aquatic humic substances. *J. Anal. Appl. Pyrolysis* 58-59: 349–359.

Hardell, H.L. and N.O. Nilvebrant. 1999. A rapid method to discriminate between free and esterified fatty acids by pyrolytic methylation using tetramethylammonium acetate or hydroxide. *J. Anal. Appl. Pyrolysis* 52:1–14.

Hatcher, P., and D.J. Clifford. 1994. Falsg pyrolysis and in situ methylation of humic acids from soil. *Org. Geochem.* 21:1081–1092.

Hatcher, P.G., H.E. Lerch III, R.K. Kotra and T.V. Verheyen. 1988. Pyrolysis G.C.-M.S. of a series of degraded woods and coalified logs that increase in rank from peat to subbituminous coal. *Fuel* 67:1069–1075.

He, Z., C.W. Honeycutt, T.S. Griffin, B.J. Cade-Menun, P.J. Pellechia, and Z. Dou. 2009. Phosphorus forms in conventional and organic dairy manure identified by solution and solid state P-31 NMR spectroscopy. *J. Environ. Qual.* 38:1909–1918.

Hedges, J.I., and D.C.Mann. 1979. The lignin geochemistry of marine sediments from the southern Washington coast. Geochim. Cosmochim. *Acta.* 43:1809–1818.

Hempfling, H., and H.R. Schulten. 1990. Chemical characterization of the organic matter in forest soils by Curie point pyrolysis-GC/MS and pyrolysis-field ionization mass spectrometry. *Organic Geochem.* 15:131–145.

Hervas, L., C. Mazuelos, N. Senesi, and C. Saiz-Jimenez. 1989. Chemical and physico-chemical characterization of vermicomposts and their humic acid fractions. *Sci. Total Environ.* 81:543–550.

Jardé, E. G. Gruau, L. Mansuy-Huault, P. Peu, and J. Martinez. 2007. Using sterols to detect pig slurry contribution to soil organic matter. *Water Air Soil Pollut.* 178:169–178.

Joll, C.A. T. Huynh, and A. Heitz. 2003. Off-line tetramethylammonium hydroxide thermochemolysis of model compound aliphatic and aromatic carboxylic acids: Decarboxylation of some ortho- and/or para-substituted aromatic carboxylic acids. *J. Anal. Appl. Pyrolysis* 70:151–167.

Jorgensen, K. and L.S. Jensen. 2009. Chemical and biochemical variation in animal manure solids separated using different commercial separation technologies. *Bioresour. Technol.* 100:3088–3095.

Kumar, K., S.C. Gupta, S.K. Baidoo, Y. Chander, and C. J. Rosen. 2005. Antibiotic uptake by plants from soil fertilized with animal manure. *J. Environ. Qual.* 34:2082–2085.

Kuroda, K. 2000. Pyrolysis-trimethylsilylation analysis of lignin: preferential formation of cinnamyl alcohol derivatives. J. *Anal. Appl. Pyrolysis* 56:79–87.

Kuroda,K., T. Ozawa, and T. Ueno. 2001. Characterization of Sago Palm (*Metroxylon sagu*) Lignin by Analytical Pyrolysis. *J. Agril. Food Chem.* 49: 1840–1847.

Lehtonen., T. 2005. Molecular composition of aquatic humic substances: analytical pyrolysis and capillary electrophoresis. *M.S. Thesis.* University of Turku, Turku, Finland.

Leinweber, P., and H.R. Schlten. 1999. Advances in analytical pyrolysis of soil organic matter. *J. Anal. Appl. Pyrolysis.* 49:359–383.

Leinweber, P., G. Jandi, K.-U. Eckhardt, A. Schlichting, D. Hofmann, and H.-R. Schulten. 2009. Analytical pyrolysis and soft-ionization mass spectrometry. p. 533-582. In: *Sensi, N., B. Xing, and P.M. Huang (eds.) Biophysico-chemical processes involving natural and non-living organic matter in environmental systems.* John Wiley & Sons, Inc., New York.

Liang, B.C., E.C. Gregorich, M. Schnitzer, and H. Schulten. 1996. Characterization of water extracts of two manures and their adsorption on soils. *Soil Sci. Soc. Am. J.* 60:1758–1763.

Mackie, R.I., P.G. Stroot, and V.H. Varel. 1998. Biochemical identification and biological origin of key odor components in livestock waste. *J. Anim. Sci.* 76:1331–1342.

McGechan, M.B. and D.R. Lewis. 2002. Transport of particulate and colloid-sorbed contaminants through soil. Part I. General principles. *Biosyst. Eng.* 83:255–273.

Meuzelaar, H.L.C., P.G. Kistemaker, and M.A. Posthumus. 1974. Recent advances in pyrolysis mass spectrometry of complex biological materials. *Biomed. Mass Spectrom.* 1:312–319.

Nagar, B.R. 1963. Examination of structure of soil humic acids by pyrolysis-gas chromatography. *Nature* 199:1213–1214.

Nakanishi, O., Y. Ishida, S. Hirao, S. Tsuge, H. Ohtani, J. Urabe, T. Sekino, M. Nakanishi and T. Kimoto. 2003. Highly sensitive determination of lipid components including polyunsaturated fatty acids in individual zooplankters by one-step thermally assisted hydrolysis and methylation-gas chromatography in the presence of trimethylsulfonium hydroxide. *J. Anal. Appl. Pyrolysis* 68-69:187–195.

Nierop, K.G., M. M. Pulleman and J.C.Y. Marinissen. 2001. Management induced organic matter differentiation in grassland and arable soil: a study using pyrolysis techniques. *Soil Biol. Biochem.* 33:755–764.

Nierop, K.G.J., and J.M. Verstraten. 2003. Organic matter formation in sandy subsurface horizons of Dutch coastal dunes in relation to soil acidification. *Organic Geochem.* 34:499–513.

Paramasivam, S., K. Jayaraman, T.C. Wilson, A.K. Alva, L.Kelson, and L.B. Jones. 2009. Ammonia volatilization loss from surface applied livestock manure. *J. Environ. Sci. Health Part B.* 44:317–324.

Plante, A.F., K. Magrini-Bair, M. Vigil, and E. A. Paul. 2009. Pyrolysis-molecular beam mass spectrometry to characterize soil organic matter composition in chemically isolated fractions from differing land uses. *Biochemistry* 92:145–161.

Saiz-Jimenez, C. 1994. Analytical pyrolysis of humic substances: pitfalls, limitations, and possible solutions. *Environ. Sci. Technol.* 29:1773–1780.

Saiz-Jimenez, C. and J. De Leeuw. 1986. Chemical characterization of soil organic matter fractions by analytical pyrolysis-gas chromatography-mass spectrometry. *J. Anal. Appl. Pyrolysis* 9:99–119.

Saiz-Jimenez, C., N. Senesi, and J. W. de Leeuw. 1989. Evidence of lignin residues in humic acids isolated from vermicomposts. *J. Anal. Appl. Pyrol.* 15:121–128.

Saiz-Jimenez,C., F. Martin, and A. Cert. 1979. Low boiling-point compounds produced by pyrolysis of fungal melanins and model phenolic polymers. *Soil Biol. Biochem.* 11(3):305–309.

Schellekens, J., P. Buurman, and X.P. Pombal. 2009. Selecting parameters for the environmental interpretation of peat molecular chemistry – A pyrolysis-GC/MS study. *Organic Geochem.* 40:678–691.

Schnitzer, M., and H.R. Schulten. 1995. Analysis of organic matter in soil extracts and whole soils by pyrolysis-mass spectrometry. *Adv. Agron.* 55:67–218.

Schnitzer, M., H. Dinel, S.P. Mathur, H.R. Schulten, and G. Owen. 1993. Determination of compost biometry. III. Evaluation of a colorimetric test by 13C-NMR spectroscopy and pyrolysis-field ionization mass spectrometry. *Biol. Agric. Hortic.* 10:109–123.

Schnitzer, M.I., C.M. Monreal, and G. Jandl 2008. The conversion of chicken manure to biooil by fast pyrolysis III. Analysis of chicken manure, biooils, and char by Py-FIMS and Py-FDMS. *J. Environ. Sci. Health Part B.* 43:81–95.

Schnitzer, M.I., C.M. Monreal, G. Jandl, P. Leinweber and P.B. Fransham. 2007. The conversion of chicken manure to biooil by fast pyrolysis II. Analysis of chicken manure, biooils, and char by curie-point pyrolysis-gas chromatography/mass spectrometry (Cp Py-GC/MS). *J. Environ. Sci. Health Part B.* 42:79–95.

Schulten, I and P. Leinweber. 1991. Influence of long-term fertilization with farmyard manure on soil organic matter: Characteristics of particle-size fractions. *Biol. Fertil. Soils* 12:81–88.

Stankiewicz, B.A., P.F. van Bergen, M.B. Smith, J.F. Carter, D.E.G. Briggs, and R.P. Evershed. 1998. Comparison of the analytical performance of filament and Curie-point pyrolysis devices. *J. Anal. Appl. Pyrolysis* 45:133–151

Stiller, A.H., D.B. Dadyburjor, J. Wann, D. Tian, and J.W. Zondlo. 1996. Co-processing of agricultural and biomass waste with coal. *Fuel Processing Technol.* 49:167–175.

Stuczynski, T. I., G. W. McCarty, J.B. Reeves, and R.J. Wright. 1997. Use of pyrolysis GC/MS for assessing changes in soil organic matter quality. *Soil Sci.* 162:97–105.

Sundrum, A. 2001. Organic livestock farming: A critical review. Livestock Production *Science* 67:207–215.

van Bochove, E., D. Couillard, M. Schnitzer, and H.R. Schulten. 1996. Pyrolysis-field ionization mass spectrometry of the four phases of cow manure composting. *Soil Sci. Soc. Am. J.* 60:1781–1786.

Vancampenhout, K., K. Wouters, B. De Vos, P. Buurman, R. Swennen, and J. Deckers. 2009. Differences in chemical composition of soil organic matter in natural ecosystems from different climatic regions – A pyrolysis–GC/MS study. *Soil Biol. Biochem.* 41:568–579.

Veeken, A.H.M., F. Adani, K.G.J. Nierop, P.A. de Jager, and H.V.M. Hamelers. 2001. Degradation of Biomacromolecules during High-Rate Composting of Wheat Straw–Amended Feces. *J. Environ. Qual.* 30:1675–1684.

Wampler, T.P. 2007. *Applied Pyrolysis Handbook.* New York Marcel Dekker, Inc., New York.

Wang, J.J., H. Zhang, J.L. Schroder, T.K. Udeigwe, Z. Zhang, S.K. Dodla, and M.H. Stietiya. 2010. Reducing potential leaching of phosphorus, heavy metals and fecal coliform from animal wastes using bauxite residues. *Water Air Soil Pollut.* DOI 10.1007/s11270-010-0420-2.

White, D.M., D.S Garland, L. Beyer, and K. Yoshikawa. 2004. Pyrolysis-GC/MS fingerprinting of environmental samples. *J. Anal. Appl. Pyrolysis* 71:107–118.

Zemany, P.D. 1952. Identification of complex organic materials by mass spectrometric analysis of their pyrolysis products. *Anal. Chem.* 24:1709–1713.

Zhang, S.Y., R.Y. Hong, J.P. Cao, and T. Takarada. 2009. Influence of manure types and pyrolysis conditions on the oxidation behavior of manure char. *Bioresour. Technol.* 100:4278–4283.

In: Environmental Chemistry of Animal Manure
Editor: Zhongqi He

ISBN: 978-1-62808-641-6
© 2013 Nova Science Publishers, Inc.

Chapter 2

STRUCTURAL AND BONDING ENVIRONMENTS OF MANURE ORGANIC MATTER DERIVED FROM INFRARED SPECTROSCOPIC STUDIES

Zhongqi He[1], Changwen Du[2] and Jianmin Zhou[2]*

2.1. INTRODUCTION

The structure of natural organic matter can be investigated using various spectroscopic methods. Infrared spectroscopy is a relative simple, yet important, technique (Hay and Myneni, 2007; He et al., 2006b; Mao et al., 2008). Infrared spectra can be obtained, often nondestructively, on samples in all three states of matter-gases, liquids, and solids, although most samples are examined in the solid form for natural organic matter studies. For a given sample, there will usually be various different sampling techniques that can be used in obtaining the spectrum (Perkins, 1993; Du and Zhou, 2009), thus permitting a researcher to choose one that may be dictated by available accessory equipment, personal preference, or the detailed nature of that particular sample (Perkins, 1993).

Infrared spectroscopy, usually in the form of Fourier transform infrared spectroscopy (FTIR), is a technique based on molecular vibrations. There are three types of motions: (i) bond stretching, (ii) bending, and (iii) tensional motions. Internal vibrational modes are usually found in the 400-4000 cm^{-1} infrared range. Several typical vibrations of C-H and oxygen-containing functional groups absorb light in the infrared region, yielding peaks (absorption bands) so that IR spectroscopy is very valuable in the identification of these functional groups and their structural arrangements in natural organic matter and other soil constituents (Hay and Myneni, 2007; Johnston and Aochi, 1996). Thus, an IR spectrum of a sample can be compared to the spectra of known reference materials or to tabulated

* Corresponding Author e-mail: zhongqi.he@ars.usda.gov
[1]USDA-ARS, New England Plant, Soil, and Water Laboratory, Orono, ME 04469, USA
[2]Institute of Soil Science, Chinese Academy of Sciences, Nanjing 210008, China

frequencies from literature (Table 2.1) so that the presence of diagnostic IR bands indicates the occurrence of particular bonding environments (components) in the sample examined. The apparent advantages of this comparative analytic approach are (i) no requirement for detailed understanding of spectroscopy, (ii) amenability to routine analysis by a nonspectroscopist, and (iii) high efficiency of spectral analysis (Johnston and Aochi, 1996).

Like in other environmental samples, organic matter in animal manures and composts has been characterized by infrared spectroscopy (Table 2.2). In this chapter, we review and discuss the structural and bonding environments of animal manure and their changes under different management practices derived from infrared spectroscopic studies. Recently, Fourier transform infrared photoacoustic spectroscopy (FTIR-PAS) has been applied in soil analyses (Du et al., 2007; 2008). In this chapter, we also present FTIR-PAS spectral data of three types of animal manure to show that this technique can also be used to characterize organic matter in animal manure.

2.2. SPECTRAL FEATURES OF ORGANIC MATTER IN ANIMAL MANURE

2.2.1. General Spectral Features

The comparative spectra from swine manure and sandy loam soil samples are shown in Figure 2.1 (He et al., 2003). The broad band at 3400 cm^{-1} is attributed to O-H and N-H stretching, and the bands at 2920 and 2856 cm^{-1} are attributed to aliphatic C-H stretching. The peaks from 1720 to 1510 cm^{-1} reflect stretching C=O (carboxylic acids, ketonic carbonyls), stretching C=C (phenyl-conjugated), stretching C=N, deforming N-H, and ring vibration of ortho-substituted aromatic compounds. The peaks around the 1400 cm^{-1} region are attributed to C-H deformation of aliphatic groups, O-H deformation and C-O stretching of phenolic O-H. Peaks around 1162-1018 cm^{-1} may be partly due to stretching C-O of polysaccharides, O-Al-OH, O-Fe-OH, Si-O, and P-O groups. The soil shows a distinct FTIR spectrum with a strong peak at 1028 cm^{-1} (Figure 2.1). Peaks around 1162-1018 cm^{-1} may be partly due to C-O of polysaccharides, O-Al-OH, O-Fe-OH, Si-O, and P-O groups (Francioso et al., 1998). However, the minor shoulders at 2929 and 2851 cm^{-1} (aliphatic) implied that organic matter was not the major contributor of the peak around 1028 cm^{-1}, which indicates the presence of a large amount of inorganic oxides and a relatively small amount of organic compounds in the sandy loam soil. On the other hand, the FTIR spectra of the swine manure showed strong peaks at 2920 and 2851 cm^{-1}, indicating that the matrix of swine manure is aliphatic. In addition, The FTIR spectrum of the swine manure is distinguished by a strong absorption band at 1650 cm^{-1}, a moderately strong band in the 1540 cm^{-1} region, and a strong band in the 1050 cm^{-1} region. Indeed, these features are general for animal manure as they have been also observed in the FTIR spectra of the solid fractions of cattle and swine slurries (Hsu and Lo, 1999; Inbar et al., 1989), poultry manure (Hachicha et al., 2009; Schnitzer et al., 2007), and dairy manure (Calderon et al., 2006).

Table 2.1.Typical bonding structures of natural organic matter identified by infrared spectroscopy [adapted from Johnston and Aochi (1996) and Tan (2003)].

Band range (cm^{-1})	Functional group	Vibration mode
3500-3200	Carboxylic acids, phenol, alcohols, amines, amides	O-H stretch, N-H stretch
3150-3000	Aromatic	C-H stretch
2970-2820	Aliphatic	C-H stretch
1725-1720	COOH groups	C=O stretch
1650-1630	Amide (I), aromatic, double bond conjugated with carbonyl, COO^- groups	C=O stretch, other vibrations
1650-1540	COO- groups	Asymmetric COO^- stretching
1450-1360	COO- groups	Symmetric COO^- stretching
1465-1440	CH2, CH3 groups	C-H bend
1250-1200	COOH groups	C-O stretch, O-H bend
1170-950	Polysaccharide	C-O stretch

2.2.2. Spectra Type

Stevenson and Goh (1971) classified the infrared spectra of humic acid (HA) and related substances to three types. Those belonging to Type I show strong bands at 3400, 2900, 1720, 1600 and 1200 cm^{-1}, with no discernible absorption being evident in the 1640 cm^{-1} region. The 1600 cm^{-1} band is about equal in intensity to the one at 1720 cm^{-1}. Spectra of Type II are characterized by a very strong 1720 cm^{-1} band, a shoulder at 1650 cm^{-1} or so, and the absence of a 1600 cm^{-1} band. Type III spectra have spectral feature similar to Type I, but show additional relatively strong bands near 1549 and 1050 cm^{-1}. Absorption between 2900 and 2840 cm^{-1} is also more pronounced. These features of Type III spectra are indicative of proteins and carbohydrates (Stevenson and Goh, 1971). Inbar et al. (1989) and Hsu and Lo (1999) observed the FTIR spectra of the solid fractions of cattle and swine slurries resemble the characteristics of Type III spectra of humic acids (Table 2.2). We further analyzed the spectral feature of other animal manures and found most of them resembling Type III spectra proposed by Stevenson and Goh (1971). It seems that the classification is not limited to humic substances, rather applicable to general natural organic matter.

2.2.3. Unique Characteristics of Animal Manure

Stevenson and Goh (1971) proposed that the humification process consisted, in part, of a loss of COOH groups and a change in the environment of C=O from the free or weakly H-bonded state to strongly chelated forms, and reflected in the spectral changes in the 1700-1600 cm^{-1}. Compared to Type III spectra of soil humic acids in literature (Stevenson and Goh, 1971), spectra of animal manure show a smaller or even no absorbance peak/shoulder at 1720 cm^{-1}. Whereas humic acids in Type III spectra are supposed to be recently formed from parent organic matter (Inbar et al., 1989), we proposed the weak absorption at 1720 cm^{-1} is an indication of more neutral (i.e. less humic acidic) organic matter in animal manure.

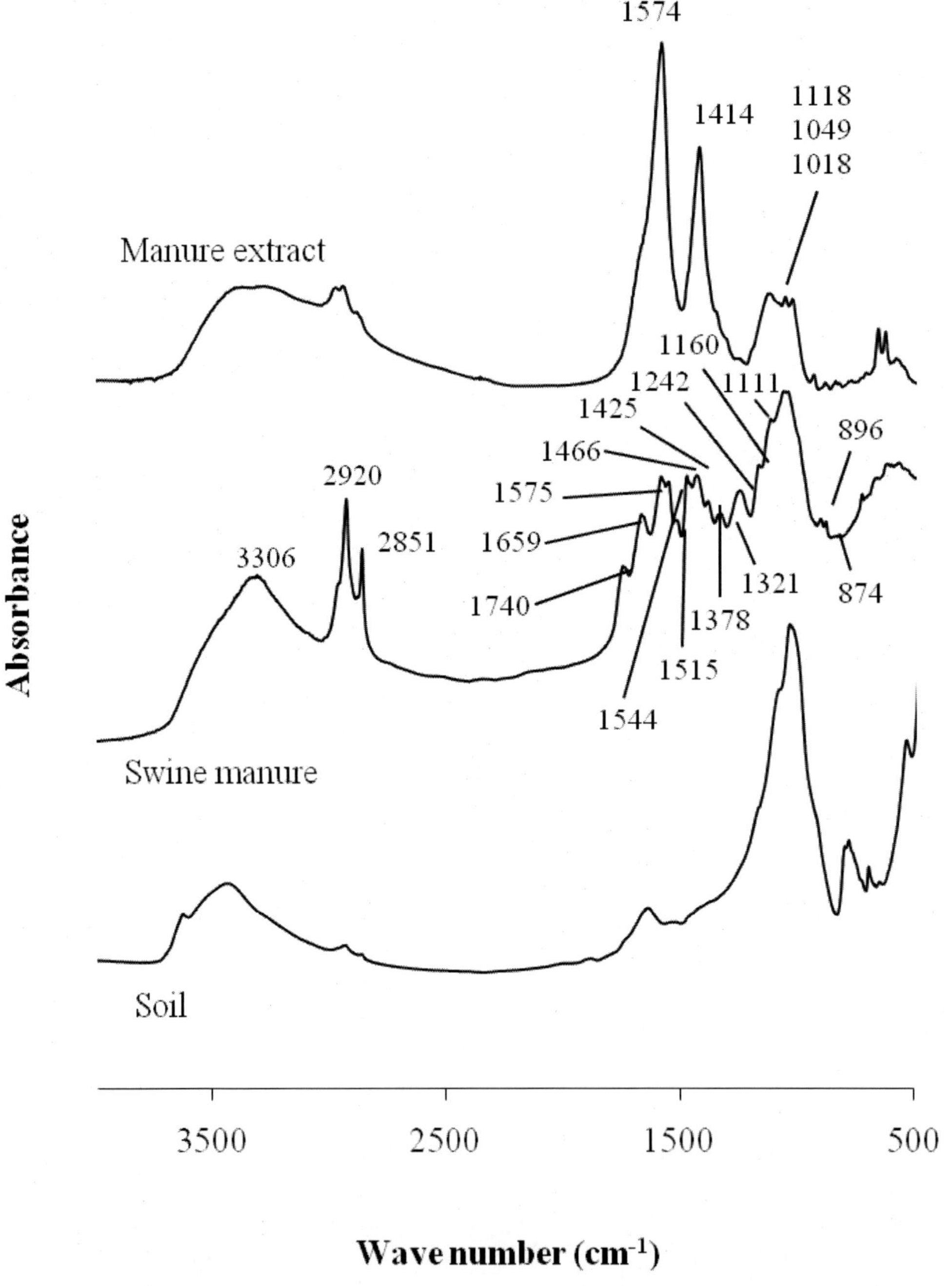

Figure 2.1. FT-IR spectra of soil and swine manure and freeze-dried water extracts of swine manure [adapted from He et al. (2003)].

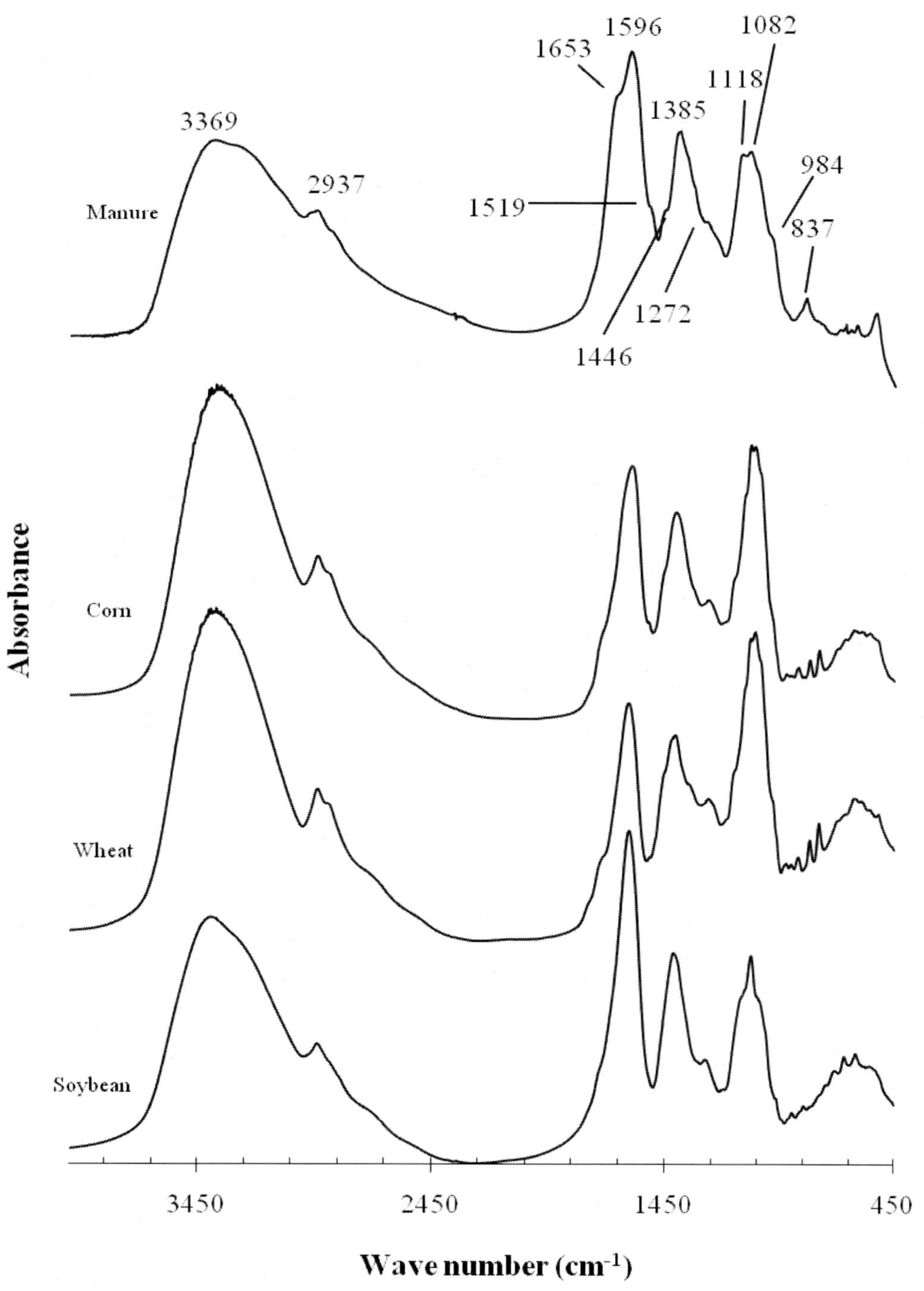

Figure 2.2. FT-IR spectra of freeze-dried water extracts of plant shoots and dairy manure [adapted from He et al. (2009)].

Table 2.2. Studies of animal manure organic matter (OM) using infrared spectroscopy

References	Samples	Goals	Spectrum type[a]
Tan et al., 1975	Water extracts of poultry litter	Adsorption with soil	III
Inbar et al., 1989	Separated (solid) cattle manure	Composting characterization	III
Dinel et al., 1998	Pig slurry colloidal fractions	OM characterization	III
Giusquiani et al., 1998	Pig slurry liquid fractions	Dissolved OM characterization	II, III
Hsu and Lo, 1999	Separated (solid) pig manure	Composting characterization	III
Plaza et al., 2002	Humic fractions of pig slurry	Soil amendment with manure	III
He et al., 2003	Swine manure and its water extracts	Comparative characterization	III
Plaza et al., 2003	Fulvic acids of pig slurry	Soil amendment with manure	II
Calderon et al., 2006	Dairy manure	Aerobic decomposition	III
Hernandez et al., 2006	Humic fractions of pig slurry	Metal-humic binding	III
Huang et al., 2006	Humic fractions of pig manure	Co-composting mechanism	III
Chien et al., 2007	Humic fractions of pig manure	Composting characterization	III
Hernadez et al., 2007	Fulvic acids of pig slurry	Metal-humic binding	II
Schnitzer et al., 2007	Chicken manure	Biooils production	III
Senesi et al., 2007	Humic fractions of pig slurry	Comparative reviews	III
Zmora-Nahum et al., 2007	Various manure composts	Comparative characterization	III
Carballo et al., 2008	Compost extracts of cattle manure	Composting characterization	III
Gibert et al., 2008	Sheep manure/lime stone	Acid mine drainage treatment	N/A
Plaza et al., 2008	Humic fractions of cattle manure	Vermicomposting	III
Hachicha et al., 2009	Poultry manure/olive mill sludge	Co-composting mechanism	III
He et al., 2009	Water extracts of dairy manure	Dissolved OM characterization	N/A
Marcato et al., 2009	Pig slurry	Anaerobic digestion	N/A

a: Spectrum types are assigned by the authors of this chapter based on Stevenson and Goh (1971) and Tan (2003). Type I spectra show equally strong bands at 1720 cm^{-1} and 1600 cm^{-1} with no discernible absorption being evident at 1640 cm^{-1}. Type spectra show a very strong 1720 cm^{-1} band, a shoulder at 1650 cm^{-1} and the absence of a 1600 cm^{-1} band. By definition, Type III spectra are similar to Type I with additional strong bands between 2900 and 2840 cm^{-1}. However, the absorbance at 1720 cm^{-1} is very weak or not observed in Type III spectra assigned to some manure samples in this table. N/A, not applicable to any of the three types.

Hsu and Lo (1999) reported that the spectra of pig manure differed from spectra of hardwood bark, pine bark, winery solid waste, and municipal solid waste composts (Niemeyer et al., 1992). The authors (Hsu and Lo, 1999) found four main differences between the spectra of the separated pig manure/compost and those of other types of composts: (i) the aromatic and polysaccharide peaks had become the main peak in the spectra of pig manure, bark, and winery waste composts as opposed to the 1450 cm^{-1} peak in the municipal solid waste spectra; (ii) a narrow sharp peak at 1800 cm^{-1} in municipal solid waste spectra that appears as a small peak at 1710 to 1720 cm^{-1} in bark and winery waste compost spectra, does not appear in the pig manure spectra; (iii) a sharp peak at 2520 cm^{-1}, which can be attributed to S-H stretch of aromatic or nonaromatic mercaptans and sulfides in the municipal solid waste spectra, does not appear in the pig manure spectra; and (iv) an intense and sharp peak at 870 cm^{-1} in the municipal solid waste spectra that does not appear in spectra of pig manure and other compost.

It is worth discussing the assignment of the peak at 1050 cm^{-1} in manure spectrum (Fig. 1) as many inorganic and organic phosphate compounds have absorbance in this region (He et al., 2006a; 2007). Bands in this region have been originally assigned to alcoholic and polysaccharide C-O stretching of humic substances (Stevenson and Goh, 1971). On the other hand, both silicate and phosphate compounds also have absorbance in this region (Johnston and Aochi, 1996). Thus, in addition to the contribution from alcoholic and polysaccharide, absorption bands in this region have also been assigned to vibrations of a SiO_2-related impurity in humic substance (Agnelli et al., 2000; Olk et al., 2000; Tan, 2003) and manure (Hsu and Lo, 1999; Inbar et al., 1989), however, rarely to indigenous phosphate groups in a sample. In a few cases, Francioso et al. (1998) attributed absorbance band in this region mainly to phosphate groups as a result of the high concentration of total P in their humic acids extracted by NaOH plus pyrophosphate although they could not exclude that the band in this region may arise from C-O stretching or an inorganic impurity. He et al. (2006b) observed that the intensity of this band changed from a shoulder or very weak shoulder band to a minor band in six humic and fulvic samples, followed the same increasing order of P content rather than of Si content in the six samples, implying a correlation between the absorbance in this region and P content. Thus, it was reasonable to assign the absorbance band at 1100-1000 cm^{-1} to phosphate groups in the humic substances (He et al., 2006b). As the level of P in animal manure is high, we proposed that the FTIR band of animal manure in this region is likely due to alcoholic, polysaccharide, inorganic and organic phosphate compounds. The contribution of Si-O should be not significant as Si content in animal manure is low (He et al., 2003).

2.3. Spectral Features of Water Extractable Organic Matter (WEOM) in Animal Manure

Tan et al. (1975) reported the infrared spectrum of the freeze-dried water extract of poultry litter. The spectral features resemble those of Type III of humic substances, as it is characterized by bands at 2920 cm^{-1} and strong bands at 1620 cm^{-1} and 1400 cm^{-1} for aliphatic C-H and carboxyl stretching vibration, respectively. Strong bands are also present in the 1130-1050 cm^{-1} region due to polysaccharides. Based on the spectral characteristics and elemental analysis, Tan et al. (1975) concluded that the dominating compound in the water extract of their poultry litter samples was polysaccharide-like materials originating probably as microbial decomposition products of the wood shavings (bedding materials), feed residues, or both, although the extract contained nitrogenous compounds and other substances.

Compared to that of the whole manure, the spectrum of the freeze-dried water extract of swine manure shows remarkable absorbance bands at 1574 cm^{-1}and 1414 cm^{-1} (Figure 2.1) (He et al., 2003) and does not fall into any of the three types of spectra. The band at 1574 cm^{-1} could be assigned to secondary amide (amide II, N-H bending vibrations) (Schnitzer et al., 2007), and the band at 1414 cm^{-1} to C-H deformation of aliphatic groups, O-H deformation, and C-O stretching of phenolic O-H (Giusquiani et al., 1998). The multiple bands at 1118, 1048, and 1018 cm^{-1} indicate the presence of different phosphate forms in addition to polysaccharides. This assignment is based on the high P:Si mole ratio (60:1) in the swine manure extract (He et al., 2003) and the fact that various P forms have been identified in the swine manure extracts (He and Honeycutt, 2001).

He et al. (2009) further comparatively investigated manure- and plant-derived water extractable organic matter by multiple spectroscopic techniques. The spectrum of freeze-dried water extract of dairy manure is more similar to that of the water extracts of poultry litter (Tan et al., 1975) than to that of swine manure (He et al., 2003). That is, the spectrum of the water extract of dairy manure resembles Type III spectra of humic substances. However, unlike HA and fulvic acid (FA) and similar to other two types of animal manure, peaks associated with COO^{-} groups and conjugated ketones, such as those in the 1720-1710 cm^{-1} and 1230-1210 cm^{-1} regions, were not apparent in the spectrum of the water extracts of dairy manure (Figure 2.2). He et al. (2009) proposed two hypotheses to explain the observation: (1) no significant abundance of COO^{-} groups in these manure water extracts, and (2) there were COO^{-} groups in these samples, but they were bonded with polyvalent ions present in the WEOM extracts. However, the relatively low concentrations of polyvalent ions in the dairy manure extract implied less possibility of the second hypothesis. In other words, even though they could not exclude the second hypothesis, the lack of these strong absorbance bands is more likely due to the first hypothesis. With the lack of significant carboxylic groups in these samples, the authors (He et al., 2009) attributed the peak at 1596 cm^{-1} in Figure 2.2 to aromatic compounds in these water extracts. A shoulder, rather a peak, at 1529 cm^{-1} (N-H stretching and NH$_2$ deformation) implied that the water extract of dairy manure contain less nitrogenous compounds than the water extracts of swine manure tested by the same research group (He et al., 2003; 2009). Another shoulder at 1653 cm^{-1} was assigned to olefine or aromatic compounds (Chang Chien et al., 2007; Francioso et al., 1996). Based on the strengths of the two bands (Figure 2.2), these olefinic and N compounds in the water extracts of dairy manure and seven plant shoots decreased in the order of lupin>manure>crimson clover>hairy vetch>corn>wheat>alfalfa>soybean (spectra of lupin, crimson clover, hairy vetch and alfalfa are not shown in Figure 2.2). On the contrary to dairy manure, the band in the 1272 cm^{-1} region is relatively strong in the plant spectra (Figure 2.2). This band was contributed by C-O stretching of esters, ethers, and phenols (Chang Chien et al., 2007). Combined with the broad bands at 1385cm^{-1} in these spectra, it appears that these plant water extracts contain more aliphatic and/or phenolic groups. From both UV-visible and FTIR spectra, He et al. (2009) concluded that the plant-derived water extractable organic matter possessed less humic-like characteristics than dairy manure-derived water extractable organic matter which itself is also less humified than soil humic substances. Thus, the differential characteristics of these samples derived from different sources may be useful for investigating the humification process of plant and manure organic matter (He et al., 2009). Previously, Plaza et al. (2008) investigated organic matter humification by vermicomposting of cattle manure alone and mixed with two-phase olive pomace. The FTIR spectra of the HA-like fractions of their cattle manure or the mixture with olive pomace resemble those of HAs from organic wastes of various origin and nature whereas they differed markedly from those of typical of soil HAs. After vermicomposting, the FTIR spectra of both cattle manure and manure/pomace humic-like fractions approach the spectral features typical of soil HAs.

Giusquiani et al. (1998) separated total dissolved organic carbon fraction (TDOC) of pig sludge into acid insoluble dissolved organic carbon (AIDOC) and acid soluble dissolved organic carbon (ASDOC) fractions. The spectra of TDOC and AIDOC are similar to Type III spectra with a shoulder band at 1714 cm^{-1} and an apparent band around 1520 cm^{-1}. The spectrum of ASDOC is similar to Type II spectra with the strong absorbance at 1714 cm^{-1}. These spectral differences suggested that the ASDOC fraction was concentrated with free carboxylic groups, which was also confirmed by the acidic functional group analysis (Giusquiani et al., 1998). On the other hand, the TDOC-AIDOC spectral difference showed a weaker absorbance of ASDOC at 1640 cm^{-1}, which was explained by the absence of the metal-carboxylic bonding in the purified ASDOC fraction. These observations suggested the involvement of dissolved organic carbon from pig slurry in the complexation of heavy metal ions (i. e. formation of DOC-metal complexes). The environmental implication is that the possible groundwater pollution hazard after pig slurry application to calcareous soils may be related to the potential mobility of the Cu-DOC complex in pig slurry (Giusquiani et al., 1998).

2.4. SPECTRAL FEATURES OF HUMIC FRACTIONS IN ANIMAL MANURE AND COMPOST

Plaza et al. (2002, 2003) investigated the spectral features of HA and FA extracted from pig slurry and soils with 3 or 4 years of pig slurry application. As the whole animal manure, the FTIR spectrum of pig slurry HA resembled Type III spectra. In contrast, the HA of the control soil without slurry application was similar to spectra of Type I. The FTIR spectra of HAs isolated from slurry-amended soils were more similar to the spectrum of control soil HA than to that of pig slurry-HA. However, the spectra of PS-amended soil HAs did differ from that of control soil-HA for (a) the stronger relative intensity of the bands at 1710 cm^{-1} and in the 1250 to 1240 cm^{-1} region, which tended to slightly decrease with the number of pig slurry applications; and (b) the slight increase of the relative intensities of the bands between 2930-2850 cm^{-1} and of the shoulder at 1040 cm^{-1}, which tended to slightly increase with the years of pig slurry applications. Compared with the corresponding HA fractions, the differences between the spectra of the pig slurry and control soil FAs are less than those observed for the corresponding HAs as both FA spectra belong to Type II spectra (Plaza et al., 2003). Only minor differences were observed as follows: (a) the slightly stronger absorptions at 2940 cm^{-1}, 1520 cm^{-1}, 1230-1220cm^{-1}, and 1050-1040 cm^{-1}, and (b) the weaker shoulder at 1660 cm^{-1}for slurry FA than for control soil FA. The spectra of FAs isolated from pig slurry-amended soils were similar to that of control soil as they were not affected by the years or the annual rate of pig slurry application. These observations imply that some HA components from pig slurry have been incorporated into soil HA fractions. However, soil FA fractions have been less impacted by pig slurry application.

Huang et al. (2006) extracted HA and FA from immature and mature pig manure composted with sawdust. The spectra of both HA and FA belong to Type II spectra. The spectra of HA at day 0 and day 63 of composting show a distinct reduction of the bands assigned to aliphatic C-H stretch occurred at 2950 and 2850 cm^{-1} by composting. Peaks in the polysaccharide region at 1160 cm^{-1} decreased, while the 1420 cm^{-1} peak became sharper, as compared to a HA spectrum from the raw compost, which indicated less $-OCH_3$ and $-OH$ polysaccharide groups in HA isolated from pig manure compost at the end of the studied process. The relative height of the aromatic region at 1650 cm^{-1} and 1250 cm^{-1} rose as the composting process proceeded. The FA spectra are similar to those of HA; However, a relatively weaker peak at 1650 cm^{-1} as compared to HA was found, indicating that fewer C=C bonds were present in FA and a lower degree of aromaticity. On the other hand, a sharp peak occurred at 1040 cm^{-1} which did not occur in the HA spectra. This peak did not change much at day 63 as compared to day 0. The FTIR spectra in this research indicated that the humification of organic matter mainly occurred in the HA fraction but little in the FA fraction during the composting process of the pig manure (Huang et al., 2006).

Chang Chien et al. (2007) extracted three fractions of humic substances, HA (MW > 1000 Da), FA (MW > 1000 Da), and FA (MW < 1000 Da) from pig manure-based compost, following the International Humic Substances Society method. The spectrum of HA showed a strong absorption band at 1657 cm^{-1}, while the spectra of both FA s (MW > 1000 and MW < 1000) showed a relatively weak absorption band and shoulder, respectively. This indicates more C=C bonds in HA than those in FA (MW > 1000) and FA (MW < 1000). The spectra of FAs showed a stronger absorption band at around 1720-1727 cm^{-1} (C=O bond), while the spectrum of HA showed a weak absorption shoulder. This clearly indicates that C=O bond contents are higher in both FA (MW > 1000) and FA (MW < 1000) than in HA. Moreover, the absorption band at 1720-1727 cm^{-1} of the FTIR spectrum of FA (MW < 1000) was much stronger than that of FA (MW > 1000), indicating more C=O bonds in FA (MW < 1000) than in FA (MW > 1000). This FTIR spectral study (Chien et al., 2007) confirmed that FA (MW < 1000) has more reactive functional groups such as C=O and carboxylic groups, which may be correlated to the higher reactivity of FA (MW < 1000) than FA (MW > 1000) and HA (MW > 1000) as revealed by a previous by the same group (Chang Chien et al., 2006). The observation of more C=O bonds in FA (a strong band at 1720−1727 cm^{-1}) contradicts the spectral data of Huang et al. (2006) who extracted HA and FA by a different sodium pyrophosphate method.

Hernandez et al. (2006) also observed that the FTIR spectra of HA from pig slurry differ markedly from those of soil HAs. Also, the spectra of pig slurry-amended soil HAs differ from those of soil HAs without pig slurry in that there is a slight increase of the relative intensities of the bands at about 2930 cm^{-1}, and of the shoulder at 1040-1080 cm^{-1}. This observation suggests a partial incorporation of the HA fraction of pig slurry into native soil HA. Whereas the authors (Hernandez et al., 2006) attributed the shoulder at 1040-1080 cm^{-1} to Si-O of silicate impurities, an apparent band in the pig slurry HA spectrum and the increasing intensity of the shoulder in soil HA spectra with the slurry application rate suggest incorporation of manure P into the soil HA fractions.

2.5. FTIR ANALYSIS OF ORGANIC MATTER TRANSFORMATION DURING COMPOSTING

In order to quantify relative changes in spectral intensity during the composting process, Inbar et al. (1989) calculated the ratios between several main peaks including 2930 cm^{-1}, 1655 cm^{-1}, 1425 cm^{-1}, 1385 cm^{-1} and 1050 cm^{-1}. The 1385/2930 (COO$^-$, CH$_3$/aliphatic C-H) and 1385/1050 (COO$^-$, CH$_3$/polysaccharides C-O) ratios increased with composting time. The 1655/2930 (aromatic C=C, COO$^-$/aliphatic C-H) and 1425/1050 (COO$^-$, CH$_2$/polysaccharides C-O) ratios increased with composting time and were linearly correlated with the cation exchange capacity (CEC). From these data, Inbar et al. (1989) concluded that the polysaccharides and aliphatic C-H stretching decreased during the composting process while the concentration of the aromatic C=C alkyl C and carboxylate ions increased. Thus, the relative intensity of the distinct FTIR peaks may serve as a semiquantitative method for the evaluation of compost maturity (Inbar et al., 1989).

Similarly, Hsu and Lo (1999) conducted FTIR analysis of organic matter transformations during composting of pig manure. They (Hsu and Lo, 1999) found distinct changes in the peak intensity of pig manure resulting from the composting process, which they quantified by calculating the ratio between the intensity of major peaks.. The 1650/2930 ratio (aromatic C/aliphatic C) increased from 1.04 to 1.68, the 1650/2850 ratio (aromatic C/aliphatic C) increased from 1.49 to 2.33, the 1650/1560 ratio (aromatic C/amide II bond) increased from 1.36 to 1.67, and the 1650/1050 ratio (aromatic C/polysaccharide) increased from 0.86 to 1.11. The changes in these ratios indicate that easily degradable OM constituents, such as aliphatic and amide components, polysaccharides, and alcohols, are chemically or biologically oxidized and, therefore, the mature compost contained more aromatic structures of higher stability. This composting mechanism is also applicable to animal manure co-composted with other organic wastes, such as poultry manure with olive mill sludge (Hachicha et al., 2009).

Zmora-Nahum et al. (2007) reported the Diffuse Reflectance Infrared Fourier Transform (DRIFT) spectra of 37 commercial composts with or without animal manure. The manure-based composts present uniform spectra with distinctive characteristics. The aliphatic peaks at 2925 cm^{-1} and in the 1450-1460 cm^{-1} range were more pronounced than in the wood composts. All spectra exhibited a small peak at 1510-1525 cm^{-1} and the 1640/1030 ratio was higher than that found for the wood composts. The DRIFT spectra of the manure composts do not resemble the spectra in the work of Inbar et al. (1989), with the 1460/1650 ratio higher than one for five of the six composts. The authors (Zmora-Nahum et al., 2007) hypothesized that the difference may be due to the fact that Inbar et al. (1989) focused on composting of the solid fraction of liquid cattle manure which is very rich in straw, whereas in the Zmora-Nahum study manure was composted without separation. Two important findings of this comparative research is that (1) the ratio of the polysaccharide to aliphatic peak (1060/2925) is distinctive of the different groups of composts, and (2) a clear grouping of each source material can be seen when placing this peak ratio vs. the %OM. For example, the grape marc and oilcake composts group at the peak ratio range of 1–2.1, and high (60–90%) OM whereas the peak ratio of the manures ranges between 2.5 and 3.6 with a moderate percentage of OM. The peak ratio of the wood composts reaches very high values (ranging from 3.2 to 9.7). These extremely high values indicate this peak originated from the minerals in the ash and not from the polysaccharides, and, at least for the wood composts, does not indicate the amount of polysaccharides. Within each group there is no trend in the peak ratio that could be expected from a degrading series. Yet, this ratio can be used as a fingerprint to the different source materials.

2.6. FTIR Analysis of Organic Matter Transformation during Decomposition

Calderon et al. (2006) analyzed FT-IR spectral changes of dairy manures during aerobic decomposition. A spectral comparison of manures incubated for 10 weeks shows the net effect of manure decomposition on the FTIR spectra. Decomposed manures had lower absorbance than the fresh manure at 3400 cm^{-1}(N-H and O-H groups) and 2870 cm^{-1} (C-H groups). Fatty acids and proteins are rich in C-H groups, while cellulose is not. These results suggest that there was a preferential utilization of fatty acids and/or proteins relative to cellulose during decomposition. In contrast, the decomposed manures had higher absorbance at the 1653 cm^{-1} (proteins), 1510 cm^{-1} (lignin) and 897 cm^{-1}(cellulose). This research demonstrated that FTIR profiles can be used to monitor chronological changes in manure decomposition, which is also supported by pyrolysis-mass spectroscopic data (Calderon et al., 2006).

Marcato et al. (2009) analyzed the FTIR spectral features of raw and digested pig slurries to evaluate the impact of anaerobic digestion on organic matter quality in pig slurry. Both raw and digested slurry FTIR spectra exhibited the same absorbance areas, but they differed in the intensity of some peaks. In digested slurry, the spectra showed a remarkable decrease of: (i) aliphatic structures and lipids (bands at about 2930–2920 cm^{-1}, 2860–2850 cm^{-1} and 1460 cm^{-1}), (ii) amides (bands at about 3330 cm^{-1}, 1665–1635 cm^{-1} and 1570 cm^{-1}), (iii) polysaccharides (1040 cm^{-1}). On the other hand, the digested slurry FTIR spectra revealed an increase in carbonates (897-875 cm^{-1}) probably due to organic matter mineralization during anaerobic digestion. These spectral changes of pig slurry during anaerobic digestion are similar to those during aerobic dairy manure decomposition (Calderon et al., 2006), indicating both aerobic and anaerobic digestion of animal manure begins with degradation of the labile lipid, protein and carbohydrate components.

2.7. Infrared Photoacoustic Study of Animal Manure

To evaluate the feasibility of FTIR-PAS in characterizing of manure organic matter, we obtained the FTIR-PAS spectra of the feed and manure of three common animals (Figure 2.3). Numerous bands can be observed in the 600-2000 cm^{-1} region, around 2900 cm^{-1}, and 3300-3600 cm^{-1}. The total spectral appearance is similar between manure and feed spectra, but many minor changes can be detected. According to the features PAS spectra these animal manures can also be assigned as Type III spectra for non-visible absorption at 1720 cm^{-1}. Comparing conventional absorption spectra, there are significant differences; more absorption peaks can be found in the 3300-3600 cm^{-1} region, i. e. 3514, 3410, and 3317 cm^{-1}; the aliphatic C-H absorption, i. e. 2931 cm^{-1}, is almost the same as conventional transmittance infrared spectrum. There is no visible band around 1596 cm^{-1} in PAS spectra, which may be overlapped by the strong band around 1651 cm^{-1}. The band intensity in 1000-1200 cm^{-1} is much stronger than that in 1500-1650 cm^{-1}, but in conventional transmittance spectrum there is inverse results. More absorption bands can be observed in the 500-1500 cm^{-1} region, which totally deprived from vibration of C-H, C-C, C-O, C=O, N-H, P-O, Si-O et al., and the specific assignments need further experiment for confirmation, which will be valuable in structure identification .

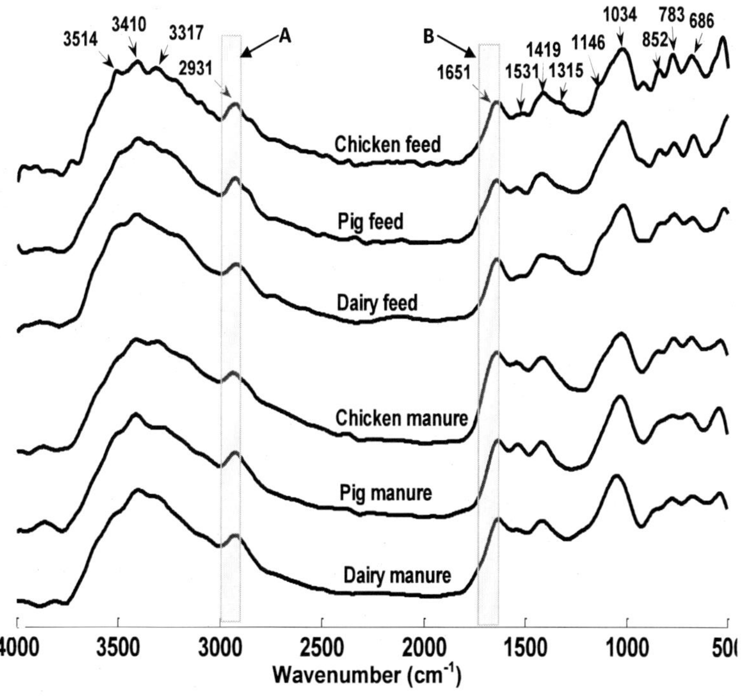

Figure 2.3. FTIR-PAS spectra of animal feeds and manures.

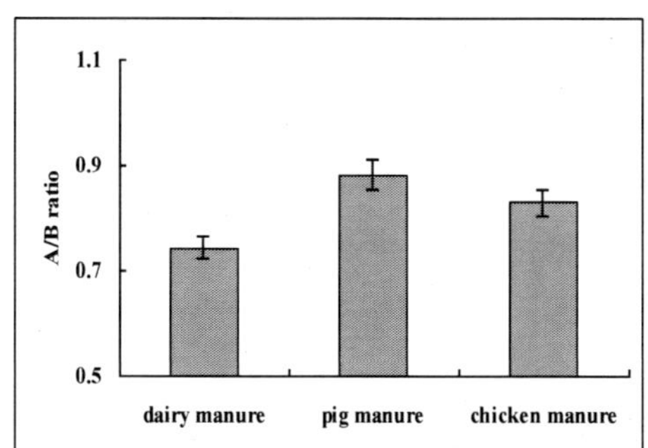

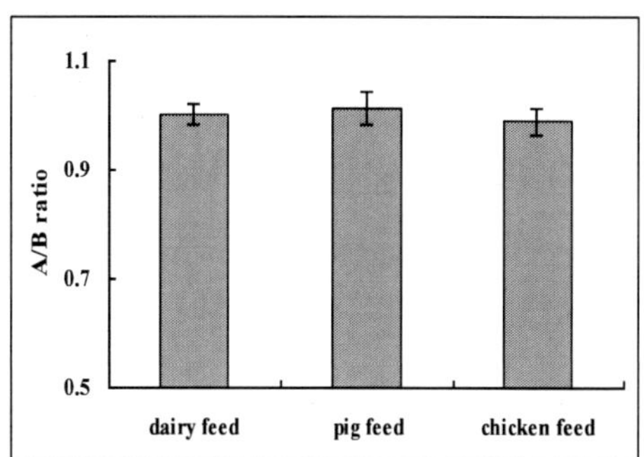

Figure 2.4. A/B ratio from FTIR-PAS spectra of manures and feeds. Intensity of 3000-2800 cm^{-1} and 1740-1600 cm^{-1} band are denoted as band A and B, respectively. Data are shown in average with standard deviations (n=5, 10, and 7 for dairy, pig, and chicken manure and feed, respectively).

To systematically analyze the FTIR-PAS spectra of these manure samples, we adapted the methods used in soil organic matter by Ellerbrock et al. (2005) and Kaiser et al. (2007).

The absorptions in the regions of 3000-2800 cm^{-1} and 1740-1600 cm^{-1} are denoted as Band A (hydrophobic) and B (hydrophilic) (Kaiser et al., 2007) and the A/B ratio calculated from the cumulative peak heights of bands A and B may be used as measure for the potential wettability (Ellerbrock et al., 2005). The ability of a liquid (e.g., water) to spread on a solid surface (i.e., the wettability) is related to the solid–water contact angles, which depends on the solid surface free energy and the surface tension. For organic matter in manures, the wettability is a relative measurement of the contents of hydrophilic C=O groups (i.e., O and N containing hydroxyl and carboxyl groups) vs. hydrophobic CH groups. Analysis of 22 manure samples indicated that the wettability of the dairy manure is significantly higher than that of chicken manure and pig manure, and the wettability of chicken is the lowest; however, no significant difference was observed among the 22 animal feeds (Figure 2.4). The A/B ratios of animal feed are significantly higher that of animal manures, thus, besides animal feed we assume that animal digesting systems should play the important role in the difference of manure wettability, and manure types could be identified by this manure property.

This new technique of FTIR-PAS has several advantages when compared to conventional absorption spectroscopy summarized below: (1) No sample preparation is required for PAS studies, and *in situ* studies can be performed. (2) Unlike conventional spectroscopy, scattering does not pose a problem with PAS, since losses due to the scattering and reflection do not produce PA signals. This aspect makes PA particularly attractive when studying strongly scattering samples. (3) The infrared photoacoustic method is capable of being used to study the mechanism of non-radioactive de-excitation; this is not possible when using conventional techniques, and this is verified by some new features in PAS spectra of manures. (4) Photoacoustic spectroscopy is suitable to study high absorbing sample, such as manure. (5) The PAS spectra of various types of manure samples can be obtained in a short time. This allows for rapid and efficient evaluation of manure quality, which may assist in application of manure. (6) Photoacoustic spectroscopy (PAS) does not require that the sample be transmitting, has low sensitivity to surface condition, and can probe over a range of selectable sampling depths from several micrometers to more than 100 μm, thus more sample information can be profiled using this feature. However, the disadvantage of FTIR-PAS is that there may be strong noise influence during PAS spectra recording; to improve the signal-to-noise ratio determination should be done in a quiet environment with enough scans.

This preliminary work demonstrates the potential application of FTIR-PAS for the characterization of organic substances in animal manures. However, currently the exploration is still very limited. In further research, specific spectral assignments of animal manure should be made and confirmed, and the step-scanning function should be applied to explore more structural information for organic substances in animal manure in deeper layers.

2.8. CONCLUSION

Infrared spectroscopy, generally in the form of FTIR, has been used to characterize the organic matter in animal manure and relevant materials or fractions. The FT-IR spectra of most manure samples resemble the Type III spectra of humic substances with strong aliphatic characters. However, the absorption band around 1720 cm^{-1} is very weak or even disappears in manure spectra, indicting fewer free carboxylic groups in animal manure than in humic substances. In addition to the contribution from alcoholic and polysaccharide, absorption bands in the 1100-1000 cm^{-1} region have been frequently assigned to vibrations of a SiO_2-related impurities in animal manure. Because the level of P in animal manure is high and the level of Si is low, it seems more reasonable to assign the FT-IR band of animal manure in this region partly to inorganic and organic phosphate compounds than to silicates. This assignment is supported by the positive correlation of peak intensity and P content in humic samples (He et al., 2006b), and better explains the observation that longer years and higher annual application rates increased this band intensity in HA fractions of slurry-amended soil (Plaza et al., 2002; Hernandez et al., 2006). FTIR spectroscopy revealed that composting and decomposition of animal manure begin by the degradation of labile organic matter components. Relative spectral intensities, such as the peak ratios of 1650 /2930, 1620 /1110, 1385 /2930, 1650 /1560 and 1511/2930 cm^{-1}, can be used to monitor the change (stability and maturity) of manure organic matter under different management practices. FTIR PAS spectroscopy offers an alternative method to characterize organic matter in animal manure. It requires minimal sample pre-treatment and allows the use of larger amounts of substance in the analysis, thus minimizing the experimental deviations due to heterogeneous characters of animal manure. Further modification and verification of this method may provide a fast and cheap method for qualitative and semiquantitative assessments of animal manure samples from different sources with a wide range of properties.

REFERENCES

Agnelli, A., L. Celi, A. Degl'Innocenti, G. Corti, and F.C. Ugolini. 2000. Chemical and spectroscopic characterization of the humic substances from sandstone-derived rock fragments. *Soil Sci.* 165:314-327.

Calderon, F.J., G.W. McCarty, and J.B. Reeves III. 2006. Pyrolysis-MS and FT-IR analysis of fresh and decomposed dairy manure. *J. Anal. Appl. Pyrolysis.* 76:14-23.

Carballo, T., M.V. Gil, X. Gomez, F. Gonzalez-andres, and A. Moran. 2008. Characterization of different compost extracts using Fourier-transform infrared spectroscopy (FTIR) and thermal analysis. *Biodegradation* 19:815-830.

Chang Chien, S.W., M.C. Wang, and C.C. Huang. 2006. Reactions of compost-derived humic substances with lead, copper, cadmium, and zinc. *Chemosphere.* 64:1353-1361.

Chang Chien, S.W., M.C. Wang, C.C. Huang, and K. Seshaiah. 2007. Characterization of humic substances derived from swine manure-based compost and correlation of their characteristics with reactivities with heavy metals. *J. Agric. Food Chem.* 55:4820-4827.

Dinel, H., M. Schnitzer, and H. Schulten. 1998. Chemical and spectroscopic characterization of colloidal fractions separated from liquid hog manures. *Soil Sci.* 163:665-673.

Du, C.W., and J.M. Zhou. 2009. Evaluation of soil fertility using infrared spectroscopy: a review. *Environ. Chem. Lett.* 7: 97-113.

Du, C.W., R. Linker, A. Shaviv, and J. M. Zhou. 2007. Characterization of soils using photoacoustic mid-infrared spectroscopy. *Appl. Spectrosc.* 61: 1063-1067.

Du, C.W., R. Linker, A. Shaviv, and J. M. Zhou. 2008. Soil identification with Fourier transform infrared photoacoustic spectroscopy. *Geoderma* 143: 85-90.

Ellerbrock, R.H., H.H. Gerke, J. Bachmann, and M.-O. Goebel. 2005. Composition of organic matter fractions for explaining wettability of three forest soils. *Soil Sci. Soc. Am. J.* 69:57-66.

Francioso, O., C. Ciavatta, V. Tuganoli, S. Sanchez-Cortes, and C. Gessa. 1998. Spectroscopic characterization of pyrophosphate incorporation during extraction of peat humic acids. *Soil Sci. Soc. Am. J.* 62:181-187.

Francioso, O., S. Sanchez-Cortes, V. Tugnoli, C. Ciavatta, L. Sitti, and C. Gessa. 1996. Infrared, Raman, and nuclear magnetic resonance (^{1}H, ^{13}C, and ^{31}P) spectroscopy in the study of fractions of peat humic acids. *Appl. Spectr.* 50:1165-1174.

Gibert, O., J. de Pablo, J.L. Cortina, and C. Ayora. 2008. Evaluation of a sheep manure/limestone mixture for in situ acid mine drainage treatment. Environ. *Eng. Sci.* 25:43-52.

Giusquiani, P.L., L. Concezzi, M. Busibelli, and A. Macchioni. 1998. Fate of pig sludge liquid fraction in calcareous soil: agricultural and environmental implications. *J. Environ. Qual.* 27:364-371.

Hachicha, S., F. Sellami, J. Cegarra, R. Hachicha, N. Drira, K. Medhioub, and E. Ammar. 2009. Biological activity during co-composting of sludge issued from the OMW evaporation ponds with poultry manure-Physico-chemical characterization of the processed organic matter. *J. Hazard. Mater.* 162:402-409.

Hay, M.B., and S.C.B. Myneni. 2007. Structural environments of carboxyl groups in natural organic molecules from terrestrial systems. Part 1: Infrared spectroscopy. *Geochim. Cosmochim. Acta* 71:3518-3532.

He, Z., and C.W. Honeycutt. 2001. Enzymatic characterization of organic phosphorus in animal manure. *J. Environ. Qual.* 30:1685-1692.

He, Z., C.W. Honeycutt, and T.S. Griffin. 2003. Comparative investigation of sequentially extracted P fractions in a sandy loam soil and a swine manure. *Commun. Soil Sci. Plant Anal.* 34:1729-1742.

He, Z., C.W. Honeycutt, T. Zhang, and P.M. Bertsch. 2006a. Preparation and FT-IR characterization of metal phytate compounds. *J. Environ. Qual.* 35:1319-1328.

He, Z., T. Ohno, B.J. Cade-Menun, M.S. Erich, and C.W. Honeycutt. 2006b. Spectral and chemical characterization of phosphates associated with humic substances. *Soil Sci. Soc. Am. J.* 70:1741-1751.

He, Z., C.W. Honeycutt, B. Xing, R.W. McDowell, P.J. Pellechia, and T. Zhang. 2007. Solid-state Fourier transform infrared and ^{31}P nuclear magnetic resonance spectral features of phosphate compounds. *Soil Sci.* 172:501-515.

He, Z., J. Mao, C.W. Honeycutt, T. Ohno, J.F. Hunt, and B.J. Cade-Menun. 2009. Characterization of plant-derived water extractable organic matter by multiple spectroscopic techniques. *Biol. Fertil. Soils.* 45:609-616.

Hernandez, D., C. Plaza, N. Senesi, and A. Polo. 2006. Detection of copper(II) and zinc(II) binding to humic acids from pig slurry and amended soils by fluorescence spectroscopy. *Environ. Pollut.* 143:212-220.

Hernandez, D., C. Plaza, N. Senesi, and A. Polo. 2007. Fluorescence analysis of copper(II) and zinc(II) binding behavior of fulvic acids from pig slurry and amended soil. *Eur. J. Soil Sci.* 58:900-908.

Hsu, J.H., and S.L. Lo. 1999. Chemical and spectroscopic analysis of organic matter transformations during composting of pig manure. *Environ. Pollut.* 104:189-196.

Huang, G.F., Q.T. Wu, J.W.C. Wong, and B.B. Nagar. 2006. Transformation of organic matter during co-composting of pig manure with sawdust. Bioresour. *Technol.* 97:1834-1842.

Inbar, Y., Y. Chen, and Y. Hadar. 1989. Solid state carbon-13 nuclear magnetic resonance and infrared spectroscopy of composted organic matter. *Soil Sci. Soc. Am. J.* 53:1695-1701.

Johnston, C.T., and Y.O. Aochi. 1996. Fourier transform infrared and Raman spectroscopy., p. 269-321, *In* D. L. Sparks, ed. Methods of soil analysis. Part 3-chemcial methods. *Soil Sci. Soc. Am.,* Madison, WI.

Kaiser, M., R.H. Ellerbrock, and H.H. Gerke. 2007. Long-term effects of crop rotation and fertilization on soil organic matter composition. *Eur. J. Soil Sci.* 58:1460-1470.

Mao, J., D.C. Olk, X. Fang, Z. He, J. Bass, and K. Schmidt-Rohr. 2008. Influence of animal manure application on the chemical structures of soil organic matter as investigated by advanced solid-state NMR and FT-IR. *Geoderma* 146:353-362.

Marcato, C.E., R. Mohtar, J.C. Revel, P. Pouech, M. Hafidi, and M. Guiresse. 2009. Impact of anaerobic digestion on organic matter quality in pig slurry. *Intern. Biodeterio. Biodegra.* 63:260-266.

Niemeyer, J., Y. Chen, and J.M. Bollag. 1992. Characterization of humic acids, compost, and peat by diffuse reflectance Fourier-transform infrared spectroscopy. *Soil Sci. Soc. Am. J.* 56:135-140.

Olk, D.C., G. Brunetti, and N. Senesi. 2000. Decrease in humification of organic matter with intensified lowland rice cropping: a wet chemical and spectroscopic investigation. *Soil Sci. Soc. Am. J.* 64:1337-1347.

Perkins, W.D. 1993. Sample handing in infrared spectroscopy-an overview., p. 11-53, *In* P. B. Coleman, ed. *Practical sampling techniques for infrared analysis.* CRC Press, Boca Raton, FL.

Plaza, C., R. Nogales, N. Senesi, E. Benitez, and A. Polo. 2008. Organic matter humification by vermicomposting of cattle manure alone and mixed with two-phase olive pomace. *Bioresour. Technol.* 99:5085-5089.

Plaza, C., N. Senesi, J.C. Garcia-Gil, G. Brunetti, V. D'Orazio, and A. Polo. 2002. Effects of pig slurry application on soils and soil humic acids. *J. Agric. Food Chem.* 50:4867-4874.

Plaza, C., N. Senesi, A. Polo, G. Brunetti, J.C. Garcia-Gil, and V. D'Orazio. 2003. Soil fulvic acid properties as a means to assess the use of pig slurry amendment. *Soil Tillage Res.* 74:179-190.

Schnitzer, M.I., C.M. Monreal, G.A. Facey, and P.B. Fransham. 2007. The conversion of chicken manure to biooil by fast pyrolysis I. Analyses of chicken manure, biooils and char by ^{13}C and ^{1}H NMR and FTIR spectrophotometry. *J. Environ. Sci. Health. Part B.* 42:71-77.

Senesi, N., C. Plaza, G. Brunetti, and A. Polo. 2007. A comparative survey of recent results on humic-like fractions in organic amendments and effects on native soil humic substances. *Soil Biol. Biochem.* 39:1244-1262.

Stevenson, F.J., and K.M. Goh. 1971. Infrared spectra of humic acids and related substances. Geochim. Cosmochim. *Acta* 35:471-483.

Tan, K.H. 2003. p. 180-188 *Humic matter in soil and the environment.* Marcel Dekker, Inc., New York, N.Y.

Tan, K.H., V.G. Mudgal, and R.A. Leonard. 1975. Adsorption of poultry litter extracts by soil and clay. *Environ. Sci. Technol.* 9:132-135.

Zmora-Nahum, S., Y. Hadar, and Y. Chen. 2007. Physico-chemical properties of commercial composts varying in their source materials and country of origin. *Soil Biol. Biochem.* 39:1263-1276.

In: Environmental Chemistry of Animal Manure
Editor: Zhongqi He

ISBN: 978-1-62808-641-6
© 2013 Nova Science Publishers, Inc.

Chapter 3

CARBON FUNCTIONAL GROUPS OF MANURE ORGANIC MATTER FRACTIONS IDENTIFIED BY SOLID STATE ^{13}C NMR SPECTROSCOPY

Zhongqi He[1][*] *and Jingdong Mao*[2]

3.1. INTRODUCTION

Similar to infrared (IR) spectroscopy (Chapter 2), nuclear magnetic resonance (NMR) spectroscopy is a non-destructive technique that uses the magnetic resonance of nuclei to investigate chemical structural environments around them. NMR was well established for organic chemical applications by 1965 (Preston, 1996). The scope of NMR applications includes solutions (liquids), solids, and intermediate physical states. The great strength of NMR in natural organic matter research is its unique ability to provide information on more complex materials which are characterized by irregular structures, and strong physical links to each other or to mineral matter (Preston, 1996). Many organic matter samples are analyzed using solid state C-13 NMR that presents the following advantages:

1) Some organic matter samples or fractions are not soluble;
2) Solid-state NMR facilitates a much larger sample than solution NMR, enhancing signal and shortening run times. In solution NMR signal intensity is dependent upon the concentration of sample in an NMR tube. Making the organic matter concentration high enough to achieve a strong signal in solution state may lead to aggregation, resulting in lower sensitivity, lower resolution and loss of structural information;

---

[*] Corresponding author: Zhongqi.He@ars.usda.gov
[1]USDA-ARS, New England Plant, Soil, and Water Laboratory, Orono, ME 04469, USA
[2]Department of Chemistry & Biochemistry, Old Dominion University, Norfolk, VA 23529, USA

3) Solid-state NMR generally involves less sample handling. The sample can be analyzed without any pretreatment and extraction, i.e. the sample can be examined as a whole and secondary reaction can be avoided;

4) Solid-state NMR avoids solvent effects on organic matter structures and solvent; artifact peaks that are common problems in solution NMR;

5) Solid-state NMR is non-intrusive, i.e., it does not consume sample. Solution NMR is not. The alteration of sample during solution NMR preparation and analysis would void valuable sample for other analyses;

6) It is easier and more straightforward to detect unprotonated carbons using solid-state techniques;

7) In solution NMR, the fast tumbling of molecules averages the anisotropic interactions, while in solid-state NMR we can take advantage of these anisotropic interactions by using specially developed pulse sequences to extract structural information not obtainable from solution NMR;

8) Solid-state techniques can identify domains and heterogeneities within organic matter structures, which solution NMR cannot (Mao et al., 2002; Mao and Schmidt-Rohr, 2006).

Perhaps, solid state ^{13}C cross-polarization /magic angle spinning (CP/MAS) NMR represents the most common approach for characterizing organic matter. Conte et al. (2004) reviewed relevant literature which applied ^{13}C CP/MAS NMR spectroscopy to the qualitative and semi-quantitative characterization of natural organic matter. Other solid-state ^{13}C NMR techniques, such as advanced spectral-editing techniques and two-dimensional ^{1}H–^{13}C heteronuclear correlation (2D HETCOR), have also been recently used on investigation of natural organic matter and its components (Mao et al., 2007a; 2007b). This chapter first reviews the spectral features and functional groups of organic matter which can be identified by these techniques, and then synthesizes and analyzes the structural information of organic matter in animal manure derived from these techniques.

3.2. SOLID STATE C-13 NMR TECHNIQUES AND STRUCTURAL INFORMATION OF ORGANIC MATTER

To conduct advanced solid-state NMR experiments, around 100 mg of samples is required to fill in a 4-mm rotor tube. Since solid-state NMR analysis is non-invasive, the sample preparations are very simple. The only requirement is that these samples be dry and in solid state. Therefore, sample-drying via freeze-drying would render any organic matter samples suitable for solid-state analysis as long as they do not contain high contents of paramagnetic materials such as Fe(III), Mn(II), and Cu(II). The most widely-used solid-state ^{13}C NMR technique in investigating natural organic matter is cross polarization/magic angle spinning (CP/MAS). This technique improves sensitivity remarkably through the transfer of magnetization from the abundant ^{1}H to dilute ^{13}C spins through cross-polarization. Anisotropic interactions such as chemical shift anisotropies and dipolar couplings lead to broad NMR spectra. In solution NMR, these anisotropic interactions are averaged due to fast tumblings of molecules. But this is not the case in solid state. Therefore, in CP/MAS, magic angle spinning is used to reduce or remove chemical shift anisotropies, and high power decoupling is employed to eliminate dipolar couplings. However, CP/MAS spectra are only semi-quantitative. There exists the reduced CP efficiency for nonprotonated carbons, mobile components, or regions having short proton rotating-frame spin-lattice relaxation time ($T_{1\rho}^{H}$). The extensive application of CP/MAS technique in the literature has led to the underestimation of the sp^2-carbon region and overestimation of the sp^3-hybridized carbons in natural organic matter.

As routine ^{13}C solid-state NMR spectra discussed above consist of broad and overlapping bands in which functional groups cannot be clearly distinguished, more advanced solid-state NMR techniques are needed to selectively retain certain peaks and eliminate others for more clearly revealing specific functional groups. For this purpose, Mao and Schmidt-Rohr developed and revised a series of spectral-editing techniques. Some of these techniques include ^{13}C chemical shift anisotropy (CSA) filter (Mao et al., 2008), dipolar distortionless enhancement by polarization transfer (DEPT) for CH selection (Schmidt-Rohr and Mao, 2002), CH_2 selection (Mao and Schmidt-Rohr, 2005), and long-range dipolar dephasing for selecting fused ring carbons (Mao and Schmidt-Rohr, 2003). These new and advanced techniques are demonstrated in a complex humic acid from a peat (Figure 3.1). It is worth noting that these techniques can be easily used for the study of manure samples although we demonstrate the results of a peat humic acid here. Also, two-dimensional ^1H-^{13}C heteronuclear correlation NMR is used to detect connectivities and proximities of different functional groups (Figure 3.2) (Mao et al., 2007b). In addition, ^1H spin diffusion can be used to detect domains or heterogeneities on a 1- to 50-nm scale (Mao and Schmidt-Rohr, 2006). Currently, these techniques have been used in only limited manure-related organic matter research (Mao et al., 2007b; 2008; He et al., 2009).

Table 3.1. Distribution of carbon composition in untreated manure and manure fractions semi-quantified by solid state C-13 CPMAS NMR spectroscopy.

C groups (chemical shifts)[a]	Untreated chicken manure[b]	Water extracts of dairy manure[c]	Water extracts of composted dairy manure[c]	Colloidal fractions of liquid hog manure[d]
	% of total C			
Alkyls, 0-40 ppm	12.4	14	14	32.4
OCH_3 and protein, peptide, amino NCH, 41-60 ppm	6.7	17	15	18.0
O-alkyls such as those of carbohydrates, 61-105 ppm	68.1	29	35	22.3
Aromatics, 106-150 ppm	3.2	22	24	15.3
Aromatic C-O, 151-170 ppm	3.1	10	7	4.5
COO and N-C=O, 171-190 ppm	6.5	8	6	7.4
Total aliphatic, 0-105 ppm	87.2	60	64	72.7
Total aromatic, 106-170 ppm	6.3	32	31	19.8
Aromaticity, (106-170)*100/(0-170)	6.7	35	32	21.4

[a] Terminology of the functional groups reported in the three papers is not exactly same in the three references. Several stating chemical shifts used in the calculation in Dinel et al. (1998) are one ppm lower than in other two papers.
[b] Adapted from Schnitzer et al. (2007).
[c] Adapted from Liang et al. (1996).
[d] Adapted from Dinel et al. (1998).

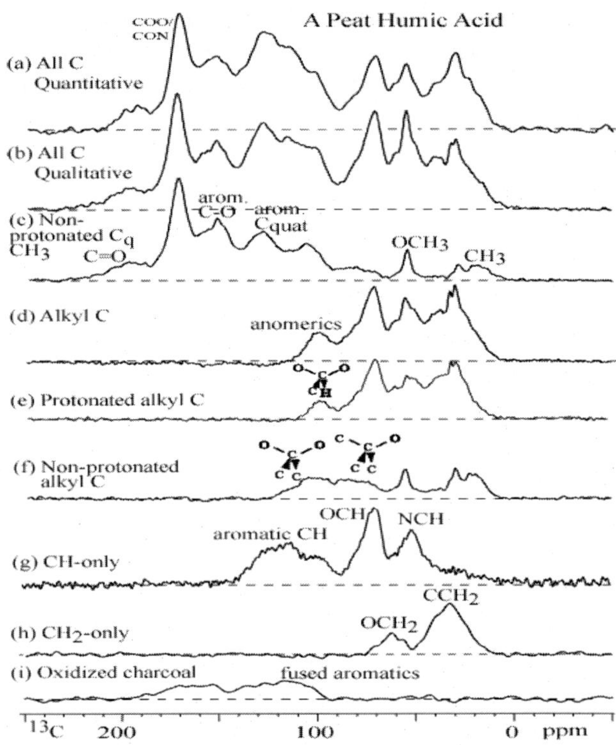

Figure 3.1. Identification of specific functionalities in a peat HA by C-H spectral editing. (a) DP/MAS showing all the quantitative structural information. (b) CP/TOSS showing all the qualitative structural information. (c) dipolar-dephased CP/TOSS selecting nonprotonated and mobile groups like CH_3. (d) sp^3-C selected by a ^{13}C CSA filter, which identifies OCO carbons. (e) protonated sp^3-C signals by a ^{13}C CSA filter plus short CP. (f) nonprotonated or mobile sp^3-C by a CSA filter and dipolar dephasing, which in particular identifies OC_qO carbons. (g) dipolar DEPT at a 4-kHz spin rate selecting CH signals. OCH and NCH bands are clearly observed. (h) CH_2-only. (i) Fused aromatics by long-range dipolar dephasing. Data reorganized from Mao and Schmidt-Rohr (2003, 2004 , 2005), Schmidt-Rohr and Mao (2002) and Schmidt-Rohr et al. (2004). All spectra were acquired at a 400 MHz instrument.

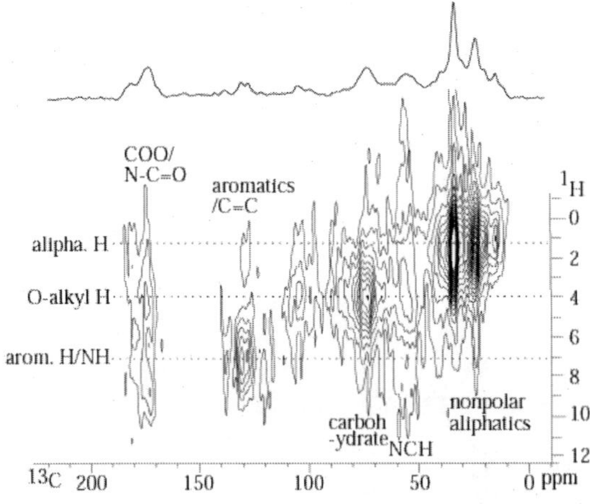

Figure 3.2. Two-dimensional ^1H-^{13}C heteronuclear correlation NMR spectrum of a pig feces sample. The spectra above the 2D HETCOR spectra are corresponding 1D ^{13}C spectra of this sample. Figure is adapted from Mao et al. (2007b). All spectra were acquired at a 400 MHz instrument.

3.3. DISTRIBUTION OF CARBON FUNCTIONAL GROUPS IN ANIMAL MANURE

3.3.1. General Features

Animal manure is generally high in organic matter intensity (Moral et al., 2005) and thus carbons so that it is well suitable for ^{13}C NMR analysis. Currently, there are only a few reports on the solid state ^{13}C NMR spectral features of untaxed animal manure for carbon composition of whole animal manure (Gomez et al., 2007; Mao et al., 2007b; Schnitzer et al., 2007). Schnitzer et al. (2007) reported the ^{13}C CP/MAS NMR spectrum of chicken manure. This spectrum shows well-defined signals at 65, 73, 75, 85, 88, and 105 ppm. The major signals at 73 and 75 ppm are assigned to C-2, C-3 and C-5 of cellulose. The signal at 105 ppm is assigned to deoxygenated and anomeric C-1 of cellulose. The peak at 65 ppm is due to crystalline components of C-6 in hexose, and the shoulders at 85 and 88 ppm arise from non-crystalline and crystalline components of C-4 of cellulose. These assignments indicate the dominance of cellulose in the poultry manure. The authors assume that the cellulose was mainly from the sawdust mixed in the manure, rather than from poultry excreta. On the other hand, the spectrum does show other organic matter components in the chicken manure. The signals at 23, 27, and 32 ppm were assigned to functional groups of CH_3, CH_2, and $(CH_2)_n$ respectively. The assignment of 27 ppm and 32 ppm signals is controversial as Hu et al. (2000) proposed that both signals are due to $(CH_2)_n$ with crystalline (32 ppm) and amorphous (27 ppm) forms, respectively. At the downfield region, the signal at 132 ppm was assigned to aromatic C of lignin or C in N-heterocyclics, and signals at 150, 156, and 160 ppm to guaiacyl and syringyl OH of lignin. The relative strong signals at 170 and 174 were attributed to C in COOH groups of uronic acids, and/or C in amides and esters. Their semi-quantitative data of the C functional groups in the chicken manure is listed in Table 3.1. It is worth poitning out that the peaks assignments in Schnitzer et al. (2007) are not complete. For example, the authors attributed the region of 151-170 ppm totally to phenolics but aromatic $C-OCH_3$ also resonates within this region (Mao et al., 2007a; 2007b).

3.3.2. Comparison of Chemical Structures of Transgenic and Conventional Pig Manures

Mao et al. (2007b) obtained detailed structural information on organic matter of eight manures samples from phytase transgenic and conventional pigs. The transgenic Yorkshire pigs, trademarked Enviropig, excrete phytase from their salivary glands in order to digest phytate (IP6, *Myo*-inositol (1,2,3,4,5,6) hexa*kis*phosphate) P in cereal grain diets (Golovan et al., 2001). These manure samples represented different treatments associated with genetics, diet, and animal growth stage. Spectral editing techniques such as dipolar dephasing, ^{13}C CSA filter, CH and CH_2 selection and quantitative DP/MAS NMR techniques were used to identify and quantify specific functional groups. Two-dimensional ^{1}H-^{13}C heteronuclear correlation NMR was used to detect their connectivity. Data from the advanced NMR analysis indicated that the chemical structures of the feces of transgenic Enviropig^{TM} pigs are different from those of conventional pigs. More carbohydrates and less nonpolar aliphatics or

COO/N-C=O were found in pig feces from transgenic pigs (Table 3.2). Also the chemical environments of nonpolar aliphatics were more diverse in the feces from transgenic pigs than those from conventional pigs, suggesting that nonpolar aliphatics are more altered in the digestive tracts of transgenic pigs. This research (Mao et al., 2007b) also found no consistent effect from diet with or without supplemental phosphate or at different growth stages.

3.3.3. Changes of the Distribution of Carbon Functional Groups in Stabilized Manure Products

The stabilization processes of animal manure refer to aerobic composting, anaerobic digestion and mixed aerobic and anaerobic treatments (Gomez et al., 2007). Stabilization is used to reduce pathogens, eliminate unpleasant smells, and decrease or eliminate the potential for putrefaction (Gomez et al., 2007). Since stability (maturity) of natural organic matter is a relative term, defining it is not a trivial challenge (Chen, 2003). Solid state ^{13}C NMR spectroscopy is one of the very useful methods to investigate the organic matter decomposition during composting (Inbar et al., 1989; Chen, 2003). Tang et al. (2006) characterized the maturing process of cattle manure mixed with rice straw using ^{13}C CP/MAS NMR spectroscopy and other techniques. Cattle manure compost was matured for up to 12-18 months after 4 days of thermophilic composting. Solid state ^{13}C CP/MAS NMR analysis was used to monitor the change of C composition during the process. Similar to other reports (Gomez et al., 2007; Schnitzer et al., 2007), Tang et al. (2006) observed strong peaks at 72 and 105 ppm and attributed them mainly to cellulose. These two peaks decreased considerably in the spectra of 8-12 month samples, suggesting that intensive degradation of cellulose occurred during the maturing process. The degradation of cellulose was further confirmed by the increase of the carboxyl peak at 174 ppm in the maturing samples as oxidative degradation of organic matter is usually accompanied by the generation of carboxyl groups (Kogel-Knabner, 1997). The relative increases in carboxyl groups observed are likely related to an accumulation or stabilization of compounds/molecules containing COOH, such as long chain fatty acids. Tang et al. (2006) further classified the C functional groups identified by ^{13}C CP/MAS into four groups: alkyl C, *O*-alkyl C, aryl-C and carbonyl C. The semi-quantitative data based on ^{13}C CP/MAS demonstrated a successive decrease in the *O*-alkyl C components during the maturing period. An increase in the ratio of alkyl C to *O*-alkyl C was also observed in matured samples, with a rapid change after 8 months of maturing. This ratio increase implies a preferential loss of carbohydrates and humification of organic matter in the compost during the process, thus confirming it is a sensitive index of extent of decomposition which was proposed by Baldock et al. (1997).

Zmora-Nahum et al. (2007) compared solid-state CPMAS ^{13}C-NMR spectra of 23 composts from various composting companies from France, Greece and the Netherlands. The uniformity of the wood-based composts was apparent in the narrow distribution of each carbon type in their NMR spectra, with moderate amounts of polysaccharide C, 36.3–46.4%; aliphatic C, 20.8–25.7%; aromatic C, 16.8–20.3%; carboxyl C, 8.5–11.7%. Animal manure based composts varied more widely. The range of their polysaccharide C contents was wider, 29.8–53.5%; aliphatic C, 15.6–22.6%; aromatic C, 15.7–22.6%; carboxyl C, 7.8–12.4%. The very high amount of polysaccharide C was found in the compost displaying a high dissolved organic carbon (DOC, 1800 mg l^{-1}), which might be due to the immature nature of these composts. More than that, Zmora-Nahum et al. (2007) found that for manure composts there were positive correlations between the ratio of polysaccharide-C/aromatic-C measured by ^{13}C-NMR and polysaccharide-C/aliphatic-C band ratios measured by Diffuse Reflectance Infrared Fourier Transform (DRIFT) . However, there were no correlations for the wood wastes composts. A weak correlation was found between the polysaccharide fraction and the DOC concentration for each wood or manure compost group, but not for both groups together. A weak negative correlation was found also for the aromatic fraction vs. the DOC concentration, but only in the manure-based composts. Thus, the spectra of the different compost groups provided evidence to the degree of maturity of these composts.

Gomez et al. (2007) investigated the transformation of cattle and poultry manure under three biological stabilization processes (anaerobic digestion, aerobic composting and a combination of aerobic and anaerobic treatments) by ^{13}C CP/MAS NMR spectroscopy and thermogravimetric analysis. The change of C composition caused by stabilization processes can be semi-quantified by integrating the peak areas of the spectra (Table 3.3). Whereas these peak assignments were basically based on plant materials, humic-like materials could be generated after digesting and composting. For this reason, we believe that, after digesting and composting, there should not be only acetyl signals around 5-30 ppm. Amino NCH group could also be present around 50-60 ppm, and the aromatics between 110-160 ppm should not be totally attributed to aromatic lignin (Mao et al., 2007a; 2007b).

Table 3.2. Relative intensity (%) of carbon functional groups of feces from pigs fed with different diets and at different growth stages. Information was derived from quantitative DP/MAS spectra (Mao et al., 2007b).

Chemical shift region	164-210 ppm	112-164 ppm	63-112 ppm	48-63 ppm	0-48 ppm
Assignment	COO/N-C=O	Arom./ Olefinic C=C	Carbo-hydrate	NCH	Nonpolar aliphatics
Conventional pigs fed a conventional diet during the "growing stage"	17.2	11.4	13.9	7.3	50.2
Transgenic(Enviropig™) pigs fed a conventional diet during the growing stage	17.8	9.5	*16.8[a]*	6.7	*49.1*
Conventional pigs fed a low-P diet during the growing stage	19.5	12.9	10.3	6.7	50.4
Transgenic pigs fed a low-P diet during the growing stage	*17.1*	10.6	*18.2*	7.8	*46.4*
Conventional pigs fed a conventional diet during the "finishing stage"	22.3	11.1	8.2	6.1	52.4
Transgenic pigs fed a conventional diet during the finishing stage	*18.4*	16.6	*11.6*	8.5	*45.0*
Conventional pigs fed a low-P diet during the finishing stage	19.6	12.5	10.2	6.7	51.0
Transgenic pigs fed a low-P diet during the finishing stage	*13.0*	12.4	*18.2*	8.5	*47.8*

[a] The values in italic highlight their differences between transgenic and conventional treatments.

As straw was used as bedding materials on the livestock farms where the samples were collected, cellulose-relevant carbohydrates were the major C components in the two types of manures. The methoxyl group in cattle manure increased in the order of Fresh <Digested < Mixed < Composted. This trend might be due to the preferential degradation in the stabilization processes of other compounds rather than methoxyl groups which are part of lignin. Indeed, a great deduction of carbohydrates was observed in stabilized cattle manures, indicating the preferential degradation of these compounds rather than methoxyl groups. Both intensities of aromatic and carboxyl/carbonyl species increase in stabilized samples when compared to fresh cattle manure. The increases are more pronounced in the aerobic and mixed stabilization processes. In the meantime, the degrees of aromaticity in these manures (Table 3.3) were also in accordance with the thermal profiles obtained from the same samples (Gomez et al., 2007). Thus, Gomez et al. (2007) proposed that the NMR signals associated with aromatic lignin and carboxyl groups could be useful for process control during stabilization processes.

The C composition of fresh poultry manure is similar to that of fresh cattle manure (Table 3.3). Digested poultry manure showed important modifications in carbon composition with increments in acetyl and lipid/protein regions (Gomez et al., 2007). The digestion of poultry manure greatly reduced the intensity of hemi-cellulose species, and removed both crystalline and amorphous cellulose as observed by the remarkable reduction of signals in the 66-106 ppm regions. Thus, the total carbohydrate intensity was reduced from 60% of the fresh poultry manure samples to 44% in the digested sample (Table 3.3). However, different from the mixedly processed cattle manure, in the stabilization of poultry manure under the mixed process, no major modifications of the C composition are observed (Table 3.3). The mixedly processed poultry manure presented total carbohydrate intensity of 62%, which is similar to that of fresh poultry manure. When compared to other stabilized samples, the mixedly processed poultry manure is characterized by only a slight increase in the aromatic and carboxyl/carbonyl functional groups. The authors (Gomez et al., 2007) attributed it to the failure of the digestion phase during the process, which impeded the solubilization of the organic matter and posterior wash-out with the percolation stream in the processing system, yielding a stabilized product with high carbohydrate intensity and low degree of aromaticity.

Table 3.3. Distribution of carbon composition in fresh and stabilized cattle and poultry manures semi-quantified by solid state C-13 CPMAS NMR spectroscopy [adapted from Gomez et al. (2007)].

C composition (chemical shifts)	Fresh[a]	Digested[b]	Mixed[c]	Compost[d]
	Cattle manure (% of total C)			
Acetyl, 5-30 ppm	12.0	13.4	10.2	11.0
Lipid, protein, 30-50 ppm	8.3	11.3	10.2	7.0
Methoxyl, 50-60 ppm	4.6	5.2	6.5	9.0
Carbohydrate I[e], 60-90 ppm	48.2	40.2	40.7	40.6
Carbohydrate II[f], 95-110 ppm	8.3	7.2	4.6	4.0
Aromatic lignin, 110-160 ppm	5.6	7.2	10.2	10.0
Carboxyl/carbonyl, 160-210 ppm	11.1	14.4	15.7	16.8
Aromaticity, (106-170)*100/(0-170)	6.2	9.5	15.2	13.0

Table 3.3. (Continued)

	Poultry manure (% of total C)			
Acetyl, 5-30 ppm	11.4	13.0	10.4	NR[g]
Lipid, protein, 30-50 ppm	10.5	15.2	7.3	NR
Methoxyl, 50-60 ppm	3.0	4.5	3.1	NR
Carbohydrate I, 60-90 ppm	49.5	35.9	52.1	NR
Carbohydrate II, 95-110 ppm	9.5	5.4	10.4	NR
Aromatic lignin, 110-160 ppm	4.8	7.6	5.2	NR
Carboxyl/carbonyl, 160-210 ppm	8.6	14.1	10.4	NR
Aromaticity, (106-170)*100/(0-170)	6.3	13.8	6.9	NR

[a]: Fresh (untreated) manure; [b]: Anaerobic digested manure; [c]: Mixedly (aerobically and anaerobically) processed manure; [d]: Aerobic stabilized manure (compost); [e]: C-2, C-3, C-4, C-5 and C-6 of cellulose and xylans; [f]: C-1 of cellulose and xylans and aliphatic lignin); [g]: Not reported.

3.4. ^{13}C NMR Characterization of Water Soluble Organic Matter of Animal Manure

3.4.1. General Features

Water extractable (soluble) organic matter (WEOM) is an important component of animal manure. It plays a significant role in the regulation of fertility and sorption properties in soil (Bolster and Sistani, 2009; Briceno et al., 2008 1227; Hunt et al., 2007). For this reason, Liang et al. (1996) characterized water extracts of stockpiled and composted dairy manures and determined the adsorption isotherms of the two WEOM fractions on three soils with different textures. The solid state ^{13}C CP/MAS NMR spectra of both extracts exhibited signals at 24, 58, 73, 105, 130, 153, 155 160, and 175 ppm. The most prominent resonances were at 58, 73, and 175 ppm. Judging from the spectra, aliphatic and aromatic C in the two water extracts were similar but the composted manure extract was richer in protein C, phenolic C, and carboxylic C and poorer in carbohydrate C (Table 3.3). These observations indicate that while the chemical composition of the two extracts was qualitatively similar, there were quantitative differences. The difference was even greater between the WEOM and the whole organic matter of animal manure as aromaticity is in a range from 6-15 for fresh and stabilized dairy manures (Gomez et al., 2007), but in a range from 32-35 in the two WEOM fractions of dairy manure (Liang et al., 1996) so that aromatic compounds in animal manure seem more soluble than aliphatic fractions which include insoluble fibrous C.

3.4.2. Comparison of Spectral Features with Plant-derived WEOM

The WEOM fractions of a dairy manure sample and seven plant shoot biomass samples were comparatively characterized by various spectroscopic techniques (He et al., 2009). The seven plant samples are field grown Alfalfa (*Medicago sativa* L.), corn (*Zea mays* L.), crimson clover (*Trifolium incarnatum* L.), hairy vetch (*Vicia villosa* L.), lupin (*Lupinus albus* L.), soybean (*Glycine max* L. Merr.), and wheat (*Triticum aestivum* L.). The ^{13}C CP/TOSS (cross polarization/total sideband suppression) spectra of all plant-derived WEOM fractions exhibited some similarities. Based on the general spectral-pattern (Figure 3.3), the spectra of plant-derived WEOM could be separated into three groups: those of (1) alfalfa, hairy vetch, and soybean, (2) crimson clover and lupin, and (3) corn and wheat. The ^{13}C CP/TOSS spectra of group 1 WEOM all showed a predominant band of OCH at 72 ppm, COO/N-C=O at ca. 178 ppm, O-C-O around 100 ppm, and OCH_2 at 62 ppm. Broad signals at around 55 ppm were assigned to NCH or OCH_3, and around 0-50 ppm to CCH, CCH_2, and CCH_3. However, these WEOM fractions contained very few aromatic functional groups. The spectra of group 2 differ from those of group 1 by additional peaks at 36, 51, and 176 ppm, which probably arose from asparagine. The spectra of group 3 WEOM were generally similar to those of group 1 WEOM. However, the peak of anomerics at 100 ppm and the peak of OCH at 72 ppm were very sharp compared with those of group 1, suggesting that some sugars be crystalline. Second, there were fewer signals in their CP/TOSS spectra and more residual peaks in their dipolar-dephased spectra around 50-60 ppm, indicating that less amino acids or peptides existed in wheat and corn than in alfalfa, hairy vetch, and soybean.

In contrast, the spectra of the dairy manure WEOM were totally different from those of plant-derived plants (Figure 3.3). Its ^{13}C CP/TOSS spectrum was very broad and especially the aromatic signals (110-160 ppm) were significant. There were also small carbonyl carbons around 200 ppm and CCH_3 around ~25 ppm which were not observed in the plant WEOM samples. Its dipolar-dephased spectrum showed rich environments of nonprotonated carbons and mobile groups, and also indicated that the carbonyl groups around 200 ppm were primarily from keto groups. The presence of OCH_3 and aromatic C-O indicated the existence of significant lignin; the N-C=O and NCH signals showed the existence of peptides; and the O-C-O functional groups were indicative of sugar rings in this manure. In summary, this work demonstrates that WEOM of dairy manure possesses significant amounts of nonprotonated carbons and lignin residues, suggesting the humification of the manure-derived WEOM. Such C structural information for different types of WEOM may be useful for understanding the effects of WEOM on soil nutrient availability to plants (He et al., 2009).

3.5. SOLID AND COLLOIDAL FRACTIONS OF ORGANIC MATTER OF MANURE SLURRY

3.5.1. Solid Fractions

Chen et al. (1989) characterized a solid fibrous fractions obtained from a slurry of cattle manure. The ^{13}C CP/MAS NMR spectra representing the fresh and composted (147 days) solid fractions were comparatively analyzed. Based on the analysis, Chen et al. (1989) semi-quantitatively reported the changes of C distribution of the solid fibrous fractions after composting. Composting process increased the alkyl groups by 42% from 13.7% to 19.5% of total C while the polysaccharides decreased by 33% from 58.4% to 39.0% of total C, probably resulted from the intensive decomposition of easily degradable compounds. Composting process also increased aromatic components by 47.3% from 14.8% to 21.8% of total C. Meanwhile, the aromaticity increased from 16% at the beginning of the composting process to 25% at the end. The relative quantity of carboxyl and ester C increased from 5.7 to

10.8% during the composting process (an increase of 90%). The least abundant C functional group, ketonic C=O, increased by 68% from 1.6% to 2.7% of total C. These data indicate that the main change during the composting process is the reduction in carbohydrates (by 33%) along with an increase in all other components (Chen et al., 1989). This observation on the separated solid fraction of cattle manure slurry is consistent with the C distribution changes during the composting of whole cattle manures (Gomez et al., 2007).

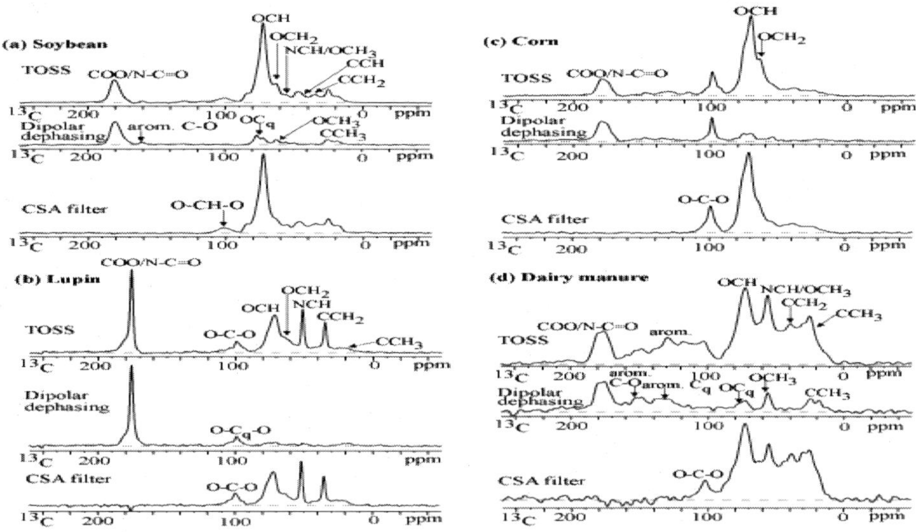

Figure 3.3. Typical solid state C-13 NMR spectra of plant- and manure-derived water extractable organic matter fractions. Spectral features of 13C CP/TOSS spectra showing qualitative structural information; corresponding dipolar-dephased (DD) ^{13}C CP/TOSS spectra showing nonprotonated carbons and mobile segments like CH_3; and selection of sp^3-hybridized carbon signals by a chemical-shift anisotropy filter, which in particular identified OCO carbons, near 100 ppm, typical of sugar rings. (a) Soybean, (b) Lupin, (g) Corn, and (d) Dairy manure. Figure is adapted from He et al. (2009). All spectra were recorded at a 400 MHz instrument.

3.5.2. Colloidal Fractions

Dinel et al. (1998) separated hog manure slurry into three fractions by mixing with $FeCl_3$ solution: floating coagulated colloids, clear solutions and fractions settling at bottom of the mixing column. As the colloidal fraction accounted for 40 to 45% of the dry matter, Dinel et al. (1998) characterized the C distribution of the colloidal fractions using ^{13}C CP/MAS NMR (Table 3.1). Similar to the whole swine manure (Mao et al., 2007), the most prominent resonances of the colloidal fractions of hog manure slurry are in the 20-60 ppm region, arising mainly from aliphatic C. In addition, they also assumed that the saturated ring carbons and carbons in side chains of sterols were major contributors to resonances in the 10-60 ppm regions whereas sterols compounds were identified in the colloidal fractions by pyrolysis-field ionization mass spectrometry (Dinel et al., 1998). The aromaticity of the colloidal fractions is higher than that of whole animal manure but lower than those of water extracts even though these manure types are different (Table 3.1). This observation seems reasonable as colloidal fractions could contain some water insoluble organic matter.

3.6. CHARACTERISTICS OF HUMIC SUBSTANCES DERIVED FROM ANIMAL MANURE

Animal manure is a valuable source of humic matter (Tan, 2003). Tan et al. (Tan et al., 1975) reported that the water soluble fraction of poultry litter carries properties similar to those of fulvic acids (FA). Although they are originally designed for the dark-colored organic materials in soil, the terminologies of humic substances have been used in organic matter fractions separated from animal manure and compost sources (Chien et al., 2007; Lin et al., 2004). Some researchers (Genevini et al., 2002b) have used the terms "humic acid (HA)-like" and "humin (HU)-like" to stress potential difference of these humic materials isolated from non-soil sources using $NaOH/Na_4P_2O_7$ extraction. These researchers further classified their organic matter materials separated in a second $NaOH/Na_4P_2O_7$ extraction after acid treatments to core-HA-like and core-HU-like fractions, and monitored modifications of these fractions during high-rate composting of pig feces amended with wheat straw using solid state ^{13}C CPMAS NMR and pyrolysis-GC/MS (Genevini et al., 2002a; Genevini et al., 2003). ^{13}C CP/MAS NMR analysis indicated the relative intensities of carbonyl, O-alkyl, methoxyl, and alkyl functional groups in HA-like fractions were 0.99, 1.10, 0.72, and 0.32 for pig feces, and 0.39, 1.27, 0.44 and 0.38 for wheat straw, respectively, compared to the intensity of aromatic group (Genevini et al., 2002a). The relative intensities of the four functional groups in core-HA-like fractions were 0.74, 0.70, 0.46, and 1.72 for pig feces, and 0.34, 0.68, 0.38 and 0.61 for wheat straw, respectively, lower than those in corresponding HA-like fractions. These data imply that the intensities of aromatic functional group in the core-HA-like fractions were relatively higher than in HA-like fractions. Compare with the structures of pig manures investigated by Mao et al. (2007b), these HA-like and core-HA like fraction of manures contain significantly more aromatics but they all contain significant lipids.

Composting changed the relative intensities of the four functional groups of HA-like and core-HA-like fractions (Figure 3.4). The biggest changes occurred with alkyl group. The abundance of O-alkyl was basically unchanged during the course. Another general observation is that the most dramatic change occurred in the first week of composting. To obtain more knowledge on the C distribution changes during composting, these researchers (Genevini et al., 2003) further characterized the insoluble HU-like and core-HU-like organic matter fractions in these pig feces-wheat straw composts. The change patterns of the four functional groups in HU- and core-HU-like fractions are different from those of HA- and core-HA-like fractions (Figure 3.4). An impressing observation is that the core-HU fractions were least in total relative intensity of the four functional groups in all the four humic-like fractions (Figure 3.4). In contrast, the O-alkyl group in the HU-like fraction was not only the most abundant group, but also decreased greatly during composting. Combined with pyrolysis-GC/MS data, the authors concluded that the core-HU-like fraction was mainly aromatic, while HU-like matter contained both core-HU-like and other types of easily degradable organic matter. Furthermore, during composting, core-HU-like matter underwent both conversion to new core-HA-like soluble matter and biodegradation to volatile products (Genevini et al., 2003).

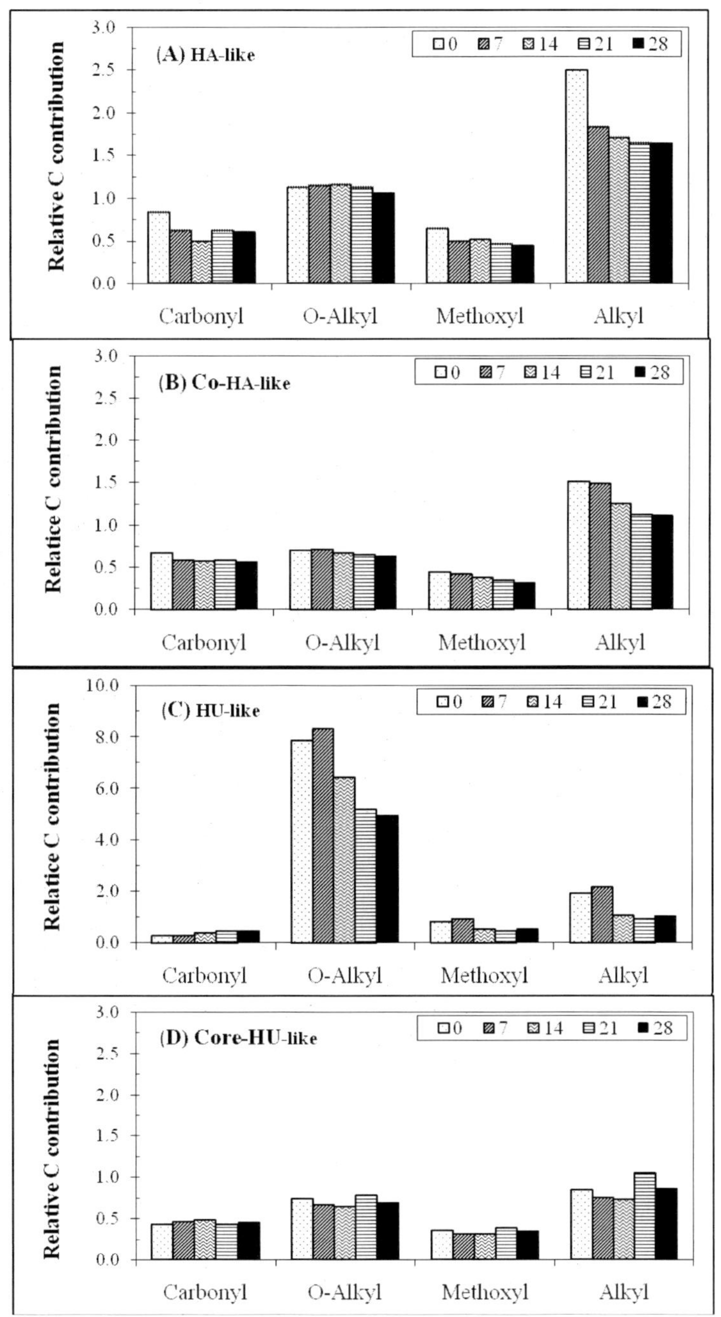

Fgiure 3.4. Change of C functional groups in humic acid (HA)-like, core-HA-like, humin (HU)-like, and core-HU-like fractions during high rate composting of pig faeces amended with wheat straw measured at day 0, 7, 14, 21 and 28. Relative C contribution was calculated from relavent CPMAS ^{13}C NMR signal areas with aromatic area=1. Figure is drawn based on data in literature (Genevini et al., 2002a; 2003).

Chien et al. (2007) extracted and characterized three fractions (HA, MW>1000; FA, MW>1000; and FA, MW<1000) of humic substances isolated from swine-based compost, assuming that these humic fractions derived from manure-based compost play a role in supplying plant nutrients and governing the reactions with organic and inorganic compounds in soil and associated environments. The ^{13}C CPMAS NMR spectra of these fractions are similar to those of humic substances extracted from soils (Conte et al., 1997). Semi-quantitatively, Chien et al. (2007) found that the sequence of the aliphatic C intensity of the humic substances extracted is FA (MW>1000)>HA (MW>1000)>FA (MW<1000), while that of the aromatic C intensity is HA (MW>1000)>FA MW<1000)>FA (MW>1000). The patterns of the carboxylic C intensity and the total functional group acidity are the same with the order of FA (MW<1000)>FA (MW>1000)>HA (MW>1000), showing the main contribution of carboxylic C to total functional group acidity. Thus, the authors concluded that the characteristic relation of carboxylic C intensities to total acidity of the humic substances derived from the swine manure-based compost is the same as the humic substances derived from natural soils. Furthermore, Chien et al. (2007) demonstrated that carboxylic C intensities, total acidities, and O intensities of the three humic fractions are in good correlation with the reactivities of the three fractions with heavy metals. Thus, data in this research confirmed that the impacts of manure or manure-based composts on mobility and bio-toxicity of heavy metals in soils are mainly through and regulated by small organic compounds in the humic fractions of the manure or compost.

3.7. CONCLUSION

Animal manure is generally high in organic matter intensity so it is well suitable for ^{13}C nuclear magnetic resonance (NMR) analysis. Solid-state ^{13}C NMR techniques used in characterizing organic matter and its components include, but are not limited, to cross-polarization /magic angle spinning (CPMAS), direct polarization/magic angle spinning (DPMAS), two-dimensional ^{1}H–^{13}C heteronuclear correlation (2D HETCOR), ^{13}C chemical shift anisotropy (CSA) filter, and saturation pulse-induced dipolar exchange with recoupling (SPIDER). Roughly, ^{13}C NMR signals can be assigned as alkyl, 0-45 ppm; NCH and OCH$_3$, 45-60 ppm; O-alkyl, 60-105 ppm; aromatics, 105-150 ppm; aromatic C-O, 150-164 ppm; N-C=O and COO, 164-190 ppm; and ketone, quinone, or aldehyde C, 190-220 ppm. Although all these C functional groups have been found in animal manure samples, solid state ^{13}C NMR studies reviewed in this chapter demonstrate that the relative intensity of these functional groups varied among types of manure, manure management practices and treatments (such as composting). Solid state ^{13}C NMR spectroscopy has also provided detailed C bonding information of water extractable organic matter, humic-like, and other fractions of animal manure under various management and environmental conditions. Information synthesized in this chapter provided current knowledge of manure organic matter derived from solid state ^{13}C NMR spectroscopy.

REFERENCES

Baldock, J.A., J.M. Oades, P.N. Nelson, T.M. Skene, A. Golchin, and P. Clarke. 1997. Assessing the extent of decompsition of natural organic materials usingsolid-state C-13 NMR spectroscopy. *Aust. J. Soil Res.* 35:1061-1083.

Bolster, C.H., and K.R. Sistani. 2009. Sorption of phosphorus from swine, dairy, and poultry manures. Commun. *Soil Sci. Plant Anal.* 40:1106-1123.

Briceno, G., R. Demanet, M.D. Mora, and G. Palma. 2008. Effect of liquid cow manure on andisol properites and atrazine adsorption. *J. Environ. Qual.* 37:1519-1526.

Chen, Y. 2003. Nuclear magnetic resonance, infra-red and pyrolysis: Application of spectroscopic methodologies to maturity determination of composts. *Compost Sci. Util.* 11:152-168.

Chen, Y., Y. Inbar, Y. Hadar, and R.L. Malcolm. 1989. Chemical properties and solid-state CPMAS ^{13}C NMR of composted organic matter. *Sci. Total. Environ.* 81/82:201-208.

Chien, S.W.C., M.C. Wang, C.C. Huang, and K. Seshaiah. 2007. Characterization of humic substances derived from swine manure-based compost and correlation of their characteristics with reactivities with heavy metals. *J. Agric. Food Chem.* 55:4820-4827.

Conte, P., R. Spaccini, and A. Piccolo. 2004. State of the art of CPMAS ^{13}C-NMR spectroscopy applied to natural organic matter. *Progr. Nucl. Magn. Reson.* Spectr. 44:215-223.

Conte, P., A. Piccolo, B. van Lagen, P. Buurman, and P.A. de Jager. 1997. Quantitative differences in evaluating soil humic substances by liquid and solid-state ^{13}C-NMR spectroscopy. *Geoderma* 80:339-352.

Dinel, H., M. Schnitzer, and H. Schulten. 1998. Chemical and spectroscopic characterization of collodial fracions separated from liquid hog manures. *Soil Sci.* 163:665-673.

Genevini, P., F. Adani, A.H.M. Veeken, K.G.J. Nierop, B. Scaglia, and C. Dijkema. 2002a. Qualitative modifications of humic acid-like and core-humic acid-like during high-rate composting of pig faeces amended with wheat straw. *Soil Sci. Plant Nutr.* 48:143-150.

Genevini, P.L., F. Adani, A.H.M. Veeken, and B. Scaglia. 2002b. Evolution of humic acid-like and core-humic acid-lik during high rate composting of pig feces mended with wheat straw. *Soil Sci. Plant Nutr.* 48:135-141.

Genevini, P.L., F. Tambone, F. Adani, A.H.M. Veeken, K.G.J. Nierop, and E. Montoneri. 2003. Evolution and qualitative modifications of humic-like matter during high rate composting of pig faces amended with wheat straw. *Soil Sci. Plant Nutr.* 49:785-792.

Golovan, S.P., R.G. Meidinger, A. Ajakaiye, M. Cottrill, M.Z. Wiederkehr, D.J. Barney, C. Plante, J.W. Pollard, M.Z. Fan, M.A. Hayes, J. Laursen, J.P. Hjorth, R.R. Hacker, J.P. Phillips, and C.W. Forsberg. 2001. Pigs expressing salivary phytase produce low-phosphorus manure. *Nat. Biotechnol.* 19:741-745.

Gomez, X., M.C. Diaz, M. Cooper, D. Blanco, A. Moran, and C.E. Snape. 2007. Study of biological stabilization processes of cattle and poultry manure by thermogravimetric analysis and 13C NMR. *Chemosphere* 68:1889-1897.

He, Z., J. Mao, C.W. Honeycutt, T. Ohno, J.F. Hunt, and B.J. Cade-Menun. 2009. Characterization of plant-derived water extractable organic matter by multiple spectroscopic techniques. *Biol. Fertil. Soils.* 45:609-616.

Hu, W.-G., J.-D. Mao, B. Xing, and K. Schmidt-Rohr. 2000. Poly(methylene) crystallites in humic substances detected by nuclear magnetic resonance. *Environ. Sci.Technol.* 34: 530-534.

Hunt, J.F., T. Ohno, Z. He, C.W. Honeycutt, and D.B. Dail. 2007. Influence of decomposition on chemical properties of plant- and manure-derived dissolved organic matter and sorption to goethite. *J. Environ. Qual.* 36:135-143.

Inbar, Y., Y. Chen, and Y. Hadar. 1989. Solid state carbon-13 nuclear magnetic resonance and infrared spectroscopy of composted organic matter. *Soil Sci. Soc. Am. J.* 53:1695-1701.

Kogel-Knabner, I. 1997. ^{13}C and ^{15}N NMR spectroscopy as a tool in soil organic matter studies. *Geoderma* 80:243-270.

Liang, B.C., E.C. Gregorich, M. Schnitzer, and H. Schulten. 1996. Characterization of water extracts of two manures and their adsorption on soils. *Soil Sci. Soc. Am. J.* 60:1758-1763.

Lin, H.T., M.C. Wang, and G.C. Li. 2004. Complexation of arsenate with humic substance in water extract of compost. *Chemosphere* 56:1105-1112.

Mao, J.-D., W.-G. Hu, G. Ding, K. Schmidt-Rohr, G. Davies, E. A. Ghabbour, and B. Xing. 2002. Suitability of Different ^{13}C solid-state NMR techniques in the characterization of humic acids. *International J. Environ. Anal. Chem.* 82: 183-196.

Mao J.-D. and K. Schmidt-Rohr. 2003. Long-range C-H dipolar dephasing in solid-state NMR, and its use for spectral selection of fused aromatic rings. *J. Magnetic Reson.* 162: 217-227.

Mao J.-D. and K. Schmidt-Rohr. 2004. Separation of acetal or ketal O-C-O ^{13}C NMR signals from aromatic-carbon bands by a chemical-shift-anisotropy filter. *Solid State NMR.* 26: 36-45.

Mao J.-D. and K. Schmidt-Rohr. 2006. Absence of mobile carbohydrate domains in dry humic substances proven by NMR, and implications for organic-contaminant sorption. Environ. *Sci.Technol.* 40: 1751-1756.

Mao J.-D. and K. Schmidt-Rohr. 2005. Methylene spectral editing in solid-state NMR by three-spin coherence selection. *J. Magnetic Reson.* 176: 1-6.

Mao, J.-D., L. Tremblay, J.P. Gagne, S. Kohl, J. Rice, and K. Schmidt-Rohr. 2007a. Humic acids from particulate organic matter in the Saguenay Fjord and the St. Lawrence Estuary investigated by advanced solid-state NMR. Geochim. *Cosmochim. Acta* 71:5483-5499..

Mao, J.-D., A. Ajakaiye, Y. Lan, D.C. Olk, M. Ceballos, T. Zhang, M.Z. Fan, and C.W. Forsberg. 2007b. Chemical structure of manure from conventional and phytase transgenic pigs investigated by advanced solid-state NMR spectroscopy. *J. Agric. Food Chem.* 56:2131-2138.

Mao, J., D.C. Olk, X. Fang, Z. He, J. Bass, and K. Schmidt-Rohr. 2008. Influence of animal manure application on the chemical structures of soil organic matter as investigated by advanced solid-state NMR and FT-IR. *Geoderma* 146:353-362.

Moral, R., J. Moreno-Caselles, M.D. Perez-Muricia, A. Perez-Espinosa, B. Rufete, and C. Pareds. 2005. Characterisation of the organic matter pool in manures. *Bioresour. Technol.* 96:153-158.

Preston, C.M. 1996. Applications of NMR to soil organic matter analysis: history and prospects. *Soil Sci.* 161:144-166.

Schmidt-Rohr, K. and J.-D. Mao. 2002. Efficient CH-group selection and identification in ^{13}C solid-state NMR by dipolar DEPT and ^{1}H chemical-shift filtering. *J. Am. Chem. Soc.* 124: 13938-13948.

Schmidt-Rohr, K., J.-D. Mao, and D.C. Olk. 2004. Nitrogen-bonded aromatics in soil organic matter and their implications for a yield decline in intensive rice cropping. *Proc. Natl. Acad. Sci. USA* 101: 6351-6354.

Schnitzer, M.I., C.M. Monreal, G.A. Facey, and P.B. Fransham. 2007. The conversion of chicken manure to biooil by fast pyrolysis I. Analyses of chicken manure, biooils and char by ^{13}C and ^{1}H MR and FTIR spectrophotometry. *J. Environ. Sci. Health. Part B.* 42:71-77.

Tan, K.H. 2003. p. 27-29. *Humic matter in soil and the environment.* Marcel Dekker, Inc., New York, N.Y.

Tan, K.H., V.G. Mudgal, and R.A. Leonard. 1975. Adsorption of poultry litter extracts by soil and clay. *Environ. Sci. Technol.* 9:132-135.

Tang, J., N. Maie, Y. Tada, and A. Katayama. 2006. Characterization of the maturing process of cattle manure compost. *Process Biochem.* 41:380-389.

Zmora-Nahum, S., Y. Hadar, and Y. Chen. 2007. Physico-chemical properties of commercial composts varying in their source materials and country of origin. *Soil Biol. Biochem.* 39:1263-1276.

In: Environmental Chemistry of Animal Manure
Editor: Zhongqi He

ISBN: 978-1-62808-641-6
© 2013 Nova Science Publishers, Inc.

Chapter 4

ULTRAVIOLET-VISIBLE ABSORPTIVE FEATURES OF WATER EXTRACTABLE AND HUMIC FRACTIONS OF ANIMAL MANURE AND RELEVANT COMPOST

Mingchu Zhang[1,], Zhongqi He[2] and Aiqin Zhao[1]*

4.1. INTRODUCTION

The absorption of electromagnetic radiation in the ultraviolet (UV, 200−400 nm) and visible (400−800 nm) regions is associated with the electronic transitions of the bonding electrons in a matter. The absorption of UV-visible radiation by organic compounds is due to the presence of specific segments or functional groups (chromospheres) which contain unbonded electrons (e. g., carbonyl groups, S, N, or O atoms, and conjugated C-C multiple bonds) (Swift, 1996). Theoretically, the UV/visible absorbance spectrum of a compound is a characteristic which can be used in its identification. However, because the peaks of UV/visible absorbance spectra of natural organic matter are broad, it is difficult to identify a particular compound in a mixture of simple molecules and practically impossible in a complicated organic matter sample such as dissolved organic matter from soils or animal manure (Swift, 1996). However, the color of natural organic matter did attract the attention of many scientists who have attempted using the UV/visible spectroscopy for organic matter characterization (Tan, 2003, Baes and Bloom, 1990; Chen et al., 2002; Domeizel et al., 2004; Ghosh and Schnitzer, 1979; Kalbitz et al., 2000; Wang and Hsieh, 2001).

Whereas the UV/visible absorbance spectra are generally broad and featureless, the absorption intensity or absorptivity at certain wavelengths of organic matter components varies with the types, sources, environmental factors, and management conditions under which samples are taken. The variations constitute the basis of UV-visible spectroscopic

---

[*] Corresponding Author: mzhang3@alaska.edu
[1] Department of High Latitude Agriculture, School of Natural Resources and Agricultural Sciences, University of Alaska Fairbanks, Fairbanks, AK 99775, USA
[2] USDA-ARS, New England Plant, Soil and Water Laboratory, Orono, ME 04469, USA

characterization of natural organic matter (Table 4.1). For example, the absorbances at 260 and 280 nm have been frequently used to monitor the dissolved organic carbon fractions eluted from size-exclusion high performance liquid chromatography (HPLC) columns (Li et al., 2007; Takeda et al., 2009). Specific absorption at 254 nm ($SUVA_{254\ nm}$; i.e., measured absorptivity divided by the dissolved organic carbon concentration) has been an important parameter to assess the aromaticity of dissolved organic matter (Jaffrain et al., 2007; Musikavong and Wattanachira, 2007). The absorbance ratios at 250 and 365 nm (E2/E3), and at 465 and 665 nm (E4/E6), can be used to characterize a variety of properties of organic matter such as molecular weight, aromaticity, and polarity (Heymann et al., 2005; Yang and Xing, 2009). In this chapter, we first review and synthesize the information on these UV-visible absorptive features of organic matter fractions of animal manure and manure-related compost. We then use two case studies to comparatively analyze the UV/visible absorptivities of the water extractable organic matter (WEOM) fraction of conventional and organic dairy manure, and hay field soil with long-term histories (0–20 years) of poultry litter application.

4.2. UV/VISIBLE SPECTRA OF MANURE FRACTIONS

He et al. (2003; 2009a) reported on the UV/visible spectra of WEOM of soil, dairy manure, and plant shoot samples (Figure 4.1). The spectrum of soil WEOM is made of a monotonically-decreasing curve with the increasing wavelength. This basically-featureless characteristic of the UV-visible spectrum of the soil WEOM sample is typical for soil humic substances (Baes and Bloom, 1990; Ghosh and Schnitzer, 1979), which indicates that there are many different chromospheres in this sample. The UV/visible spectrum of dairy manure WEOM shows an absorbance shoulder around 280 nm. In the spectra of plant-derived WEOM fractions, the absorbance shoulder is more obvious, or becomes an absorbance peak between 260 and 280 nm. The absorbance peak or shoulder is due to aromatic and/or phenolic compounds with conjugated C=C and C=O double bonds which have strong absorbance in the range of 200 to 300 nm (Abbt-Braun et al., 2004; Baes and Bloom, 1990). An apparent second absorbance shoulder or peak between 300–350 nm appeared in the spectra of the WEOM fractions of corn, hairy vetch, and alfalfa, indicating the presence of some ring-fission products of phenolic carboxylic compounds in these WEOM fractions as strong absorbance of these ring-fission intermediates has been frequently observed during microbial metabolism of phenolic compounds (He and Spain, 2000). From the comparison, it is obvious that the UV/visible spectrum of manure WEOM is more complicated than that of soil sample, but simpler than plant WEOM samples. The order seems reasonable considering the abundance and complicacy of organic matter in the three types of materials.

He et al. (2003) also compared the UV/visible spectra of swine manure fractions sequentially extracted by H_2O, 0.5 M $NaHCO_3$ (pH 8.5), 0.1 M NaOH, and 1 M HCl (Figure 4.2). Although the UV/visible absorbance may arise from both organic (such as carbonyl group, conjugated C-C multiple bonds, hetero atoms) and inorganic (such as Fe, Cu, or Mn complexes) chromospheres, dissolved organic matter should be the major contributor as the HCl fraction contains the highest concentrations of Fe and Mn but shows the weakest absorbance. Whereas the authors (He et al., 2003) visually observed that the first three fractions of the swine manure were brown, and the fraction extracted by 1 M HCl was clear, the UV/visible spectroscopy disclosed clearly the differences among these fractions. Although the spectra of these fractions in the 400–800 nm are featureless, differences in the UV (< 400 nm) absorbance between the fractions are evident. For example, the shapes, but not the strengths, of the spectra between H_2O and $NaHCO_3$ fractions, and between NaOH and HCl fractions, are similar (Figure 4.2).

Yin et al. (2010) reported on UV/visible spectroscopic characteristics of pyrolysis bio-oil derived from cattle manure. They identified the conjugate structures of benzene, carbonyl group, and polycyclic aromatic hydrocarbon in the bio-oil product by the wavelengths of 250 – 290, 290 –300 and >300 nm, respectively. The absorbance in the UV/visible spectrum shows the presence of aromatic hydrocarbons, which imply that the bio-oil might have high density and viscosity (Yin et al., 2010).

4.3. SPECIFIC ULTRAVIOLET ABSORBANCE WAVELENGTHS ASSOCIATED WITH PROPERTIES OF MANURE ORGANIC MATTER

4.3.1. E2/E3 Ratio

The E2/E3 ratio is a bulk spectroscopic parameter that has been widely related to the molecular weight property of dissolved organic matter. The ratio is the quotient of absorbance at 250 nm to that at 365 nm (Guo and Chorover, 2003; Peuravuori and Pihlaja, 1997; Yang and Xing, 2009), or to that at 254 and 365 nm (Hunt et al., 2007b; Ohno et al., 2005). Generally, a low E2/E3 ratio reflects a high average molecular weight (Peuravuori and Pihlaja, 1997). Similar to the E2/E3 ratio, Sellami et al. (2008) reported on E2/E4 and E2/E6 ratios. They proposed that the ratio E2/E4 be used as an indicator of the relative abundance of lignin at the beginning of humification, and that the ratio E2/E6 reflects the relation between non-humified and highly humified materials. However, it is noteworthy that the authors measured E2, E4, and E6 at 280, 472, and 664 nm, respectively, which differ from the wavelengths used in most of other studies.

Hunt et al. (2007b) reported that E2/E3 ratios of the water extracts of fresh and 10-day incubated dairy manure were 3.16 and 2.94, respectively. However, E2/E3 ratios of the water extracts of plant residues either decreased or increased after the same 10-day decomposition incubation. Hunt et al. (2007a) separated both fresh and decomposed WEOM of animal manure and plant residues into hydrophilic and hydrophobic fractions. The E2/E3 ratios of the hydrophilic and hydrophobic fractions were 64.0 and 6.7 for fresh poultry manure, respectively, and 3.2 and 3.7 for fresh dairy manure, respectively. After decomposition, the E2/E3 ratios of the two fractions decrease to 6.0 and 4.7 for poultry manure, respectively, and to 1.9 and 2.1 for dairy manure, respectively. Of total 10 samples, the E2/E3 ratio of seven hydrophilic, but four hydrophobic fractions are significantly impacted by decomposition. The result supports the conclusion that the hydrophobicity of natural dissolved organic matter generally increases with decomposition. Hunt and Ohno (2007) further determined the E2/E3 ratio of fresh and decomposed WEOM derived from 10 plant biomass materials and three manure (dairy, swine, and poultry) sources and analyzed the relationship between the E2/E3 ratio and the inherent fluorescent components. The tryptophan- and tyrosine-like components showed a strong correlation with E2/E3 ratio, suggesting a relationship between these two fluorphores and the low molecular weight compounds in these dissolved organic matter samples.

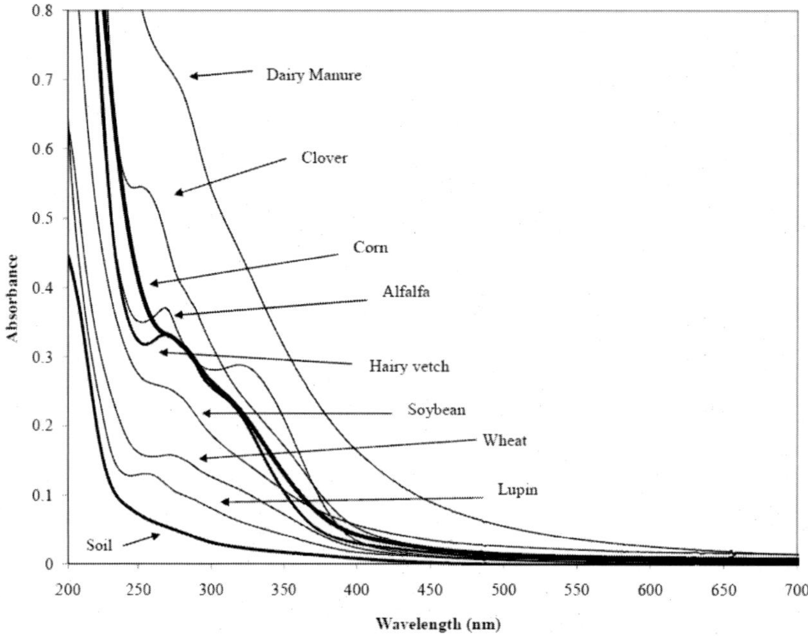

Figure 4.1. UV-visible spectra of water extracts of dairy manure, soil, and plant shoots. Data are adapted from He et al. (2003, 2009a).

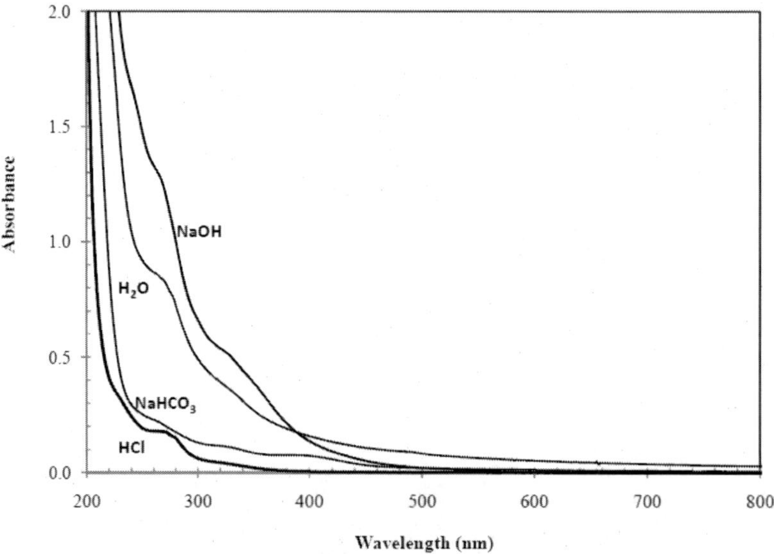

Figure 4.2. UV-visible spectra of sequentially-extracted H_2O, 0.5 M $NaHCO_3$ (pH 8.5), 0.1 M NaOH and 1 M HCl fractions of swine manure. Data are adapted from He et al. (2003).

Table 4.1. UV/visible wavelengths used in dissolved organic matter characterization.

Wavelength (nm)	Property
250	Aromaticity, apparent molecular weight
254	Aromaticity
260	Hydrophobic C content
265	Relative abundance of functional groups
272	Aromaticity
280	Hydrophobic C content, humification index, apparent molecular size
285	Humification index
300	Humic substance characterization
340	Humic substance characterization
350	Color
360	Humic substance characterization
365	Apparent molecular size
400	Color, relative abundance of functional groups
436	Quality indicator
465	Relative abundance of functional groups
472	Relative abundance of functional groups
600	Color, relative abundance of functional groups
665	Relative abundance of functional groups
664	Relative abundance of functional groups

Compiled from Jaffrain et al. (2007), He et al. (2006), Sellami et al. (2008) and references therein.

4.3.2. SUVA

Specific absorption or absorptivity at 254 nm (i.e., L mg^{-1} m^{-1}) is the absorbance divided by the dissolved organic carbon (DOC) concentration. It is frequently referred to as specific ultraviolet absorbance (SUVA) (Zmora-Nahum et al., 2007), SUVA$_{254}$ (Shao et al., 2009), or SUVA$_{254\ nm}$ (Jaffrain et al., 2007). This parameter has been used to assess the aromaticity of DOC from various sources (Embacher et al., 2008; Jaffrain et al., 2007; Musikavong and Wattanachira, 2007). Shao et al. (2009) measured SUVA values of WEOM during biostabilization of municipal solid waste (MSW). They found that the SUVA value increases from 0.04 to 3.13 m^{-1} mg^{-1} C L during the active stage of decomposition coupling with a decrease in DOC concentration and an increase in relatively resistant aromatic fraction released during MSW biostabilization. After day 30, the SUVA value remained steady at around 3.00 m^{-1} mg^{-1} C L, indicating the stabilization of the MSW. Embacher et al. (2008) characterized the WEOM of the soils under three fertilization treatments in a Haplic Chernozem differing in fertilization intensity for over 90 years: (i) no fertilization (Control), (ii) mineral fertilization (NPK), and (iii) mineral plus additional farmyard manure fertilization (NPK+FYM). Their data demonstrated that fertilization increases SUVA value and thus, the aromatic content of WEOM. The SUVA value increased following the order: Control < NPK < (NPK + FYM). The authors also noticed that under all fertilization schemes, the SUVA values were significantly lower in the samples from below the plough depth than within the plough depth, indicating a selective retardation of WEOM consisting of relatively condensed and aromatic compounds.

Zmora-Nahum et al. (2007) measured the absorbance at 254 and 465 nm and reported the SUVA values of water extracts of 37 commercial composts based on wood, green, animal manure, grape marc (GM), oilcake, spent mushroom substrate (SMS), and municipal solid waste (MSW). The absorbance was divided into two groups: high for samples containing high concentrations of humified materials and low for samples containing high concentrations of low molecular weight compounds. The absorbance of extracts from wood, green, manure, MSW and SMS-based composts belonged to the first group whereas the GM and oilcake compost extracts fell within the second group. The only exception was the SMS sample that demonstrates an exceptionally low absorbance in the visible range at a high DOC concentration. For the first group, the absorbance of the compost extracts in the UV range (254 nm) correlates well with DOC concentration (R^2 =0.86). Except for the SMS sample, all other samples show high correlation (R^2 = 0.84) between the DOC concentration and the absorbance at 465 nm. The low SUVA values of the second group suggest that these oilcake and GM composts were rich in soluble aliphatic compounds (Zmora-Nahum et al., 2007).

4.3.3. Absorptivity at 280 nm

The absorptivity at 280 nm, also referred to $SUVA_{280}$ (Cook et al., 2009), is a parameter to estimate the aromaticity and the molecular size of DOM (Table 4.1). Using absorptivity at 280 nm, Ohno and Crannell (1996) estimated the average molecular weights of two green (vetch and clover) manure WEOMs at ~710 and ~ 850, respectively, and of cattle and dairy manure WEOMs at ~2000 and ~2800, respectively. The higher molecular weights are thus used to explain the inability of the animal manure extracts to inhibit P sorption. Hunt et al. (2007a) further measured the absorptivity at 280 nm of the hydrophilic (HPL) and hydrophobic (HPB) fractions of WEOM of these two animal manures before and after a 10-d laboratory microbial decomposition. The HPL fractions have absorptivity values equal to or greater than the corresponding HPB fractions. The molar absorptivity was 432 L mol^{-1} cm^{-1} for the fresh poultry manure-derived HPL fractions, and 225 L mol^{-1} cm^{-1} for the fresh HPB fractions. The authors (Hunt et al., 2007a) attributed the difference to the presence of high concentrations of relatively polar N-containing compounds such as polypeptides and nucleic acids in the HPL fraction. The low molar absorptivity (33.8 L mol^{-1} cm^{-1}) of the decomposed poultry manure-derived HPL fraction suggests the decomposition of these UV absorbing organic N compounds in the fresh poultry manure-derived WEOM fraction. Whereas a high initial molar absorptivity (117 L mol^{-1} cm^{-1}) of the dairy manure-derived HPL WEOM was observed, the value subsequently increased to 304 L mol^{-1} cm^{-1} following decomposition. The contrary observation suggests the presence of humified organic carbon compounds in the dairy manure WEOM fractions. Indeed, nine of the 10 HPL fractions (both manures plus shoots and roots of corn, soybean, hairy vetch and crimson clover) demonstrated significantly higher absorptivity values after decomposition, indicating greater aromatic content of the post-decomposition WEOM. Following decomposition, four of the 10 HPB fractions significantly increased in absorptivity relative to their fresh values, whereas six decreased. The results illustrate the extent and complexity of the changes that can occur to WEOM during even a short-term decomposition of its source material (Hunt et al., 2007a).

Mathur et al. (1993) determined the UV/visible absorbance of water extracts of four types of compost containing various fresh animal manures and shredded waste paper. During the 59-day composting, the absorption at 280, 465, and 665 nm of these compost water extracts rises sharply after start of composting, but declined by day 20. It further declines and levels until day 59 for nearly all feedstocks. At the end of composting, the absorption values at 280 and 465 nm were not different from day zero, although the soluble C concentration had decreased. As such, Mathur et al. (1993) claimed that the absorbance at neither wavelengths can be used as a desirable parameter for determining compost biomaturity. However, the absorbance at 665 nm was well below the absorbances at other test wavelengths for all feedstocks at time zero and thus, the author proposed to use the absorbance at 665 nm as a reliable, scientific sound, valid test method for compost maturity for regulators because of its high sensitivity. Schnitzer et al. (1993) evaluated the proposed test method by solid state ^{13}C-NMR spectroscopy and pyrolysis-field ionization mass spectrometry. Their results corroborate the conclusion of Mathur et al. (1993) that absorption at 665 nm of compost water extracts could serve as a suitable test for compost maturity.

4.4. VISIBLE E4/E6 RATIOS OF MANURE ORGANIC MATTER FRACTIONS

4.4.1. Definition of E4/E6

The absorbance ratio, E4/E6, was first used as an index for the rate of light absorption of dilute humic substance solutions in the visible range (Tan, 2003). The E4/E6 ratio was originally measured at 400 and 600 nm (Tan, 2003), and has still been reported in literature (Desalegn et al., 2008; He et al., 2006). The ratio of absorbance has recently been measured at 465 and 665 nm in many studies (Chen et al., 1977; Kang and Xing, 2008). Occasionally, the E4/E6 ratio is measured at 472 and 664 nm (Sellami et al., 2008). A high E4/E6 ratio, 7–8 or higher, corresponds to curve with steep slopes and is usually observed for fulvic or humic acids of relatively low molecular weights. On the other hand, a low E4/E6 ratio, 3–5, corresponds to less steep curves that are observed for humic acids and other related high molecular weight compounds (Tan, 2003). As aliphatic molecules are larger in molecular size than aromatic molecules, the lower E4/E6 ratios reflect an increase in aliphaticity of humic acid fractions (Heymann et al., 2005). Kang and Xing (2008) proposed that lower E4/E6 ratio reflects well-decomposed soil organic matter which has a relatively lower polarity.

4.4.2. Measuring Conditions of E4/E6

Chen et al. (1977) showed that the E4/E6 ratios of humic and fulvic acids could be affected by pH, but was independent of humic and fulvic acid concentrations, at least in the 100–500 mg L^{-1} range. Thus, they proposed a procedure to measure the E4/E6 ratio of humic and fulvic acids. First, 2–4 mg of sample are dissolved in 10 mL of 0.05 M NaHCO$_3$ (pH 8.3) solution. Due to the acidic property of the sample, the resulting pH of the sample solution is near 8.0. Then, the absorbance of the sample solution is measured at 465 and 665 nm with

0.05 M NaHCO$_3$ (pH 8.3) solution as a blank reference. Whereas it has been adopted by some researchers (Giusquiani et al., 1998; Heymann et al., 2005), this procedure is not always followed for various practical reasons. In addition, it seems more feasible sometimes to measure the E4/E6 ratio in other solutions (e.g. water extracts, Mathur et al., 1993) or at a pH value that is also suitable for other characterization purposes (Grassi, et al., 2010). There are also some other papers with no information on measuring conditions reported (Li et al., 2001). Due to these reasons, we do not put the measuring conditions of E4/E6 ratios when we list these values in Table 4.2.

Table 4.2. E4/E6 ratio of animal manure and related compost components.

Sample	Fraction	E4/E6	Reference
Cattle farmyard manure, poultry manure, pig slurry, and other organic wastes	Humic acid Fulvic acid	2.6-5.2 2.4-3.3	Riffaldi et al., 1983
Mixtures of dairy, beef, pig and sheep manure with or without shredded paper	Water extract	3.7 -5.8 (B)[1] 7.8-9.2 (A)	Mathur et al., 1993
Pig sludge	Acid soluble Acid insoluble	11.6 4.4	Giusquiani et al., 1998
Pig manure, sewage sludge, and sawdust	Humic+ fulvic Humic acid Fulvic acid	1.69, 6.23, and 4.94 4.66, 7.16, and 10.97 7.33, 6.62, and 3.06	Li et al., 2001
Mixture of pig manure, sewage, and sawdust	Humic+ fulvic Humic acid Fulvic acid	2.41(B) and 1.37 (A) 5.99 (B) and 3.66 (A) 3.66 (B) and 26.59 (A)	Li et al., 2001
Pig slurry	Humic acid Fulvic acid	4.0 14.1	Plaza et al., 2002; 2003
Mixture of pig manure and sawdust	Humic acid Fulvic acid	3.56 (B) and 2.50 (A) 7.69 (B) and 4.83 (A)	Huang et al., 2006
Cattle manure	Humic acid	8.8 (B) and 5.1 (A)	Plaza et al., 2008
Olive pomace and cattle manure	Humic acid	8.8 (B) and 4.8 (A)	Plaza et al., 2008
Mixtures of horse manure and biowaste	Humic acid	8.5-11.5 (B) 3.8-9.5 (A)	Desalegn et al., 2008
Cattle manure compost	Humic acid	2.35	Grassi and Rosa, 2010

[1] B=before composting, A=after composting.

4.4.3. Absorbance at 465 and 665 nm of Water Extracts of Manure and Compost

Mathur et al. (1993) found that the E4/E6 ratio of water extracts of four animal manure mixtures changed from 3.7 to 9.2 over the 59-day composting time (Table 4.2). The increase implied that the water extracts contained less condensed aromatic substances at later stage composting, in contrast to the authors' expectation. Thus, absorbance at 665 nm as a parameter for compost biomaturity (refer to discussion in section 4.3.3) was recommended (Mathur et al., 1993). Using the absorbance at 665 nm of water extracts, Charest and Beauchamp (2002) evaluated the maturity of composts of de-inking paper sludge, poultry manure, and chicken broiler floor litter that contains 0.6%, 0.7%, and 0.9% of N, respectively. After 24 weeks of composting, the absorbance at 665 nm of hot water extracts from all these three feedstocks was still higher than 0.008, a value indicating maturity as suggested by Mathur et al. (1993). Based on this test, the authors (Charest and Beauchamp, 2002) concluded that none of their compost piles could be considered mature after 24 weeks of composting.

Similarly, Zmora-Nahum et al. (2005) measured absorbance at 465 nm of the water extracts of three composts from three types of source materials (municipal solid waste, separated cow manure, and biosolids). The organic carbon (OC) concentration of water extracts from all the composts decreases rapidly within the first month and then stabilized at 4 g kg^{-1} towards the end of the composting process. The absorbance of the compost water extracts at 465 nm correlated highly with the OC concentrations, with the regression coefficient (R^2) values ranging from 0.82 to 0.99. With the high correlation, Zmora-Nahum et al. (2005) suggested that the absorbance at 465 nm may be used as a simple and cost-effective substitute for OC determination, which can also serve as a tool for compost producers after calibration.

4.4.4. E4/E6 Ratio of Manure and Compost Fractions

Unlike the water extracts of compost (Mathur, 1993), E4/E6 ratios of humic acids extracted from manures alone or with other materials all decrease after composting (Table 4.2). These lower E4/E6 ratios reflect relatively large particle size, molecular weight, and humification degree of humic acids in these materials after composting. The consistent decreases imply that the E4/E6 ratio of compost humic acid fractions is a meaningful indicator for evaluating the maturity of compost (Huang et al., 2006). However, the observation on the E4/E6 ratios of fulvic acids of two composting samples is not consistent (Table 4.2). The E4/E6 ratio of fulvic acids in the mixture of pig manure, sewage, and sawdust had a markedly increase from day-0 to day-100 of composting (Li et al., 2001), whereas the value of the fulvic acids in the mixture of pig manure and sawdust decreased after a 63-day composting (Huang et al., 2006). Apparently, more research is needed to understand how the fulvic acid fraction changes during composting.

The spectral ratio of E4/E6 has also been used as a parameter for organic waste characterization. Riffaldi et al. (1983) measured E4/E6 ratios of humic and fulvic acids of various organic waste samples including rye-straw, cattle farmyard manure, aerobic sewage sludge, poultry manure, municipal refuse compost, and pig slurry. Generally, the E4/E6 ratios were significantly higher, with the exception of the pig slurry sample, in the humic acid fractions than in the fulvic acid fractions (Table 4.2). Giusquiani et al. (1998) measured E4/E6 ratios of acid soluble OC (ASDOC) and acid insoluble dissolved OC (AIDOC) fractions of pig sludge. The E4/E6 ratio (11.6) of ASDOC was higher than typical values of soil fulvic acids, suggesting the degree of molecular complexity in this fraction may be considered lower than soil fulvic acids. The E4/E6 ratio (4.4) of AIDOC is comparable to typical values of soil humic acids. Thus, the ASDOC and AIDOC fractions of pig sludge were similar to fulvic and humic acids, respectively. Plaza et al. (2002; 2003) reported on the E4/E6 ratios of the fulvic acids (14.1) and humic acids (4.0) of pig slurry (Table 4.2). They found that both values were smaller than those of soil fulvic acids (18.0) and humic acids (4.9), respectively. However, all E4/E6 values of both fractions of soils amended with pig slurry were smaller than those of unamended soil, but greater than those of pig slurry. This observation suggests the contribution of pig slurry humic substances to soils.

4.5. SPECTRAL MODELING

4.5.1. Spectral Slope

In addition to use of SUVA at different wavelengths and absorptive ratios (E2/E3, and E4/E6) to characterize dissolved organic matter, spectral slopes have been used as a proxy for quantity and quality of water soluble organic compounds in fresh or sea waters. There are several forms of spectral models in which the spectral slope can be developed from, but the single exponential model (SEM) is often used because the slope can easily be derived from a linear regression after a log transformation of following equation.

$$a_\lambda = a_{\lambda ref}e^{-S(\lambda - \lambda_{ref})}$$

Eq. 4.1

where a = absorptivity (m^{-1}); λ = wavelength (nm); and λ_{ref} = reference wavelength.

The slope simulated from Eq. 4.1 has been used to determine the ratio of humic and fulvic acids in a sample (Carter et al., 1989) and the nominal molecular weight of fulvic acids (Hayase and Tsubota, 1985), and to differentiate seasonal and spatial variation and sources of dissolved organic matter in sea water (Stedmon and Markager 2001; Del Vecchio and Blough 2004). However, the value for the slope varies, depending on the range of wavelength that is used for simulation. Sarpal et al. (1995) obtained a better simulation result for Antarctic seawater when narrow wavelength intervals (260 to 330 nm and 330 to 410 nm) were used in comparison with broad wavelength intervals (260 to 410 nm). In addition, statistical approaches (log transformation and linear regression vs. non-linear regression) used to derive the slope S can also lead to variations of the results. Nevertheless, progress has been achieved that different spectral slopes is related to different portion of chromophores of dissolved organic matter in water (Twardowski et al., 2004).

In addition to SEM, there are other models used to derive the spectral slopes. These are SEM + constant (Stedmon and Markager, 2001), hyperbolic, double exponential with second slope fixed, and double exponentials (Twardowski et al. 2004). Twardowski et al. (2004)

showed that these models can better characterize the absorbance of dissolved organic matter in water than SEM.

There is no report on spectral slope values of water soluble organic matter in animal manures, compost, and manure/compost amended soils. Using SEM + constant, Zhang et al. (2006) compared the spectral slopes of water extractable organic matter from soils of three different land uses. They found that the slope value is higher for water extractable organic matter (molecular weight less than 1 kD) from forest soil (0.025) than from agricultural soil (0.018). For large size fractions of water extractable organic matter (> 1 kD), the spectral slope value is similar between forest and agricultural soils. With an increasing trend of organic amendment in agricultural soils, the change of water soluble soil organic matter needs to be well understood because water soluble organic matter in soil not only affects plant nutrients, but also directly contributes to dissolve organic matter in fresh water and sea water. The work by Zhang et al. (2006) suggests that the spectral slope may be a useful tool for agricultural scientists to understand soil soluble organic matter in addition to the spectral ratios that researchers currently use.

4.5.2. Spectral Deconvolution

The above mentioned modeling approaches are all based on the shape of absorbance curve against wavelength. However, another fundamentally different approach, UV spectral deconvolution, has been also reported. Through this approach, the overlapped signals can be deconvoluted from the known reference signals. The mathematical principle for such deconvolution is that an overlapping signal of unknown compounds in a solution is a series of known reference signals in a linear combination as follow:

$$P = \sum_{i=1}^{q} a_i\, PR_i \pm r_p \qquad\qquad\qquad \text{Eq. 4.2}$$

where P is unknown spectrum, a_i and r are respective coefficients from the ith reference spectrum PR_i. This modeling approach is originally developed as a method to quickly estimate dissolved organic carbon in waste water (Gallot and Thomas, 1993). Using such relationship, Thomas et al. (1993) determined organic carbon, suspended solid, and nitrate in wastewater. Domeizel et al. (2004) used such method to monitor compost maturity. They separated different stages of compost into three components (i. e. non-humified fraction, fulvic acids, and humic acids), and found the ratio of humification (C_{HA}/C_{FF}, HA= humic acids, FF=fulvic fraction) in compost is related to the modeling parameters of three reference coefficients. Hassouna et al. (2007) proposed this method as a quick method to estimate nitrate and total and fractional water extractable organic carbon from soil. They found a strong relationship between estimated and measured organic fractions (hydrophobic, transphilic and hydrophilic), and between estimated and measured nitrate.

There are many other methods of deconvolution used for signal decomposition process. The mathematical approaches to deconvolute overlapped signals are often complicated. The linear combination approach proposed by Gallot and Thomas et al. (1993) is developed to deconvolute dissolved organic compounds in waste water based on the assumption of spectral signal from each compound in a mixture is a vector, and can be linearly combined to constitute the spectral signal for a mixture in solution. The underlined fundamental for such method is that no interactions exist among compounds in the solution. This approach has been successfully used to deconvolute chemicals with no interactions in solution (Gonen and Rytwo, 2009). However, for complex solutions like WEOM from manure, compost or soil, one can not warrant that there is no interaction among the components that constitute the WEOM. In fact, conjugation among different components (due to existence of lone pairs of electrons) can always occur, resulting in a shift of maximum absorbance to longer wavelengths (bathochromic). From that perspective, the deconvolution approach that researchers used may not be a valid way to theoretically deconvolute WEOM spectral signals from soil and organic wastes for the purpose of identifying their components. As such, no attempt is made for use of the deconvolution method in the case studies in this chapter to characterize WEOM from either manure or manure amended soils.

4.6 CASE STUDIES

4.6.1. Background

Case studies are presented herein on using above mentioned UV-Vis spectral parameters to characterize WEOM in 1) dairy manure samples with organic and conventional feed sources and 2) pasture soil samples with 0–20 years of poultry litter application. Dairy manure samples were collected in commercial farms of Maine, USA. Of the samples, fifteen were from organic, and four from conventional dairy farms. Poultry litter-amended soil samples were collected from five grass hay fields with three replicates in Sandy Mountain region of north Alabama. The annual poultry litter application for each of the field was 0, 11.4, 22.7, 54.5, and 27.2 Mg ha^{-1}, respectively. The soil samples (0–20 cm) were taken from poultry liter treated slope (3–8%) area in 0, 5, 10, 15, and 20 years after application. The soil is a Hartselle (find sandy loam, siliceous, thermic, Typic Hapludults). Relevant soil properties are thoroughly discussed in separated publications (He et al., 2008; 2009b). Samples were extracted by deionized water with a sample:solution ratio of 1:10 for soil samples and 1:100 for dairy manure samples. Solution samples were filtrated through 0.45 μm Millipore filter paper and total C and N were determined. Solution samples from dairy cow manure were further diluted five times with deionized water for determination of UV-Vis spectra at 250, 254, 280, 365, 465, and 665 nm. Spectral slope was modeled from 300 to 375 nm using Eq. 4.1. Absorbance from each sample was converted to absorptivity using Eq. 4.3 as follow:

$$a = \frac{A}{l} \qquad\qquad\qquad \text{Eq. 4.3}$$

where a = absorptivity (m^{-1}); A = absorbance, and l = path length (m). The SUVA$_{254}$, SUVA$_{280}$ were then normalized using solution total OC concentration. Unpaired student t test was used for mean comparison of spectral parameters of organic dairy and conventional manures. Analysis of variance (completely randomized design) was conducted for poultry amended soils, and mean comparisons among five different poultry litter application histories for spectral properties were made using Least Significant Difference (LSD) at 95% confidence.

4.6.2. Case I. SUVA at 254, 280 nm, Spectral Ratios of E2/E3 and E4/E6, and Spectral Slopes (300 – 375 nm) of Dairy Manure from Organic and Conventional Feed Sources

Total C and N, water extractable OC and N differed from the organic feed to the conventional feed, indicating the dietary effect on diary manure properties (Table 4.3). For the two different feed sources, WEOM from the conventional manure was smaller in $SUVA_{254}$ and $SUVA_{280}$ values, showing that dairy manure from the organic feed source was higher in aromaticity and larger in molecular size than the dairy manure from the conventional feed sources. Shao et al. (2009) showed an increase in SUVA 254 and 280 nm in WEOM after decomposition of municipal solid waste. Similarly, Akagi et al. (2007) demonstrated an increase in SUVA at 254 and 280 nm of soil WEOM after incubation. As such, the manure from the organic feed source might be more resistant to further decomposition in comparison with the manure of the conventional source. Hunt et al. (2007a) found that high $SUVA_{280}$ is associated with higher hydrophilic fractions in WEOM from manures. They attributed that to the existence of polar N compounds. The C:N ratio for the total C and N was 26:1 for the conventional feed source in contrast to the 24:1 of the organic source. But the C:N ratio of the two for the WEOM was rather similar (12:1). The higher SUVA values both at 254 and 280 nm for the manure of the organic feed source indicated that the feed sources may influence WEOM components of dairy manure even their C:N ratio was similar. There was no difference for the E2/E3 ratio between these two manures showing they had a similar degree of humification. The E4/E6 (465 and 665 nm) ratio, on the other hand, was lower with the conventional feed source as compared to the organic feed source (Table 4.4). The lower E4/E6 ratio indicated a shift of absorbance to higher wave length (increase in absorbance in 665 nm), showing the existence of long chain organic compounds in the manure from the conventional feed source, or conjugation of organic compounds in water solution due to the existence of lone pairs of electrons. The spectral slopes from 300 to 375 nm showed no statistical difference between the two sources (Table 4.4). Although spectral slope has been used to differentiate sources of DOC in sea water (Stedmon and Markager, 2001; Helms et al., 2008), there are no reports on using spectral slope to characterize WEOM from manure. Therefore, more slope modeling work is needed for manure studies so that meaningful comparisons can be made.

4.6.3. Case Study II. SUVA at 254, 280 nm, Spectral Ratios of E2/E3, E4/E6, and Spectral Slopes (300 – 375 nm) of Poultry Litter-Amended Soil with Different Application Histories

Soil total carbon and total N increased as increase in poultry litter application rate and history (Table 4.5). Coupling with this increase was the increase in water extractable C and N (Table 4.5). Soil samples with 15-year poultry litter application history received the highest cumulative amount of poultry litter application. However, its water extractable OC was not significantly different ($p > 0.05$) from that of the soils of 20-year application history, which received about half the amount of the application rate. Also, the water extractable C from the 15-year soil samples was not significantly different ($p > 0.05$) from that of the 10-year soil samples. In contrast, the water extractable N was significantly different ($p < 0.05$) in every 10 year interval (Table 4.5), linearly related to the year of application, but not to the amount of poultry litter application. Apparently, the highest poultry litter application rate in 15-year soil did not yield neither high water extractable OC nor N in soil.

Soil samples from the fields with different poultry litter application history tended to be different in their spectral properties. Although the differences are not statistically significant ($p > 0.05$), the trend of change in absorptivity is clearly demonstrated in Table 4.6. For all poultry litter-amended soils, the SUVA at both 254 and 280 nm showed an increase in absorptivity compared to the soils with no poultry litter application. The absorptivity of 254 and 280 nm of WEOM from the poultry litter-amended soils had a tendency of decrease as the litter application history was extended, indicating decreases in aromaticity. As discussed earlier, absorptivity at 254 and 280 nm increase after decomposition of WEOM (Shao et al., 2009; Akagi et al., 2007). Continuous poultry litter application might have resulted in accumulation of undecomposed poultry litter in the soil. This undecomposed poultry litter can contribute directly to pools of WEOM, diluting aromaticity of WEOM derived from the decomposed poultry litter. In determination of amino-acid nitrogen (AC-N) from the same soil, He and Senwo (2009) found that AC-N are partially from the residue and partially from the fresh applied poultry litter in the soil with application history of 10 to 15 years. Increasing poultry litter application history led to an increase of WEOM concentration in soil (Table 4.5), and both decomposed and undecomposed poultry litters in soil contributed to such increase. Over the years of application, the spectral ratio of E2/E3 was not changed as obvious as the E4/E6 ratio. In comparison, the E4/E6 ratio increased as increase of poultry litter application history, showing the existence of long chain compounds. These compounds with long chains can be either aromatic or aliphatic. But with the fact of thedecrease in SUVA 254 and 280 nm, the increase in the E4/E6 ratio might have largely come from the aliphatic long chain compounds.

Table 4.3. Total and water extractable C and N in dairy manure used in case studies.

Materials	Source of feed	Total C	Total N	Total water extractable C	Total water extractable N
		g C kg^{-1} manure	g N kg^{-1} manure	g C kg^{-1} dry manure	g N kg^{-1} dry manure
Dairy manure (n=15)	Organic	403.0 a[1]	17.1 a	19.0 a	1.69 a
Dairy manure (n=4)	Conventional	395.4 b	14.7 b	15.5 b	1.27 b

[1] Values in the same column with different letters are significant with *student t test* at $p = 0.05$.

Table 4.4. UV-Vis spectral properties of dairy manure from organic and conventional feed sources.

Materials	SUVA$_{254}$	SUVA$_{280}$	E2/E3	E4/E6	Spectral slope (300-375 nm)
	m^{-1} mg^{-1} of C L	m^{-1} mg^{-1} of C L			nm^{-1}
Organic dairy manure					
Average	13.60 a[1]	12.04 a	3.05 a	3.14 a	0.0113 a
Standard error (n=15)	0.43	0.39	0.0062	0.0388	0.00003
Conventional dairy manure					
Average	8.37 b	7.29 b	3.03 a	2.85 b	0.0112 a
Standard error (n=4)	1.30	1.11	0.052	0.073	0.00009

[1] Values with different letters in the same column are significant with *student t test* at $p \leq 0.05$.

Table 4.5. Total C and N, total water extractable C and N from poultry litter amended soils.

Application history	Annual application rate	Cumulative application	Total C	Total N	Total water extractable C	Total water extractable N
year	Mg ha^{-1} yr^{-1}	Mg ha^{-1}	mg C kg^{-1} soil	mg N kg^{-1} soil	mg C kg^{-1} dry manure	mg N kg^{-1} dry manure
0	0	0	4.33 a[1]	0.49 a	88 a	19 a
5	2.27	11.4	5.54 ab	0.71 ab	131 ab	25 ab
10	2.27	22.7	9.21 bc	1.07 bc	199 bc	51 bc
15	3.63	54.5	11.73 c	1.55 c	216 c	79 cd
20	1.36	27.2	10.68 c	1.34 c	218 c	89 d
Probability (*F test*)			0.037	0.019	0.012	0.002

[1] Values with different letters in the same column are significant with Least Significant Difference (LSD) at $p \leq 0.05$.

Absorptivity of the soil WEOM at 300 to 375 nm were used for modeling the spectral slope (Eq. 4.1) (Figure 4.3). Statistically, there was still 90% confidence to show the differences among the spectral slope values over the years of poultry litter application (Table 4.6). The spectral slopes of WEOM from the 0- and 5-year treatments were identical. In contrast, the spectral slopes of the 15- and 20-year treatments were nearly identical, but higher than those of the 0- and 5-year treatments. In between was the value of the 10-year treatment (Table 4.6). Apparently, within 5-year of poultry litter application, applied poultry litter WEOM might have been decomposed and become part of soil WEOM, resulting in a similar spectral slope to that of WEOM from the soil with no history of poultry litter application. With increase in years of the litter application, there was more WEOM from the applied poultry litter accumulated in soil without decomposition, which might have contributed to a large proportion of WEOM from the soil samples. As such, the 15- and 20-year spectral slope values were higher than those from the 0- and 5-year WEOM. Whereas spectral slope has been used for identifying sources of DOC in sea water, the slope differences in these poultry litter-amended soils may be worked as an indicator for differentiating sources of WEOM in soil. Based on the results, we can conclude that beyond 5 years, decomposition of applied poultry litters had slowed down, and applied poultry litter beyond that time contributed directly to soil WEOM. This may suggest the upper time limits in which a soil can receive poultry litter application.

4.7. CONCLUSIONS

UV-vis spectroscopy is a useful tool for characterizing water extractable or humic fractions of natural organic matter. Whereas the whole UV-visible spectra of these fractions are more or less featureless, the SUVA at 254 and 280 nm as well as spectral E2/E3 and E4/E6 ratios have been used for characteristic parameters of dissolved organic matter fractions. Similar to other organic matter research, these spectroscopic parameters are used to describe molecular weight, aromaticity, and polarity of water soluble organic matter fraction from animal manure. More than that, in manure-related studies, these parameters are also used to monitor the decomposition and humification of manure, composting process, and compost maturities and quality. Another characteristic, the spectral slope in the UV region has been used in dissolved organic carbon studies in fresh and sea waters, but not in manure or soil WEOM characterization. More work is needed in this area to characterize soil or organic wastes derived water soluble organic C.

In the case study I, we found that dairy manure from organic feed source has a higher SUVA at 254 and 280 nm than that from the conventional feed source, indicating high aromaticity of WEOM from the manure of the organic feed source. In the case study II, we found that the SUVA at 254 and 280 nm of soil WEOM increased in the first 10 years of poultry litter application, but then decreased. We attributed this observation to the accumulation of undecomposed water soluble organic matter from the applied poultry litter in soil. Whereas the E2/E3 ratio has not shown that it is an effective tool to differentiate either manure from different feed sources or soils with various years of poultry litter application, the ratio of E4/E6 appeared to be effective to show the difference among dairy cow manures with different feed sources and among soils with different poultry litter application histories. The simulated spectral slope values from 300 nm to 375 nm were not statistically different between the WEOM of the two types of dairy manures. However, they appeared to differ in the WEOM from soils with different histories of poultry litter application. Future studies in this area should be emphasized on optimal wavelength for simulation, and evaluation various spectral models so that sources of soil WEOM can be determined.

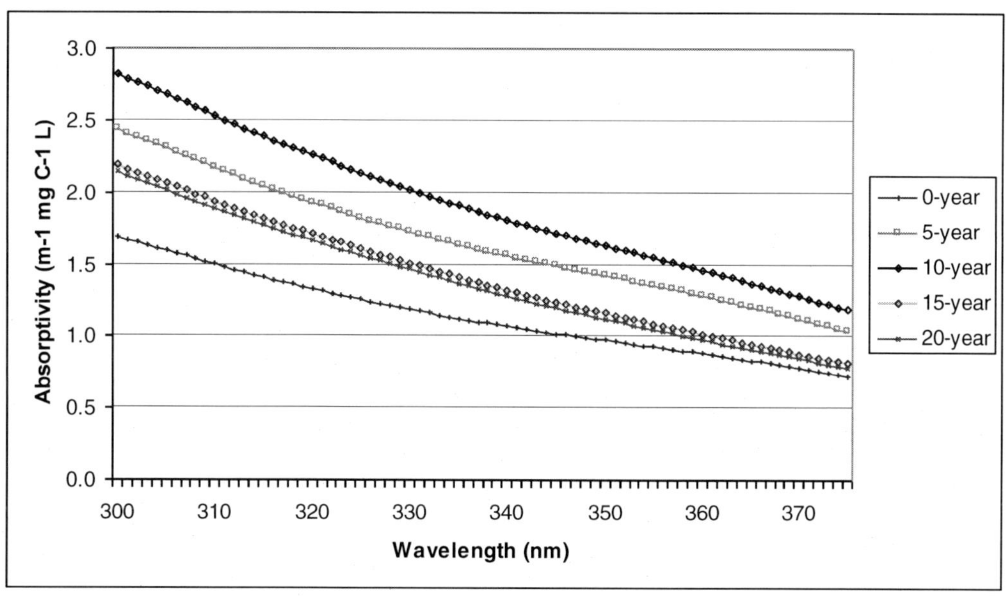

Figure 4.3. Absorptivity (300 – 375 nm) of water extractable organic matter from soil with different poultry litter amendment history, from which spectral slope (S) was simulated by the single exponential equation $a_\lambda = a_{\lambda_{ref}} e^{-S(\lambda - \lambda_{ref})}$.

Table 4.6. UV-visible spectral properties of poultry litter amended soils.

Litter application history	SUVA$_{254}$	SUVA$_{280}$	E2/E3	E4/E6	Spectral slope (300-375 nm)
	m^{-1} mg C^{-1} L	m^{-1} mg C^{-1} L			nm^{-1}
0 year	2.61 a[1]	2.18 a	3.42 a	4.13 a	0.0109 a
5 years	3.72 a	3.07 a	3.33 a	4.48 a	0.0109 a
10 years	4.27 a	3.49 a	3.28 a	5.28 a	0.0112 a
15 years	3.52 a	2.84 a	3.96 a	6.32 a	0.0131 a
20 years	3.48 a	2.78 a	4.00 a	6.02 a	0.0132 a
Probability (*F test*)	0.170	0.218	0.134	0.062	0.093

[1] Values with different letters were significant with Least Significant Difference (LSD) at $p \leq 0.05$.

REFERENCES

Abbt-Braun, G., U. Lankes, and F.H. Frimmel. 2004. Structural characterization of aquatic humic substances-the need for a multiple method approach. *Aquat. Sci.* 66:151-170.

Akagi, J. A. Zsolnay, F. Bastida. 2007. Quantity and spectroscopic properties of soil dissolved organic matter (DOM) as a function of soil sample treatments: Air-drying and pre-incubation. *Chemosphere,* 69:1040-1046.

Baes, A.U., and P.R. Bloom. 1990. Fulvic acid ultraviolet-visible spectra: influence of solvent and pH. Soil *Sci. Soc. Am. J.,* 54:1248-1254.

Carder, K.L., R.G., Steward, G.R. Harvey, and P.B. Ortner. 1989. Marine humic and fulvic acids: Their effects on remote sensing of ocean chlorophyll, *Limnol. Oceanogr.,* 34:68-81.

Charest, M.H., and C.J. Beauchamp. 2002. Composting of de-inking paper sludge with poultry manure at three nitrogen levels using mechanical turning: behavior of physico-chemical parameters. *Bioresour. Technol.,* 81:7-17.

Chen, J., B. Gu, E.J. Le Boeuf, H. Pan, and S. Dai. 2002. Spectroscopic characterization of the structural and functional properties of natural organic matter fractions. *Chemosphere* 48:59-68.

Chen, Y., N. Senesi, and M. Schnitzer. 1977. Information provided on humic substances by E4/E6 ratios. *Soil Sci. Soc. Am. J.,* 41:352-357.

Chin, Yu-Ping, G. Alken, and E. O'Loughlin. 1994. Molecular weight, polydlspersity, and spectroscopic properties of aquatic humic substances. *Environ. Sci. Technol.,* 28:1853-1858.

Cook, R.L., J.E. Birdwell, C. Lattao, and M. Lowry. 2009. A multi-method comparison of Atchafalaya Basin surface water organic matter samples. *J. Environ. Qual.* 38:702-711.

Del Vecchio, R., and N.V. Blough. 2004. Spatial and seasonal distribution of chromophoric dissolved organic matter and dissolved organic carbon in the Middle Atlantic Bight. *Marine Chem.* 89:169-187.

Desalegn, G., E. Binner, and P. Lechner. 2008. Humification and degradability evaluation during composting of horse manure and biowaste. *Compost Sci. Util.* 16:90-98.

Domeizel, M., A. Khalil, and P. Prudent. 2004. UV spectroscopy: a tool for monitoring humification and for proposing an index of the maturity of compost. *Bioresour. Technol.,* 90:177-184.

Embacher, A., A. Zsolnay, A. Gattinger, and J.C. Munch. 2008. The dynamics of water extractable organic matter (WEOM) in common arable topsoils: II. Influence of mineral and combined mineral and manure fertilization in a Haplic Chernozem. *Geoderma,* 148:63-69.

Gallot, S. and O. Thomas. 1993. Fast and easy interpretation of a set of absorption spectra: Theory and qualitative applications for UV examination of waters and wastewaters. Fresenius. *J. Anal Chem.,* 346:976-983.

Ghosh, K., and M. Schnitzer. 1979. UV and visible absorption spectroscopic investigations in relation to macromolecular characteristics of HUMIC substances. *J. Soil Sci.,* 30:735-745.

Giusquiani, P.L., L. Concezzi, M. Busibelli, and A. Macchioni. 1998. Fate of pig sludge liquid fraction in calcareous soil: agricultural and environmental implications. *J. Environ. Qual.* 27:364-371.

Gonen, Y. and G. Rytwo. 2009. Using a Matlab implemented algorithm for UV-vis spectral resolution for pKa determination and multicomponent analysis . *Analytical Chemistry Insights,* 4:21-27.

Grassi, M., and M. Rosa. 2010. Humic acids of different origin as modifiers of cadmium-ion chemistry: A spectroscopic approach to structural properties and reactivity. *Inorg. Chim. Acta, 363:*495-503.

Guo, M., and J. Chorover. 2003. Transport and fractionation of dissolve organic matter in soil column. *Soil Sci.,* 168:108-118.

Hassouna, M., F. Theraulaz, and C. Massiani. 2007. Direct estimation of nitrate, total and fractional water extractable organic carbon (WEOC) in an agricultural soil using direct UV absorbance deconvolution. *Talanta,* 71:861-867.

Hayase, K., and H. Tsubota. 1984. Sedimentary humic-acid and fulvic-acid as fluorescent organic materials. Geochim. *Cosmochim. Acta,* 49:159-163.

He, Z., and J.C. Spain. 2000. Reactions involved in the lower pathway for degradation of 4-nitrotoluene by *Mycobacterium* strain HL 4-NT-1. *Appl. Environ. Microbiol.,* 66:3010-3015.

He, Z., C.W. Honeycutt, and T.S. Griffin. 2003. Comparative investigation of sequentially extracted P fractions in a sandy loam soil and a swine manure. *Commun. Soil Sci. Plant Anal.,* 34:1729-1742.

He, Z., T. Ohno, B.J. Cade-Menun, M.S. Erich, and C.W. Honeycutt. 2006. Spectral and chemical characterization of phosphates associated with humic substances. *Soil Sci. Soc. Am. J.,* 70:1741-1751.

He, Z., J. Mao, C.W. Honeycutt, T. Ohno, J.F. Hunt, and B.J. Cade-Menun. 2009a. Characterization of plant-derived water extractable organic matter by multiple spectroscopic techniques. *Biol. Fertil. Soils,* 45:609-616.

He, Z., Z.N. Senwo, I.A. Tazisong, and D.A. Martens. 2009b. Amino compounds in poultry litter, litter-amended soil and plants. *The Proceedings of the International Plant Nutrition Colloquium XVI.,* Vol. http://repositories.cdlib.org/ipnc/xvi/1048. On-line publication, Sacramento, CA.

He, Z. I.A. Tazisong, Z.N. Senwo, and D. Zhang. 2008. Soil properties and macro cations status impacted by long-term applied poultry litter. *Comm. Soil Sci. and Plant Anal.,* 39:858-872.

Helms, J.R., A. Stubbins, J.D. Ritchie, E.C. Minor, D.J. Kieber, and K. Mopper. 2008. Absorption spectral slopes and slope ratios as indicators of molecular weight, source, and photobleaching of chromophoric dissolved organic matter. *Limonol. Oceanogr.,* 53:955-969.

Heymann, K., H. Mashayekhi, and B.S. Xing. 2005. Spectroscopic analysis of sequentially extracted humic acid from compost. *Spectr. Lett.,* 38:293-302.

Huang, G.F., Q.T. Wu, J.W.C. Wong, and B.B. Nagar. 2006. Transformation of organic matter during co-composting of pig manure with sawdust. *Bioresour. Technol.,* 97:1834-1842.

Hunt, J.F., and T. Ohno. 2007. Characterization of fresh and decomposed dissolved organic matter using excitation-emission matrix fluorescence spectroscopy and multiway analysis. *J. Agric. Food Chem.,* 55:2121-2128.

Hunt, J.F., T. Ohno, Z. He, C.W. Honeycutt, and D.B. Dail. 2007a. Influence of decomposition on chemical properties of plant- and manure-derived dissolved organic matter and sorption to goethite. *J. Environ. Qual.,* 36:135-143.

Hunt, J.F., T. Ohno, Z. He, C.W. Honeycutt, and D.B. Dail. 2007b. Inhibition of phosphorus sorption to goethite, gibbsite, and kaolin by fresh and decomposed organic matter. *Biol. Fertil. Soils,* 44:277-288.

Jaffrain, J., F. Gerard, M. Meyer, and J. Ranger. 2007. Assessing the quality of dissolved organic matter in forest soils using ultraviolet absorption spectrophotometry. *Soil Sci. Soc. Am. J.*, 71:1851-1858.

Kalbitz, K., S. Geyer, and W. Geyer. 2000. A comparative characterization of dissolved organic matter by means of original aqueous samples and isolated humic substances. *Chemosphere*, 40:1305-1317.

Kang, S., and B. Xing. 2008. Humic acid fractionation upon sequential adsorption onto goethite. *Langmuir*, 24:2525-2531.

Li, F., A. Yuasa, Y. Katamine, and H. Tanaka. 2007. Breakthrough of natural organic matter from fixed bed adsorbers: investigations based on size-exclusion HPLC. *Adsorpt.-J. Int. Adsorpt. Soc.*, 13:569-577.

Li, G.X., F.S. Zhang, Y. Sun, J.W.C. Wong, and M. Fang. 2001. Chemical evaluation of sewage sludge composting as a mature indicator for composting process. *Water Air Soil Pollut.*, 132:333-345.

Mathur, S.P., H. Dinel, G. Owen, M. Schnitzer, and J. Dugan. 1993. Determination of compost biomaturity .2. Optical-density of water extracts of composts as a reflection of their maturity. *Biol. Agric. Hortic.*, 10:87-108.

Musikavong, C., and S. Wattanachira. 2007. Reduction of dissolved organic matter in terms of DOC, UV-254, SUVA and THMFP in industrial estate wastewater treated by stabilization ponds. *Environ. Monit. Assess.*, 134:489-497.

Ohno, T., and B.S. Crannell. 1996. Green and animal manure-derived dissolved organic matter effects on phosphorus sorption. *J. Environ. Qual.*, 25:1137-1143.

Ohno, T., T.S. Griffin, M. Liebman, and G.A. Porter. 2005. Chemical characterization of soil phosphorus and organic matter in different cropping systems in Maine, USA. *Agric. Ecosyst. Environ.*, 105:625-634.

Peuravuori, J., and K. Pihlaja. 1997. Molecular size distribution and spectroscopic properties of aquatic humic substances. *Anal. Chim. Acta*, 337:133-149.

Plaza, C., R. Nogales, N. Senesi, E. Benitez, and A. Polo. 2008. Organic matter humification by vermicomposting of cattle manure alone and mixed with two-phase olive pomace. *Bioresour. Technol.*, 99:5085-5089.

Plaza, C., N. Senesi, J.C. Garcia-Gil, G. Brunetti, V. D'Orazio, and A. Polo. 2002. Effects of pig slurry application on soils and soil humic acids. *J. Agric. Food Chem.*, 50:4867-4874.

Plaza, C., N. Senesi, A. Polo, G. Brunetti, J.C. Garcia-Gil, and V. D'Orazio. 2003. Soil fulvic acid properties as a means to assess the use of pig slurry amendment. *Soil Tillage Res.*, 74:179-190.

Riffaldi, R., R. Levi-Minzi, and A. Saviozzi. 1983. Humic fractions of organic wastes. *Agric. Ecosyst. Environ.*, 10:353-359.

Sarpal, R. S., K. Mopper, and D.J. Keiber. 1995. Absorbance properties of dissolved organic matter in Antarctic sea water. *Antarc., J.* 30:139-140.

Schnitzer, M., H. Dinel, S.P. Mathur, H.R. Schulten, and G. Owen. 1993. Determination of compost biometry. III. Evaluation of a colorimetric test by ^{13}C-NMR spectroscopy and pyrolysis-field ionization mass spectrometry. *Biol. Agric. Hortic.*, 10:109-123.

Sellami, F., S. Hachicha, M. Chtourou, K. Medhioub, and E. Ammar. 2008. Maturity assessment of composted olive mill wastes using UV spectra and humification parameters. *Bioresour. Technol.*, 99:6900-6907.

Shao, Z.H., P.J. He, D.Q. Zhang, and L.M. Shao. 2009. Characterization of water-extractable organic matter during the biostabilization of municipal solid waste. *J. Hazard. Mater.,* 164:1191-1197.

Stedmon, C.A. and S. Markager. 2001. The optics of chromophoric dissolved organic matter (CDOM) in the greenland sea: An algorithm for differentiation between marine and terrestrially derived organic matter. *Limnol Oceanogr.,* 46:2087-2093.

Swift, R.S. 1996. Organic matter characterization., p. 1011-1084, *In* D. L. Sparks, ed. Methods of Soil Analysis. Part 3-Chemical Methods. *Soil Sci. Soc. Am.,* Madison, WI.

Takeda, A., H. Tsukada, Y. Takaku, and S. Hisamatsu. 2009. Fractionation of metal complexes with dissolved organic matter in a rhizosphere soil solution of a humus-rich Andosol using size exclusion chromatography with inductively coupled plasma-mass spectrometry. *Soil Sci. Plant Nutr.,* 55:349-357.

Tan, K.H. 2003. *Humic matter in soil and the environment.* p. 176-180 Marcel Dekker, Inc., New York, N.Y.

Thomas, O., F. Theraulaz, M. Domeizel, and C. Massiani. 1993. UV spectral deconvolution: A valuable tool for waste water quality determination. *Environ. Tech.,* 14:1187-1192.

Twardowski, M.S., E. Boss, J.M. Sullivan, and P.L. Donaghay. 2004. Modeling the spectral shape of absorbing chromophoric dissolved organic matter. *Marine Chem.,* 89:313-326.

Wang, G.S., and S.T. Hsieh. 2001. Monitoring natural organic matter in water with scanning spectrophotometer. *Environ. Int.,* 26:205-212.

Yang, K., and B.S. Xing. 2009. Adsorption of fulvic acid by carbon nanotubes from water. *Environ. Pollut.,* 157:1095-1100.

Yin, S.D., R. Dolan, M. Harris, and Z.C. Tan. 2010. Subcritical hydrothermal liquefaction of cattle manure to bio-oil: Effects of conversion parameters on bio-oil yield and characterization of bio-oil. *Bioresour. Technol.,* 101:3657-3664.

Zhang, M., S. D. Sparrow, and S. Seefeldt. 2006. Spectral characteristics of water extractable organic matter from soils of different land uses in a subarctic Alaska environment. *Abstract in Soil Science Society of America Annual Conference.* Nov. 12-16. 2006 Indianapolis Indiana, USA

Zmora-Nahum, S., Y. Hadar, and Y. Chen. 2007. Physico-chemical properties of commercial composts varying in their source materials and country of origin. *Soil Biol. Biochem.,* 39:1263-1276.

Zmora-Nahum, S., O. Markovitch, J. Tarchitzky, and Y.N. Chen. 2005. Dissolved organic carbon (DOC) as a parameter of compost maturity. *Soil Biol. Biochem.,* 37:2109-2116.

In: Environmental Chemistry of Animal Manure
Editor: Zhongqi He

ISBN: 978-1-62808-641-6
© 2013 Nova Science Publishers, Inc.

Chapter 5

FLUORESCENCE SPECTROSCOPIC ANALYSIS OF ORGANIC MATTER FRACTIONS: THE CURRENT STATUS AND A TUTORIAL CASE STUDY

Tsutomu Ohno[1,] and Zhongqi He[2]*

ABBREVIATIONS

C_{TS},	total soluble C;
DI-H_2O,	deionized-distilled water;
DOM,	dissolved organic matter,
EEM,	excitation-emission matrix;
EM,	emission;
EX,	excitation
PARAFAC,	parallel factor analysis

5.1. INTRODUCTION

Fluorescence spectroscopy has been one approach to chemically characterize organic matter (OM) from various sources (Senesi et al., 2007; He et al., 2008; Murphy et al., 2008; Macalady and Walton-Day, 2009; Santin et al., 2009). One of the chief advantages of fluorescence spectroscopy is its high sensitivity which can provide information on the chemical properties of OM fractions (i.e. water extractable organic matter and humic

---

[*] Corresponding author: ohno@maine.edu
1 Department of Plant, Soil, and Environmental Sciences, University of Maine, 5722 Deering Hall, Orono, ME 04469-5722, USA.
2 USDA-ARS, New England Plant, Soil, and Water Laboratory, Orono, ME 04469, USA.

substances) without any pretreatment. This reduces substantially concerns regarding the introduction of artifacts due to processing steps to concentrate the organic matter (Senesi, 1990; 1992). Fluorescence analysis for the characterization of OM fractions has undergone substantial advancement recently with adoption of two techniques. First, it has become common to use excitation-emission matrix (EEM) spectroscopy to generate three-dimensional fluorescence spectra. The EEM method measures the emission (EM) spectra over a range of excitation (EX) wavelengths resulting in a landscape surface defined by the fluorescence intensity at EX and EM wavelength pairs. The EEM approach has been used to characterize OM fractions extracted from a variety of sources relevant to agronomic nutrient management: crop residues (Merritt and Erich, 2003; Ohno and Cronan, 1997); manures (Ohno and Bro, 2006; Hunt and Ohno, 2007), wastewater treatment residues (Westerhoff et al., 2001) and soils (Ohno et al., 2009; He et al., 2010). Although the EEM has greater spectral information density than the traditional fluorescence approaches, the EEM landscape has been typically characterized by "peak picking" the locations of one or more peaks visually observable in the fluorescence intensity landscape. Two fluorophores frequently observed in DOM samples are located near EX 270~280 nm and EM 335~350 and the other at EX 310~325 nm and EM 420~445. These have been characterized as "protein-like" and "humic-like", respectively (Coble et al., 1990; Merritt and Erich, 2003). Chen et al. (2003) have quantified the EEM spectra by operationally delineating the EEM landscape into five regions and calculating the integrated volume under each region to characterize the DOM. The regions are characterized as aromatic protein-like (two regions), fulvic acid-like, microbial by-product-like, and humic acid-like.

Secondly, the use of the statistical parallel factor analysis (PARAFAC) method has been demonstrated to decompose a suite of EEM landscapes into *chemically meaningful* spectral components (Bro, 1997; Andersen and Bro, 2003; Smilde et al., 2004). PARAFAC provides a direct estimate of the relative concentration of the OM components present in the data set as well as the excitation and emission spectra of the components. Thus, PARAFAC can be seen as providing the spectral signatures of the individual fluorophores present in complex and heterogeneous OM mixture without needing any kind of separation methodology. Application of PARAFAC to OM extracted from plant biomass, soil, and manures revealed that five fluorophores could be present (Ohno and Bro, 2006). Seven components were identified by Hunt and Ohno (2007) on a set of DOM isolated from crop residues and manures which were subjected to microbial decomposition in a laboratory incubation. The potential of application of the EEM-PARAFAC method in characterizing manure OM fractions and their impacts on soil OM composition has not been fully explored.

The objectives of this chapter are: (1) review and discuss fluorescence spectroscopic studies of OM fractions of animal manure; (2) to provide an introduction to PARAFAC for the characterization of fluorescence spectra of OM; and (3) to present as a case study an example workflow of PARAFAC analysis of a set of EEM spectra of dissolved organic matter (DOM, i.e. water extractable OM) samples derived from six animal manures and three relevant soil samples to promote the application of this method in manure OM-relevant studies.

5.2. FLUORESCENCE FEATURES
OF HUMIC FRACTIONS OF ANIMAL MANURE

Plaza et al. (2002; 2003) compared the fluorescence spectra of humic (HA) and fulvic (FA) acids extracted from pig slurry and soils amended with pig slurry (PS). The main feature of the EM spectra was a unique, typical, broad band with the maximum centered at a wavelength that is much shorter (450 nm) for PS-HA than for any soil HA (513-519 nm). The PS-amended soil HAs feature the EM maximum at a slightly longer wavelength than that of HA in control soil without PS amendment and a broad shoulder that extends to shorter wavelengths. The EX spectrum of PS-HA is very different from those of soil HAs and is characterized by a prominent band in the intermediate-wavelength region (391 nm) and some small bands and shoulders at short and long wavelengths (337 and 438 nm) . The EX spectrum of HA in control soil without PS amendment features two prominent bands of almost equal relative intensity at long wavelength (453 and 464 nm) and an intense shoulder at 394 nm. The EX spectra of PS amended soil HAs are similar to one to another and to that of control soil HA. However, with respect to C-HA, PS-amended soil HAs exhibit a less intense shoulder at intermediate wavelength and the band at 464-465 nm is slightly more intense than that at 451-453 nm. The authors (Plaza et al. 2002) assumed that the large overall fluorescence intensity and the short wavelengths measured for the main fluorescence peaks of PS-HA were due to the presence of simple structural components of wide molecular heterogeneity and low molecular weight, degree of aromatic polycondensation, level of conjugated chromophores, and humification degree. On the contrary, the small fluorescence intensities and long wavelengths of major peaks of soil HAs may be ascribed to the presence of an extended, linearly condensed aromatic ring network and other unsaturated bond systems capable of a great degree of conjugation in large molecular weight units of great humification degree. Compared with the corresponding HA fractions, the FAs feature markedly different fluorescence spectra (Plaza et al. 2003). The main feature of EM spectra of FAs is a unique, typical, broad band with the maximum centered at a wavelength that was shorter (453 nm) for PS-FA than for control soil-FA (459 nm). The PS-amended soil FAs feature the EM maximum at a slightly shorter wavelength than that of control soil-FA. The EX spectra of all FAs were characterized by a prominent peak in the intermediate-wavelength region (392 nm) which was accompanied by a shoulder extending to longer wavelengths. The excitation spectrum of PS-FA was only slightly different from that of any soil FA. Thus, the differences in fluorescence features of FAs between PS and soils were smaller than those between the two types of HAs (Platza et al. 2003). The authors (Plaza et al. 2002; 2003) also measured synchronous scan excitation spectra by scanning simultaneously both the excitation (varied from 300 to 550 nm) and the EM wavelengths while maintaining a constant, optimized 18 nm wavelength difference between Ex and EM. The synchronous spectral changes of soil HA and FA impacted by PS-amendment are similar to those of the EX spectral change under the same conditions.

Hernandez et al. (2006; 2007) measured fluorescence EEM spectra of HAs and FAs in the absence and presence of either Cu(II) or Zn(II) as fluorescence spectroscopy is a well-established, reliable and invaluable means for examining interactions of humic fractions with metal ions (e.g. Senesi,1992; Wu et al., 2004). The EEM spectrum of PS-HA was characterized by a unique fluorophore centered at an excitation/emission wavelength pair

(EEWP) of 325/437 nm. The main features of the EEM spectra of soil HAs were a typical broad band with the maximum centered at an EEWP value of 440/508 nm, and a wide shoulder at shorter wavelengths whose relative intensity, with respect to the main peak, tends to increase with increasing amendment rate. Addition of Zn(II) and Cu(II) caused a marked decrease of the fluorescence intensity of all HAs, especially soil HAs. A marked decrease of the EX and EM wavelength maxima was observed for Cu(II)-treated HAs, which was explained by the large quenching effect of Cu(II) ion on HA fluorescence at long EX and EM wavelengths (Hernandez et al., 2006) . On the contrary, the EX and EM peaks remain almost constant when Zn(II) was added. These results are indicative of a marked modification of the electronic structure of HAs upon interaction with Zn(II), and especially Cu(II). The different extent of the modifications observed can be possibly ascribed to the different strength of bonding achieved between Cu(II) or Zn(II) and the various HAs, on dependence of the nature of ligand atoms, the metal:HA ratio, the type of HA groups involved in metal complexation, and the three-dimensional arrangement of HA macromolecules. Hernandez et al. (2006) further analyzed experimentally-determined values and model-derived nonlinear regressions of fluorescence intensity of the main peak in fluorescence EEM spectra of HAs isolated from PS, control soil , and soils amended with PS at different rates as a function of increasing total concentration of Cu(II) and Zn(II). This modeling analysis indicated that PS application decreased Cu(II) and Zn(II) complexing capacities and binding affinities of soil HA. These effects increased with increasing the rate per year of PS application to soil, and are expected to have a large impact on bioavailability, mobilization, and transport of Cu(II) and Zn(II) ions in PS-amended soils (Hernandez et al., 2006).

Similarly, the EEM spectra of FAs in the absence of Cu(II) or Zn(II) consisted of a unique broad band centered at an EEWP value that was a little shorter for PS-FA (325/430 nm) than for soil FAs (325–330/435–438 nm) (Hernandez et al. 2007). Whereas the EEWP values of these FAs differed from those of corresponding HAs (Hernandez et al. 2007), addition of Zn(II) and Cu(II) caused a shift to shorter values, i.e. a blue shift, of the EX and EM wavelengths of the main peak of PS-FA as observed in HA samples. These results suggest the occurrence of marked modifications of the electronic structure of FAs upon interaction with Cu(II) and

Zn(II). However, compared to the corresponding soil HA fractions, the chemical and structural characteristics and Cu(II) and Zn(II) binding behavior of soil FAs were less affected than those of soil HAs from PS amendment (Hernandez et al., 2007).

Huang et al. (2006) characterized the fluorescence spectral changes of HA and FA extracted from pig manure during co-composting with sawdust up to 63 days. As the composting time increased, the relative fluorescence intensity of HA decreased, and the peak of EM spectra shifted from 440 nm at day 0 to the longer wavelength at 484 nm at day 63. This indicated less fluorescing but more aromatic structures in HA at day 63. The relative fluorescence intensity of FA increased with composting time, but the peak remained at 440 nm for the EM spectra from day 0 and day 63, which indicated the increased fluorescing structure in FA, but no great change of aromatic structure after composting. The EM spectra, although broad, showed maximum intensity wavelengths that were longer for HA than for fulvic acid, indicating a greater degree of condensed, aromatic character in humic acid in pig manure compost. EX spectra of HA confirm the increased humification with the age of composting by showing an increase in the intensity level of the peaks at an intermediate wavelength of 398 nm relative to peaks and shoulders at longer wavelength of HA. Only a slight increase in intensity was observed at the same wavelength for FA on day 63 as compared to day 0. The relative fluorescence intensity of the excitation peaks changed with composting time for HA, showing an increase in the peak intensity at 398 nm at day 63 with respect to 390 nm at day 0. The general observations of Huang et al. (2006) are similar to those of Plaza et al. (2002; 2003), indicating a similar mechanism of organic matter transformation in composting and soils.

The fluorescence EEM spectra of HA extracted from cattle manure (CM) and products vermicomposted with olive pomace (OP) (Plaza et al. 2008) are different from those of PS-HA (Hernandez et al., 2006). The unique EEWP main band was at 325/441 nm for CM-HA, and it was slightly shorter for OP + CM-HA (310/405 nm) . After vermicomposting, the EEWP band shifted to 395/487 nm and 400/503 nm for HAs of CM and OP+CM, respectively. Thus, vermicomposting made the EEWP value to increase slightly longer for OP + CM HA than for CM-HA. Plaza et al. (2008) concluded that fluorescence features of the HAs isolated from CM and OP + CM before vermicomposting were similar to those of HAs isolated from other types of untreated pig slurry and sewage sludge HAs (Hernández et al., 2006; Plaza et al., 2006). However, fluorescence spectra of the HA-like fractions after the vermicomposting process closely approach those typical of soil HAs (Senesi et al., 1991). Combined with other elemental and spectroscopic analyses, the authors (Plaza et al. 2008) concluded that vermicomposting was thus able to promote organic matter humification in both CM alone and in the mixture OP + CM, thus enhancing the quality of these materials as soil organic amendments.

5.3. FLUORESCENCE FEATURES OF WATER EXTRACTABLE ORGANIC MATTER OF ANIMAL MANURE

Ohno and Bro (2006) characterized water extractable organic matter (WEOM) of plant materials and four types of animal manures (i.e. beef, dairy, poultry, and swine) using EEM spectroscopy and multiway analysis. At first, the authors presented the emission spectra structure for WEOM derived from wheat residue, poultry manure, and soil samples. As these emission spectra were obtained from a single emission scan at a fixed excitation wavelength of 254 nm, full-scan EEM fluorescence spectroscopy collects emission spectra from a range of excitation wavelengths, allowing a complete profile of fluorescence intensity response along both excitation and emission wavelength variables. The relative concentrations of the identified fluorophores for the samples were then modeled by the PARAFAC approach. Ohno and Bro (2006) found that whereas the soil WEOM was well modeled by three fluorophore components, five-component models fitted the individual samples in the manure and plant WEOM sets. More than that, the animal manure WEOM distribution was fairly even across the five modeled components. The crop- and wetland plants-derived WEOM component profiles were similar, and they differed from the tree-leaves-derived WEOM. Thus, this study demonstrates the ability of EEM-PARAFAC to yield information about the composition of WEOM from different sources (Ohno and Bro, 2006).

Hunt and Ohno (2007) then compared fresh and decomposed WEOM derived from some of these plant and manure samples using the same way. The visual "peak picking" approach suggested the potential presence of four fluorophores in the fresh corn WEOM centered at

EEWP of <240/440 nm, 250/300 nm, 270/355 nm, and 320/440 nm. Three fluorophores can be visually identified for the fresh (<240/440 nm, 270/350 nm and 330/430 nm) and the decomposed poultry manure WEOM (355/440 nm, 260/345 nm, and 250/ 465 nm). The EEPP values of manure WEOM samples are apparently different from those of humic fractions of pig and cattle manure (Hernandez et al., 2006; 2007; Plaza et al. 2008). The PARAFAC analysis modeled seven fluorescence components: tryptophan-like, tyrosine-like, and five humic substance-like components. Hunt and Ohno (2007) found that for most of the plant-derived WEOM solutions, decomposition significantly affected the concentration of three humic substance-like-associated components, increasing two and decreasing one. The effect of decomposition upon WEOM derived from swine, poultry, and dairy manures was dependent on the manure source.

Ohno et al. (2008) further characterized WEOM derived from soil, plant, and animal manure samples , including those used in previous studies (Ohno and Bro, 2006; Hunt and Ohno, 2007), using time-resolved fluorescence spectroscopy. Fluorescence decay measurements were carried out using a conventional time-correlated single-photon counting (TCSPC) apparatus, and the decay profiles were fit to multi-exponential functions using the multi-way PowerSlicing method. Their data indicated that the lifetime means of the three WEOM components are in general agreement with the three component lifetimes reported for a soil-derived FA. Ohno et al. (2008) noted that the WEOM fraction is a highly soluble and labile fraction of OM, while fulvic acids are refractory components of organic matter extracted using strong base. Similarly, EEM spectral analysis of reference humic substances (He et al., 2006; 2008) and these WEOM samples (Ohno and Bro, 2006; Hunt and Ohno, 2007; Ohno et al., 2009) has demonstrated that they all contain similar PARAFAC-derived components. Thus, Ohno et al. (2008) concluded that the similarity in the fluorescence properties of these two seemingly different organic matter fractions (i.e. the very labile WEOM and the highly refractory FA and HA) reflects a new paradigm for the structure of soil humic substances which should be further explored.

5.4. PARAFAC OF FLUORESCENCE EEM SPECTRA

The PARAFAC model which was initially developed by Harshman (1970) has been thoroughly detailed for use in analytical chemistry in a book (Smilde et al., 2004). Briefly, for a simple case where only one fluorophore exists, the emission intensity at a specific wavelength, j, when excited at wavelength k, can ideally be approximated as:

$$x_{jk} = ab_jc_k \qquad [1]$$

where x_{jk} is the intensity of the light emitted at emission wavelength j at excitation wavelength k, a is the concentration (in arbitrary scale) of the analyte, b_j is the relative emission emitted at wavelength j, and c_k is the relative amount of light absorbed at the excitation wavelength k. However, for DOM which is composed of a complex and heterogeneous mixture of fluorophores, Eqn. 1 can be expressed for F , the number of analytes (i.e. fluorophores), present in the mixture as:

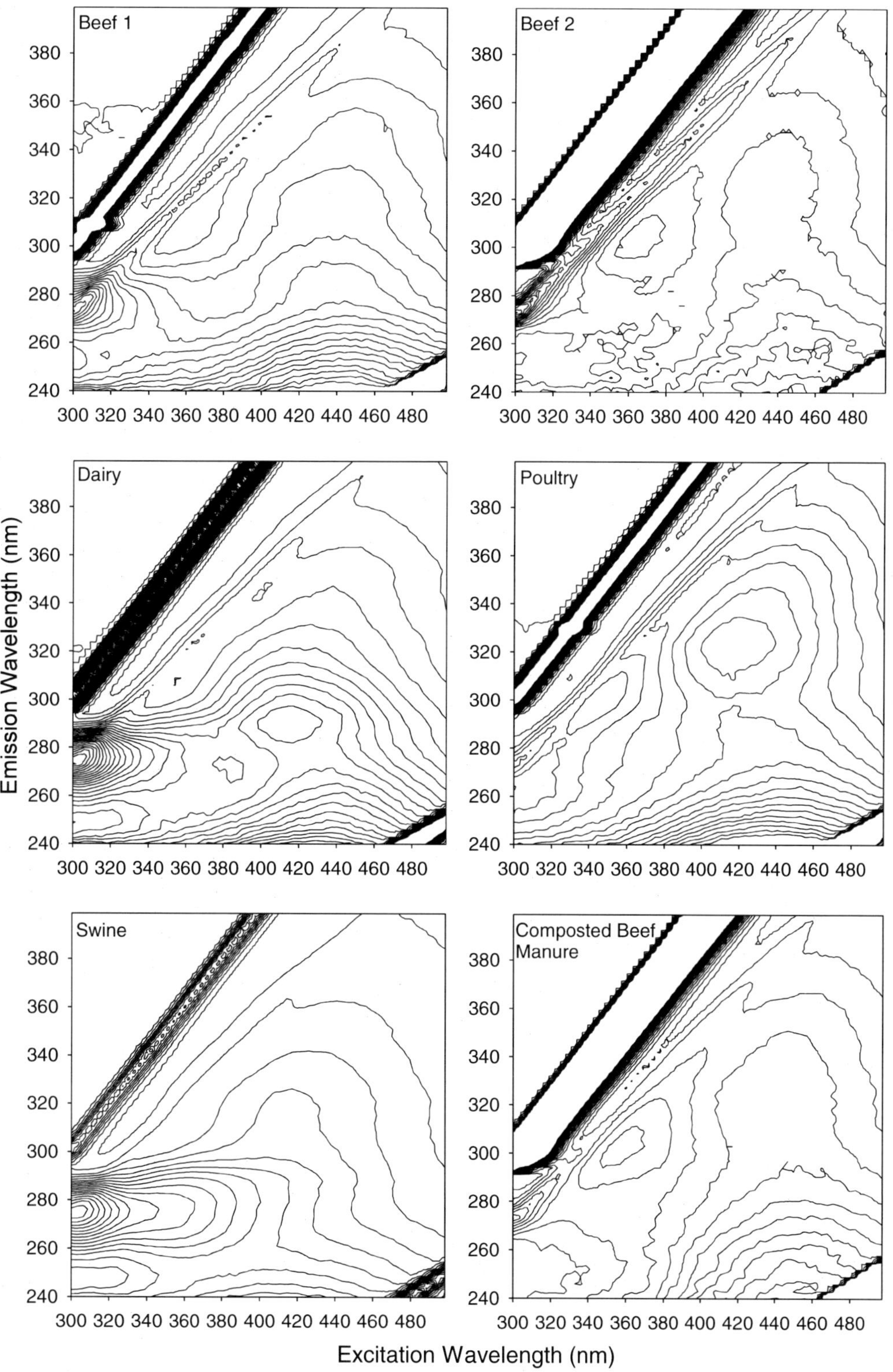

Figure 5.1. Excitation-emission matrix fluorescence spectra for the six aqueous manure extract dissolved organic matter used in this study.

$$x_{jk} = \sum_{f=1}^{F} a_f b_{jf} c_{kf}$$

[2]

Eqn. 2 implies that the contribution to the emission from each analyte is independent of the contributions of the remaining analytes. In the above equation, the relative emission of analyte f at emission j is b_{jf}, the relative absorption at excitation k is c_{kf}, and the concentration of analyte f is a_f. For the case where there is more than one sample, and a_{if} being the concentration of the fth analyte in the ith sample, the model then becomes:

$$x_{ijk} = \sum_{f=1}^{F} a_{if} b_{jf} c_{kf}$$
$$, i=1,..I;j=1,...J;k=1,..K$$

[3]

This model of several samples is exactly the same as the underlying PARAFAC model shown in Eqn. 2. Thus it can be seen that the PARAFAC model takes input data (i.e. measured EEMs) and the number of components to provide estimates of the concentrations and the excitation and emission spectra. The elements x_{ijk} can be held in a three-way array X of size $I \times J \times K$ where I is the number of samples, J the number of emission wavelengths and K the number of excitation wavelengths as indicated in Figure 5. 1. Thus, by fitting a least squares PARAFAC model with the right number of components, the relative concentrations, emission spectra and excitation spectra can be determined.

There are some pre-processing steps which are required to be performed on the EEM spectra prior to PARAFAC analysis. Light scattering from interaction of the incident beam with the solvent molecules results in diagonal lines in the fluorescence EEM even in the absence of fluorescing molecules. The intense elastic first order (EM $\square$= EX) and second order (EM = 2 EX) Rayleigh lines are typically clearly visible. In addition, a less intense inelastic Raman scattering is often evident with the scatter line shifted to longer emission wavelengths from the first order Rayleigh scatter line due to partial non-radiative energy conversion prior to photon emission. If a fluorophore is located on the EEM landscape on or near a scatter line, the PARAFAC analysis may be influenced by the presence of the scatter lines (Andersen and Bro, 2003). Thus, steps must betaken to minimize the influence of scatter lines and other attributes of the EEM landscape prior to PARAFAC modeling. Subtractions of a blank EEM containing only the solvent from each sample EEM will remove the lower intensity Raman scatter effects effectively. The intense Rayleigh scatter will likely to remain partially after the blank subtraction (Christensen et al., 2005). Thus, the Rayleigh scatter lines and the region immediately adjacent to the scatter lines are mathematically removed by setting the fluorescence intensity values as missing. The EEM has a triangular shaped region where the EM is less than the EX where fluorescence is physically not possible. These EX:EM data pairs are set to zero.

For researchers who would like to start using EEM/PARAFAC to analyze their organic matter samples, several papers exist in addition to the book already cited (Smilde, 2004). Bro (1997) and Andersen and Bro (2003) provide a useful discussion of on how to use PARAFAC to analyze fluorescence data. In addition, Stedmon and Bro (2008) provides a more recent and detailed tutorial with links to a MATLAB Toolbox written to run the PARAFAC analysis.

These references will be valuable references to researchers who are new to this aspect of fluorescence spectroscopy.

5.5. CASE STUDY: MANURE AND MANURE-AMENDED SOIL ANALYSIS

5.5.1. Step 1: Aqueous Extracts

Six manure samples and three manure-amended soil samples were used in this illustrative study (Table 5. 1). All samples were air-dried and sieved through a 2-mm sieve. Deionized-distilled water (DI-H$_2$O) was used for all solutions and extractions. The manure samples were extracted at a 40:1 water (volume) to sample (mass) ratio at 4 °C for 16 h with periodic hand shaking, centrifuged at 900 X g for 25 min prior to vacuum filtering through 0.4 ☐m polycarbonate filters. The water: sample ratio selection is not a crucial factor and a variety of ratios have been used by researchers. The manure identified as Beef 2 and the composted manure were incorporated using standard recommendations based on N content of the manure into field plots at the Rogers Sustainable Agricultural Research Farm in early June, 2009. Replicated samples of the plots which received the two manures and control plots which did not receive the amendments were sampled four weeks after incorporation. Soil DOM was extracted by adding 10.0 mL of deionized-H$_2$O to 1.00 g of soil in a 15-mL centrifuge tube. The suspensions were shaken on an orbital shaker for 30 min at room temperature (22 ± 1 °C), centrifuged at 900 X g for 30 min, and filtered through 0.45 μm Acrodisk syringe filters. Extraction at 4 °C is recommended to minimize microbial alteration during the extraction process (Zhou and Wong, 2000).

5.5.2. Step 2: Inner-Filtration Effects

Fluorescence spectrometric measurements need careful attention of inner-filtering effects which is the attenuation of fluorescence signal by the solution itself prior to detection (MacDonald et al., 1997). Primary inner-filtering is the absorption of the excitation beam prior to reaching the analyte in the measurement zone and secondary inner-filtering is the absorption of the emitted fluorescence photon in the solution. Although soil DOM extracts usually have sufficiently low DOC concentration so that inner-filtration effects are minimal, animal manure extracts at 40:1 ratio typically contain 350 to 700 mg dissolved organic carbon (DOC) L^{-1} in solution (Ohno and Crannell, 1996). These high concentrations of DOC will lead to absorbance in the UV excitation wavelengths used in EEM to be offscale. There are two routine approaches for minimizing inner-filtering effects in DOM analysis. One technique is the dilution of the solution to a fixed absorbance value at the lowest EX wavelength used in the EEM scan. The fixed value typically ranges between 0.1 and 0.3 absorbance units (Cox et al., 2000; Ohno, 2002). The other approach is to dilute the solution to a fixed concentration, typically around 25 ppm DOC (Yang et al., 1994). In both of these approaches, the concept is to suppress the amount of inner-filtering to negligible level and is appropriate for most routine analysis.

There is a more rigorous, exact correction for inner-filtration that can also be used. This requires an absorbance measurement of the solution at all wavelengths present in the EEM scan. The exact correction can be calculated as: $I = I_o (10^{-b(A_{ex} + A_{em})})$ where I is the instrument detected fluorescence intensity, I_o is the fluorescence intensity without inner-filtration, b is the assumed path length of both excitation and emission beams (typically 0.5 cm with 1.0 cm fluorescence cells), and A_{EX} and A_{EM} are absorbance values at the EX and EM wavelengths (Ohno, 2002). This exact correction is quite time-consuming and experience has shown that the approximate correction is usually appropriate in most settings since by fixing either absorbance for DOC concentration, results are obtained without a concentration bias.

For this case study, the solution was diluted to 0.10 absorbance units at 240 nm. This wavelength was the lowest EX wavelength in the EEM and would be the case where inner-filtration would be at a maximum. An absorbance value at 240 nm of the 40:1 extract was initially obtained to suggest how much dilution is needed. Using this initial absorbance value, a first dilution is made to obtain solution with absorbance of ~0.25. An exact absorbance value of this first dilution solution is made and used to calculate the exact dilution needed to bring the absorbance down to 0.100 units. This can be best done gravimetrically using the relationship: $(A_i)(volume_i) = (A_f)(volume_f)$ where i and f refer to initial and final solutions. Thus, one can solve for the volume of the solution needed by dividing the product of A_f (usually 0.1) and volume of sample needed (usually 10 mL) by the absorbance value of the solution for this final dilution step. Using a balance sensitive to 0.01 g units and assuming that 1 mL of solution is equivalent to 1 g makes the dilution step rapid and accurate to ensure proper dilution to the desired absorbance value.

5.5.3. Step 3: Obtaining the EEM Scan

There are a variety of fluorescence instruments capable of producing of EEM rapidly. Even an instrument not designed for EEM spectroscopy can be used by manually combining a series of EX scans at different EX wavelengths into a single EEM. The specific instrument conditions should be set based on the instrument itself and experience of the operator. The EEM wavelength used by DOM researchers has been quite variable: EX, 200-390 nm & EM, 300-600 nm (Westerhoff et al., 2001); EX, 220-450 nm & EM, 230-600 nm (Stedmon et al., 2003); EX, 220-493 nm & EM, 300-600 nm (Alberts and Takacs, 2004); and EX, 240-400 nm & EM, 300-500 nm (Ohno and Bro, 2006). Signal noise can be a significant problem in the <230 nm range on some instruments and may pose a lower EX wavelength limit. Although there are DOM related fluorophores in this <230 nm range, it is not recommended to extend the EEM range into wavelength regions where instrument noise becomes significant. The EEM spectra of the six manure samples are shown in Figure 5. 1. All samples have a peak with EX <240 nm and a broad EM peak in the 380-460 nm range. The beef 1, dairy, and swine manure spectra display a sharp peak at EX 275 nm and EM 305 nm. Both of the beef manures, swine, and compost EEM spectra have peaks at approximate EX 320 nm and EM 440 nm. The dairy and poultry have an EM peak at 420 nm with the EX peak at 290 nm for dairy manure and 320 nm for poultry manure. Using visual 'peak picking', each EEM spectra has two or three identifiable peaks. The EEM spectra of the soil DOM of beef and composted beef manure amended soils, as well as the control soil are shown in Figure 5. 2. Unlike the manure EEM spectra, the soil EEM spectra for all three treatments are similar with the broad peak at EX <240 nm and EM peak in the 380-460 nm range, and a EX 320 nm and EM 440 nm peak.

5.5.4. Step 4: PARAFAC Modeling

The PARAFAC models can be run with PLS_Toolbox (Eigenvector Research Inc., Mason, WA) or the routines provided in the DOMFluor Toolbox (Stedmon and Bro, 2008). Both of these are designed to work in the MATLAB (Mathworks, Natic, MA) computing environment. The subtraction of the blank EEM from all the samples can be done as a simple matrix subtraction within MATLAB. The routines provided in the DOMFluor toolbox will allow for removal of the Raman scatter replacing the scatter intensities with a NaN (not a number) in the data array. A non-negativity constraint is applied to the parameters to allow only chemically relevant results since negative concentrations and fluorescence intensities are chemically impossible. PARAFAC modeling requires that the user select the number of components (i.e. fluorophores) to model. These PARAFAC-derived components can be viewed as being 'quasiparticles' in the sense defined by Sposito and Blaser (1992). PARAFAC components are not real molecules, but rather they are mathematical entities representing non-interacting ligands whose modeled parameters closely mimics the actual mixture of fluorescing organic matter components present in the DOM extracts. With DOM being composed of a suite of compounds with subtle chemical differences, PARAFAC modeling is useful because it provides an objective description of the average behavior of the discrete, diverse classes of compounds present in samples. For PARAFAC analysis of DOM, the modeling is repeated for number of components present from 2 to ~10.

5.5.5. Step 5: Determining Proper Component Number

PARAFAC models with two to seven components were computed for this animal manure and soil data set. Although there are a variety of ways to determine the correct number of components present in the data set such as the percentage of variation modeled and the number of iterations required for model fitting, the core consistency diagnostic score which should be close to 100 % for appropriate models is approach most frequently used. The core consistency provides an estimate of how well the model captures trilinear information, and if the consistency diagnostic value turns low it is a strong indication that the model is invalid (Bro and Kiers, 2003). The core consistency diagnostic scores for the two to seven component PARAFAC models are shown in Figure 5. 3. There is a sharp drop off between the five-component and six component models indicating that the five component model provides the greatest spectral resolution of components and is likely the most appropriate model for this manure data set. The five component PARAFAC model explained 99.99 % of the variability in the data set. The choice of the appropriate number of components in using PARAFAC for modeling EEM data is a critical step of the method and has been described in detail in several papers (Andersen and Bro, 2003; Stedmon and Markager, 2005).

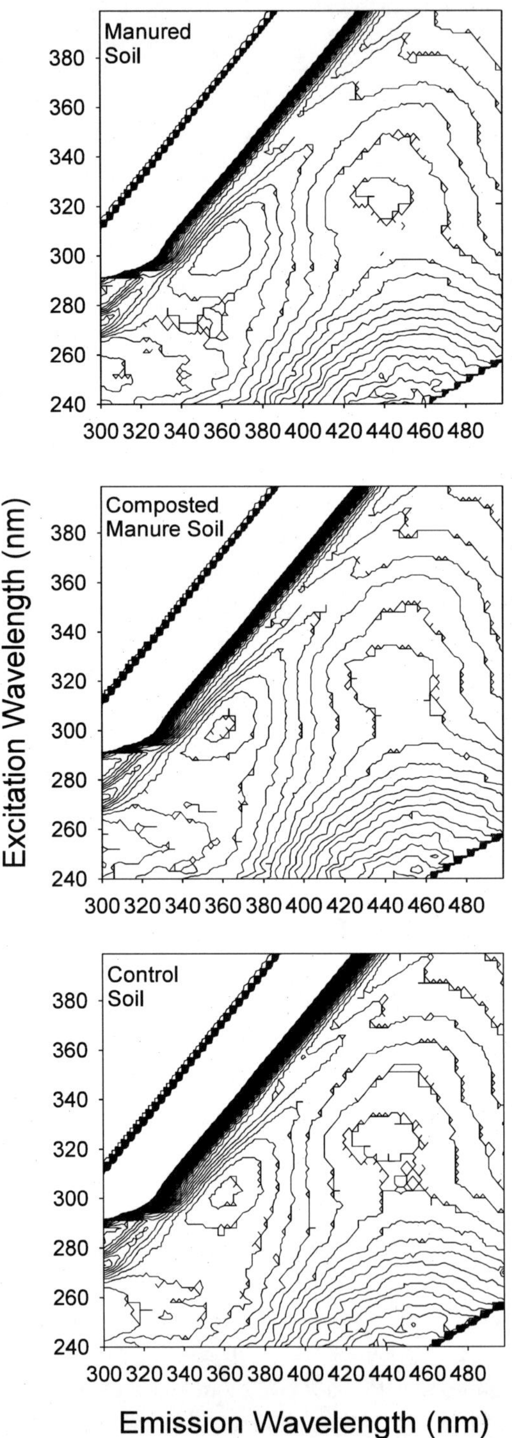

Figure 5.2. Excitation-emission matrix fluorescence spectra the manure amended, composted manure-amended, and control soil dissolved organic matter.

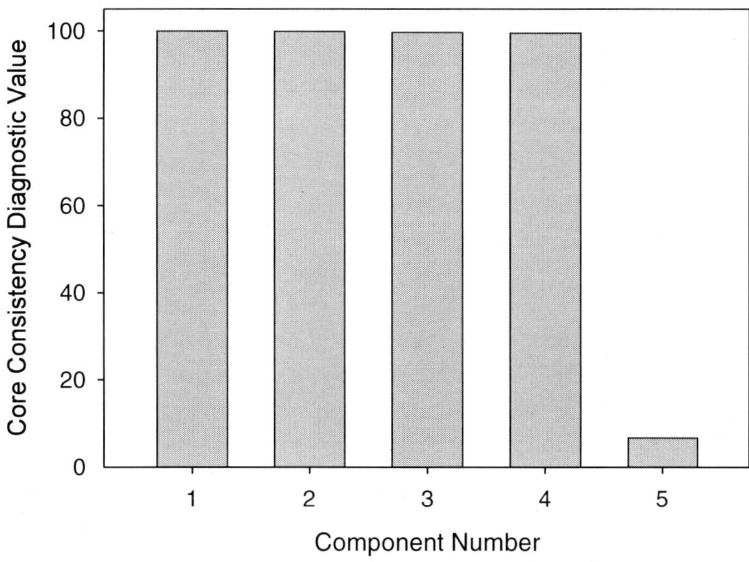

Figure 5.3. Core consistency diagnostic values for the PARAFAC models having two to six components for the manure and manure-amended soil DOM data set.

5.5.6. Step 6. Details of the Selected PARAFAC Model

After the selection of the proper component number (i.e. number of fluorphores present), the details of the model can be studied to obtain the chemical characteristics of the DOM under investigation. The two loadings representing the EX spectra and the EM spectra, as well as the "concentration" score of the components can be obtained from the PARAFAC model. The EX and EM spectra of the five components are shown in Figure 5. 4a and 5.4b as contour and X-Y plots, respectively. The contour plot provides an efficient visualization of the component as an EEM spectrum, while the X-Y plot provides a more traditional view of the separate EX and EM spectra . The X-Y plotted spectra also allow easier presentation of the intensities of the EX and EM spectra as compared to the contour plot. Both types of presentations have been used in literature (e. g. He et al., 2008; 2010; Hunt and Ohno, 2007; Ohno and Bro, 2006; Ohno et al., 2009). Component 1 has been identified as being associated with proteins containing tyrosine (Stedmon and Markager, 2005). Components 2 and 3 have two EX spectral peaks and a single EM peak which is often found in humic substances (Sierra, 2005). Component 4 is represents a fluorophore of unknown chemical provenance. Component 5 has an EX maxima <240 nm and a EM that encompasses the entire 300-500 nm range which is in a range that corresponds to an often observed, but poorly defined peak classified as the "A" peak (Yamashita e t al., 2003).

The relative concentration scores are shown in Table 5.1. It is stressed that the relative fraction distributions presented in Table 5.1 are based on the relative fluorescence signal intensities of the sample components, rather than on their chemical concentrations. Calculation of their true chemical concentration would require knowledge of the fluorescence quantum efficiencies of the individual components which are currently unknown. The component distribution between the DOM extracted from these three soils was similar to each other. This suggests that the addition of manure or composted manure does not lead to prolonged changes in the chemical nature of DOM. There is a possibility that organic amendment alters the chemical composition of DOM, but the change occurs in a part of the DOM molecule that is not fluorescence-active. The soil DOM was dominated by components 2 and 4 which accounted for about 70% of the fluorescent signal indicating that the red-shifted humic-like and the unknown humic-like fraction govern the chemical characteristic of the DOM. The low amount of the protein-like component indicate that these proteins or polypeptides do not remain in a water-soluble form after animal manure is incorporated into the soil most likely due to microbial uptake as a source of N, or due to soil sorption reactions.

In contrast to the soil samples, all five component contents of the individual samples in the manure samples were highly variable indicating that composition of the DOM varied appreciably among manures (Table 5.1). Interestingly, the protein-like component 1 was highly variable ranging from 47% in the swine manure to 8% in the composted beef manure. This variability may be due to differing feed sources which have different levels of protein content. The manure DOM, like those of soil, has the red-shifted humic-like component 2 in the greatest amount which suggests that this component is found in all DOM regardless of source. Component 3 which also is characterized as humic-like was in general found at a higher concentration than for the soil DOM. The uncharacterized component 4 was a minor fraction of the manure DOM, as was component 5 in both soil and manure DOM.

Table 5.1. Relative percentage distribution of the five PARAFAC components in water extracted organic matter (OM) samples from the soils and manure amendments used in this case study.

OM Type	Component 1	Component 2	Component 3	Component 4	Component 5
Manure-Amended Soil	7.8	42.2	14.7	31.1	4.2
Composted Manure Amended Soil	7.8	40.3	13.4	33.5	5.0
Control Soil	7.8	42.2	14.0	32.3	5.1
Beef Manure 1	18.8	34.0	12.7	7.8	26.6
Beef Manure 2	9.9	57.2	28.1	4.8	0.0
Dairy Manure	29.4	25.5	18.5	7.2	19.5
Poultry Manure	12.4	32.8	30.8	10.2	13.7
Swine Manure	47.6	16.8	20.8	5.5	9.3
Composted Beef Manure	8.0	45.4	16.5	24.6	5.5

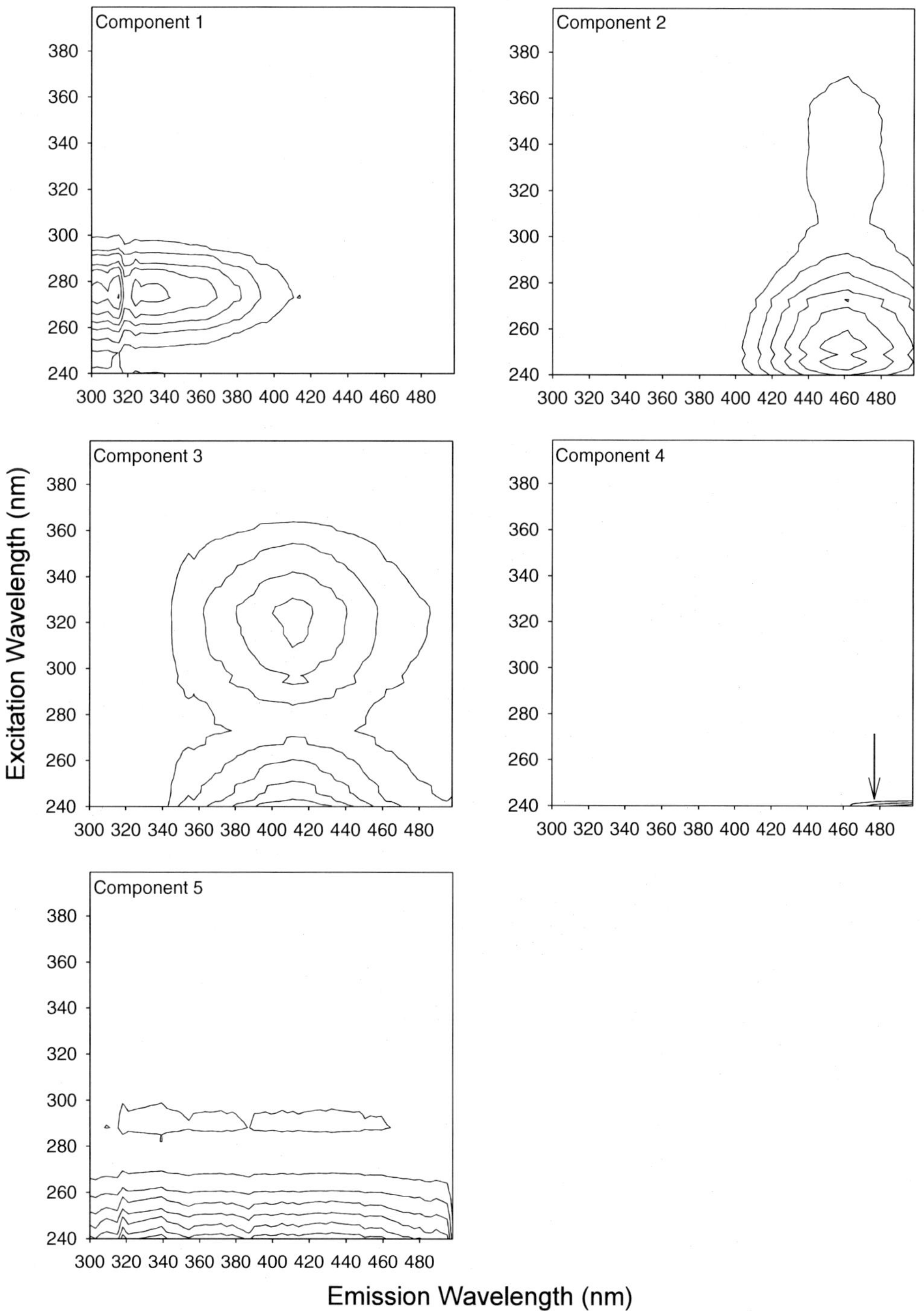

Figure 5.4. (Continued on Next Page).

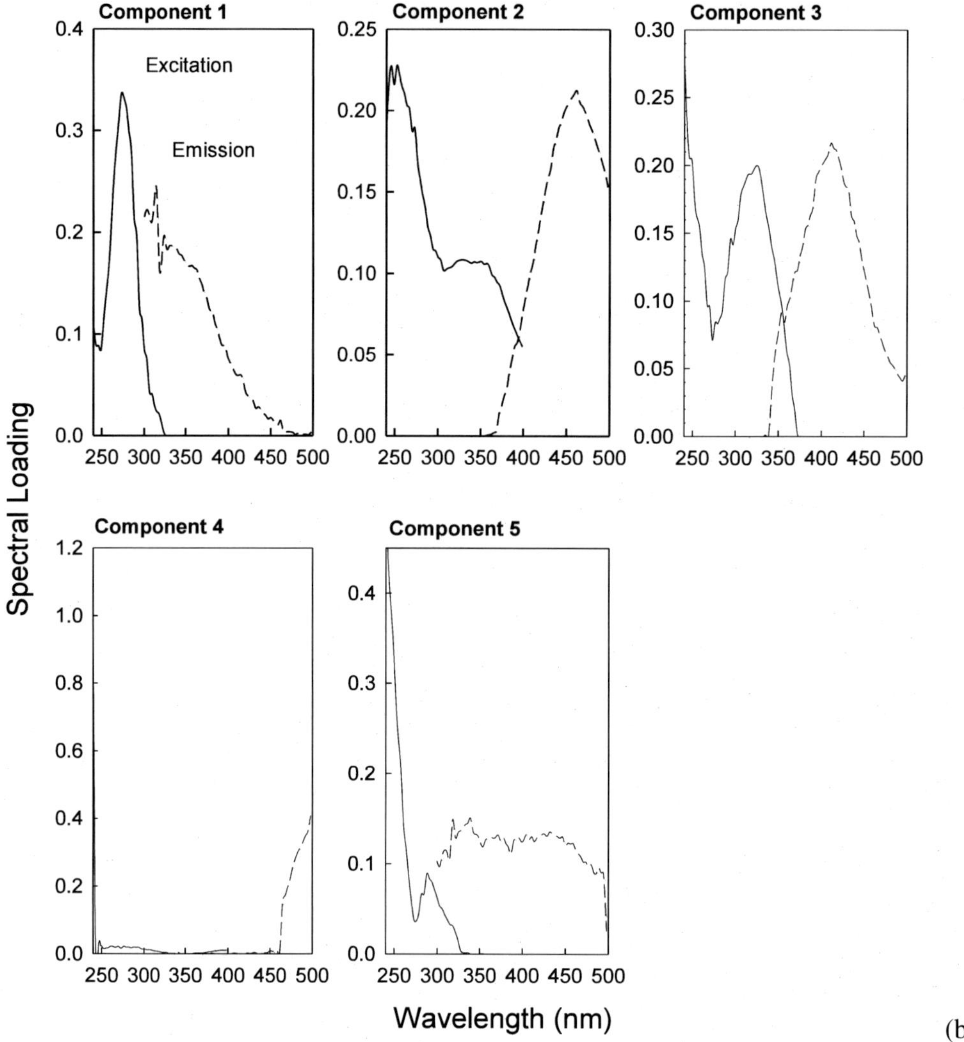

Figure 5.4. Excitation-emission matrix fluorescence spectra for the five PARAFAC components modeled for the manure and manure-amended soil DOM set. (a). In contour formats; (b) in x-y formats.

Principal components analysis of the component relative distribution was conducted to show the relationships of the component contents amongst the soil and manure DOM sources (Figure 5.5). The first two principal components captured 84 % of the variability and two interpretations of the DOM data set are easily apparent. First, the strong self-similarity for the soil DOM is shown with a tight clustering of the points. Secondly, the strong dissimilarity for the animal manure DOM is evident with the distance variability between their points on the plot. Beef 1, dairy, and swine manures are relatively similar with a cluster of points, but differ from beef manure 2, poultry, and the composted manure DOM. The composted manure is in the cluster of points of soils which suggests that the microbial alterations during the composting process results in DOM which is quite similar to soil DOM.

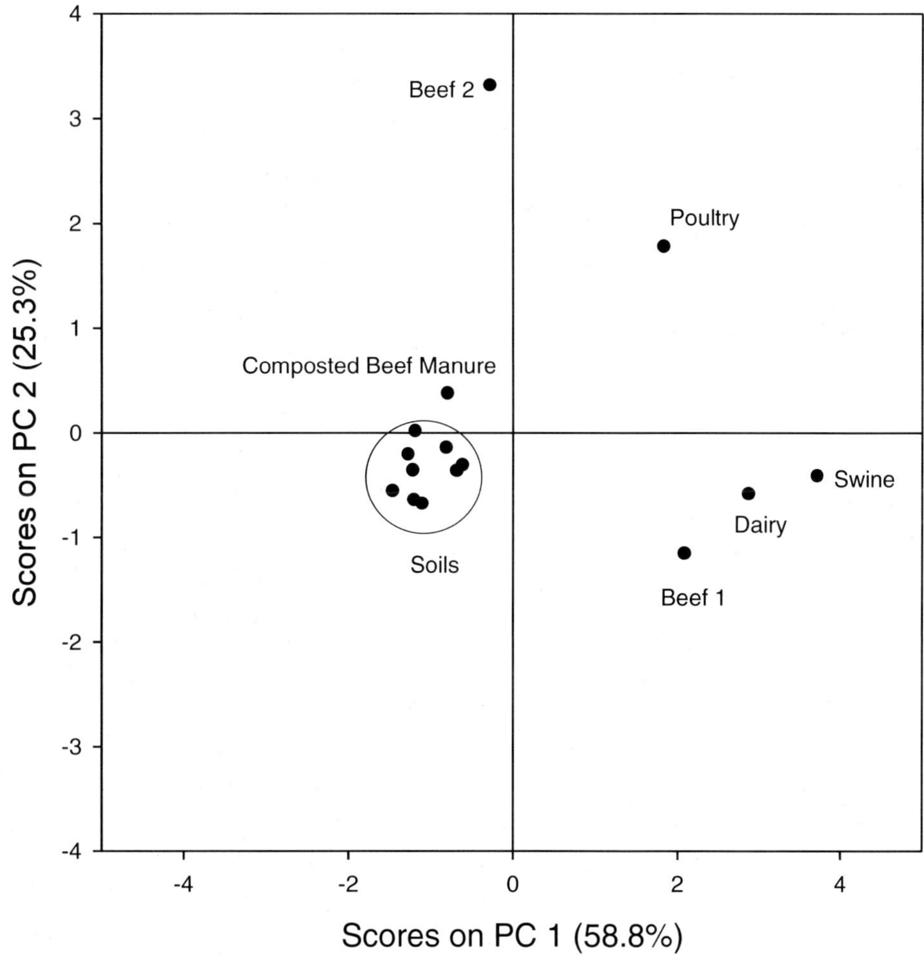

Figure 5.5. Principal component analysis of the concentration scores for the five component PARAFAC model of the manure and manure-amended soil DOM set.

5.6. CONCLUSION

The application of animal manures to agricultural soils is an approach to biocycling nutrients which can eliminate or reduce the need for purchased fertilizes. Animal manures, in addition to crop nutrients, also contains significant amounts of dissolved organic carbon which can affect important soil chemical processes such as altering the bioavailability of crop nutrients, solubilization and transport of metals, and sequestration of soil carbon. The involvement of the manure-derived organic matter in these soil reactions would be expected to depend on its chemical structure. Fluorescence spectroscopy is a sensitive and rapid analytical method to probe the chemical properties of organic matter and thus is ideally suited to investigate how manure amendment affects soil processes that are of relavence to the sound management of animal manures in agricultural system.

This tutorial provides step-by-step details involved in using PARAFAC to analyze fluorescence EEM spectra. PARAFAC provides a rigorous method to obtain objective quantitative chemical information from fluorescence spectra. This method provides a statistical approach for spectral interpretation. With this chemometric approach, EEM fluorescence can be easily used to follow organic matter dynamics in soil systems. In the past few years since the demonstration of the use of PARAFAC on organic matter characterization, it has been successfully used to gain a better understanding of how organic matter is involved in many soil ecosystem processes. The use this approach in studies involving manures and landspreading of manures should significantly increase our understanding of the importance of the high variability of manure characteristics and the transformation of the manure-derived organic matter upon incorporation into soils.

ACKNOWLEDGMENTS

This project was supported in part by USDA-National Research Initiative Competitive Grant no. 2008-35107-04480 from the USDA Cooperative State Research, Education, and Extension Service, and Hatch funds provided by the Maine Agricultural and Forest Experiment Station.

REFERENCES

Alberts, J.J., and M. Takacs. 2004. Total luminescence sprectra of IHSS standard and reference fulvic acids, humic acids and natural organic matter: comparison of aquatic and terrestrial source terms. *Org. Geochem.* 35:243-256.

Andersen, C.M., and R. Bro. 2003. Practical aspects of PARAFAC modeling of fluorescence excitation-emission data. *J. Chemometrics* 17:200-215.

Bro, R. 1997. PARAFAC: Tutorial and applications. *Chemometrics Intelligent Lab. Syst.* 38:149-171.

Bro, R., and H.A.L. Kiers. 2003. A new efficient method for determining the number of components in PARAFAC models. *J. Chemometrics* 17:274-286.

Chen, W., P. Westerhoff, J.A. Leenheer, and K. Booksh. 2003. Fluorescence excitation-emission matrix regional integration to quantify spectra for dissolved organic matter. *Environ. Sci. Technol.* 37:5701-5710.

Christensen, J.H., A.B. Hansen, J. Mortensen, and O. Andersen. 2005. Characterization and matching of oil samples using fluorescence spectroscopy and parallel factor analysis. *Anal. Chem.* 77:2210-2217.

Coble, P.G., S.A. Green, N.V. Blough, and R.B. Gagosian. 1990. Characterization of dissolved organic matter in the Black Sea by fluorescence spectroscopy. *Nature* 348:432-435.

Cox, L., R. Celis, M.C. Hermosin, J. Cornejo, A. Zsolnay, and K. Zeller. 2000. Effect of organic amendments on herbicide sorption as related to the nature of the dissolved organic matter. *Environ. Sci. Technol.* 34:4600-4605.

Harshman, R.A. 1970. Foundations of the PARAFAC procedure: Models and conditions for an 'explanatory' multi-modal factor analysis. *UCLA Working Papers in Phonetics* 16:1-84.

He, Z., T. Ohno, B.J. Cade-Menun, M.S. Erich, and C.W. Honeycutt. 2006. Spectral and chemical characterization of phosphates associated with humic substances. *Soil Sci. Soc. Am. J.* 70:1741-1751.

He, Z., T. Ohno, D.C. Olk, and F. Wu. 2010. Capillary electrophoresis profiles and fluorophore components of humic acids in Nebraska corn and Philippine rice soils. *Geoderma* 156:143-151.

He, Z., T. Ohno, F.C. Wu, D.C. Olk, C.W. Honeycutt, and M. Olanya. 2008. Capillary electrophoresis and fluorescence excitation-emission matrix spectroscopy for characterization of humic substances. *Soil Sci. Soc. Am. J.* 72:1248-1255.

Hernandez, D., C. Plaza, N. Senesi, and A. Polo. 2006. Detection of copper(II) and zinc(II) binding to humic acids from pig slurry and amended soils by fluorescence spectroscopy. *Environ. Pollut.* 143:212-220.

Hernandez, D., C. Plaza, N. Senesi, and A. Polo. 2007. Fluorescence analysis of copper(II) and zinc(II) binding behavior of fulvic acids from pig slurry and amended soil. *Eur. J. Soil Sci.* 58:900-908.

Huang, G.F., Q.T. Wu, J.W.C. Wong, and B.B. Nagar. 2006. Transformation of organic matter during co-composting of pig manure with sawdust. *Bioresour. Technol.* 97:1834-1842.

Hunt, J.F., and T. Ohno. 2007. Characterization of fresh and decomposed dissolved organic matter using excitation-emission matrix fluorescence spectroscopy and multi-way analysis. *J. Agric. Food Chem.* 55:2121-2128.

Macalady, D.L., and K. Walton-Day. 2009. New light on a dark subject: On the use of fluorescence data to deduce redox states of natural organic matter (NOM). *Aquat. Sci.* 71:135-143.

MacDonald, B.C., S.J. Lvin, and H. Patterson. 1997. Correction of fluorescence inner filter effects and the partitioning of pyrene to dissolved organic carbon. *Analytica Chimica Acta* 338:155-162.

Merritt, K.A., and M.S. Erich. 2003. Influence of organic matter decomposition on soluble carbon and its copper-binding capacity. *J. Environ. Qual.* 32:2122-2131.

Murphy, K., C.A. Stedmon, D. Waite, and G. Ruiz. 2008. Distinguishing between terrestrial and autochthonous organic matter sources in marine environments using fluorescence spectroscopy. *Mar. Chem.* 108:40-58.

Ohno, T. 2002. Fluorescence inner-filtering correction for determining the humification index of dissolved organic matter. *Environ. Sci. Technol.* 36:742-746.

Ohno, T. and B.S. Crannell. 1996. Green and animal manure-derived dissolved organic matter effects on phosphorus sorption. *J. Environ. Qual.* 25:1137-1143.

Ohno, T., and C.S. Cronan. 1997. Comparative effects on ionic-and nonionic-resin purification treatments on the chemistry of dissolved organic matter. *Intern. J. Environ. Anal.* 66:119-136.

Ohno, T., and R. Bro. 2006. Dissolved organic matter characterization using multi-way spectral decomposition of fluorescence landscapes. *Soil Sci. Soc. Am. J.* 70:2028-2037.

Ohno, T., Z.M. Wang, and R. Bro. 2008. PowerSlicing to determine fluorescence lifetimes of water-soluble organic matter derived from soils, plant biomass, and animal manures. *Anal. Bioanal. Chem.* 390:2189-2194.

Ohno, T., Z. He, I.A. Tazisong, and Z.N. Senwo. 2009. Influence of tillage, cropping, and nitrogen source on the chemical characteristics of humic acid, fulvic acid, and water-soluble soil organic matter fractions of a long-term cropping system study. *Soil Sci.* 174:652-660.

Plaza, C., N. Senesi, J.C. Garcia-Gil, G. Brunetti, V. D'Orazio, and A. Polo. 2002. Effects of pig slurry application on soils and soil humic acids. *J. Agric. Food Chem.* 50:4867-4874.

Plaza, C., N. Senesi, A. Polo, G. Brunetti, J.C. Garcia-Gil, and V. D'Orazio. 2003. Soil fulvic acid properties as a means to assess the use of pig slurry amendment. *Soil Tillage Res.* 74:179-190.

Plaza, C., G. Brunetti, N. Senesi, and A. Polo. 2006. Molecular and quantitative analysis of metal ion binding to humic acids from sewage sludge and sludge-amended soils by fluorescence spectroscopy. *Environ. Sci. Technol.* 40:917-923.

Plaza, C., R. Nogales, N. Senesi, E. Benitez, and A. Polo. 2008. Organic matter humification by vermicomposting of cattle manure alone and mixed with two-phase olive pomace. *Bioresour. Technol.* 99:5085-5089.

Santin, C., Y. Yamashita, X.L. Otero, M.A. Alvarez, and R. Jaffe. 2009. Characterizing humic substances from estuarine soils and sediments by excitation-emission matrix spectroscopy and parallel factor analysis. *Biogeochemistry* 96:131-147.

Senesi, N. 1990. Molecular and quantitative aspects of the chemistry of fulvic acid and its interactions with metal ions and organic chemicals. Part II. The fluorescence spectroscopy approach. *Anal. Chem. Acta.* 232:77-106.

Senesi, N., T.M. Miano, M.R. Provenzano, and G. Brunett. 1991. Characterization, differentiation, and classification of humic substances by fluorescence spectroscopy. *Soil Sci.* 152:259-271.

Senesi, N. 1992. Application of electron spin resonance and fluorescence spectroscopies to the study of soil humic substances. p. 13-52. *In*: J. Kubat (ed) *Humic Substances in Soil, Sediment, and Water. John Wiley and Sons, New York.*

Senesi, N., C. Plaza, G. Brunetti, and A. Polo. 2007. A comparative survey of recent results on humic-like fractions in organic amendments and effects on native soil humic substances. *Soil Biol. Biochem.* 39:1244-1262.

Sierra, M.M.D., M. Giovanela, E. Parlante, and E.J. Soriano-Sierra. 2005. Fluorescence fingerprint of fulvic and humic acids from varied origins as viewed by single-scan and excitation/emissiojn matriz techniques. *Chemosphere* 58:715-733.

Smilde, A.K, R. Bro, and P. Geladi. 2004. Multi-way analysis. *Applications in the chemical sciences.* Wiley, New York.

Sposito, G., and P. Blaser. 1992. Revised quasiparticle model of protonation and metal complexation reactions. Soil *Sci. Soc. Am. J.* 56:1095-1099.

Stedmon, C.A., and S. Markager. 2005. Resolving the variability in dissolved organic matter fluorescence in a temperate estuary and its catchment using PARAFAC analsysis. *Limnol. Oceanogr.* 50:686-697.

Stedmon, C.A., and R. Bro. 2008. Characterizing dissolved organic matter fluorescence with parallel factor analysis: a tutorial. *Limnol. Oceanogr. Methods* 6:572-579.

Stedmon, C.A., S. Markager, and R. Bro. 2003. Tracing dissolved organic matter in aquatic environments using a new approach to fluorescence spectroscopy. *Marine Chem.* 82:239-254.

Westerhoff, P., W. Chen, and M. Esparza. 2001. Fluorescence analysis of a standard fulvic acid and tertiary treated wastewater. *J. Environ. Qual.* 30:2037-2046.

Wu, F.C., R.B. Mills, R.D. Evans, and P.J. Dillon. 2004. Kinetic of metal-fulvic acid complexation using a stopped-flow technique and three-dimensional excitation emission fluorescence spectrophotometer. *Anal. Chem.* 76:110-113.

Yang, A., G. Sposito, and T. Lloyd. 1994. Total luminescence spectroscopy of aqueous pine litter (O horizon) extracts: organic ligands and their Al or Cu complexes. *Geoderma* 62:327-344.

Yamashita, Y., and E. Tanoue. 2003. Chemical characterization of pretein-like fluorophores in DOM in relation to aromatic amino acids. *Marine Chem.* 82:255-271.

Zhou, L.X., and J.W.C. Wong. 2000. Microbial decomposition of dissolved organic matter and its control during a sorption experiment. *J. Environ. Qual.* 29:1852-1856.

PART II. NITROGEN AND VOLATILE COMPOUNDS

In: Environmental Chemistry of Animal Manure
Editor: Zhongqi He

ISBN: 978-1-62808-641-6
© 2013 Nova Science Publishers, Inc.

Chapter 6

AMMONIA EMISSION FROM ANIMAL MANURE: MECHANISMS AND MITIGATION TECHNIQUES

Pius M. Ndegwa[1,], Alexander N. Hristov[2] and Jactone A. Ogejo[3]*

6.1. INTRODUCTION

Ammonia (NH_3) volatilization is one of the most important pathways through which nitrogen (N) is lost from animal manures. Ammonia volatilization is a critical issue because its loss not only reduces the fertilizer-value of the manure but, in most cases, also has negative impacts on the environment. Agriculture is believed to be the largest source of global NH_3 emission, with the majority of emissions (~80%) estimated to originate from animal manures. Potential adverse consequences associated with NH_3 emission to the environment include: respiratory diseases caused by exposure to high concentrations of secondary fine particulate aerosols formed from NH_3 (commonly referred to as $PM_{2.5}$); nitrate contamination of drinking water; eutrophication of surface water bodies manifested in harmful algal blooms and decreased water quality; vegetation and ecosystem changes caused by excess N deposition; and soil acidification through nitrification and leaching. Reducing NH_3 emissions from animal manures can ultimately mitigate these environmental impacts. In order to design effective techniques for reducing NH_3 loss from animal manures, it is important to first understand the fundamental mechanisms involved in such emissions. This Chapter is divided into two sections. Section 6.2 presents processes and mechanisms leading to NH_3 production and volatilization from animal manures. Section 6.3, on the other hand, summarizes basic principles of the techniques and case-studies, either already developed or in development, for minimizing NH_3 emissions from animal manures.

[*] Corresponding author: ndegwa@wsu.edu; Phone: 509.335.8167; Fax: 509.335.2722.
[1] Biological Systems Engineering, Washington State University PO Box 646120, Pullman, WA 99164, USA
[2] Dairy and Animal Science Department, Pennsylvania State University, University Park, PA 16802, USA
[3] Biological Systems Engineering, Virginia Tech. 212 Seitz Hall, Blacksburg, VA 24061, USA

6.2. MECHANISMS OF AMMONIA EMISSIONS

6.2.1. Processes Responsible for Ammonia in Manure

In monogastric animals; dietary N concentration is the primary factor determining N excretions and NH_3 volatilization losses from manure. The efficiency of utilization of dietary N in pigs, for example, ranges from 30 to 50% (Jongbloed and Lenis, 1992; Arogo et al., 2001), which is comparable to the efficiency of utilization of dietary N in the ruminants (see following paragraphs). In general, overfeeding of protein and amino acids imbalance are the main reasons for this low efficiency. Numerous studies have demonstrated reduced NH_3 emissions from pig manure with reduced dietary crude protein (CP) intake (Le at al., 2009). Meticulous balance of animal needs and amino acids supply can dramatically improve N utilization efficiency and consequently reduce NH_3 emissions from manure. For example, feeding low protein diets supplemented with individual amino acids to meet, but not exceed, animal requirements has produced remarkable improvements in the efficiency of dietary N utilization for production purposes in pigs (Baker, 1996).

Nitrogen metabolism in ruminants is a more complex process than in monogastric animals because of extensive breakdown and modification of proteins in the reticulo-rumen. The ruminant animal is unique in its ability to convert feed N into microbial protein. The metabolizable protein needs of the ruminant are met primarily from two sources of amino acids: microbial protein synthesized in the rumen, and un-degraded feed protein in the rumen; with a small contribution from endogenous protein secretions. Microbial protein amino acid composition is very similar to the amino acid composition of tissue and milk protein (NRC, 2001; Lapierre et al., 2006), which makes it an ideal source of amino acids for the animal. Feed protein that by-pass ruminal degradation, however, may not provide digestible essential amino acids in the quantity or ratios sufficient for maintenance or production needs. Even if the amino acids absorbed in the gut match closely the amino acid requirements, the liver significantly modifies the amino acid profile of metabolizable protein available to the animal (Blouin et al., 2002). During this metabolism, amino acids are deaminated and NH_3-N, being a tissue toxin, is detoxified to urea by the liver. Urea synthesized in the liver is transported by the blood, partially reabsorbed in the kidneys, and can be recycled back to the digestive tract (Stewart and Smith, 2005), contributing to the ruminal NH_3-N pool. A large portion of the dietary proteins and non-protein compounds entering the rumen are degraded by the ruminal microorganisms to peptides, amino acids, and eventually to NH_3-N (Hristov and Jouany, 2005). NH_3-N is absorbed into the blood stream through the rumen wall or other sections of the gastrointestinal tract (Reynolds and Kristensen, 2008) and also contributes to urea synthesis in the liver.

The mechanisms controlling urea recycling in ruminants are complex and, in spite of intensive research in the past couple of decades, are not completely understood (Reynolds and Kristensen, 2008). Recent studies have reemphasized the importance of these mechanisms in preserving N and potential to provide available N for microbial growth when dietary protein may be deficient (Lapierre and Lobley, 2001; Reynolds and Kristensen, 2008). The level of dietary CP is one of the most important factors determining urea recycling rate to the gut and utilization by the microbes in the rumen (Reynolds and Kristensen, 2008). A series of classic experiments from the Research Center for Animal Production in Dummerstorf-Rostock have demonstrated, for example, that the efficiency of supplemental urea utilization for microbial protein synthesis in the rumen is sharply decreased (respectively, urinary N losses are increased) as dietary or plant protein availability increases (Voigt et al., 1984; Piatkowski and Voigt, 1986). Growing cattle (Wickersham, 2006), or dairy cows (Ruiz et al., 2002) fed low-CP diets have the ability to recycle to the gut virtually all urea synthesized in the liver, with very little being lost in urine. As pointed out by Reynolds and Kristensen (2008), ruminants will still excrete N in urine, but it will predominantly be N other than urea. Urea transferred to the gastrointestinal tract will be utilized for anabolic purposes, i.e. microbial protein synthesis, at a much greater rate in ruminants fed low-CP diets (Reynolds and Kristensen, 2008). Even at high levels of dietary CP intake, cattle will recycle a significant proportion of urea to the gut. Gozho et al. (2008) studied urea recycling in dairy cows fed diets with 17 to 17.4% CP and reported urea entry rate to the gastrointestinal tract of around 62%. Urea utilization for microbial protein synthesis in the rumen, however, was only about 20%.

Current feeding systems for ruminants (NRC, 1996, NRC, 2001) do not account for urea recycling and likely overestimate the protein (particularly ruminally-degradable protein, RDP) requirements of the animal. Meta-analysis of a large (1,734 diets) dataset demonstrated that among several dietary and animal performance variables, dietary CP was the most important factor determining milk N efficiency in dairy cows (Huhtanen and Hristov, 2009). Variability in milk yield may explain some of the variability in milk N efficiency when included in a model with dietary CP, but was insignificant as a stand-alone prediction variable. Other more recent estimations have shown that increasing dietary CP concentration with 1%-unit may increase milk protein N yield by approximately 2.8 g d^{-1}, but will result in 35.7 g d^{-1} dietary N not being utilized for milk protein synthesis (Hristov and Huhtanen, 2008). A major fraction of this unaccounted N will be excreted in urine, which is more susceptible for leaching and evaporative losses than fecal N (Bussink and Oenema, 1998). Huhtanen et al. (2008), for example, using a dataset of mainly grass silage-based diets estimated that 84% of the incremental N intake at constant dry matter (DM) intake is excreted in urine. Therefore, a better control of ruminal N, particularly NH_3-N, metabolism is an obvious way to achieve an improvement in the efficiency of N utilization by ruminants and to limit the excretion of nitrogenous compounds resulting in environmental pollution around animal production areas (Hristov and Jouany, 2005).

Urea is the main constituent of ruminant urine. Bristow et al. (1992), among others, reported that urea N represented from about 60 to 90% of all urinary N in cattle, with similar proportions for sheep and goats. Other significant nitrogenous compounds are hippuric acid, creatinine, and metabolites of purine bases catabolism, such as allantoin, uric acid, xanthine, and hypoxanthine. Bussink and Oenema (1998) summarized existing literature and gave a range of urinary urea as proportion of total N of 50 to 90%. In the urine of high-producing dairy cows, urea represents 60 to 80% or more of the total urinary N (Reynal and Broderick, 2005; Vander Pol et al., 2007) and this proportion gradually increases as dietary CP level and intake increase (Colmenero and Broderick, 2006). Urea is the main source of NH_3 volatilized from cattle manure (Bussink and Oenema, 1998). These authors indicated that 4 to 41% of the urinary N may be volatilized, while N volatilization from feces is considerably less, 1 to 13%. Urea is not volatile, but once mixed with feces; it is rapidly hydrolyzed to NH_3-N and carbon dioxide by the abundant urease activity in fecal matter (Bussink and Oenema, 1998). The following steps in NH_3 volatilization have been identified (Monteny and Erisman, 1998): (1) urea hydrolysis mediated by urease); (2) dissociation (governed by pH and temperature); and

(3) volatilization (which is a function of temperature and air velocity). Miner et al. (2000) demonstrated the relationship between concentration of NH_3-N (as percent of the total ammoniacal nitrogen, TAN) and pH and temperature. Irrespective of ambient temperature, less than 1% of TAN is in form of NH_3-N at pH 7. As pH increases, however, up to 10, 50, and 85% of TAN is in NH_3-N form at pHs of 8, 9, and 10, respectively; with substantial differences between 10°C and 30°C. Air velocity over the liquid manure surface has been identified as the main factor determining NH_3 release rates (Ni, 1999).

A series of experiments were recently conducted to quantify the relative contribution of urinary vs. fecal N to NH_3 volatilization losses from cattle manure (Lee and Hristov, 2010a). Feces and urine of lactating dairy cows were labeled separately with ^{15}N through labeling of ruminal microbial protein (see Hristov et al., 2005 for labeling protocol), combined in a 1:1 ratio, and incubated for 10 d in a laboratory scale, closed-chamber system. Ammonia emitted was captured in an acid trap and analyzed for ^{15}N. As either feces or urine were the only sources of ^{15}N above background enrichment, the origin of NH_3 volatilized during the incubation could be traced and quantitatively determined. Results from this study are presented in Figures 6.1 and 6.2. The proportion of NH_3 originating from fecal N (Figure 6.1) was negligible in the first 48 h of the incubation, represented 0.04 ± 0.006 by d 5, and then gradually increased to 0.11 ± 0.019 of the emitted NH_3 by d 10. The proportion curve fitted well a logistic regression model (adjusted $R^2 = 0.91$; P < 0.001). The proportion of NH_3 originating from urinary N represented 0.94 ± 0.027 at 24 h, 0.97 ± 0.002 at 48 h, 0.91 ± 0.004 at 72 h, and gradually decreased to 0.87 ± 0.005 at incubation d 10. The curve fitted well an exponential decay regression model (adjusted $R^2 = 0.92$; P < 0.001). The average recovery of NH_3 by this approach was 0.95 ± 0.011 for the 10 d of manure incubation. This study also clearly demonstrated that the main source of NH_3 volatilized from cattle manure during the initial 10 d of storage is urinary N, representing on average 90% of the emitted NH_3. The contribution of fecal N was relatively low, but gradually increased to about 10% by d 10, as mineralization of fecal N progressed. Using a similar approach, Thomsen (2000) estimated that urinary N accounted for 79% of the total N losses from sheep manure after 7 d of composting, but only for 64% at the end of the 86-d storage period. If manure was stored anaerobically, urinary N accounted for 94% of the total N losses after 28 d and for 68% at 86 d.

Reducing ration CP (Frank et al., 2002) or RDP concentration (Van Duinkerken et al., 2005) effectively reduces volatile N losses from manure. Metabolizable protein supply has similar effect. Weiss et al. (2009) reported that increasing dietary metabolizable protein increased NH_3 produced per gram of manure mainly because of increased urinary N excretion with a significantly smaller contribution of fecal N. Recent studies also show a remarkable effect of decreasing dietary CP on NH_3 emitting potential of dairy manure. In one study, a replicated Latin square design with 6 ruminally-cannulated cows, 3 diets varying in CP (HighCP, 15.4; MedCP, 13.4, and LowCP, 12.9% CP, DM basis) and RDP, but having similar metabolizable protein were fed to lactating dairy cows (Agle et al., 2010). Both MedCP and LowCP resulted in lower ruminal NH_3-N pool size and absolute and relative excretion of urinary N compared with the HighCP diet. Excretion of fecal N and milk yield and composition were not affected by diet. As a result of the greater urinary N excretion with the HighCP diet, cumulative (15 d) NH_3 emissions from manure were significantly greater (P < 0.001) for HighCP compared with MedCP and LowCP (2,278, 1,673, and 1,418 mg, respectively; Figure 6.3). The rate of NH_3 emission was also considerably greater for HighCP compared with the low-CP diets (138, 98, and 83 mg NH_3 d^{-1}, respectively; $P < 0.001$). These observations were confirmed in a follow-up experiment (Lee and Hristov, 2010b) with high-producing dairy cows (average milk yield was 38 kg d^{-1}) fed high (16% CP, HighCP) or low-CP (14% CP, LowCP) diets. Ammonia emission rates were 202 and 132 mg N h^{-1} for HighCP and LowCP manure, respectively ($P < 0.001$). Cumulative NH_3 emission was 45% less ($P < 0.001$; Figure 6.4) for LowCP compared with HighCP manure. Manure produced from these diets was applied to 61 × 61 × 61 cm lysimeters collected from a Hagerstown silt loam (fine, mixed, mesic Typic Hapludalf) in order to determine NH_3 emissions from soil amended with manure from low- and high-CP diets. Manure application rate was 9.3 g of N lysimeter^{-1}, corresponding to a field application rate of 300 kg N ha^{-1}, and was identical for the 2 types of manure. The HighCP manure had higher N content (4.4 vs. 2.8%, DM basis) and proportion of NH_4^+-N and urea-N in total manure N (51.4 vs. 30.5%) than the LowCP manure and as a result, more LowCP than HighCP manure (2.36 vs. 1.65 kg) was applied to each lysimeter. After manure application, NH_3 emissions were measured using a photoacoustic infrared gas analyzer at 3, 8, 23, 28, 50 and 100 h. As Figure 6.5 shows, NH_3 emission was significantly greater ($P < 0.05$) from HighCP- than from LowCP-manure amended soil. The area under the cumulative (100 h) NH_3 emission curve for LowCP was smaller ($P < 0.05$) than the area for HighCP manure (56.8 and 114.8 mg NH_3 m^{-2} min^{-1} × h, respectively). This experiment clearly demonstrated that manure from dairy cows fed reduced CP diets had decreased NH_3 emitting potential and would result in significantly lower NH_3 volatilization when applied to soil, compared to manure from cows fed a high-CP diet.

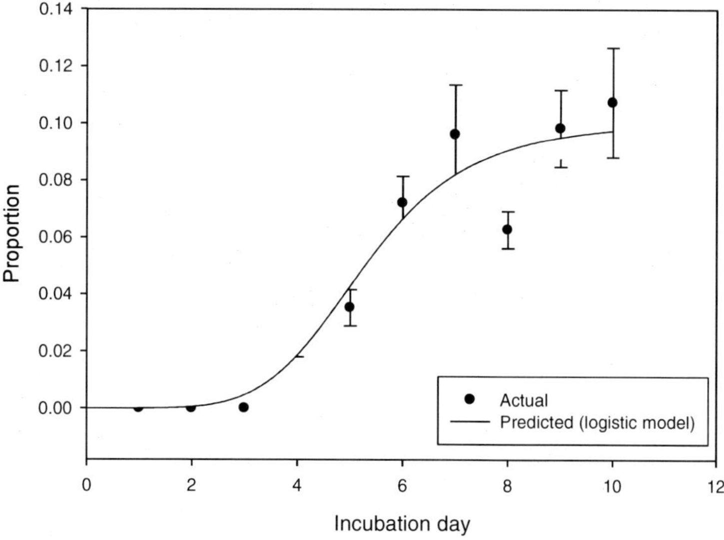

Figure 6.1. Proportion of ammonia originating from fecal N in dairy manure incubated for 10 d in a closed-chamber system (from Lee and Hristov, 2010a). Symbols are measured (means ± SE) and lines are predicted values (logistic regression model).

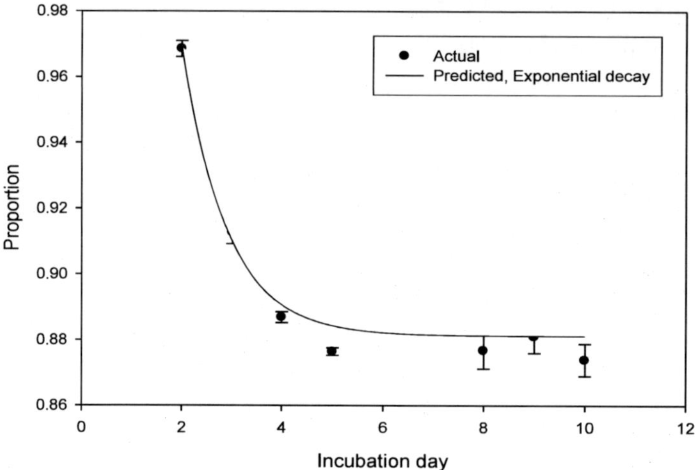

Figure 6.2. Proportion of ammonia originating from urinary N in dairy manure incubated for 10 d in a closed-chamber system (from Lee and Hristov, 2010a). Symbols are measured (means ± SE) and lines are predicted values (exponential decay regression model).

With high-producing cows lowering dietary CP may, in certain situations, result in decreased milk yield (Broderick, 2003), which would be unacceptable to most producers and nutritionists in the field. These performance effects in most cases stem from the complex interactions of protein with DM and energy intake (Huhtanen and Hristov, 2009). In the study mentioned above (Lee and Hristov, 2010b), for example, milk yield was significantly reduced (by 3 kg d^{-1}; P < 0.04) by the low-CP diet. This effect was clearly a result of essential amino acid deficiency and could be avoided by providing sufficient metabolizable protein, or supplementing the diet with synthetic, ruminally-protected amino acid limiting milk production (Broderick et al., 2008).

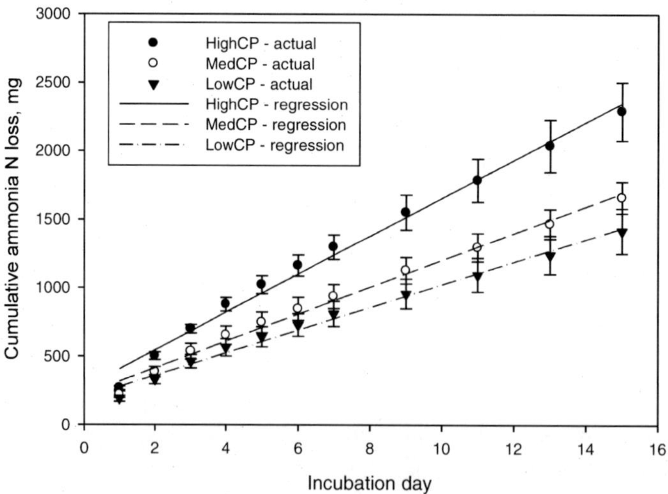

Figure 6.3. Effect of dietary crude protein concentrations on cumulative ammonia losses from dairy manure (from Agle et al., 2010). Symbols are measured (means ± SE) and lines are predicted values (linear regression). HighCP, MedCP, and LowCP are diets with crude protein concentration (DM basis) of 15.4, 13.4, and 12.9%, respectively.

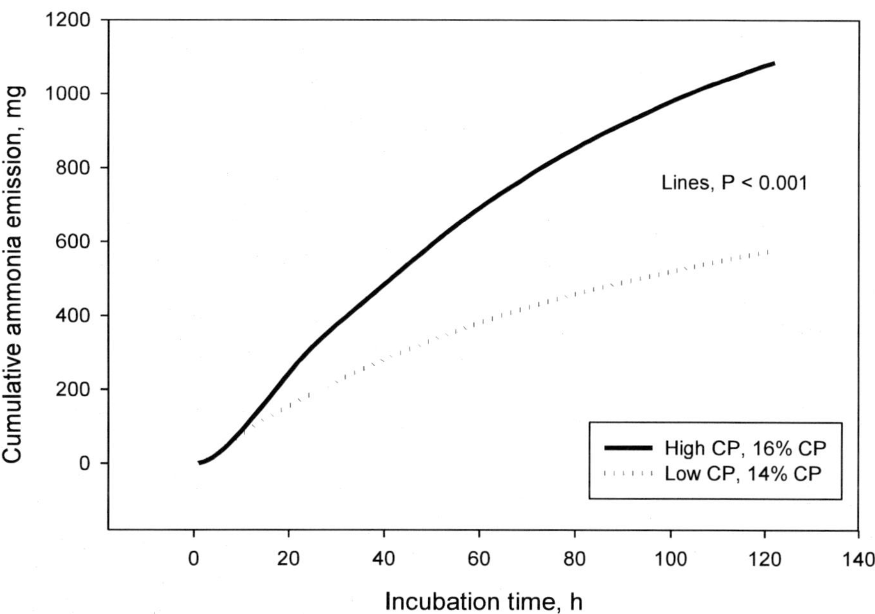

Figure 6.4. Effect of dietary crude protein concentrations on cumulative ammonia losses from dairy manure (from Lee and Hristov, 2010b). HighCP and LowCP are diets with crude protein concentration (DM basis) of 16 and 14%, respectively.

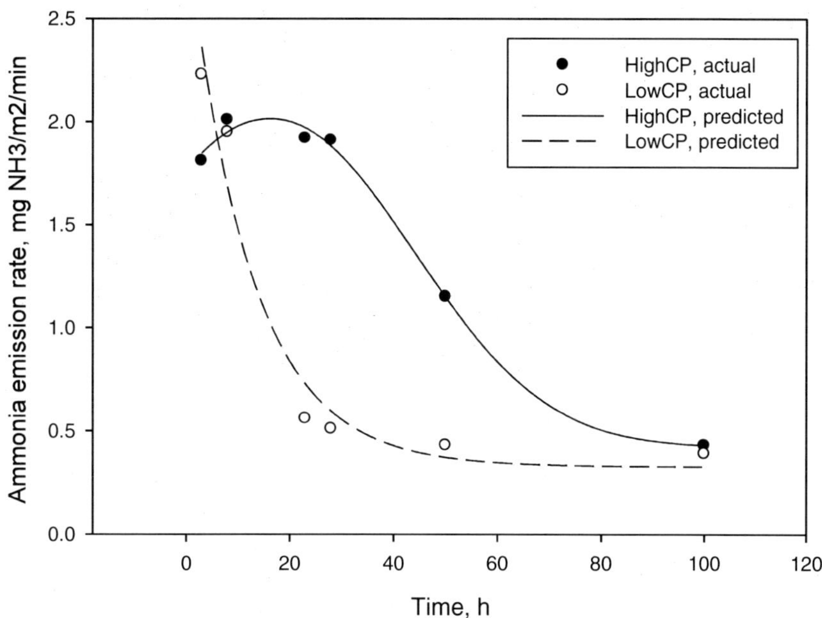

Figure 6.5. Ammonia volatilization from soil amended with manure from dairy cows fed high- (HighCP, 16%), or low-crude protein (LowCP, 14%) diets (from Lee et al., 2010). Symbols are measured and lines are predicted values (peak, modified Gaussian and exponential decay regression models, respectively).

6.2.2. Ammonia Release Mechanisms

As noted in the previous subsection, livestock manures contain N in both organic and forms. Excess N in the feed and inefficient utilization of CP or amino acids in diets is the source of this N in excreted urine and feces. Most of the N (up to 97%) is excreted as urea in the urine of sheep, cows, and pigs; while the rest is excreted as undigested organic N in the feces (McCrory and Hobbs, 2001; Varel, 1997; Mobley et al., 1995). Within hours to a few days, urea is hydrolyzed to NH_4^+-N in a process catalyzed by the microbial enzyme urease originating mainly from feces (Beline et al., 1998). In contrast, the microbial breakdown of organic N in feces into NH_4^+-N in a process referred to as ammonification or mineralization requires months or even years to effect. The NH_4^+-N resulting from either urea hydrolysis or organic N decomposition, or both, is the one susceptible to volatilization from manure depending on pH and temperature conditions.

The hydrolysis of urea to NH_3-N (or to NH_4^+-N) in aqueous environments, which is catalyzed by the enzyme urease (with nickel as the co-factor in the urease active sites) occurs in two steps (Kaminskaia and Kostic, 1997; Udert et al., 2003; Banini et al., 1999; Todd and Hausinger, 1989). In the first step (depicted in Equation 6.1), a mole of urea is hydrolyzed into a mole of NH_3-N (or NH_4^+-N depending on pH conditions) and a mole of the unstable carbamic acid. The mole of unstable carbamic acid then spontaneously decomposes into another mole of NH_3-N and a mole carbon dioxide in a second step (presented in Equation 6.2). A mole of urea, therefore, produces two moles of NH_3-N. There are no documented cases of uncatalyzed hydrolysis of urea in aqueous solutions (Kaminskais and Kostic, 1997), which demonstrates the importance of urease in the formation process of NH_3-N from urea.

$$NH_2(CO)NH_2 + H_2O \rightarrow NH_3 + NH_2(CO)OH \tag{6.1}$$

$$NH_2(CO)OH \rightarrow NH_3 + CO_2 \tag{6.2}$$

The conversion of organic-N (proteins, amino polysaccharides, and nucleic acids) to NH_3-N or NH_4^+-N, on the other hand, is mediated by a host of enzymes produced by heterotrophic microbes (Vavilin et al., 2008; Zhang et al., 2007; Horton et al., 1992). The process also takes place in two distinct stages. First, extracellular enzymes (e.g. proteases, peptidases, chitinase, chitobiase, lyzosyme, ribonucleases, deoxyribonucleases, exonucleases, and endonucleases) break down organic-N polymers into monomers (amino acids, amino sugars, and nucleic acid). Second, the monomers then pass across the microbial cell membrane and are further metabolized by intracellular enzymes (e.g. dehydrogenases, oxidases, and kinases) into NH_4^+-N (Barracklough, 1997; Barak et al., 1990). Some of the NH_4^+-N is incorporated into the microbial biomass in a process referred to as assimilation; the excess or surplus NH_4^+-N is released back into the bulk manure. The mineralization of protein N to NH_4^+-N, for example, involves: (i) the formation of intermediate amino acid N from protein N which is catalyzed by proteases and (ii) hydrolysis of this amino acid N to NH_4^+-N, which is catalyzed by either amino acid dehydrogenases or amino acid oxidases (Nannipieri and Eldor, 2009).

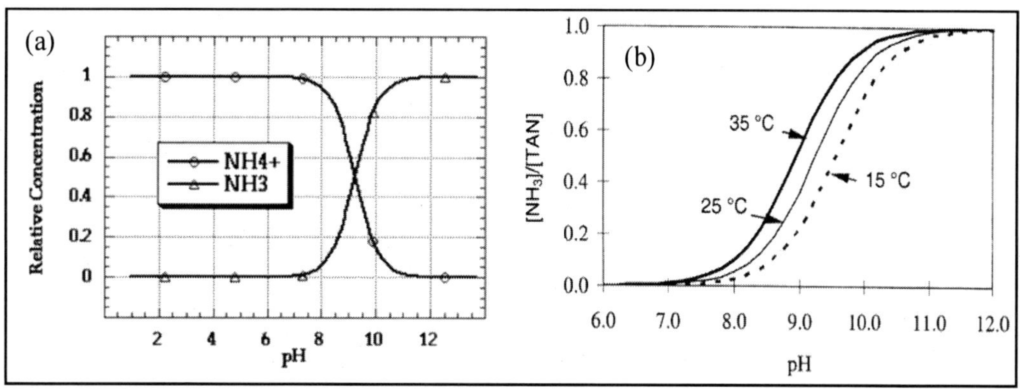

Figure 6.6. Equilibrium between NH_4^+-N and NH_3-N in aqueous solutions as a function of pH and temperature.

As mentioned earlier, NH_4^+-N itself is not volatile but it is amenable to volatilization once it is dissociates into the NH_3-N. In aqueous environments, NH_4^+-N and NH_3-N exist in equilibrium that is governed by both pH and temperature conditions. At constant temperature, for example, the pH of manure determines the equilibrium between NH_4^+-N and NH_3-N (see Equation 6.3). Lowering the pH of manure results in an equilibrium shift that favor the NH_4^+-N form, which effectively lowers the potential of NH_3 volatilization. On the other hand, raising the pH pushes the equilibrium towards the NH_3-N form, which exacerbates NH_3 volatilization. In general, NH_3 volatilization is directly proportional to the proportion of NH_3 in the total ammoniacal nitrogen (TAN = NH_4^+-N + NH_3-N) in the aqueous solutions such as the manure slurries. The influence of pH and temperature on the dissociation of NH_4+-N is shown in Figure 6.6. The fraction of TAN present as NH_3-N increases with increase in the pH of the manure (Figure 6.6a). For a given pH, on the other hand, the fraction of TAN present as NH_3-N increases with increase in temperature (Figure 6.6b). At pH values lower than 8.3 (at 25C), the proportion of TAN present as NH_3-N is less than or equal to 0.5. Decreasing proportion of NH_3-N in liquid manure results in a lower potential of **NH_3** to volatilize (Figure 6.6a: Sawyer and McCarty, 1978). The greatest change in the proportion of TAN present as **NH_3-N** occurs between pH 7 and pH 10. Below pH of 7 **NH_3** volatilization decreases progressively to about pH 4.5 where there is almost no measurable **NH_3-N** (Ndegwa et al., 2008; Hartung and Phillips, 1994; Sawyer and McCarty, 1978). The influence of temperature on ammonium-ammonia equilibrium, on the other hand is shown in Figure 6.6b (Loehr, 1974). Increasing temperature increases dissociation of NH_4^+-N to NH_3-N and thus also enhances NH_3 volatilization.

$$NH_4^+ \overset{pH}{\Longleftrightarrow} NH_3 + H^+ \qquad\qquad (6.3)$$

Theoretically, the process of NH_3 volatilization involves movement of NH_3-N to the manure surface followed by its subsequent release into the ambient air (Teye and Hautala, 2008; Ni, 1999). A conceptual model of NH_3 formation and volatilization is presented in Figure 6.7. Transfer of NH_3-N to the manure surface is achieved through diffusion mass transfer because of concentration gradient, while the release of NH_3 from manure surface to ambient air is mainly through convective mass transfer (Ni, 1999; Kirk and Nye, 1991; van der Molen et at., 1990; Olesen and Sommer, 1993). The resistance of NH_3-N transfer to the surface is relatively small compared to the resistance of its release into the ambient air. The latter process is thus more significant to the overall NH_3 volatilization process. In general, NH_3 volatilization increases with an increase in: concentration of NH_3-N in the manure near the surface of manure containment, air velocity and turbulence, manure temperature, and manure pH (Teye and Hautala, 2008; Arogo et al., 1999; Sommer et al., 1991; Olesen and Sommer, 1993; Vlek and Stumpe, 1978). There are, however, emerging thoughts that the concentration of NH_3-N at the surface of the manure containment where volatilization occurs may be different from the bulk liquid concentration. De Visscher et al. (2002) reported that, although suspended solids are involved in the transport of the NH_3-N to the liquid interface, their effect may not be accounted for in the estimation of NH_3 volatilization using bulk liquid concentrations only. Ro et al. (2007) observed that gas bubbles, methane in particular, may also increase NH_3 volatilization via entraining of NH_3 from the sludge layer where anaerobic digestion occurs. The NH_3 entrained in the bubbles is also not accounted for in the estimation of volatilization using bulk liquid concentration.

6.3. AMMONIA EMISSION MITIGATION TECHNIQUES

Techniques for reducing NH_3 losses from animal manure may be put into three broad categories: (i) those that minimize N in the animal manure prior to its excretion, (ii) those that reduce volatile N species (NH_3-N or NH_4^+-N) in the excreted manure, and (iii) those that physically contain and treat NH_3-N or NH_4^+-N species after they have already been formed. The basic principles behind each of these NH_3 mitigation strategies are discussed in this section. An overview of the performance of each technique is presented following the discussion of the principle or strategy in question.

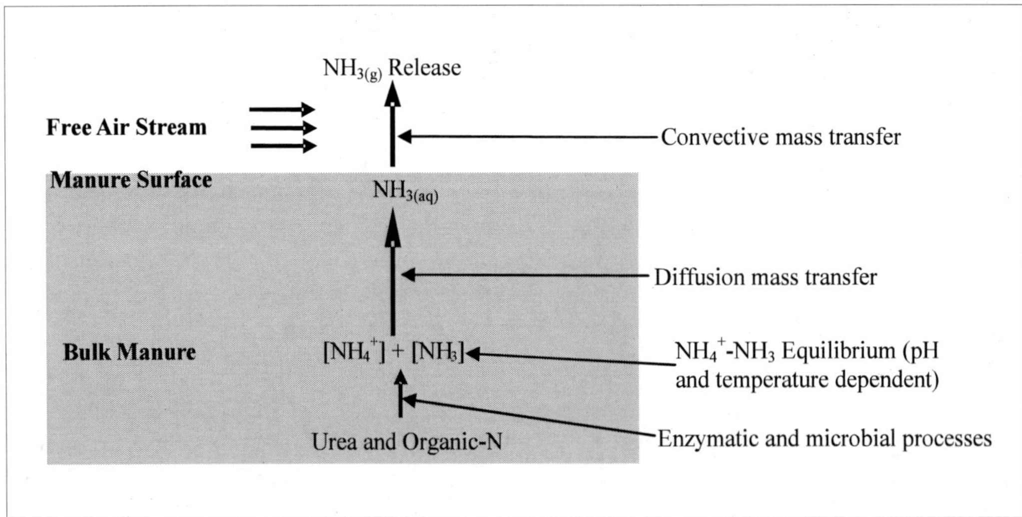

Figure 6.7. A conceptual model of ammonia formation and volatilization from manures.

6.3.1. Reduction of N in Manure

Minimizing N excretion, which can be achieved through dietary-modifications, is naturally the first line of defense in curbing NH_3 emissions from livestock operations (Satter et al., 2002). Available research data indicate that diets fed to animals have profound effects on NH_3 emissions from excreted manure. Overfeeding dietary protein, imbalanced amino acid supply, and reduced energy availability for ruminal fermentation result in increased urinary and fecal N losses and consequently increased NH_3 emissions from manure.

In non-ruminants (for example; pigs), NH_3 losses have been reduced by either shifting N excretion from urine to feces by increasing fiber in the feed or reducing the N content in the diet (Canh et al., 1997; Canh et al., 1998b). Several reports indicate that reducing CP in pig diets and supplementing with amino acids can reduce N excretion by 28-79 % in the manure based on an average of 8% reduction in N excretion per unit of CP reduction (Kerr, 1995; Canh et al., 1998a). Panetta et al. (2006) reported decreased NH_3 emission rates from 2.46 to 1.05 mg min^{-1} with decreasing dietary CP levels from 17.0 to 14.5%. Similarly, O'Connell et al. (2006) observed increased NH_3 emissions from manure slurries of pigs fed a 22% CP diet compared with 16% diet. For broiler and layer chickens, reduced protein diets have resulted in reduced N excretion (Jacob et al., 2000). Thus, with some few notable exceptions (McGinn et al., 2002; Clark et al., 2005), reduction of dietary CP result in significant reduction in NH_3 loss from pig facilities (Otto et al., 2003; Hayes et al., 2004; Velthof et al., 2005) and poultry operations (Ferguson et al., 1998; Nahm, 2003). Other strategies such as supplementing diet with zeolite (Kim et al., 2005), antibiotics and probiotics (Han and Shin, 2005), vegetable oil (Leek et al., 2004), plant extracts (rich in tannins and saponins; Colina et al., 2001; Vliwisli et al., 2002), and exogenous enzymes (Smith et al., 2004; O'Connell et al., 2006) have been used with varied success to reduce NH_3 losses from pig and cattle manure. In practice, efforts to reduce NH_3 emissions must be balanced with animal performance in determining optimal protein concentrations and forms in the diet (Cole et al., 2005; Panetta et al., 2006).

In ruminants (cattle; for example), diet composition can have significant effects on urinary urea excretion and consequently NH_3 losses from manure and overall efficiency of utilization of dietary N (Klopfenstein et al., 2002; Satter et al., 2002). Generally, ruminants are relatively inefficient utilizers of dietary N. The efficiency of transfer of feed N into milk protein N (MNE; found as milk protein N ÷ N intake × 100) is on average 25±0.1%, with a min and max of 14 and 40%, respectively (Hristov et al., 2004a); the bulk of the remaining N being lost to the environment via urine and feces. Within limit, urinary N losses in dairy cows linearly decrease with decreasing dietary CP levels without affecting milk and milk protein yields and composition; MNE of 36% was achieved with the lowest CP (13.5%; Olmos Colmenero and Broderick, 2006). Cows fed 15.0 to 18.5% CP diets produced similar milk yields (32 to 39 kg d^{-1}) while simultaneously increasing N excretion and urinary N proportion (Groff and Wu, 2005). Reduction of urinary N excretion from dairy cows can be achieved mainly by the reduction of N intake in the form of ruminally degradable protein (Kebreab et al., 2002). Utilizing a combination of prediction equations (urine volume) and actual analyses (urine composition), de Boer et al. (2002) demonstrated the importance of the ruminal N balance (OEB) in reducing N losses in dairy cows. Increasing OEB from 0 to 1,000 g cow^{-1} d^{-1} resulted in linear increase in urinary N excretions. Feeding excess RDP resulted in greater ruminal N and milk urea N concentrations and increased urinary N losses (by 27%; Hristov et al., 2004). Decreasing CP in diets fed to cows in mid [17 to 15% CP, ruminally undegradable protein (RUP) of 5.5 to 7.3], or late lactation (14 to 12.5% CP) can reduce the cost of the diet and waste N excreted from the cow. However, early lactating dairy cows need sufficient dietary RUP. After peak milk and dry mater intake (DMI), CP and especially RUP requirements decline with declining milk production (Kalscheur et al., 1999). Using ruminally-protected amino acids enables an efficient use of low CP diets for production purposes. With ruminally-protected methionine (up to 25 g d^{-1}), milk yield was maintained and MNE increased from 26 to 34% as dietary CP decreased from 18.6 to 14.8% (Broderick, 2005). Methionine supply to low (13%) CP diets decreased proportion of urinary N in the total excreta N (Krober et al., 2000). Carbohydrate level and availability in the diet can also have a significant effect on ruminal N utilization and consequently urinary urea output. Increasing dietary net energy of lactation concentration from 1.55 to 1.62 Mcal kg^{-1} decreased urinary urea N excretion and increased MNE (from 25 to 30%, respectively), while increasing dietary CP level from 15.1 to 18.4% had an opposite effect by increasing urinary urea N excretion and decreasing MNE (Broderick, 2003).

Dietary CP levels and effects on urinary urea excretion are directly related to NH_3 emissions from cattle manure. Smits et al. (1995) fed dairy cows two diets differing in OEB (40 vs. 1,060 g d^{-1}) and reported a significant increase in urinary urea-N concentrations and NH_3 emissions from manure (by 39%) with the high-OEB diet. Külling et al. (2001) demonstrated that at 17.5% CP in the diet, N losses from manure after 7 weeks of storage were from 21 (slurry) to 108% (urine-rich slurry: urine:feces ratio of 9:1) greater than the N losses from manure from cows fed 12.5% CP, with respective NH_3 emissions rates of 163 and 42 $\mu g\ m^{-2}\ s^{-1}$. Low protein diets (13.5-14% CP) fed to dairy cows resulted in significantly lower NH_3 release from manure compared with the high CP (15-19%) diets (Frank and Swensson, 2002; Frank et al., 2002). Similar results were reported for feedlot cattle (Cole et al., 2005; Todd et al., 2006). For example, decreasing CP content of finishing cattle diets from 13 to 11.5% reduced daily NH_3 flux by 28% (Todd et al., 2006). In summary, reducing CP in beef cattle diets is a practical and cost-effective way of reducing NH_3 emissions from feedlots.

Ammonia volatilization is directly related to the proportion of aqueous NH_3 in the total ammoniacal-N (TAN). In general, at constant temperature pH determines the equilibrium between NH_4^+-N and NH_3-N with a lower pH favoring the NH_4^+-N form and hence lower potential of NH_3 volatilization. Thus, low urinary pH may be a key factor for reducing NH_3 emissions from cattle manure. Various dietary treatments can decrease urinary pH (Stockdale, 2005). Anionic salts (Tucker et al., 1991; Bowman et al., 2003; Mellau et al., 2004) and high fermentable carbohydrates levels (Mellau et al., 2004; Andersen et al., 2004) can reduce urinary pH to below 6.0. In non-ruminants, diet acidification with organic (benzoic) acids (Martin, 1982) or Ca and P salts (Kim et al., 2004) reduced urinary pH and NH_3 emissions from pig manure (Canh et al., 1997, Canh et al., 1998a, b).

6.3.2. Minimizing Volatile N-Species

6.3.2.1. *Urine-feces Segregation*

In cows and pigs, urea is mainly excreted in the urine while the enzyme urease, on the other hand, is predominantly found in the feces. Urease is known to catalyze urea hydrolysis into NH_4^+-N. This is precisely the basis of the segregation of feces and urine immediately upon excretion of either so as to reduce their contact time. In general, the urine-feces segregation concept can be seen in two main design systems. One design (Figure 6.8a) uses a conveyor-belt or a net to separate urine and feces, with urine flowing into one pit, while feces left on the belt or the net are conveyed into a separate collection-pit. In the other design (Figure 6.8b), appropriate floor designs ensure that urine drains away from feces into a urine-pit immediately after discharge and the feces are then scraped or washed into a separate pit.

The available information indicates that segregation of urine from feces has achieved as much as 99% reduction in NH_3 emissions in laboratory studies (Panetta et al., 2004). Pilot and full-scale urine-feces segregation systems, however, have not been equally effective. Several researchers have evaluated conveyor belt systems (Lachance et al., 2005; Stewart et al., 2004). Lachance et al. (2005) evaluated the performances of two urine-feces separation systems (a belt and a net) in pig grower-finisher housing. A 49% reduction in NH_3 emissions was obtained with urine-feces separation directly under the slat-floor using either a belt or a net. Stewart et al. (2004) evaluated an inclined conveyor belt used directly as a dunging area in a swine barn. This system's average NH_3 emission was 47% lower than a conventional pit pull-plug design system.

A few studies evaluating the feasibility of feces-urine separation concept via various floor designs have also been conducted. Swierstra et al. (2001) evaluated a pre-cast concrete floor with perforated grooves in the pit under a conventional slatted floor in a mechanically ventilated cow barn. Urine drained along the grooves and through the grooves' perforations while feces were scraped to one end of the alley. In comparison to the control compartment, NH_3 emissions reductions were 46 and 35% in the test compartment with grooved pre-cast floor with perforations open and closed, respectively. A v-shaped solid floor with a gutter at the bottom of the v-groove to drain urine in cow houses was evaluated by Braam et al. (1997a) for the mitigation of NH_3 emission in the barn. Ammonia emission in this system was on average 50% lower than in the control barn. Swierstra et al. (1995), on the hand, compared NH_3 emissions from a solid sloping floor with a central urine gutter with a slatted floor in a cow barn. Ammonia emission was 50% lower with the inclined solid floors than with the conventional slatted floor. A similar study by Braam et al. (1997b) also evaluated a traditional slatted floor and two solid floor systems; one of the latter was sloped (3%) and drained urine into a gutter, while the other was not inclined at all. The solid floors were either scraped 4 times every hour or once every two hours (normally-scraped). The non-sloped solid floor normally-scraped had the same NH_3 emission as the slatted floor, while the sloped normally-scraped solid floor, further reduced NH_3 emission by 21% over the other two systems. A marginal 5% further reduction in emission was achieved by increasing the scraping by four fold; which probably does not economically justify the extra scraping efforts.

The urine-feces segregation methods reviewed in this chapter reduced NH_3 emissions from livestock barns by about 50% compared to the conventional manure handling systems (co-mingled or mixed urine-feces systems). This observation suggests that one could choose any of the methods by a casual toss of the coin. In reality, this is far from truth. The following critical factors, among others, need to be considered in making the choice of the method for separating urine from feces: cost of installing the system, maintenance, simplicity of operation, and cost of operation.

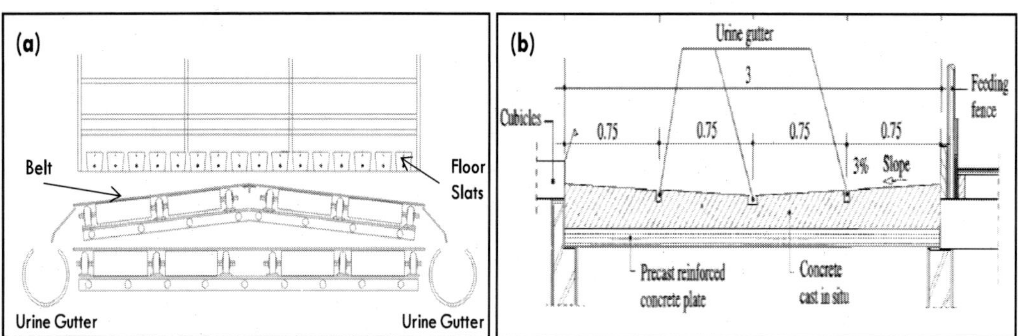

Figure 6.8. Cross-sections of Conceptual Urine-Feces Segregation Systems (a: Inclined or convex belt adapted from Funk and Polakow, 2004c; b: Slopped floor adapted from Braam et al., 1997b).

6.3.2.2. Urease Inhibitors

The enzyme urease found in the feces catalyzes urea (in cattle and swine manures) hydrolysis into NH_4^+-N when urine mixes with feces (Beline et al., 1998). One method to block urea hydrolyzes is to segregate urine from feces as described in the preceding section. Another approach is blocking urea hydrolysis using urease inhibitors. Inhibiting the enzyme activity, in general, has the same result as preventing urine from coming into contact with feces.

In laboratory studies, two urease inhibitors; cyclohexylphosphoric triamide (CHPT) and phenyl phosphorodiamidate (PPDA), successfully controlled urea hydrolyses in typical cattle and swine slurries (Varel, 1997). At dosages of 10 mg L^{-1}, both inhibitors delayed hydrolysis of urea in cattle waste and swine waste for 4-11 d. In contrast, complete hydrolyses of urea in control cattle or swine manure were observed in one day. Weekly addition of the inhibitors was the most effective method of preventing urea hydrolysis. Weekly additions of 10, 40, and 100 mg of PPDA per liter of cattle waste (5-6 g urea L^{-1}) prevented 38, 48, and 70% of the urea, respectively, from being hydrolyzed during a period of 28 d. For the swine waste (2·5 g urea L^{-1}), the same PPDA concentrations prevented 72, 92, and 92%, respectively, of the urea from being hydrolyzed during the same study period.

Another laboratory study evaluated the effects of rate and frequency of urease inhibitor application on NH_3 emissions from simulated beef cattle feed-yard manure surfaces (Parker et al., 2005). The urease inhibitor N-(n-butyl) thiophosphoric triamide (NBPT) was applied at rates of 0, 1, and 2 kg ha^{-1}; at every 8, 16, and 32 d. Synthetic urine was added every two days to the manure surface. This urease inhibitor applied every 8 d was most effective, with the 1 and 2 kg NBPT ha^{-1} treatments resulting in 49% to 69% reduction in NH_3 emission rates, respectively. According to the authors, the 8-day, 1 kg NBPT ha^{-1} treatments had the most promising benefit to cost ratios ranging between 0.48 and 0.60. Although the technical and economic potentials of use of NBPT for reducing NH_3 emissions in beef cattle feed yard was demonstrated, the authors cautioned that higher NBPT application rates may be necessary with time because of possible buildup of urea in the pen surfaces. In an earlier study, Varel et al. (1999) reported accumulation of urea, less concentration of TAN, and more concentrations of total-N in cattle feedlot manure when 20 mg [NBPT] kg^{-1} of manure was applied weekly for six weeks compared with control.

The effectiveness of urease inhibitors in the mitigation of NH_3 emissions from animal manures has been well established at the laboratory-scale. Adoption of this method in full-scale livestock operations is, however, yet to be realized. The slow acceptance of this technique may be attributed to the unknown effects of these chemicals (i.e. urease inhibitors) on the crops or pastures where the manure is eventually applied as fertilizer; in addition to increased labor costs associated with repeated application of the inhibitors.

6.3.2.3. Lowering Manure pH

As already discussed in section 6.2.2, NH_3 volatilization is directly proportional to the proportion of NH_3-N in the total ammoniacal nitrogen (TAN) in the manure slurries. At constant temperature, the dissociation constant (K_d), which is function of media pH, determines the equilibrium between NH_4^+-N and NH_3-N in aqueous systems. The lower the pH of manure the lower the proportion of NH_3-N and, therefore, the lower the potential of NH_3 volatilization. Acidification of animal manure for mitigation of NH_3 loss relies on this fundamental principle.

A review of past studies demonstrates the effectiveness of pH reduction in the mitigation of NH_3 volatilization from livestock manure (Table 6.1). Gradual acidification of pig and cattle slurries with sulfuric acid (H_2SO_4) from a pH of 8 progressively reduced NH_3 emissions and completely stopped NH_3 volatilization at pHs of 5 and 4 in pig slurries and in cattle slurries, respectively (Molloy and Tunney, 1983). In another study, Jensen (2002) maintained a pH of 5.5 using H_2SO_4 in swine manure in full-scale swine buildings with slatted floors and under-floor manure pits. This treatment reduced ambient NH_3 concentrations by 75-90%, in addition to increasing the weight of pigs by 1074 g d^{-1} during the study period; compared to the pigs in the control buildings. In a similar study, Stevens et al. (1989) used H_2SO_4 to acidify cow and pig slurries to pHs of 5.5 and 6.0. At these pH conditions, NH_3 volatilizations were effectively reduced by 95% in the lab and by 82% in the field. Similar studies (Frost et al., 1990) using H_2SO_4 to acidify cattle slurry to a pH of 5.5 reduced NH_3 loss by 85%. Al-Kanani et al. (1992) similarly reported NH_3 loss reduction of 75% when H_2SO_4 was applied to swine manure in lab experiments. Somewhat lower NH_3 loss reductions (14-57%) from cattle slurry were reported by Pain et al. (1990) on lowering the pH to 5.5 with H_2SO_4.

Husted et al. (1991) investigated use of hydrochloric acid on the acidification of stored cattle slurry, and reported a potential NH_3 loss reduction of as much as 90% compared to the control. On the other hand, Safley et al. (1983) achieved ~50% reduction in NH_3 loss using phosphoric acid within 28 d of dairy cattle manure storage. Al-Kanani et al. (1992) reported significantly less (~90%) loss of NH_3 with the same phosphoric acid on swine manure. Phosphoric acid, however, adds P concentration in the manure, which is undesirable if manure will be applied on soils with adequate or elevated P levels. Some of the weaker acids like propionic and lactic acids were also found to be equally as effective as the strong acids because they reduced NH_3 emissions by as much as 90% when pH of the manure was maintained at 4.5 (Parkhust et al., 1974).

Other studies on the effect of pH reduction on mitigation of NH_3 emissions from livestock manure have investigated the use of other types of acidifying agents such as aluminum potassium sulfate or alum, ferric chloride, sodium hydrogen sulfate, and calcium chloride. Although most of these chemicals effectively reduce manure pH, they are generally not as effective in reducing NH_3 loss as the strong acids because they can't maintain stable pH conditions like their counterparts.

Table 6.1. Summary of ammonia emissions reduction from manure storages by lowering pH.

Animal Species	Agent or substance	Emissions reduction (%)	References
Cattle and pig	Sulfuric acid	14-100	Molloy and Tunney, 1983; Jensen, 2002; Stevens et al., 1989; Frost et al., 1990; Al-Kanani et al., 1992; Pain et al., 1990
Cattle	Hydrochloric acid	90	Husted et al., 1991;
Cattle and pig	Phosphoric acid	50	Safley et al., 1983
Pig	Phosphoric acid	90	Al-Kanani et al., 1992
Broiler	Alum	89	Li et al., 2006;
Cattle	Alum	91-98	Shi et al., 2001
Cattle	Calcium chloride	71-78	Shi et al., 2001; Witter, 1991
Poultry and cattle	Calcium chloride	10-15	Kithome et al., 1999; Husted et al., 1991;
Cattle	Monocalcium phosphate monohydrate	87	Al-Kanani et al., 1992

Li et al. (2006) reported 89% reduction in NH_3 volatilization when alum was applied at the rate of 2 kg per m^2 [surface area]. Armstrong et al. (2003) observed that application of liquid alum equivalent of 45, 90, and 135 kg per 93 m^2 of broiler litter surface was effective at maintaining in-house NH_3 concentrations at below 25 ppm for two weeks, three weeks, and three weeks of the grow-out, respectively. Shi et al. (2001) investigated the efficacy of alum on beef cattle manure. Compared to the control, NH_3 emissions reduction during 21 d of monitoring were 91.5% at 4500 kg ha^{-1} alum and 98.3% at 9000 kg ha^{-1} alum. The advantage of alum use in the reduction of NH_3 emissions is reduction of soluble phosphorus and thus the potential for phosphorus runoff or leaching once manure is land applied. A thorough review of the use of alum to treat animal manure to mitigate ammonia emissions is presented later in Chapter 15.

Addition of calcium and magnesium salts during aerobic treatment of livestock manure reduced NH_3 emissions by 85 to 10 0% within 2-3 weeks and by 23-52% in the seventh week of treatment (Witter and Kirchmann, 1989a). In another study, Shi et al. (2001) evaluated the efficacy of $CaCl_2$ on reducing NH_3 emissions from beef cattle manure in the laboratory. Compared to the control, NH_3 emissions 21 d after application were reduced by 71.2 and 77.5% at 4500 and 900 kg ha^{-1} $CaCl_2$ application. In general, $CaCl_2$ was less effective than alum at the same application rates. Witter (1991) studied the effect of pre-treating fresh and anaerobically digested manure on mitigating NH_3 loss upon land application of the respective slurries. Ammonia losses were reduced by 73 and 8% with respect to the fresh and anaerobically digested manures, respectively, within 2 d after application. In other studies,

Kithome et al. (1999) noted a 10% decrease in NH_3 loss after addition of 20% $CaCl_2$ to poultry manure. The latter result was similar to the maximum 15% NH_3 emission reduction obtained by Husted et al. (1991) after addition of 300-400 meq l^{-1} of $CaCl_2$ to cattle slurry. Another pH reducing agent that has received attention is monocalcium phosphate monohydrate (MCPM). Al-Kanani et al. (1992) observed a significant reduction in both pH and NH_3 emission (87%) when MCPM was applied to cattle manure. In similar research, Mackenzie and Tomar (1987) investigated addition of MCPM to aerated and non-aerated liquid swine manures. The pH of manure decreased with addition MCPM, but when addition of salt was discontinued the pH started to increase. During subsequent aeration, however, total nitrogen (TN) decreased significantly in the control manure while no significant change was observed in the TN in the manure treated with MCPM.

Overall, strong acids are more cost-effective at reducing manure pH than the weaker acids or the acidifying salts. Strong acids, however, are more hazardous for use on the farm. Therefore, although the acidifying salts and other weaker acids may be less effective than strong acids, they are non-hazardous and relatively low cost; which increases their suitability for on-farm use. A notable negative consequence of manure acidification is the enhanced volatilization of malodorous volatile fatty acids and hydrogen sulfide.

6.3.2.4. Ammonium Binding Agents

This category of substances has a high affinity for adsorbing NH_4^+-N and NH_3-N, thus reducing their volatilization potential. The methods of NH_4^+-N and NH_3-N binding in some cases are not well understood. In general, however, most of these substances reduce NH_3 loss by either trapping it in microscopic pits present in their structures or adsorbing it on their surfaces (as is in the case of zeolite and Sphagnum peat moss, for example). Clinoptilolite zeolite is a naturally occurring mineral with a high cation-exchange capacity and specific affinity for NH_4^+ ions (Mumpton and Fishman, 1977). Two other additives in this category evaluated for abatement of NH_3 emissions in livestock manures are Sphagnum peat moss (*Sphagnum fuscum* peat) and yucca plant extracts (saponins). In contrast to clinoptilolite zeolite, sphagnum peat moss exhibits high adsorptive capacity for NH_3-N rather than NH_4^+-N. Additionally, sphagnum peat moss, which is derived from decayed plant materials has a large surface area that offers extensive adsorbing sites. The mechanism by which yucca extracts reduce NH_3 loss from animal manures is currently not clear although it is generally believed to be through binding or transformation of NH_4^+-N. A summary of the performance of these substances is provided in Table 6.2.

A layer of 38% zeolite placed on the surface of composting poultry manure reduced NH_3 loss by 44% (Kithome et al., 1999). An earlier study by Witter and Kirchamann (1989b) investigating the efficacy of zeolite on the reduction of NH_3 loss from poultry manure during aerobic incubation reported an insignificant 1.5% reduction in NH_3 loss when mixed with manure in the ratio of 1:4. Nakaue et al. (1981) observed a reduction of up to 35% NH_3 loss by the addition of 5 kg m^{-2} of zeolite to broiler litter. Portejoie et al. (2003) investigated reduction of NH_3 loss in pig manure during storage and land application using zeolite, and reported a 71% reduction in NH_3 emissions. Li et al. (2006) evaluated the efficacy of zeolite in reducing NH_3 emissions from fresh poultry manure in laboratory experiments. Application of typical medium rates of 5% (w/w) zeolite reduced NH_3 emission by 81%. Zeolite seems to be more effective for reduction of NH_3 emission in animal slurries and liquid manures than in solid poultry manures.

Table 6.2. Summary of ammonia emissions reduction from manure storages using ammonium binders.

Animal Species	Binding Agent	Emissions reduction (%)	References
Poultry	Zeolite	1.5-96	Kithome et al., 1999; Witter & Kirchamann, 1989b; Nakaue et al., 1981; & Li et al., 2006
Pig	Zeolite	71	Portejoie et al., 2003
Pig	Sphagnum peat moss	80-99	Al-Kanani et al., 1992; & Barrington and Moreno, 1995
Poultry	Sphagnum peat moss	24	Witter & Kirchamann, 1989b
Pig	Saponins (yucca extract)	23	Kemme et al., 1993
Pig	Alliance®	24	Heber et al., 2000
Poultry	De-Odorase®	50	Amon et al., 1997

Al-Kanani et al. (1992) compared the efficacy of several amendments on liquid hog manure and concluded that Sphagnum peat moss was just as effective as the strong acids (reduced NH_3 volatilization by as much as 99%), although it did not drop the pH to the same levels as the acids. Barrington and Moreno (1995) observed that a 2-cm cover of floating Sphagnum reduced NH_3 loss by as much as 80%. Similar results were reported by other researchers (Al-Kanani et al., 1992), but Witter and Kirchamann (1989b) reported a somewhat lower (24%) reduction in NH_3 emissions when sphagnum peat, mixed in the ratio of 1:4, was used in poultry manure during aerobic incubation. This product also seems to be more effective on the liquid slurries than on the solid poultry manure in the same way as zeolite. Kemme et al. (1993) reported NH_3 loss reduction of 23% when saponins were applied to pig slurries. Panetta et al. (2004) reports similar results when these extracts were applied to swine slurry in laboratory studies. In this category, saponins seem to be less effective in mitigating NH_3 emissions than either zeolite or peat moss.

A host of other brand name additives have also been evaluated. Heber et al. (2000) evaluated a commercial manure additive (Alliance®) developed by Monsato EnvironChem (St. Louis, MO.) to improve air quality in swine buildings. Alliance® was sprayed into the manure stored in pits underneath slatted floors. Compared to the control, this additive reduced NH_3 emissions by 24%, but also further diluted the manure by 20%. The authors estimated the cost of this additive at $1.38 pig^{-1} year^{-1} or $0.50 hog^{-1} based on 135-d growth cycles, and a product cost of $3.43 L^{-1}. They also noted that because of the modest reduction in NH_3 emission, this additive may not be cost-effective to most producers. Amon et al. (1997) compared the effectiveness of another commercial additive (De-Odorase®) to a control (no additive) in broiler production. This product (De-Odorase®) significantly reduced NH_3 emission by 50% over the control. It is important for producers to ensure effectiveness of the respective additives has been scientifically verified by independent and reputable institutions before they adopt them for use in their facilities.

6.3.2.5. Biological Treatments

In general, biological treatments either assimilate and immobilize volatile N in non-volatile organic forms or transform volatile N into non-volatile inorganic N and benign gaseous N_2. The former processes remove NH_4^+-N from animal manure, via microbial and plant (duckweed, algae, etc) uptake, which is then immobilized in some form of organic N. If the microbial or plant biomass is subsequently removed from the system, this removal is permanent. The converse is also true. If the microbial or plant biomass is not removed from the system, the immobilized organic N could eventually be mineralized back into inorganic forms. In spite of the important roles that microbial assimilation and plant uptake of NH_4^+-N play in mitigating NH_3 emissions, these processes will not be discussed further in this chapter as they are deemed to lean more towards N recovery or alternative uses of animal manures. The conversion of volatile N species to non-volatile species or into benign gaseous N_2 is thus the most common biological treatment system. This system invariably consists of a nitrification process in series with a denitrification process.

During the nitrification phase, ammonia-oxidizing bacteria oxidize TAN (NH_4^+-N and NH_3-N) to nitrite in step 1 (Equations 6.4 and 6.5). In step 2, nitrite-oxidizing bacteria oxidize nitrite to nitrate as shown in Equation 6.6. Although the products (nitrite and nitrate) of nitrification process are not volatile, which effectively mitigates NH_3 emission, they are however, not absolutely environmentally friendly. If such treated manure were applied on land, nitrite and nitrate could leach to ground waters or run-off to surface waters. In the denitrification phase, these compounds (nitrite and nitrate) are biologically reduced to rather environmentally benign N_2 by denitrifying bacteria according to Equations 6.7 and 6.8. In general, the reaction rate of nitrification is extremely low compared to that of denitrification; consequently, nitrification is the rate-limiting step. Nitrification is the more critical step, and usually receives more attention in biological treatment of wastewaters for reduction of NH_3 emissions. Common biological treatment systems consist of either single or two bioreactors. The single-reactor-systems are either run alternately in aerobic and anaerobic modes, or have both aerobic and anoxic zones in the same reactor to effect nitrification and denitrification, respectively. In contrast, these processes take place in separate reactors in the two-reactors-systems. To enhance the nitrification kinetics in particular, other features such as cell immobilization on inert materials or other methods of biomass enrichment are usually incorporated.

$$NH_4^+ + 1.5O_2 \rightarrow NO_2^- + 2H^+ + H_2O \qquad (6.4)$$

$$NH_3 + 1.5O_2 \rightarrow NO_2^- + H^+ + H_2O \qquad (6.5)$$

$$NO_2^- + 0.5O_2 \rightarrow NO_3^- \qquad (6.6)$$

$$2NO_3^- + 10e^- + 12H^+ \rightarrow N_2 + 6H_2O \qquad (6.7)$$

$$2NO_2^- + 6e^- + 8H^+ \rightarrow N_2 + 4H_2O \qquad (6.8)$$

Hu et al. (2003) studied a continuous-flow intermittent aeration (IA: an aerobic phase in series with an anaerobic phase) process for N removal from anaerobically pre-treated swine wastewater at the laboratory scale. Nitrification-denitrification was achieved in this reactor at 3 d hydraulic retention time (HRT) and 20 d solids retention time (SRT). Nitrogen removal rates exceeded 80%; and nitrite and nitrate were less than 20 mg L^{-1} in the effluents. A similar system was evaluated by Zhang et al. (2006) for treating swine manure rich in N. In this study, a bench-scale sequencing batch reactor (SBR) was operated in a cyclic aerobic-anoxic mode at low aeration of 1.0 L[air] m^{-3} [wastewater] s^{-1} during the aerobic cycle. Approximately 97.5% of the TN in the treated **manure** was removed, with only 15 mg L^{-1} of the oxidized N left in the effluent. Luostarinen et al. (2006) evaluated a single-moving bed bioreactor (MBBR) for treatment of anaerobically pre-treated **dairy** parlor wastewater and a mixture of kitchen waste and black water. The effect of intermittent aeration and continuous versus sequencing batch operation was also studied. The MBBRs removed 50-60% of N irrespective of the operational mode. Complete **nitrification** was achieved, but denitrification was impeded by insufficient carbon.

A continuous-flow two-reactor (aerobic and anoxic) system for treatment of swine wastewater was evaluated by Pan and Drapcho (2001). The aerobic reactor was maintained at 5mg L^{-1} dissolved oxygen. This system was operated at 35 h HRT in the anoxic and at 36 h HRT in the aerobic. At steady state, TAN in the effluent was reduced by about 85%, of which 51% was retained as nitrate in the effluent. A similar bench-scale system was evaluated by Ten-Have et al. (1994) for treatment of supernatant from settled sow **manure.** This system also consisted of separate reactors for **nitrification** and denitrification and a recycle of mixed liquor from former to the latter. More than 99% of the TAN was converted to nitrate. Complete **denitrification** was not accomplished because of inadequate fermentable carbon in the manure supernatant. Molasses was added to provide the extra carbon needed. Shin et al. (2005) investigated a slightly different two-reactor system for biological removal of N from **swine** wastewater rich in organic matter and N. This system consisted of a submerged membrane bioreactor (MBR) for nitrification followed by an anaerobic up-flow bed filter (AUBF) reactor for denitrification. Total N removal efficiency of 60% was achieved at an internal recycle ratio of three times flow-rate. Complete **nitrification** of the ammoniacal-N was achieved in the process.

Vanotti and Hunt (2000) evaluated an immobilized-cell (encapsulated in polymer resin) system for enhanced nitrification of TAN in swine wastewater. This system was evaluated for treatment of high-strength swine lagoon wastewaters containing about 230 mg $[NH_4^+\text{-}N]$ L^{-1} and 195 mg $[BOD_5]$ L^{-1}. A culture of acclimated lagoon nitrifying sludge immobilized in 3 to 5 mm polyvinyl alcohol polymer pellets was used for this experiment. Alkalinity was maintained with inorganic carbon to ensure a liquid pH within the optimum range (7.7-8.4). In batch treatment, only 14 h were needed for nitrification of NH_4^+-N. In contrast, it took 10 d for a control (no-pellets) aerated reactor to start nitrification, while as much as 70% NH_3 was lost via air stripping. In continuous flow treatment, nitrification efficiencies of 95% were obtained with NH_4^+-N loading rates of 418 mg[N] L^{-1} [reactor] d^{-1} at 12 h HRT. In all cases, the NH_4^+-N removed was entirely recovered in oxidized N forms. The immobilized-cell technology thus further enhanced TAN removal from anaerobic swine lagoons wastewater.

An 8 m^3 d^{-1} pilot scale two-reactor system was evaluated by Westerman et al. (2000) for treatment of supernatant from settled flushed swine wastewater. The main system consisted of two up-flow aerated biofilters connected in series. The aerated biofilters operated at ~27°C removed about 84% of the TKN, 94% of the TAN, and 61% of the TN. A significant portion of the TAN was converted to nitrite and nitrate. The TKN, TAN, and TN removal averaged 49%, 52%, and 29%, respectively, when the reactors were ran at ~10°C. The unaccounted N of about 24% could have been lost through NH_3 volatilization or through denitrification within the biofilm. Westerman and Bicudo (2002) later evaluated a full-scale nitrification-denitrification system for biological treatment of flushed swine manure in a 3,000 finishing-swine facility. The system consisted of a pond with a mixing zone for denitrification, and an aeration zone for nitrification, with recirculation from aeration zone to mixing zone, and a recycle from aeration zone to the barns for flushing. Nitrogen reduction in the effluent ranged from 65 to 90%, with more than 90% of the N being inorganic N. In addition, significant reduction in odor perception, irritation, and unpleasantness were achieved in this treatment system. Another full-scale nitrification-denitrification system for swine wastewater on a 52,800 grow-finish facility was reported by Townsend et al. (2003). In this system, nitrification and denitrification occurred in a single aerobic-anoxic tank centrally located on the farm reducing TN by an average of 87%.

It is clear that these systems can effectively mitigate NH_3 emissions from animal manures. A major hindrance in their adoption is probably the cost of initial installation and subsequent operation. A key element of completing biological N removal from animal manure with nitrification-denitrification system is the carbon source to drive the denitrification process. This is especially important in the two-reactor systems in which most of the organic matter may be consumed in the nitrification reactor resulting in inadequate carbon to power the denitrification process in the denitrification reactor.

6.3.3. Building and Manure Managements

Urine and feces build-up on the floor is the main source of NH_3 volatilization within livestock buildings. The longer the manure is on either the floor or in the building the greater the NH_3 volatilization. Moreover, the manure is also thinly spread-out further exacerbating NH_3 volatilization because spread-out manure has increased surface areas per unit volume. In general, therefore, manure removal frequency is a key element to the mitigation of NH_3 emission within the building. Scraping, flushing, slatted floors, conveyor belts or combinations of these systems are the most common methods of removing manures from the floors or buildings. Several case studies are presented in the following paragraphs to illustrate the effectiveness of manure management and building-design in the mitigation of NH_3 emissions and concentrations within buildings.

Source: http://www.makingenergy.com.

Figure 6.9. A typical manure flushing system.

Source: http://www.menschmfg.com

Figure 6.10. Manure scraping.

Several manure management studies indicate that flushing manure from floors (Figure 6.9) with water every 2 to 3 h can achieve 14 to 70% reduction in NH_3 loss compared to use of conventional slatted floors in dairy barns. Increasing flushing frequency, increasing the amount of water, and use of fresh water (rather than recycled water) can further improve the performance of manure flushing system in regard to reducing NH_3 emission from buildings. Because these practices may also increase both the volume of the slurry to be handled and the cost of slurry utilization, however; a compromise between flushing frequency, amount of water, use of fresh water and the respective additional reduction of NH_3 losses needs to be established.

Figure 6.11. Typical slatted floor system (Courtesy: http://www.depts.ttu.edu/porkindustryinstitute/ environmental_enrichment_for_pigs.htm).

Kroodsma et al. (1993) investigated the effects of different manure managements on NH_3 emissions from freestall dairy houses. Scraping manure (see Figure 6.10) every 3.5 h did not significantly decrease NH_3 emissions while flushing with water every 3.5 h decreased the emissions by up to 70%. In addition, more frequent flushing (every 1-2 h) over shorter periods (2 s) was more effective than longer (3-6 s) but less frequent flushing (every 3.5 h). Ogink and Kroodsma (1996) evaluated two cattle manure management systems for reduction of NH_3 emissions from cow houses with partially slatted floors. In one method, manure was scraped from the slats and was subsequently flushed with water every 2 h, using 20 L[water] d^{-1} cow^{-1}. The other method was similar, except that 4 g of formalin per liter of flushing-water was added. Method one reduced emission by 14% while method two reduced emissions by 50% in comparison to a control (no scraping or flushing). Misselbrook et al. (2006) reported that pressure washing and the use of a urease inhibitor in addition to yard scraping were more effective means of reducing emissions compared with yard scraping alone. Additionally, the latter studies noted that reduced yard area per animal also effectively reduced total emissions.

In slatted floor systems (Figure 6.11), the frequency of manure removal from the pits under the slats is also an important component in the management of NH_3 emissions within the building. In one study, Hartung and Phillips (1994) compared four different manure removal strategies: a partially slatted floor (PSF) with a slurry pit emptied every two weeks; a PSF with a sloped slurry channel beneath that is flushed several times a day, a PSF floor with a continuous re-circulatory flushing; and a PSF floor with a continuous re-circulatory flushing combined with basin and plug. The control was a PSF with slurry pit underneath providing storage for six months. Respective NH_3 volatilizations were 20, 60, 40, and 80% less than in the control. In a similar study, Lachance et al. (2005) reported ~ 46% reduction in NH_3 emissions when manure was removed every 2-3 d, compared to the 8 weeks removal schedule in the control. Lim et al. (2004) evaluated several manure management strategies on reduction of NH_3 emissions in finishing-pigs confinements. The strategies included daily flushing and static pits with 7, 14, and 42 d manure accumulation cycles with and without pit recharge (with secondary lagoon effluent after emptying). Lower NH_3 emissions were observed with flushing and static pit recharge systems. Mean NH_3 emissions were 15, 27, and 25g d^{-1} AU^{-1} for the 1, 7, and 14 d cycles without pit recharge, and 10, 12, and 11 g d^{-1} AU^{-1} for the 7, 14, and 42 d cycles with pit recharge, respectively. Mean daily NH_3 emissions from the rooms with static pits were 51 to 62% lower with recharge than without recharge. In general, less NH_3 emissions occurred with pit-recharge at more frequent pit-emptying.

Ammonia volatilization within the buildings is also known to be a function of the building ventilation. Increasing ventilation increases NH_3 loss because of reduced resistance to NH_3 transfer from the manure into the surrounding air. For example, a common practice to reduce elevated NH_3 levels in poultry houses is to increase ventilation rates above the values needed for proper litter moisture control. Although increased ventilation rates successfully reduce NH_3 concentration in the house, it also translates directly into higher NH_3 loss from the manure as well higher operational costs.

6.3.4. Emissions Capture and Treatment

This category of NH_3 emissions mitigation strategies involves capturing or trapping the vent-air. In most cases the captured or trapped gases are subsequently treated and transformed into less environmentally harmful forms. Broadly, technologies behind these strategies can be put into two main groups: (i) filters and biofilters, and (ii) permeable and impermeable covers.

6.3.4.1. Filters and Biofilters

Conventional filters remove NH_3 from vent-air either physically or chemically. Biofilters, on the other hand, not only physically trap NH_3 but also utilize microorganisms to biologically degrade or convert trapped NH_3 into benign N-species. Removing NH_3 from vented air using water and acid filters or scrubbers, for example, is feasible where barns are mechanically ventilated (Sommer and Hutchings, 1995; Groot Koerkamp, 1994). In these types of systems (Figure 6.12), NH_3 gas is absorbed by a trickling strong acid. The inert packed media in the absorption tower provides more surface area and the necessary turbulence, which enhance NH_3 absorption process. Dry filters are similar but remove NH_3 by either trapping it in microscopic pits present in the packed media (such as carbon and zeolite) or adsorbing it on the surface of the media. In most cases, practical applications of filters and scrubbers are limited in livestock operations due to their relatively high cost and technical problems due to dust, especially in poultry and swine houses. For this reason, biofilters (Figure 6.13) are more common on livestock operations for removing NH_3 from vent-air prior to exhausting clean air back into the atmosphere. This sub-section, therefore, focuses more on biofilters than on filters and scrubbers. A biofilter, in general, is basically a layer of organic material (such as wood chips, compost, peat, and straw) which supports a population of microorganisms that degrade adsorbed and trapped, or NH_3 moving through the bed.

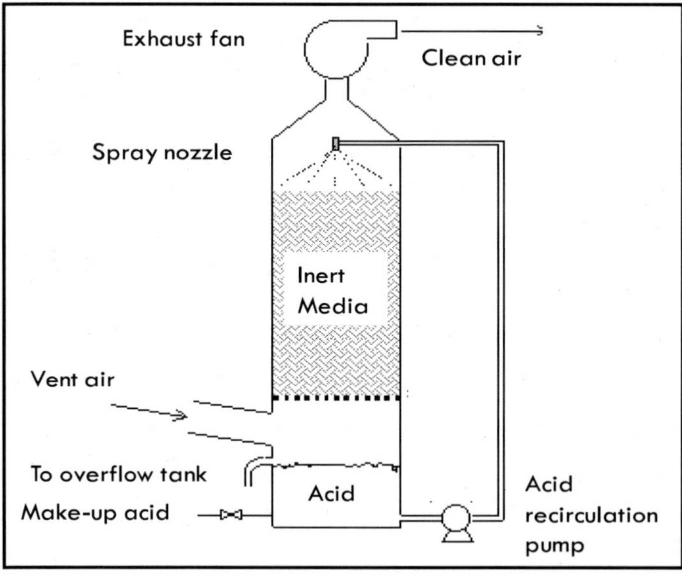

Figure 6.12. A schematic of Ammonia Scrubbing Trickling Strong Acid (Adapted from: http://www.cheresources.com/biofilters.shtml).

Figure 6.13. A typical woodchip biofilter at a large Wastewater Treatment Plant (Courtesy: Franz B. Frechen/University Kassel).

A 20-cm deep biofilter consisting of a mixture of compost and woodchips tested for removal of NH_3 from swine housing ventilation air was evaluated by Sun et al. (2000). This biofilter removed ~83% of NH_3 in the carrier air at the biofilter-media moisture content of 50% and residence time of 20 s. Tanaka et al. (2003) also reported a reduction of 94% in NH_3 from composting vent-air in a biofilter consisting of finished compost (of cattle manure and sawdust) within the first 72 h of treatment. Hong and Park (2005) reported 100% NH_3 removal efficiency from air from a dairy manure-crop residues compost pile in a 50-cm deep, 50:50 manure compost and coconut peels biofilter. Sheridan et al. (2002) evaluated a pilot scale wood chip biofilter for reducing NH_3 from exhaust air from a pig-finishing building. A 50 cm deep biofilter made from 20 mm screen size wood chips efficiently removed between 54 and 93% NH_3 depending on volumetric loading rate. A filter bed moisture level of 63% or greater was recommended to maintain the biofilter efficiency. A biofilter consisting of a mixture of pine and perlite removed 95.6% NH_3 from ventilation air from a swine facility in a pilot-scale system (Chang et al., 2004). Kastner et al. (2004) reported that a biofilter made of pre-screened yard waste compost reduced NH_3 by 25 to 95% in ventilation air from a modern 2400-sow farrow-to-wean unit, depending on residence time and inlet NH_3 concentration. Martinec et al. (2001.) evaluated several biofilter materials (biochips, coconut peels, bark-wood, pellets+bark, and compost) for reduction of NH_3 from swine operations. Ammonia reduction with these materials ranged between 9 and 26%.

Evidently, the performance of biofilters in removing NH_3 from vent-air is very broad. This wide range of performances (9-100%) is attributed to the wide range of biofilter-materials, filter-bed moisture content, the residence time of the air in the biofilter, NH_3 load in the incoming air, and the degree of establishment of the microbial community in the biofilter. Well designed and properly maintained biofilters are in general effective in mitigating NH_3 emissions from livestock manure.

Source: Funk et al. 2004a, 2004b.

Figure 6.14. Impermeable Covers: (a) A floating plastic lagoon cover and (b) an inflatable cover.

For the readers interested in more details on acid scrubbers and trickling filters, a thorough overview of these technologies treating exhaust air from pig and poultry houses is given by Melse and Ogink (2005). In general, NH_3 removal in acid scrubbers ranges from 40-100%, with an overall average of 96%. On the other hand, NH_3 removal efficiency in biotrickling filters ranges from -8 to 100% with an overall average of 70%.

6.3.4.2. Impermeable and Permeable Covers

The simplest control method to mitigate NH_3 emissions from storage and treatment systems (of animal manure) open to the atmosphere is to use a physical cover to contain the emissions. Impermeable covers (Figure 6.14), which are meant to trap and maintain gases released from such systems, are regularly used in conjunction with scrubbers or biofilters described in the preceding subsection. The effectiveness of these covers, therefore, depends on both their trapping efficiency and on the effectiveness of the scrubber or biofilter they are coupled with. Permeable covers (Figure 6.15), on the other hand, do not contain NH_3 like the impermeable covers, but impede, trap, and bio-transform NH_3 just like biofilters, and include materials such as straw, cornstalks, peat moss, foam, geotextile fabric, and Leca® rock. The performances of impermeable and permeable covers are summarized in Table 6.3.

Source: Burns and Moody, 2008.

Figure 6.15. Permeable covers: (a) Covering lagoon with barley straw (Source: Duratech) and (b) LECA covered lagoon.

In comparison to an uncovered control, two impermeable covers: a floating film (two 2-mm thick polyethylene film layers glued together) and a tarpaulin, effectively reduced NH_3 emissions from swine manure lagoons by 99.7 and 99.5%, respectively (Funk et al., 2004a,b). Scotford and Williams (2001) reported nearly 100% NH_3 loss reduction from a pig slurry lagoon covered with a floating 0.5-mm thick reinforced ultraviolet light-stabilized opaque polyethylene cover. Funk et al. (2004b) reported effective control of NH_3 emission using an air-supported 0.35-mm vinyl coated fabric cover installed on an earthen-embanked swine lagoon, but experienced major challenges in controlling gas leakage. Ammoniacal-N is not soluble in oil; therefore, thin layers of oil (oil-films) can also create impermeable covers over stored manure slurries. Heber et al. (2005) evaluated the efficacy of soybean oil sprinkling on NH_3 emission mitigation in tunnel-ventilated swine finishing barns. The oil treated barn resulted in 40% less NH_3 emission than the control barn. Better results have been reported when a layer of vegetable oil was placed on the surface of manure liquid/slurry. Guarino et al. (2006) reported a reduction of NH_3 emissions between 79 and 100% when 3 and 9-mm layers of vegetable oil were applied on stored pig and cattle slurries. Portejoie et al. (2003) reported similar NH_3 emission reductions (93%) with a 10-mm oil layer. Other laboratory and on-farm studies with a 6 mm rapeseed oil layer indicated control of NH_3 emissions by up to 85%, while a thinner 3-mm layer was not effective (Hornig et al., 1999).

Table 6.3. Summary of the performances of permeable and impermeable covers in abating ammonia emissions from livestock manure storages.

Cover Type (s)	Emissions reduction (%)	References
Polyethylene	80-100	Funk et al., 2004a; Scotford and Williams, 2001; Miner et al., 2003
Tarpaulin	99.5	Funk et al., 2004b
Oil films	40-100	Heber et al., 2005; Guarino et al., 2006; Portejoie et al., 2003; Hornig et al., 1999
Geotextile cover	44	Bicudo et al., 2004
Straw covers	37-90	Clanton et al., 2001; Sommer et al., 1993; Horning et al., 1999; Guarino et al., 2006; Xue et al., 1999; Miner & Pan, 1995
Surface crust, peat, & PVC foil	24-32	Sommer et al., 1993;
Leca rock	14-87	Sommer et al., 1993; Balsari et al., 2006
Polymer composite	17-54	Zahn et al., 2001
Pegulit	91	Horning et al., 1999
Wood chips	17-91	Guarino et al., 2006
Corn stalks	37-60	Guarino et al., 2006
Zeolite on permeable cover	90	Miner & Pan, 1995
Polystyrene foam	45-95	Miner and Suh, 1997

A permeable geotextile cover installed on swine manure lagoon provided 44% reductions in NH_3 emissions, although the cover performance deteriorated after one year (Bicudo et al., 2004). Clanton et al. (2001) reported 37, 72, and 86% reduction in NH_3 emissions from swine manure lagoons with 10, 20, and 30 cm-thick straw covers, respectively, supported on a geotextile fabric. The permeable geotextile fabric itself did not have significant effect on NH_3 emissions in the absence of an overlaying straw layer. Compared to uncovered cattle and pig slurry, surface crust, peat, straw, PVC foil, and Leca® rock achieved approximately 24, 32, 60, 26, and 14% NH_3 emission reductions, respectively (Sommer et al., 1993). Zahn et al. (2001) reported a 54% reduction of NH_3 emissions from a lagoon covered with an acclimated proprietary polymer composite bio-cover. Relative to an uncovered control, Hornig et al. (1999) reported NH_3 emissions reduction of 80 to 91% with straw and Pegulit (a natural mineral buoyant material) covers. Development of a surface crust in stored cattle manure was as effective as a 15 cm layer of straw and reduced NH_3 emissions by as much as 20% (Sommer et al., 1993). Guarino et al. (2006) reported effective NH_3 emission reduction from pig and cattle slurry with adequate cover thickness of wheat straw, wood chips, and corn stalks. With 14 cm thick straw, wood chips, and corn stalks covers, NH_3 emissions reductions were 100, 91, and 60%, respectively. However, by using 7 cm thick covers, the respective NH_3 emission reductions were only 59, 17, and 37%. In laboratory studies, Xue et al. (1999) reported that 5 to 10 cm straw covers reduced NH_3 emissions by 90% from dairy manure storages. Miner and Pan (1995) reported permeable covers configured with straw, zeolite, or a combination of both, effectively reduced NH_3 emissions by 90% from manure storages. A permeable polystyrene foam cover was reported to reduce NH_3 emissions by 45 to 95% in manure lagoons (Miner and Suh, 1997). In other laboratory and field studies, Miner et al. (2003) reported NH_3 emission reductions from swine slurries of about 80% using a 5 cm thick permeable polyethylene foam lagoon cover. Balsari et al. (2006) evaluated a low cost cover (Leca® balls layer) for NH_3 emission abatement from swine slurry storage and observed a significant NH_3 emission reduction (up to 87%) with a 10 cm layer of Leca® balls.

Impermeable covers are generally more effective (up to 100%) than permeable covers in mitigating NH_3 emissions from stored manure. The costs for covers, however, vary widely depending on the material used, the geographical region, and the method of application. The length of the time the covers will be in place is an important consideration. Removal and clean-up of the material left behind when the useful life of the covers is over is equally important. In addition, if no biofilters are used to treat the trapped gases under impermeable covers, excessive NH_3 and other gaseous emissions may occur during land application. Massey et al. (2003) evaluated the economics of installing impermeable lagoon covers on swine farms, and showed that at \$0.72 -\$3.41 cwt^{-1} of hog marketed, the initial purchase price of the cover was the biggest hurdle. The second major hurdle is the availability of more land base to receive the conserved N. As high as 3.5 times larger land base may be required to responsibly utilize the conserved N in such covered manure lagoons compared to open lagoons.

6.3.4.3. Land Application

Depending on local environmental conditions, significant NH_3 volatilization can occur following manure surface-spreading or broadcasting (Figure 6.16) to fertilize crop and pasture fields. Minimizing time manure is left exposed on the ground surface is, therefore, the soundest strategy for reducing NH_3 emissions during and after field application of manure.

Some of the common methods available to limit this exposure time include: direct injection; prompt plowing-in; increased infiltration; and washing-in after applications of manure. Coupling these manure application techniques with use of NH_4^+ binding agents or acidifiers can, in addition, improve NH_4^+ uptake by crops and pastures further decreasing NH_3 loss. A summary of the effectiveness of various manure application strategies in reducing NH_3 emissions is given in Table 6.4, while brief narratives are presented in the remaining paragraphs of this sub-section.

Generally, direct manure injection (Figure 6.17) and immediate incorporation of manure into the soil reduce NH_3 losses better than other application methods. Direct manure injections to 3-30 cm depths reduced NH_3 volatilization by 47-98% compared to surface applications (Hoff et al., 1981; Thompson et al., 1987; van der Molen et al., 1990; Svensson, 1994; Rubaek et al., 1996; Morken and Sakshaug, 1998; Smith et al., 2000; Sommer and Hutchings, 2001). Ammonia volatilization, in general, decreases progressively with increased depth of manure injection. Where direct manure injection or immediate incorporation is not possible, other surface placement methods such as band spreading, trailing shoe, and shallow slot injection are more effective than surface broadcasting. These practices have been reported to reduce NH_3 losses by 39-83% compared with surface broadcasting (Thompson et al., 1990a; Svensson, 1994; Frost, 1994; Smith et al., 2000). In the long run, however, some of the studies) did not indicate any advantage of band spreading over broadcasting (Thompson et al., 1990a; Svensson, 1994.

Table 6.4. Summary of livestock manure application strategies for abatement of ammonia emissions.

Application Strategy	Emissions reduction (%)	References
Direct injection	47-100	Hoff et al., 1981; Thompson et al. 1987; Rubaek et al., 1996; Morken & Sakshaug, 1998; Smith et al., 2000; Thompson & Meisinger, 2002; Svensson, 1994; Van der Molen et al., 1990; Huijsmans et al., 2003
Slot injection	80-92	Morken & Sakshaug, 1998; Frost, 1994; Huijsmans et al., 2001
Band application	0-65	Thompson et al., 1990a; Smith et al., 2000; Morken & Sakshaug, 1998; Huijsmans et al., 2001
Trailing shoe	43	Smith et al., 2000
Slurry dilution	44-91	Morken & Sakshaug, 1998; Frost, 1994; Sommer & Olesen, 1991;
Low soil water content	70	Sommer & Jacobsen, 1999
Soil surface cultivation	40-90	Sommer & Thomsen, 1993; Van der Molen et al., 1990; Huijsmans et al., 2003

Source: flickr.com.

Figure 6.16. Manure broadcasting on pasture.

Source: flickr.com.

Figure 6.17. Direct manure injection.

Manure infiltration studies have shown NH_3 losses from surface applied slurry are inversely related to infiltration. One method of increasing manure infiltration into the soil is manure dilution with water. Manure slurry diluted ~100% with water reduced NH_3 losses by 44-91% (Sommer and Olesen, 1991; Stevens et al., 1992; Frost, 1994; Morken and Sakshaug, 1998). Another method of increasing infiltration is cultivating the soil surface or increasing the surface roughness. Cultivating the soil surface before surface application of slurry reduced NH_3 losses by 40-90% (Sommer and Thomsen, 1993). A similar method of increasing infiltration is cultivating the top 6 cm of the soil to mix applied slurry with soil. This manure-soil mixing reduces NH_3 loss by ~60% compared to surface application (Van der Molen et al., 1990). Other studies have shown that infiltration is also higher at low soil moisture contents, and slurry application during periods of low soil moisture can reduce NH_3 loss by as high as 70% (Sommer and Jacobsen, 1999). An inverse relationship is observed between NH_3 loss and the rate of slurry application (volume time^{-1} area^{-1}). This observation suggests that intermittent slurry application might also reduce NH_3 loss because intermittent application improves infiltration (Thompson et al., 1990b).

Ammonia losses from manure applied during crop growth periods may be reduced by use of trailing hoses which apply the slurry onto the soil between rows of plants (Bless et al. 1991; Holtan-Hartwig and Bockman, 1994). The reduction in NH_3 loss is attributed to the immediate absorption of NH_4^+-N by plant leaves and roots and reduced slurry exposed surface. In addition, the canopy-modified microclimate is also not favorable for NH_3 volatilization (Holtan-Hartwig and Bockman, 1994; Thompson et al., 1990).

Atmospheric conditions play an important role in NH_3 loss reduction during slurry application. Sommer et al. (1991) reported a linear increase in NH_3 volatilization between 0 and 19°C during a 24 h period. In the same study, NH_3 loss increased significantly when wind speed increased to 2.5 m s^{-1}. No consistent increase in NH_3 loss was recorded between 2.5 and 4.0 m s^{-1} wind speeds. In an earlier study, increasing wind speed from 0.5 to 3.0 m s^{-1} increased NH_3 loss by about 29% in 5 d (Thompson et al., 1990b). These observations suggest manure applications should be scheduled preferably during non-windy, colder periods.

Arguably, direct manure injection and manure incorporation into the soil adds to the costs of manure application. The cost of injection or manure incorporation into the soil during land application to reduce NH_3 emissions may, however, be recaptured in terms of better crop yields due to a more efficient utilization of the applied manure. If other environmental benefits emanating from reduced NH_3 loss, as well as costs that may be incurred in legal conflicts due to NH_3 emissions are considered, these manure application practices are economically justifiable.

6.4. SUMMARY AND CONCLUSION

Ammonia emissions from livestock manure is associated with excess N (or unutilized N) excreted by the animal or bird with urine and feces. Diet formulation or constitution, especially with respect to N composition, is thus an important element determining N excretion and NH_3 emissions from manure. In general, the level of dietary CP is one of the most important factors determining N utilization and thus excretion in farm animals. Increasing CP concentration in the diets of cows, goats, and sheep, for example, may increase milk protein N yield but it also reduces efficiency of N utilization for milk protein synthesis. Moreover, most of the dietary N not going into milk protein synthesis is excreted in urine, which is more susceptible to volatilization than fecal N. Therefore, the easiest way to control or limit excretion of nitrogenous compounds from dairy cattle is to enhance the efficiency of N utilization for milk production through meticulous dietary CP balance. Reduced ration CP or RDP concentration results in reduced volatile N in excreted manure. On the other hand, increased dietary metabolizable protein increases volatile N excretion in the manure.

Most of the N (~90%) is excreted as urea in the urine of cows, goats, sheep, or pigs, and as undigested organic N in the feces. Both urea and organic N are eventually transformed, through enzymatic actions of microbial enzymes, into either NH_4^+-N at low pH or NH_3-N at high pH. The hydrolysis of urea to NH_3-N (or NH_4^+-N) is catalyzed by the enzyme urease. The conversion of organic-N to NH_4^+-N or NH_3-N, on the other hand, is mediated by a host of degradative enzymes produced by heterotrophic microbes. The NH_4^+-N species itself is not volatile but it is susceptible to volatilization when it dissociates to NH_3-N. In general, the dissociation of NH_4^+-N to NH_3-N increases with increase in both pH and temperature, which increases the potential for NH3 volatilization.

Techniques for reducing NH_3 losses from animal manure may be put into three broad categories: (i) those that minimize N in the animal manure prior to its excretion, (ii) those that reduce volatile N species (NH_3-N or NH_4^+-N) in the excreted manure, and (iii) those that physically contain and treat NH_3-N or NH_4^+-N species after they have already been formed. The first category mainly involves dietary changes with respect to feed N. In general, reducing N excretion through dietary changes can effectively mitigate NH_3 emissions from livestock operations. In ruminants, reducing the CP intake by as little as 5% can reduce NH_3 emissions by as much as 74% from excreted manure. For non-ruminants, similar NH_3 emission reductions have been observed by replacing CP intake with intake of amino acids.

Strategies towards reducing volatiles N species include: use of enzyme inhibitors, urine-feces segregation, pH reduction, binding of ammonium ion, and conversion of NH_3-N or NH_4^+-N into other non-volatile N species (NO_3^- or NO_2^-) or into environmentally benign N_2. The effectiveness of urease inhibitors in the mitigation of NH_3 emissions from animal manures has been demonstrated in the laboratory. Adoption of this method in full-scale livestock operations, however, is yet to be realized. The slow acceptance of this technique may be attributed to the unknown effects of these chemicals (i.e. urease inhibitors) on the crops or pastures where the manure is eventually applied as fertilizer; as well as additional labor costs associated with repeated application of inhibitors. Reported urine-feces segregation methods reduced NH_3 emissions from livestock barns by about 50% compared to the conventional manure handling systems. Reduction of manure pH to abate NH_3 emissions is effective. In general, strong acids can reduce slurry pH more cost-effectively than the weaker acids and acidifying salts to hold NH_3 in ammonium form. However, strong acids are more hazardous for use on the farm. Therefore, although the acidifying salts and other weaker acids may be less effective than the strong acids, their non-hazardous nature increases their suitability for reducing NH_3 loss from livestock manures. Among NH_4^+-N and NH_3-N binding amendments, zeolite and sphagnum are more effective for reduction of H_3 loss in manure slurries or liquid than in solid poultry manures. Yucca extract (saponins) does not seem to as effective as either zeolite or peat moss in mitigating NH_3 emissions.

There are a number of other amendments with various brand names, but their mode of operations is not known. It is important for producers to ensure that the effectiveness of these additives have been scientifically verified by independent and reputable institutions before they can adopt them for use in their facilities. Often, large amounts of the product are required and in most cases such as the use of acids or acidic salts, precautions must be taken to safeguard the safety of livestock and farm workers. In addition use of acids may result not only in an undesirable increase in the mineral content of the manure or litter, but also in the corrosion of equipment and structures. Selection of appropriate application methods for effective use of these additives is very important. Currently, there is a lack of standardized application and evaluation protocols for these additives. Biological transformation of NH_4^+-N or NH_3-N into NO_2^-, NO_3^-, or N_2 are fairly effective (up to 99%) in mitigating emissions from livestock manure. It appears that the major hindrance is the economics of installing and operating the systems.

The third category of techniques that involve capturing and treatment of volatile species include: biofilters, permeable and impermeable covers, and injection of manure during land applications. The performance of biofilters in removing NH_3 from exhaust air is wide; ranging from 9 to 100%. The wide range of biofilters performance is attributed to the variety of biofilter materials, maintenance of optimum moisture in the filter bed, residence time of the air in the biofilter, NH_3 load in the incoming air, and how well the microbial community is established in the biofilter. With respect to manure storage covers, impermeable covers are, in general, more effective than permeable covers in mitigating NH_3 emissions from manure storages. The important factors to consider when selecting permeable covers are the cost, method of application or installation, length of the time the covers will be in place, and removal and clean-up of the material left behind when the useful life of the covers is over. In contrast, use of biofilters must be considered to treat the trapped gases under impermeable covers to avoid excessive NH_3 and other gaseous emissions during manure removal for land application. The other factors to consider when choosing impermeable over permeable covers are the high initial purchase price of these covers and requirement of more land base to receive the conserved N. Besides the cost of manure application, direct manure injection or manure incorporation into the soil are the most effective (up to 98%) methods for mitigating NH_3 emissions during land application of livestock manure. In most cases, the costs of injection or incorporating manure is recoverable from improved crop yields resulting from more efficient utilization of the applied manure N. Other than the direct economic gains from improved crop yields, the latter practices result in additional environmental benefits accruing from reduced NH_3 release into the environment, which protect livestock operations from potential lawsuits associated with the NH_3 releases.

REFERENCES

Al-Kanani, T., E. Akochi, A.F. Mackenzie, I. Alli, and S. Barrington. 1992. Organic and inorganic amendments to reduce ammonia losses from liquid hog manure. *J. Environ. Qual.* 21:709-715.

Andersen, J. B., N.C. Friggens, T. Larsen, M. Vestergaard, and K.L. Ingvartsen. 2004. Effect of energy density in the diet and milking frequency on plasma metabolites and hormones in early lactation dairy cows. J. *Vet. Med. Ser.* A 51:52-57.

Amon, M., M. Dobeic, R.W. Sneath, V.R. Phillips, T.H. Misselbrook, and B.F. Pain. 1997. A farm-scale study on the use of clinoptilolite zeolite and De-Odorase® for reducing odour and ammonia emissions from broiler houses. *Bioresour. Technol.* 61(3):229-237.

Agle, M., A. N. Hristov, S. Zaman, C. Schneider, P. Ndegwa, and V. K. Vaddella. 2010. Effects of ruminally degraded protein on rumen fermentation and ammonia losses from manure in dairy cows. *J. Dairy Sci.* 93(4):1625-1637.

Armstrong, K.A., R.T. Burns, F.R. Walker, L.R. Wilhelm, and D.R. Raman. 2003. Ammonia concentrations in poultry broiler production units treated with liquid alum, in: *Proc. of the Air Pollution from Agr.* Operations III, Res. Triangle Park, North Carolina, pp. 116-122.

Arogo, J., P.W. Westerman, A.J. Heber, W.P. Robarge, abd J.J. Classen. 2001. *Ammonia emissions from animal feeding operations.* Report: White Papers. National Center for Manure & Animal Waste Management, Raleigh, NC.

Arogo, J., R. H. Zhang, G. L. Riskowski, L. L. Christianson, and D. L. Day. 1999. Mass transfer coefficient of ammonia in liquid swine manure and aqueous solutions. *J. Agri. Eng. Res.* 73(1), 77-86.

Baker, D. H. 1996. *Advances in amino acid nutrition and metabolism of swine and poultry. Pages 11-22 in.Nutrient management of food animals to enhance and protect the environment.* E. T. Kornegay, ed. CRC Lewis Publishers, Boca Raton, FL.

Banini, S., W.R. Rypniewski, K.S. Wilson, S. Miletti, S. Ciurli, and S. Mangani. 1999. A new proposal for urease mechanism based on the crystal structures of the native and inhibited enzyme from *Bacillus pasteurii*: why urea hydrolysis costs two nickels. *Structures* 7(2):205-216.

Balsari P; Dinuccio E; Gioelli F (2006). A low cost solution for ammonia emission abatement from slurry storage. *Int Congr* 1293:323-326.

Barak, P., J.A.E. Molina, A. Hadas, and C.E. Clapp. 1990. Mineralization of amino acids and evidence of direct assimilation of organic nitrogen. *Soil Sci. Soc. Am. J.* 54:769-774.

Barraclough, D. 1997. The direct or MIT route for nitrogen immobilization: A ^{15}N mirror image study with leucine and glycine. *Soil Biol.* 29(1):101-108.

Barrington, S.F., and R.G. Moreno. 1995. Swine manure N conservation in storage using sphagnum moss. *J. Environ. Qual.* 24:603-607.

Beline, F., J. Martinez, C. Marol, and G. Guiraud. 1998. Nitrogen transformations during anaerobically stored ^{15}N-labeled pig slurry. *Bioresour. Technol.* 64:83-88.

Bicudo, J.R., C.J. Clanton, D.R. Schmidt, W. Powers, L.D. Jacobson, C.L. Tengman. 2004. Geotextile covers to reduce odor and gas emissions from swine manure storage ponds. *Appl. Eng. Agr.* 20(1): 65-75.

Bless, H.G., R. Beinhauer, and B. Sattelmacher. 1991. Ammonia emission from slurry applied to wheat stubble and rape in North Germany. *Camb. J. Agr. Sci.* **117:**225-231.

Blouin, J. P., J. F. Bernier, C. K. Reynolds, G. E. Lobley, P. Dubreuil, and H. Lapierre. 2002. Effect of supply of metabolizable protein on splanchnic fluxes of nutrients and hormones in lactating dairy cows. *J. Dairy Sci.* 85:2618–2630.

Bowman, G.R., K.A. Beauchemin, and L.M. Rode. 2003. Acidification of urine by feeding anionic products to non-lactating dairy cows. *Can. J. Anim. Sci.* 83:319-321.

Braam, C.R., M.C.J. Smits, H. Gunnink, and D. Swierstra. 1997a. Ammonia emission from a double-sloped solid floor in a cubicle house for dairy cows. *J. Agr. Eng. Res.* 68(4):375-386

Braam, C.R., J.J.M.H. Ketelars, and M.C.J. Smits. 1997b. Effects of floor design and floor cleaning on ammonia emission from cubicle houses for dairy cows. *Neth. J. Agr. Sci.* 45:49-64.

Bristow, A. W., D. C. Whitehead, and J. E. Cockburn. 1992. Nitrogenous constituents in the urine of cattle, sheep, and goats. *J. Sci. Food Agric.* 59:387-394.

Broderick, G. A., M. J. Stevenson, R. A. Patton, N. E. Lobos, and J. J. Olmos Colmenero. 2008. Effect of supplementing rumen-protected methionine on production and nitrogen excretion in lactating dairy cows. *J. Dairy Sci.* 91:1092–1102.

Broderick, G.A. 2005. Protein and carbohydrate nutrition of dairy cows. *Proceedings, Pacific Northwest Animal Nutrition Conference,* Boise, Idaho, pp. 35-56.

Broderick, G.A. 2003. Effects of varying dietary protein and energy levels on the production of lactating dairy cows. *J. Dairy Sci.* 86:1370-1381.

Burns, R., and L. Moody. 2008. A review of permeable cover options for manure storages. In: *Proc. of Mitigating Air Emissions from Animal Feeding Operations.* May 19-21, Des Moines, Iowa.

Bussink D.W. and O. Oenema. 1998. Ammonia volatilization from dairy farming systems in temperate areas: a review. *Nutrient Cyc. Agroecosys.* 51:19–33.

Bussink D.W. and O. Oenema. 1998. Ammonia volatilization from dairy farming systems in temperate areas: a review. *Nutrient Cyc. Agroecosys.* 51:19–33.

Canh, T.T., A.J.A. Aarnink, J.B. Schulte, A. Sutton, D.J. Langhout, and M.W.A. Verstergen. 1998a. Dietary protein affects nitrogen excretion and ammonia emission from slurry of growing-finishing pigs. *Livest. Prod. Sci.* 56:181-191.

Canh, T.T., A. Sutton, A.J.A. Aarnink, M.W.A. Verstergen, J.W. Schrama, and J.W. Bakker. 1998b. Dietary carbohydrates alter fecal composition and pH and the ammonia emission from slurry of growing pigs. *J. Anim. Sci.* 76:1887-1895.

Canh, T.T., M.W.A. Verstergen, A.J.A. Aarnink, and J.W. Schrama. 1997. Influence of dietary factors on nitrogen partitioning and composition of urine and feces of fattening pigs. *J. Anim. Sci.* 75:700-706.

Clark, O.G., S. Moehn, I. Edeogu, J. Price, and J. Leonard. 2005. Manipulation of dietary protein and nonstarch polysaccharide to control swine manure emissions. *J. Environ. Qual.* 34:1461-1466.

Chang, D.I., S.J. Lee, and W.Y. Choi. 2004. *A pilot-scale biofilter system to reduce odor from swine operation.* ASAE paper number 044056.

Clanton, C.J., D.R. Schmidt, R.E. Nicolai, L.D. Jacobson, P.R. Goodrich, K.A. Janni, and J.R. Bicudo. 2001. Geotextile fabric-straw manure storage covers for odor, hydrogen sulfide and ammonia control. *Appl. Eng. Agr.* 17(6):849-858.

Cole, N.A., R.N. Clark, R.W. Todd, C.R. Richardson, A. Gueye, L.W. Greene, and K. McBride. 2005. Influence of dietary crude protein concentration and source on potential ammonia emissions from beef cattle manure. *J. Anim. Sci.* 83:722-731.

Colina, J.J., A.J. Lewis, P.S. Miller, and R.L. Fischer. 2001. Dietary manipulation to reduce aerial ammonia concentrations in nursery pig facilities. *J. Anim. Sci.* 79:3096-3103.

Colmenero, J. J. and G. A. Broderick. 2006. Effect of dietary crude protein concentration on milk production and nitrogen utilization in lactating dairy cows. *J. Dairy Sci.* 89:1704-1712.

de Boer, I.J.M., M.C.J. Smits, H. Mollenhorst, G. van Duikerken, and G.J. Monteny. 2002. Prediction of ammonia emissions from dairy barns using feeds characteristics Part I: Relation between feed characteristics and urinary urea concentration. *J. Dairy Sci.* 85:3382-3388.

De Visscher, A., L.A. Harper, P.W. Westerman, Z. Liang, J. Arogo, R.R. Sharpe, and O. Van Cleemput. 2002. Ammonia emissions from anaerobic swine lagoons: Model development. *J. Appl. Meteor.* 41:426-433.

Ferguson, N.S., R.S. Gates, J.L. Taraba, A.H. Cantor, A.J. Pescatore, M.L. Straw, M.J. Ford, and D.J. Burnham. 1998. The effect of dietary protein and phosphorus on ammonia concentration and litter composition in broilers. *Poultry Sci.* 77:1085-1093.

Frank, B., and C. Swensson. 2002. Relationship between content of crude protein in rations for dairy cows and milk yield, concentration of urea in milk and ammonia emissions. *J. Dairy Sci.* 85:1829-1838.

Frank, B., M. Persson, and G. Gustafsson. 2002. Feeding dairy cows for decreased ammonia emission. *Livest. Prod. Sci.* 76:171-179.

Frost, J.P. 1994. Effect of spreading method, application rate and dilution on ammonia volatilization from cattle slurry. *Grass and Forage Sci.* 49:391–400

Frost, J.P., R.J. Stevens, and R.J. Laughlin. 1990. Effect of separation and acidification on ammonia volatilization and on the efficiency of slurry nitrogen for herbage production. *J. Agr. Sci.* 115:49-56.

Funk, T., A. Mutlu, Y. Zhang, and M. Ellis. 2004a. Synthetic covers for emissions control from earthen embanked swine lagoon Part II: negative pressure lagoon cover. *Appl. Eng. Agr.* 20(2): 239-242.

Funk, T.L., R. Hussey, Y. Zhang, and M. Ellis. 2004b. Synthetic covers for emissions control from earthen embanked swine lagoons part 1: positive pressure lagoon cover. *Appl. Eng. Agr.* 20(2):233-238.

Funk, T.L., and J. Polakow. 2004c. *Manure liquid-solid separation factors that determine system choice.* SOWN papers: University of Illinois Extension.

Gozho, G. N., M. R. Hobin, and T. Mutsvangwa. 2008. Interactions between barley grain processing and source of supplemental dietary fat on nitrogen metabolism and urea-nitrogen recycling in dairy cows. *J. Dairy Sci.* 91:247–259.

Groff, E.B., and Z. Wu. 2005. Milk production and nitrogen excretion of dairy cows fed different amounts of protein and varying proportions of alfalfa and corn silage. *J. Dairy Sci.* 88:3619-3632.

Groot Koerkamp, P.W. G. 1994. Review on emissions of ammonia from housing systems for laying hens in relation to sources, processes, building design and manure handling. *J. Agr. Eng. Res.* 59(2):73-87.

Guarino, M., C. Fabbri, M. Brambilla, L. Valli, and P. Navarotto. 2006. Evaluation of simplified covering systems to reduce gaseous emissions from livestock manure storage. *Trans.* ASAE 49(3):737-747.

Han, Y.K., and H.T. Shin. 2005. Effects of antibiotics, copper sulfate and probiotics supplementation on the performance and ammonia emission from slurry in growing pigs. *J. Anim. Sci. Technol.* 47:537-546.

Hartung, J., and V.R. Phillips. 1994. Control of gaseous emissions from livestock buildings and manure stores. *J. Agr. Eng. Res.* 57:173-189.

Hayes, E.T., A.B.G. Leek, T.P. Curran, V.A. Dodd, O.T. Carton, V.E. Beattie, J.V. O'Doherty. 2004. The influence of diet crude protein level on odour and ammonia emissions from finishing pig houses. *Bioresour. Technol.* 91:309-315.

Heber, A.J., P.C. Tao, J.Q. Ni, T.T. Lim, and A.M. Schmidt. 2005. Air emissions from two swine finishing building with flushing: ammonia characteristics. *Proceedings of the Seventh International Symposium,* Beijing, China.

Heber, A.J., J.Q. Ni, T.T., Lim, C.A. Diehl, S.L. Sutton, R.K. Duggirala, B.L. Haymore, D.T. Kelly, and V.I. Adamchuk. 2000. Effect of a manure additive on ammonia emission from swine finishing buildings. *Trans. ASAE* 43(6):1895-1902.

Hoff, J.D., D.W. Nelson, and A.L. Sutton. 1981. Ammonia volatilization from liquid swine manure applied to cropland. *J. Environ. Qual.* 10(1):90-95.

Holtan-Hartwig, L., and O.C. Bockman. 1994. Ammonia exchange between crops and air. Norweg. *J. Agr. Sci.* (suppl.) 14:5-40.

Hong, J.H., and K.J. Park. 2005. Compost biofiltration of ammonia gas from bin composting. *Bioresour. Technol.* 96(6):741-745.

Hornig, G., M. Turk, and U. Wanka. 1999. Slurry covers to reduce ammonia emission and odor nuisance. *J. Agr. Eng. Res.* 73:151-157.

Horton, H.R., L.A. Moran, R.S. Ochs, J.D. Rawn, & K.G. Scrimgeour. 1992. *Principles of biochemistry (International edition).* Neil Patterson Publishers Prentice Hall, Englewood Cliffs, NJ 07632, pp. 17.1-19.26.

Hristov, A.N. and P. Huhtanen. 2008. Nitrogen efficiency in Holstein cows and dietary means to mitigate nitrogen losses from dairy operations. Pages 125-136 in Proceedings, *Cornell Nutrition Conference, Syracuse,* NY, Oct. 21-23.

Hristov, A. N. and Jouany, J.-P. 2005. Factors affecting the efficiency of nitrogen utilization in the rumen. In: Pfeffer, E. and Hristov, A. N. (eds.) *Nitrogen and Phosphorus Nutrition of Cattle: Reducing the Environmental Impact of Cattle Operations. CAB International,* Wallingford, U.K., pp. 117-166.

Hristov, A.N., J.K. Ropp, K.L. Grandeen, S. Abedi, R.P. Etter, A. Melgar, and A.E. Foley. 2005. Effect of carbohydrate source on ammonia utilization in lactating dairy cows. *J. Anim. Sci.*83:408-421.

Hristov, A.N., W.J. Price, and B. Shafii. 2004a. A meta-analysis examining the relationship among dietary factors, dry matter intake, and milk yield and milk protein yield in dairy cows. *J. Dairy Sci.* 87:2184-2196.

Hristov, A.N., R.P. Etter, J.K. Ropp, and K.L. Grandeen. 2004b. Effect of dietary crude protein level and degradability on ruminal fermentation and nitrogen utilization in lactating dairy cows. *J. Anim Sci.* 82:3219-3229.

Hu, Z. C.R. Mota, F.L. de los Reyes III, and J. Cheng. 2003. Optimization of nitrogen removal from anaerobically pretreated swine wastewater in intermittent aeration tanks, in: *The Proc. of 9th Inter. Anim., Agr. and Food Proc. Wastes Symp.,* Res. Triangle Park, NC, pp. 590-595.

Huhtanen, P. and A. N. Hristov. 2009. A meta-analysis of the effects of protein concentration and degradability on milk protein yield and milk N efficiency in dairy cows. *J. Dairy Sci.* 92:3222–3232.

Huhtanen, P. and A. N. Hristov. 2009. A meta-analysis of the effects of protein concentration and degradability on milk protein yield and milk N efficiency in dairy cows. J. *Dairy Sci.* 92:3222–3232.

Huhtanen, P., J.I. Nousiainen, M. Rinne, K. Kytölä, and H. Khalili. 2008. Utilization of and partitioning of dietary nitrogen in dairy cows fed grass silage based diets. J. *Dairy Sci.* 91:3589–3599.

Huijsmans, J.F.M., J.M.G. Hol, and G.D. Vermeulen. 2003. Effect of application method, manure characteristic, weather and field conditions on ammonia volatilization from manure applied to arable land. *Atmos. Environ.* 37:3669-3680.

Huijsmans, J.F.M., J.M.G. Hol, and M.M.W.B. Hendriks. 2001. Effect of application technique, manure characteristic, weather and field conditions on ammonia volatilization from manure applied to grassland. *Neth. J. Agr. Sci.* 49:323-342.

Husted, S., L.S. Jensen, and S.S. Jorgensen. 1991. Reducing ammonia loss from cattle slurry by the used of acidifying additives: the role of the buffer system. *J. Sci. Food Agr.* 57:335-349.

Jacob, J.P., S. Ibrahim, R. Blair, H. Namkung, and I.K. Paik. 2000. Using enzyme supplemented, reduced protein diets to decrease nitrogen and phosphorus excretion of white leghorn hens. *Asian Australas J. Anim. Sci.* 13:1743-1749.

Jensen, A.O. 2002. Changing the environment in swine buildings using sulfuric acid. *Trans. ASAE* 45(1):223-227.

Jongbloed, A.W. and N. P. Lenis. 1992. Alteration of nutrition as a means to reduce environmental pollution by pigs. *Livest. Prod. Sci.* 3 1:75-94.

Kalscheur, F.F., J.H. Vandersall, R.A. Erdman, R.A. Kohn, and E. Russek-Cohen. 1999. Effect of dietary crude protein concentration and degradability on milk production responses of early, mid, and late lactation dairy cows. *J. Dairy Sci.* 82:545-554.

Kaminskaia, N.V. and N.M. Kostic. 1997. Kinetics and mechanism of urea hydrolysis catalyzed by palladium (II) complexes. *Inorg. Chem.,* 36: 5917-5926.

Kirk, G.J.D. and P.H. Nye. 1991. A model of ammonia volatilization from applied urea. V. The steady-state drainage and evaporation. *J. Soil Sci.* 42:103-113.

Kastner, J.R., K.C. Das, and B. Crompton. 2004. Kinetics of ammonia removal in a pilot-scale biofilter. *Trans. ASAE* 47(5):1867-1878.

Kebreab, E., J. France, J.A.N. Mills, R. Allison, and J. Dijkstra. 2002. A dynamic model of N metabolism in the lactating dairy cow and an assessment of N excretion on the environment. *J. Anim. Sci.* 80:248-259.

Kemme, P.A., A.W. Jongbloed, B.M. Dellaert, and F. Krol-Kramer. 1993. The use of a Yucca Schidigera extract as 'urease inhibitor' in pig slurry. In: *Proc. First Inter. Symp. Nitrogen Flow in Pig Production and Environ Consequences,* Wageningen, NL, pp, 330-335.

Kerr, B.J. 1995. Nutritional strategies for waste reduction-management: nitrogen. In: *Proc. New Horizons in Animal Nutrition and Health. Raleigh,* NC.

Kim, J.H., S.C. Kim, and Y.D. Ko. 2005. Effect of dietary zeolite on the performance and carcass characteristics of finishing pigs. *J Anim Sci Technol* 47, 555-564.

Kim, I.B., P.R. Ferket, W.J. Powers, H.H. Stein, and T.A.T.G. van Kempen. 2004. Effects of different dietary acidifier sources of calcium and phosphorus on ammonia, methane and odorant emission from growing-finishing pigs. *Asian Australas J. Anim. Sci.* 17:1131-1138.

Kithome, M., J.W. Paul, and A.A. Bomke. 1999. Reducing nitrogen losses during simulated composting of poultry manure using adsorbents or chemical amendments. *J. Environ. Qual.* 28(1):194-201.

Klopfenstein, T., R. Angel, G.L. Cromwell, G.L. Erickson, D.G. Fox, C. Parsons, L.D. Satter, and A.L. Sutton. 2002. Animal diet modification to decrease the potential for nitrogen and phosphorus pollution. *Council for Agricultural Science and Technology* 21:1-16.

Krober, T.F., D.R. Kulling, H. Manz, F. Sutter, and M. Kreuzer. 2000. Quantitative effects of feed protein reduction and methionine on nitrogen use by cows and nitrogen emission from slurry. *J. Dairy Sci.* 83:2941-2951.

Kroodsma, W., J.W.H. Huis in 't Veld , and R. Scholtens. 1993. Ammonia emission and its reduction from cubicle houses by flushing. *Livest. Prod. Sci.* 35(3-4):293-302

Kulling, D.R., H. Menzi, T.F. Krober, A. Neftel, F. Sutter, P. Lischer, and M. Kreuzer. 2001. Emissions of ammonia, nitrous oxide and methane from different types of dairy manure during storage as affected by dietary protein content. *J. Agric. Sci. Camb.* 137:235-250.

Lachance Jr, I., S. Godbout, S.P. Lemay, J.P. Larouche, and F. Pouliot. 2005. *Separation of pig manure under slats: to reduce releases in the environment.* ASAE paper 054159.

Lapierre, H., and G. E. Lobley. 2001. Nitrogen recycling in the ruminant: A review. *J. Dairy Sci.* 84(Suppl. E):E223–E236.

Lapierre, H., D. Pacheco, R. Berthiaume, D. R. Ouellet, C. G. Schwab, P. Dubreuil, G. Holtrop, and G. E. Lobley. 2006. What is the true supply of amino acids for a dairy cow? *J. Dairy Sci.* 89(E. Suppl.):E1–E14.

Le, P. D., A.J.A. Aarnink, and A.W. Jongbloed. 2009. Odour and ammonia emission from pig manure as affected by dietary crude protein level. *Livest. Sci.* 121:267–274.

Lee, C. and A. N. Hristov. 2010a. Origin of ammonia nitrogen volatilized from dairy manure. *J. Dairy Sci.* 92 (in press).

Lee, C. and A. N. Hristov. 2010b. Effects of dietary protein concentration and coconut oil supplementation on nitrogen utilization in dairy cows. *J. Dairy Sci.* 92 (in press).

Lee, C., A. N. Hristov, C. Dell, G. Feyereisen, J. Kaye, and D. Beegle. 2010. Effect of dietary protein on ammonia emission from dairy manure. *J. Dairy Sci.* 92 (in press).

Leek, A.B.G., V.E. Beattie, and J.V. O'Doherty. 2004. The effects of dietary oil inclusion and oil source on apparent digestibility, faecal volatile fatty acid concentration and manure ammonia emission. *Anim. Sci.* 79:155-164.

Li, H., H. Xin, and R.T. Burns. 2006. Reduction of ammonia emission from stored poultry manure using additives: zeolite, Al$^+$ clear, Ferix-3, and PLT. *ASAE paper No. 064188,* 2006 ASABE Annual International Meeting, Portland, Oregon.

Lim, T.T., A.J. Heber, J.Q. Ni, D.C. Kendall, and B.T. Richert. 2004. Effects of manure removal strategies on odor and gas emissions from swine finishing. *Trans. ASAE* 47(6):2041-2050.

Luostarinen, S., S. Luste, L. Valentín, and J. Rintala. 2006. Nitrogen removal from on-site treated anaerobic effluents using intermittently aerated moving bed biofilm reactors at low temperatures. *Water Res.* 40(8):1607-1615.

Mackenzie, A.F., and J.S. Tomar. 1987. Effect of added monocalcium phosphate monohydrate and aeration on nitrogen retention by liquid hog manure. *Can. J. Soil Sci.* 67:687-692.

Martin, A.K. 1982. The origin of urinary aromatic compounds excreted by the ruminants 2. The metabolism of phenolic cinnamic acids to benzoic acid. *Br. J. Nutr.* 47:155-164.

Massey, R.E., J.E. Zulovich, J.A. Lory, and A.M. Millmier. 2003. Farm level economic impacts of owning and operating impermeable lagoon covers. In: T*he Proceedings of Air Pollution from Agricultural Operations III Conference,* North Carolina, USA, pp. 346-353.

Martinec, M., E. Hartung, T. Jungbluth, F. Schneider, and P.H. Wieser. 2001. Reduction of gas, odor and dust emissions from swine operations with biofilters. *ASAE paper number* 014079.

McCrory, D.F. and P.J. Hobbs. 2001. Additives to reduced ammonia and odor emissions from livestock wastes: *A review. J. Environ. Qual.* 30:345-355.

McGinn, S.M., K.M. Koenig, and T. Coates. 2002. Effect of diet on odourant emissions from cattle manure. *Can. J. Anim. Sci.* 82:435-444.

Mellau, L.S.B., R.J. Jorgensen, P.C. Bartlett, J.M.D. Enemark, and A.K. Hansen. 2004. Effect of anionic salt and highly fermentable carbohydrate supplementation on urine pH and on experimentally induced hypocalcaemia in cows. *Acta Vet. Scand.* 45:139-147.

Melse, R.W., and N.W.M. Ogink. 2005. Air scrubbing techniques for ammonia and odor reduction at livestock operations: review of on-farm research in the Netherlands. *Trans. ASAE* 48(6):2303-2313.

Miner, J.R., F.J. Humenik, J.M. Rice, D.M.C. Rashash, C.M. Williams, W. Robarge, D.B. Harris, and R. Sheffield. 2003. Evaluation of a permeable, 5 cm thick, polyethylene foam lagoon cover. *Trans. ASAE* 46(5):1421-1426.

Miner, J. R., F. J. Humenik, and M. R. Overcash. 2000. *Managing Livestock Waste to Preserve Environmental Quality.* Iowa State University Press, Ames, Iowa.

Miner, J.R.,and K.W. Suh. 1997. Floating permeable covers to control odor from lagoons and manure storages, in: *Proc. Of Ammonia and Odour Emissions from Animal Production Facilities,* eds. J.A.M. Voermans and G. Monteny, Rosmalen, Netherlands: Dutch Society of Agr. Eng. pp. 435-440

Miner, J.R. and H. Pan. 1995. A floating permeable blanket to prevent odor escape, in: *Proc. of the Intl. Livestock Odor Conf.,* Ames, Iowa, pp. 28-34.

Misselbrook, T.H., J. Webb, and S.L. Gilhespy. 2006. Ammonia emissions from outdoor concrete yards used by livestock-quantification and mitigation. *Atmos. Environ.* 40(35):6752-6763.

Mobley, H.L.T., M.D. Island, and R.P. Hausinger. 1995. Molecular biology of microbial ureases. *Microbiology Reviews,* 59:451-480.

Molloy, S.P. and H. Tunney. 1983. A laboratory study of ammonia volatilization from cattle and pig slurry. *Irish J. Agr. Res.* 22:37-45.

Monteny, G. J. and J. W. Erisman. 1998. Ammonia emission from dairy cow buildings: a review of measurement techniques, influencing factors and possibilities for reduction. *Neth. J. Agric. Sci.* 46:2258-247.

Morken, J., and S. Sakshaug. 1998. Direct ground injection of livestock waste slurry to avoid ammonia emission. *Nutr. Cycl. Agroecosyst.* 51:59-63.

Mumpton, F.A., and P.H. Fishman. 1977. The application of natural zeolites in animal science and aquaculture. *J. Anim. Sci.* 45(5):1188-1203.

Nahm, K.H. 2003. Current pollution and odor control technologies for poultry production. *Avian Poultry Biol. Rev.* 14:151-174.

Nakaue, H.S., J.K., Koelliker, and M.L. Pierson. 1981. Effect of feeding broilers and the direct application of clinoptilolite (zeolite) on clean and re-used broiler litter on broiler performance and house environment. *Poultry Sci.* 60:1221.

Ndegwa, P.M., A.N. Hristov, J. Arogo, R.E. Sheffield. 2008. A review of ammonia emissions mitigation techniques for concentrated animal feeding operations. *Biosyst. Eng.* 100(4):453-469.

Nannipieri, P. and P. Eldor. 2009. The chemical and functional characterization of soil N and its biotic components. *Soil Biol. Biochem.* 41:2357-2369.

Ni, J. 1999. Mechanistic models of ammonia release from liquid manure: a review. *J. Agric. Eng. Res.* 72:1-17.

NRC (National Research Council). 1996. *National Research Council. Nutrient requirements of beef cattle. Seventh Revised Edition.* National Academy Press, Washington, D.C.

NRC (National Research Council). 2001. *National Research Council. Nutrient requirements of dairy cattle*. Seventh Revised Edition. National Academy Press, Washington, D.C.

O'Connell, J.M., J.J. Callan, and J.V. O'Doherty. 2006. The effect of dietary crude protein level, cereal type and exogenous enzyme supplementation on nutrient digestibility, nitrogen excretion, faecal volatile fatty acid concentration and ammonia emissions from pigs. *Anim. Feed Sci. Technol.* 127:73-88.

Ogink, N.W.M., and W. Kroodsma. 1996. Reduction of ammonia emission from a cow cubicle house by flushing with water or a formalin solution. *J. Agr. Eng. Res.* 63(3):197-204.

Olesen, J.E. and S.G. Sommer. 1993. Modeling effects of wind speed and surface cover on ammonia volatilization from stored pig slurry. *Atmos. Environ. A-Gen.* 27(16):2567-2574.

Olmos Colmenero, J.J., and G.A. Broderick. 2006. Effect of dietary crude protein concentration on milk production and nitrogen utilization in lactating dairy cows. *J. Dairy Sci.* 89:1704-1712.

Otto, E.R., M. Yokoyama, S. Hengemuehle, R.D. Bermuth, T. von Kempen, and N.L. van Trottier. 2003. Ammonia, volatile fatty acids, phenolics, and odor offensiveness in manure from growing pigs fed diets reduced in protein concentration. *J. Anim. Sci.* 81:1754-1763.

Pain, B.F., R.B. Thompson, Y.J. Rees, and J.H. Skinner. 1990. Reducing gaseous losses of nitrogen from cattle slurry applied to grassland by the use of additives. *J. Sci. Food Agr.* 50:141-153.

Pan, P.T., and C.M. Drapcho. 2001. Biological anoxic/aerobic treatment of swine waste for reduction of organic carbon, nitrogen, and odor. *Trans. ASAE* 44(6):1789-1796.

Panetta, D.M., W.J. Powers, H. Xin, B.J. Kerr, and K.J. Stalder. 2006. Nitrogen excretion and ammonia emissions from pigs fed modified diets. *J. Environ. Qual.* 35:1297-1308.

Panetta, D. M., W.J. Powers, and J.C. Lorimor. 2004. Direct measurement of management strategy impacts on ammonia volatilization from swine manure. *ASAE paper number* 044107.

Parker. D.B., S. Pandrangi, L.W. Greene, L.K. Almas, N.A. Cole, M.B. Rhoades, and J.A. Koziel. 2005. Rate and frequency of urease inhibitor application for minimizing ammonia emissions from beef cattle feedyards. *Trans. ASAE* 48:787-793.

Parkhust, C.R., P.B. Hamilton, and G.R. Baughman. 1974. The use of volatile fatty acids for the control of micro-organisms in pine sawdust litter. *Poultry Sci.* 53:801.

Piatkowski, B. and J. Voigt. 1986. Studies on the intestinal protein supply in the cow. *Arch. Anim. Nutr.* 36:222-226.

Portejoie, S., J. Martinez, F. Guiziou, and C.M. Coste. 2003. Effect of covering pig slurry stores on the ammonia emission processes. *Bioresour. Technol.* 87(3):199-207.

Reynal, S. M. and G. A. Broderick. 2005. Effect of dietary level of rumen-degraded protein on production and nitrogen metabolism in lactating dairy cows. *J. Dairy Sci.* 88:4045–4064.

Reynolds, C. K. and N. B. Kristensen. 2008. Nitrogen recycling through the gut and the nitrogen economy of ruminants: An asynchronous symbiosis. *J. Anim. Sci.* 2008. 86(E. Suppl.):E293–E305.

Ro, K.S., A.A. Szogi, M.B. Vanotti, and K.C. Stone. 2007. Process model for ammonia volatilization from anaerobic swine lagoons incorporating varying wind speeds and gas bubbling. *Trans. ASABE* 51(1):259-270.

Rubaek, G.H., K. Henriksen, J. Petersen, B. Ramussen, and S.G. Sommer. 1996. Effects of application technique and anaerobic digestion on gaseous nitrogen loss from animal slurry applied to rye grass (*Lolium perene*). *J. Agr. Sci.* 126:481-492.

Ruiz, R., L. O. Tedeschi, J. C. Marini, D. G. Fox, A. N. Pell, G. Jarvis, and J. B. Russell. 2002. The effect of a ruminal nitrogen (N) deficiency in dairy cows: Evaluation of the Cornell Net Carbohydrate and Protein System ruminal nitrogen deficiency adjustment. *J. Dairy Sci.* 85:2986–2999.

Safley Jr, L.M., D.W. Nelson, and P.W. Westerman. 1983. Conserving manurial nitrogen. Trans. ASAE 30:1166-1170.

Satter, L.D., T.J. Klopfenstein, and G.E. Erickson. 2002. The role of nutrition in reducing nutrient output from ruminants. *J. Anim. Sci.* 80(2):E143-E156.

Sawyer, C.N., and P.L. MaCarty. 1978. *Chemistry for Environmental Engineering* (3rd edition), McGraw-Hill Inc., USA.

Scotford, I.M., and A.G. Williams. 2001. Practicalities, costs and effectiveness of a floating plastic cover to reduce ammonia emissions from a pig slurry lagoon. *J. Agr. Eng. Res.* 80(3):273-281.

Sheridan, B., T. Curran, V. Dodd, and J. Colligan. 2002. Biofiltration of odour and ammonia from a pig unit—a pilot-scale Study. *Biosyst. Eng.* 82(4):441-453.

Shi, Y., D.B. Parker, N.A. Cole, B.W. Auvermann, and J.E. Mehlhorn. 2001. Surface amendments to minimize ammonia emissions from beef cattle feedlots. *Trans. ASAE* 44(3):677–682.

Shin, J.H., S.M. Lee, J.Y. Jung, Y.C. Chung, and S.H. Noh. 2005. Enhanced COD and nitrogen removals for the treatment of swine wastewater by combining submerged membrane bioreactor and anaerobic upflow bed filter reactor. *Process Biochem.* 40(12):3769-3776.

Smith, D.R., P.A. Moore, B.E. Haggard Jr., C.V. Maxwell, T.C. Daniel, K. vanDevander, and M.E. Davis. 2004. Effect of aluminum chloride and dietary phytase on relative ammonia losses from swine manure. *J. Anim. Sci.* 82:605-611.

Smith, K.A., D.R. Jacson, T.H. Misselbrook, B.F. Pain, and R.A. Johnson. 2000. Reduction of ammonia emission by slurry application techniques. *J. Agr. Eng. Res.* 77(3):277-287.

Smits, M.C.J., H. Valk, A. Elzing, and A. Keen. 1995. Effect of protein nutrition on ammonia emission from a cubicle house for dairy cattle. *Livest. Prod. Sci.* 44:147-156.

Sommer, S.G., and O.H. Jacobsen. 1999. Infiltration of slurry liquid and volatilization of ammonia from surface applied pig slurry as affected by soil water content. *J. Agr. Sci.* 132:297-303.

Sommer, S.G., and N.J. Hutchings. 2001. Ammonia emission from field applied manure and its reduction – invited paper. *Eur. J. Agron.* 15(1):1-15.

Sommer, S.G. and N. Hutchings. 1995. Techniques and strategies for the reduction of ammonia emission from agriculture. *Water Air Soil. Pollut.* 85:237-248.

Sommer, S.G., B.T. Christensen, N.E. Nielsen, and J.K. Schjorring. 1993. Ammonia volatilization during storage of cattle and pig slurry: Effect of surface cover. *J. Agr. Sci.* 121:63-71.

Sommer, S.G. and J.E. Olesen. 1991. Effects of dry matter content and temperature on ammonia loss from surface-applied cattle slurry. *J. Environ. Qual.* 20:679–683.

Sommer, S.G., J.E. Olesen, and B.T. Christensen. 1991. Effects of temperature, wind speed and air humidity on ammonia volatilization from surface applied cattle slurry. *J. Agr. Sci.* 117:91-100.

Stevens, R.J., R.J. Laughlin, and J.P. Frost. 1992. Effects of separation, dilution, washing and acidification on ammonia volatilization from surface applied cattle slurry. J. *Agr. Sci.* 1119:383-389.

Stevens, R.J., R.J. Laughlin, and J.P. Frost. 1989. Effect of acidification with sulphuric acid on the volatilization of ammonia from cow and pig slurries. *Camb. J. Agr. Sci.* 113:389-395.

Stewart, G. S. and C. P. Smith. 2005. Urea nitrogen salvage mechanisms and their relevance to ruminants, non-ruminants, and man. *Nitr. Res. Rev.* 18:49-62.

Stewart, K.J., S.P. Lemay, E.M. Barber, C. Laguë, and T. Crowe. 2004. *Experimental manure handling systems for reducing airborne contamination of fecal origin.* ASAE Paper No. 044132. St. Joseph, Michigan.

Stockdale, C.R. 2005. Investigating the interaction between body condition at calving and pre-calving energy and protein nutrition on the early lactation performance of dairy cows. *Aust. J. Exp. Agr.* 45:1507-1518.

Sun, Y., C.J. Clanton, K.A. Janni, and G.L. Malzer. 2000. Sulfur and nitrogen balances in biofilters for odorous gas emission control. *Trans. ASAE* 43(6):1861-1875.

Svensson, L. 1994. Ammonia volatilization following application of livestock manure to arable land. *J. Agr. Eng. Res.* 58(4):241-260.

Swierstra, D., C.R. Braam, and M.C. Smits. 2001. Grooved floor system for cattle housing: ammonia emission reduction and good slip resistance. *Appl. Eng. Agr.* 17(1):85-90.

Swierstra, D., M.C.J. Smits, and W. Kroodsma. 1995. Ammonia emission from cubicle houses for cattle with slatted and solid floors. *J. Agr. Eng. Res.* 62(2):127-132.

Tanaka, A., K. Yakushido, and C. Shimaya. 2003. Adsorption process for odor emission control at a pilot scale dairy manure composting facility. In: *Proc. of Air Pollution from Agricultural Operations III Conference,* North Carolina, USA, pp. 189-196.

Ten-Have, P.J.W., H.C. Willers, and P.J.L. Derikx. 1994. Nitrification and denitrification in an activated-sludge system for supernatant from settled sow manure with molasses as an extra carbon source. *Bioresour. Technol.* 47(2):135-141.

Teye, F.K. and M. Hautala. 2008. Adaptation of an ammonia volatilization model for a naturally ventilated dairy building. *Atmos. Environ.* 42: 4345-4354.

Thompson, R.B., and J.J. Meisinger. 2002. Management factors affecting ammonia volatilization from land-applied cattle slurry in the Mid-Atlantic USA. J. *Environ. Qual.* 31:1329-1338.

Thompson, R.B., B.F. Pain, and D.R. Lockyer. 1990a. Ammonia volatilization from cattle slurry following surface application to grassland. I. Influence of mechanical separation, changes in chemical composition during volatilization, and the presence of grass sward. *Plant Soil* 125, 109-117.

Thompson, R.B., B.F. Pain, and Y.J. Rees. 1990b. Ammonia volatilization from cattle slurry following surface application to grassland. II. Influence of application rate, wind speed and applying slurry in narrow bands. *Plant Soil* 125:119-128.

Thompson, R.B., J.C. Ryden, and D.R. Lockyer. 1987. Fate of nitrogen in cattle slurry following surface application or injection to grassland. *J. Soil Sci.* 38:689-700.

Thomsen, I. K. 2000. C and N transformations in 15N cross-labeled solid ruminant manure during anaerobic and aerobic storage. *Bioresour. Technol.* 72:267-274.

Todd, R.W., N.A. Cole, and R.N. Clark. 2006. Reducing crude protein in beef cattle diet reduces ammonia emissions from artificial feedyard surfaces. *J. Environ. Qual.* 35:404-411.

Todd, M.J. and R.P. Hausinger. 1989. Competitive inhibitors of *Klebsiella aerogenes* Urease: Mechanisms of interaction with the nickel active site. *Biol. Chem.* 264(27):15835-15842.

Townsend, D.A., B.A. Paulsen, and W.J. Wells. 2003. Large scale nitrification denitrification of swine waste. In: *The Ninth International Animal, Agric. & Food Processing Wastes Proc.,* North Carolina, USA, pp. 576-583.

Tucker W B; Hogue J F; Waterman D F; Swenson T S; Xin Z; Hemken R W; Jackson J A; Adams G D; Spicer L J (1991). Role of sulfur and chloride in the dietary cation-anion balance for lactating dairy cows. *J Anim Sci* 69, 1205-1213.

Udert, K.M., T.A. Larsen, M. Biebow, and W. Gujer. 2003. *Urea hydrolysis and precipitation dynamics in a urine-collecting system.*

Van Duinkerken, G., G. André , M. C. J. Smits, G. J. Monteny, and L. B. J. Šebek. 2005. Effect of rumen-degradable protein balance and forage type on bulk milk urea concentration and emission of ammonia from dairy cow houses. *J. Dairy Sci.* 88:1099–1112.

Vander Pol, M., A. N. Hristov, S. Zaman, N. Delano. 2007. Peas can replace soybean meal and corn grain in dairy cow diets. *J. Dairy Sci.* 91:698-703.

Van Der Molen, J., A.C.M. Beljaars, W.J. Chardon, W.A. Jury, H.G. Vanfaasssen. 1990. Ammonia volatilization from arable land after application of cattle slurry. 2. Derivation of a transfer model. Netherlands *J. Agr. Sci.* 38(3):239-254.

Van der Molen, J., H.G. van Fassen, M.Y. Leclerc, R. Vriesema, W.J. Chardon. 1990. Ammonia volatilization from arable land after application of cattle slurry. 1. Field estimates. *Neth. J. Agr. Sci.* 38:145-158.

Vanotti, M.B., and P.G. Hunt. 2000. Nitrification treatment of swine wastewater with acclimated nitrifying sludge immobilized in polymer pellets. *Trans. ASAE* 43(2):405-413.

Varel, V.H., J.A. Nienaber, and H.C. Freetly. 1999. Conservation of nitrogen in cattle feedlot waste with urease inhibitors. *J. Anim. Sc.* 77:1162-1168.

Varel, V.H. 1997. Use of urease inhibitors to control nitrogen loss from livestock waste. *Bioresour. Technol.* 62(1):11-17.

Vavilin, V.A., B. Fernandez, J. Palatsi, and X. Flotats. 2008. Hydrolysis kinetics in anaerobic degradation of particulate organic material: An overview. *Waste Management* 28:939-951.

Velthof, G.L., J.A. Nelemans, O. Oenema, and P.J. Kuikman. 2005. Gaseous nitrogen and carbon losses from pig manure derived from different diets. *J. Environ. Qual.* 34 :698-706.

Vlek, P.L.G. and J.M. Stumpe. 1978. Effects of solution chemistry and environmental conditions on ammonia volatilization losses from aqueous systems. *Soil Sci. Soc. Am. J.* 42:416-421.

Voigt, J., B. Piatkowski, Z. Ceresnakova, R. Krawielitzki, and K. Adam. 1984. *Arch. Anim. Nutr.* 84:387-395.

Voorburg, J.H., and W. Kroodsma. 1992. Volatile emissions of housing systems for cattle. *Livest. Prod. Sci.* 31(1-2):57-70.

Weiss, W. P., L. B. Willett, N. R. St-Pierre, D. C. Borger, T. R. McKelvey, and D. J. Wyatt. 2009. Varying forage type, metabolizable protein concentration, and carbohydrate source affects manure excretion, manure ammonia, and nitrogen metabolism of dairy cows. J. *Dairy Sci.* 92:5607–5619.

Westerman, P.W., and J.R. Bicudo. 2002. Application of mixed and aerated pond for nitrification and denitrification of flushed swine manure. *Appl. Eng. Agr.* 18(3):351–358.

Westerman, P.W., J.R. Bicudo, and A. Kantardjieff. 2000. Upflow biological aerated filters for the treatment of flushed swine manure. *Bioresour. Technol.* 74(3):181-190.

Wickersham, E. A. 2006. *Effect of supplemental protein on nitrogen recycling in beef cattle consuming low-quality forage.* PhD Diss. Kansas State Univ., Manhattan.

Witter, E. 1991. Use of $CaCl_2$ to decrease ammonia volatilization after application of fresh and anaerobic chicken slurry to soil. *J. Soil Sci.* 42:369-380.

Witter, E., and H. Kirchmann. 1989a. Effects of addition of calcium and magnesium salts on ammonia volatilization during manure decomposition. *Plant Soil* 1154:53-58.

Witter, E. and H. Kirchmann. 1989b. Peas, zeolite and basalt as adsorbents of ammoniacal nitrogen during manure decomposition. *Plant Soil* 1154: 43-52.

Xue, S.K., S. Chen, and R.E. Hermann. 1999. Wheat straw-cover for reducing ammonia and hydrogen sulfide emissions from dairy manure storage. Trans. *ASAE* 42(4):1095-1101.

Zahn, J.A., A.E. Tung, B.A. Roberts, and J.L. Hatfield. 2001. Abatement of ammonia and hydrogen sulfide emissions from a swine lagoon using a polymer biocover. *J. Air Waste Manag. Assoc.* 51:562-573.

Zhang, B., P. He, F. Lu, L. Shao, and P. Wang. 2007. Extracellular enzyme activities during regulated hydrolysis of high-solid organic wastes. *Water Res.* 41:4468-4478.

Zhang, Z., J. Zhu, J. King, and W. Li. 2006. A two-step fed SBR for treating swine manure. *Process Biochem.* 41(4):892-900.

In: Environmental Chemistry of Animal Manure
Editor: Zhongqi He

ISBN: 978-1-62808-641-6
© 2013 Nova Science Publishers, Inc.

Chapter 7

ORIGINS AND IDENTITIES OF KEY MANURE ODOR COMPONENTS

Daniel N. Miller and Vincent H. Varel*

7.1. MANURE ENVIRONMENTAL ISSUES

Significant increases in animal and crop production efficiency in the last century have radically changed the way that food is produced, yielding an abundance of nutritious and inexpensive food for consumers, but not without new challenges on the farm. Historically, agriculture was a highly integrated production system where crops raised on the farm were fed to the animals, and the valuable manure nutrients needed for crop production were returned to the same fields that produced the feed. Agriculture now has become much more specialized with crop production largely dependent upon chemical inputs on one farm and transport of feedstuffs to another farm specializing in concentrated animal production. Manure is viewed more frequently as a waste and not a valuable fertilizer resource. To handle the larger numbers of livestock, specialized housing and production areas (confined animal feeding operations or CAFO) have been developed and become a standard production practice. The CAFO has been utilized for many familiar animal species, including beef cattle, dairy, swine, and poultry, and for the less familiar such as buffalo, elk, and alligator. The design and size of CAFO ranges greatly and is influenced by species, climatic region, local practices, and regulations (from local to federal).

One challenge common to all CAFO is how to best manage the large volumes of concentrated manure, which is a complicated mixture of urine and feces that potentially contains bedding material, spilled feed, soil, insects, and excess water. Typically, the manure is in a highly concentrated form and accumulates in either a manure storage structure or on the soil surface, as seen with cattle feedlots. During storage, manure composition changes, either purposefully through engineered treatment systems or naturally through the activity of manure and soil microorganisms. A wide range of manure treatment systems are available.

---

* Email: dan.miller@ars.usda.gov USDA-ARS, Lincoln, NE 68583, USA

Most commonly in the Midwest, the treated manure or effluent is utilized for crop or forage production, based upon its nitrogen (N) or phosphorus (P) content. In other areas of the U.S., land is unavailable for manure application, and therefore animal producers rely more heavily upon manure treatment to remove excess N and P nutrients prior to land application in order to stay within prescribed land application guidelines for N and P.

A variety of environmental issues, such as gas emissions, pathogens, nutrients, and pharmaceutically active compounds, can arise at any point during the production, handling, storage, and application of manure. Gas emissions from the manure or manure/soil surface are a prominent issue due to their effects at local and global scales and include greenhouse gases (CO_2, CH_4, and N_2O), volatile nitrogen compounds (NH_3 and methylamine), volatile sulfur compounds (H_2S, and various methyl sulfides), and a multitude of volatile organic compounds (VOC). Zoonotic pathogens capable of infecting animals and humans are often detected in manures, and therefore proper precautions involving manure application to crops need to be taken to avoid human and livestock exposure. Similarly, when manure is applied to fields, nutrient losses by runoff after an extreme rainfall event occurs could affect local surface and ground waters leading to eutrophication. Recent attention to the trace levels of hormones and antibiotics in manure is also a concern given the large manure volumes generated in a typical CAFO. A common thread connecting all of these manure environmental issues is the role of microorganisms. Microbes can be directly involved through potential exposure of manure pathogens to crops, or they may be involved by virtue of their activities, as in the production of greenhouse gases during manure decomposition or nutrient transformations. The scope of this chapter is by necessity narrow and focuses exclusively on manure odor compounds, but many obvious intersections with other environmental issues will be evident, particularly with greenhouse gas production and other gaseous (potentially odorous) emissions.

7.2. ODOR SOURCES IN ANIMAL AGRICULTURE

There are a variety of sources in the animal production environment that are associated with odors, yet not all of them are directly related to animal excreta. One large, non manure source occurs during the storage and preparation of feedstuffs. Silage is a common cattle feed made from a variety of forages and grains that undergo a limited acid fermentation during storage. This fermentation preserves the feed and enables cattle feeders to provide feed long after the fresh feed would have spoiled. The odor associated with silage is often described as a sharp, sour odor, but not highly objectionable. Another odor source that is feed-related, but on the opposite end of the moisture spectrum are odor issues associated with the preparation of dried feed ingredients. Soybeans and grains need to be fragmented into smaller particles for animals to get the most nutrition and energy out of the feed. In the process dust can be generated, which when inhaled will produce an odor response by the human nose. Further processing of the ground feed stocks by cooking or addition of other ingredients, such as fats and oils, to bind the feed will yield additional odors. The particular odors of grain and pellet processing are again distinctive, but not considered highly offensive.

The most objectionable odor sources in the animal production environment are areas and structures associated with fresh and stored manure. Manure storage and treatment structures seem to be the primary source and come in all shapes and sizes with multiple points (both physical and temporal) where odors may be emitted during animal production. Taking one example, a swine production system, there are several options for manure handling that generally vary in the amount of water used to dilute the manure. Deep pit manure storage is at one end of the spectrum and dilutes manure very little. On the other end of the spectrum are flush systems where a considerable amount of water is used to move manure to an outside storage structure. In the deep pit system, feces and urine make their way through slatted floors into a pit beneath the production area, where several fans pull air off the slurry surface and vent the gases outside the building. Periodically, the pits are mixed and the slurry is pumped into large tanker trucks or applied via a hose-drag system which transports the manure to crop fields. Taking this particular example, manure odors are emitted in several distinct stages. First, feces and urine are voided from the animal, but not necessarily at the same time and on the same spot. The urine, which already contains some odorous nitrogen and organic compounds, is briefly in contact with the slatted floor where microbes present on the surface begin breaking down substrates in the urine. As the fresh manure works its way through the slatted floor (a much more time-consuming process compared to the urine), gases and odor compounds generated inside the digestive system of the hog during normal digestion processes are emitted to the atmosphere. The proportion of odor emitted in this stage, relative to whole barn emission, is unknown, but it may make up a large fraction of the emission. Once in the slurry pit, the manure undergoes further degassing and decomposition in a protected, warm enclosure with odorous compounds volatilizing at a relatively constant rate. After a swine production cycle, the aged manure slurry is mixed in the pit with the result being a very large out-gassing of odor compounds. Out-gassing of H_2S during this period is a real danger and has been known to asphyxiate workers. A priority of the swine industry is to educate workers about this risk. Finally, the choice of field application method will significantly impact manure odor emissions. Surface application by spraying exposes the manure to the air and produces the largest odorous compound emissions, both during the spraying event and as the manure dries on the soil surface. Another less offensive method is subsurface slurry incorporation where manure is applied beneath the soil surface.

In the swine deep pit production example, there were multiple points where odors were emitted (excretion, storage, mixing prior to field application, and finally field application). Many variations of this deep pit production system are available, each slightly changing the potential for odor emission. Additionally, other swine production systems are available and utilize different manure handling, treatment, and application systems, (i.e., a flush system to an anaerobic lagoon with center pivot spray application or an engineered anaerobic digester followed by solids composting, or low intensity aeration of slurry in lagoons), and hence the potential sources of odor emission quickly become even more complex. Finally, if the multiple livestock species reared on CAFO are considered, solutions to every possible combination of sites emitting odor becomes very challenging. Some progress towards finding odor solutions for the wide variation in potential sources can be made by understanding the process at deeper mechanistic and biochemical levels, but an understanding of odor, how it's measured, and its chemical basis, is necessary.

7.3. ODOR—DESCRIPTION, QUANTIFICATION, ANALYSIS

The first generation of methods developed to measure and describe odor used a number of parameters. The basis of these methods relies upon the noses of human panelists to detect and characterize the odor. Even today when the science of olfactometry (i.e., measurement of the response of human assessors to olfactory stimuli) is well developed, the human nose is often considered the final authority for odor research with a litany of standard practices and terminology. At its most basic level, odor is described using four major descriptors, which are frequency, intensity, duration and offensiveness (FIDO). Frequency relates to how often a particular odor is detected, intensity relates to the perceived magnitude of an odor stimulus, duration is the period of time that the odor is experienced, and offensiveness relates to how unpleasant the odor is sensed. Frequency and duration are measured in terms of time scales, whereas intensity and offensiveness are often rated on subjective scales. Hedonic tone is a property of odor similar to offensiveness, but it is slightly broader in scope. Rated on a similar arbitrary scale, offensive odors would be rated with large negative values and very pleasant odors would be rated with large positive numbers. Thus, scaling various odor attributes assigned by human panels is one method to measure and compare odor.

An alternative to arbitrary scaling methods are methods that rely upon (and quantify) how much dilution with odor-free air is needed to either render an odorous air sample unrecognizable (recognition threshold) or to dilute the odorous air sample to a level where no odor is detected (detection threshold). Like the various scaling methods, these dilution methods also rely upon humans as the sensor and have been developed for laboratory and field applications. Typically laboratory dilution olfactometry utilizes forced-choice olfactometry. In this method, a bag of odorous air is mixed with charcoal-scrubbed (odor-free) air and presented to a panelist as one of three possible choices (the two other choices consist of just odor-free air. The panelist is then tasked with identifying which of the three samples contains the odor sample. As the dilution increases, a threshold is reached where the panelists are no longer able to accurately pick the odorous air sample from the odor-free air samples, and the detection threshold is recorded.

Both scaling and dilution olfactometry rely upon the human nose as a detector, but not all noses are the same and individual perception of odors change from day-to-day. Age, sex, health, personal habits (smoking, for example), and personal history all affect the perception of odor and what may be easily detectible by a young non-smoker may not be detectible by an older smoker suffering from a head cold! One important practice to manage nose-to-nose variability is through the use of reference odorants. Typically *n*-butanol is used, since it is widely available, stable, and relatively nontoxic at the concentrations used in olfactometry. A range of concentrations of *n*-butanol in water can be used to train panelist to a specific reference scale. For dilution olfactometry, a target *n*-butanol concentration for detection may be used to screen panelists for selection. Beyond the human variation in nose acuity, other factors including olfactory saturation and fatigue can affect odor measurements during an olfactory panel. Recognizing and minimizing all the day-to-day and nose-to-nose variation is no easy task, thus olfactometry is expensive and typically conducted in specialized laboratories that contract with research and regulatory entities to measure specific samples.

Over the past several years, field and laboratory techniques have been developed to sample and quantify specific odor compounds in manure, manure slurry, soils, compost, and air. Most sample matrices (except air) are amenable to direct extraction, provided that the target odor compounds are abundant enough. Typical manure extractions involve water, alcohols, ether, or a variety of other organic solvents. For air, a concentration step during sampling is usually necessary. Anyone working in animal production environments is familiar with the odors that linger on clothing. In fact, an early method to capture and analyze odor compounds involved exposing cloth swatches to odorous environments followed by laboratory olfactometry or extraction and/or chemical analysis. This method has seen recent modification with the use of steel plates and solid phase microextraction (Bulliner et al., 2006). Today, sorption methods have become more refined utilizing specific adsorbants (Tenax, carboxen, divinylbenzene, polydimethylsiloxane, etc.) and controlled exposures to sample odorous air or manure samples (Loughrin et al., 2008; Miller and Woodbury, 2006; Shabtay et al., 2009).

Sorption tubes or bars and solid phase micro extraction (SPME) are currently the best options to concentrate diffuse odor compound molecules to the level where the sample can be introduced to laboratory equipment for identification and quantification. Gas chromatography is most commonly used to separate and quantify specific odorant compounds in an odor sample. A variety of chromatography columns over a range of hydrophobicity is usually used to separate individual odor compounds based upon their chemical properties. Several choices available for detecting the separated compounds include thermal conductivity, flame ionization, electron capture, and mass selective detectors. Each detector has its strengths and weaknesses, but one of the most versatile detectors employed is the mass selective spectra detector. This detector enables not only detection and quantification of odor compounds but also identification or confirmation of the odor compound based upon the fragmentation of the parent molecule in the detector. One recent innovation in odor compound detection and quantification merges laboratory gas chromatography and human-based olfactometry (Lo et al., 2008; Wright et al., 2005). This technique places a sniffing port downstream of the chromatography column and splits the stream in two with one portion going to a standard detector and the other to the sniffing port. This enables an operator to rate individual odorant molecules on their intensity, offensiveness, or odor character.

Although laboratory olfactometry and gas chromatography instruments provide a versatile way to analyze odor in a controlled environment, there is a strong need for field instrumentation. The need for field instruments is based upon the reactive nature of specific odor compounds, which react with plastic in Tedlar bags (Keener et al., 2002; Trabue et al., 2006) or react with oxygen (or other constituents) in the atmosphere. However, environmental variations in temperature, humidity, and wind speed may induce some variation between otherwise identical odor measurements, which are controlled in the laboratory. Field instruments utilize similar detection methods. For human olfactometry, ambient air is diluted and mixed with charcoal-scrubbed air in a facemask, and odor is detected by a trained human nose. Typically, a dial is used to regulate the ratio of ambient air to clean air to a level where the odor is undetectable, and thus a detection threshold is obtained. Field biochemical methods utilize hand-held electronic detectors (most commonly ionization) that can be coupled to a miniature gas chromatograph. Electronic nose technology is an offshoot of the biochemical detector technology that utilizes an array of sensors to provide a signal that can be compared to existing signal patters. Utilizing neural networks, the electronic nose can 'learn' various odors (swine versus cattle) and in theory provide much more repeatable measurements of odor compared to human panelists. Initial poor capabilities of the electronic nose to predict manure odor concentration, as determined by human panelists, have shown considerable predictive improvement (Gralapp et al., 2001; Qu et al., 2008). However, additional obstacles affecting the sensors in electronic noses, such as contamination, fatigue, and life span, need to be overcome before the technology becomes a routine method for odor measurement and characterization.

7.4. ODOR CHEMISTRY

Manure odors are a complex mixture of chemicals possessing properties that would not normally be predicted based upon the concentration and properties of the individual chemicals. Even simple mixtures of two or three chemicals may behave in unpredictable ways. Two very offensive odor molecules may either cancel out each other, or drastically enhance the perceived offensiveness, or even produce a pleasing fragrance. Trying to predict perceived odor properties based upon a known mixture of gases is exceedingly difficult. Nevertheless, research into the types of chemicals, their abundances, and the human response has provided some insights into how chemistry and perceived odor are related to one another. To understand this relationship, it is necessary to understand the chemicals encountered in manure and odor.

A wide variety of chemicals have been detected in solid, liquid, and gaseous odorous samples associated with livestock production. Lists of compounds and the locations they were detected have been published elsewhere (Hartung and Phillips, 1994; Le et al., 2005b; O'Neill and Phillips, 1992; Schiffman et al., 2001), but new odor compounds continue to be identified (Lo et al., 2008). Major constituents can be grouped into five general groups including volatile fatty acids (VFAs), alcohols, aromatic compounds, sulfides, and amines/ammonia. Minor constituents include other VOC, such as ketones, esters, ethers, and aldehydes. The most offensive odor compounds arise within the animal during feed digestion and subsequently during manure storage and are the product of an incomplete anaerobic microbial fermentation of substrates (Le et al., 2005b; Mackie et al., 1998; Rappert and Müller, 2005) by a very diverse group of microorganisms (Cotta et al., 2003; Ouwerkerk and Klieve, 2001). During complete anaerobic fermentation, organic matter is converted to CO_2, CH_4, H_2S, NH_3, H_2O and new microbial biomass in a three step process (Figure 7. 1). In the first step, complex organic matter, composed of polymers, is first broken down by extracellular enzymes into smaller monomers and dimers, which can then be transported inside microbial cells for further use in metabolic (energy harvesting and cell building) activities. Many of the microbes involved in this first step would be grouped as acidogens (acid-formers) based upon their fermentation products. In the second step, end products from acidogens (and other types of microorganisms) are then utilized as substrates by another guild of microbes, H_2-utilizing acetogenic bacteria (acetogens). Finally, in the third step, very simple compounds like alcohols, formate, acetate, H_2, and CO_2 are used by strictly anaerobic, methanogenic Archaea to produce CH_4 gas. To a very limited extent, this process occurs in nearly every digestive tract (from termites to dairy cows) and manure-impacted environments (feedlot surfaces to swine manure pit) as evidenced by the emission of CH_4 from these systems. The problem of incomplete anaerobic fermentation arises in that the activity of some guilds of microorganisms (acidogens) outpace the capacity of another group (methanogens) to consume the first group's end products (volatile malodorous compounds, according to the human nose), leading to an accumulation of products and a potential for emission. Why methanogens activity or abundance is limited in some manure-impacted environments may be due to a combination of factors including their slow growth rates and environmental conditions. In cattle feedlots, the environmental conditions are very dynamic with fluctuating oxygen, pH, and moisture regimes that limit the growth and activity of specialized, strictly anaerobic microorganisms like methanogens. In swine manure pits and lagoons, the conditions are not as dynamic, and a large population of methanogens will develop that consumes a large fraction of the VFAs and produces copious methane.

7.4.1. Volatile Fatty Acids

Volatile fatty acids (VFAs) are defined as short-chain (C_2 through C_6) carboxylic acids and are produced from carbohydrates during acidogenesis and acetogenesis. Both straight-chain and branched-chain VFAs are consistently detected in fresh excreta and aged manure. Strictly speaking, acetogenesis involved the transformation of fatty acids greater than two carbon chains long (i.e., propionate), alcohols, and other organic acids, such as lactic acid, into acetate, whereas a wide variety of fermentation products, including VFAs, are produced during acidogenesis. To some extent amino acid fermentation can also contribute to VFAs formation through a deamination reaction that yields VFAs, NH_3, and H_2. Lactic acid is one of the principal fermentation products found in beef cattle manures. Although it is not strictly classified as a volatile fatty acid, it is produced during acidogenesis largely from carbohydrate fermentation. Functionally similar to propionate, it is a C_3 carboxylic acid containing a hydroxyl group at the C_2 carbon. It has a much lower pK_a compared to VFAs, and hence, it is not considered volatile under normal pH ranges experienced in the digestive tract or in manure-impacted environments.

7.4.2. Alcohols

A number of different alcohols are formed during manure fermentation, including both straight-chain and branched-chain varieties. The principal alcohols include ethanol and propanol with traces of butanol. Isobutanol has also been detected, but at very low abundances. In general, alcohol is more abundant in beef cattle manure than in swine manure when slurried and incubated.

7.4.3. Aromatic Compounds

A wide variety of compounds containing aromatic benzene rings have been detected in manures. These compounds are divided into two groups, phenols with a single six-carbon benzene ring bearing a hydroxyl group and indoles that contain a benzene ring fused to a nitrogen heterocyclic pyrrole ring. Several variations on the structures include the addition of methyl or ethyl groups (*p*-cresol, 4-ethyl phenol, and skatole), or carboxylic acids (benzoic acid, phenylacetate, and phenylpropionate) to the base aromatic ring structure. Most of these compounds have very strong, distinct odors and are strongly lipophilic. All aromatic odor compounds share a similar origin—fermentation of aromatic amino acids. The three parent aromatic amino acids are tyrosine, phenylalanine, and tryptophan.

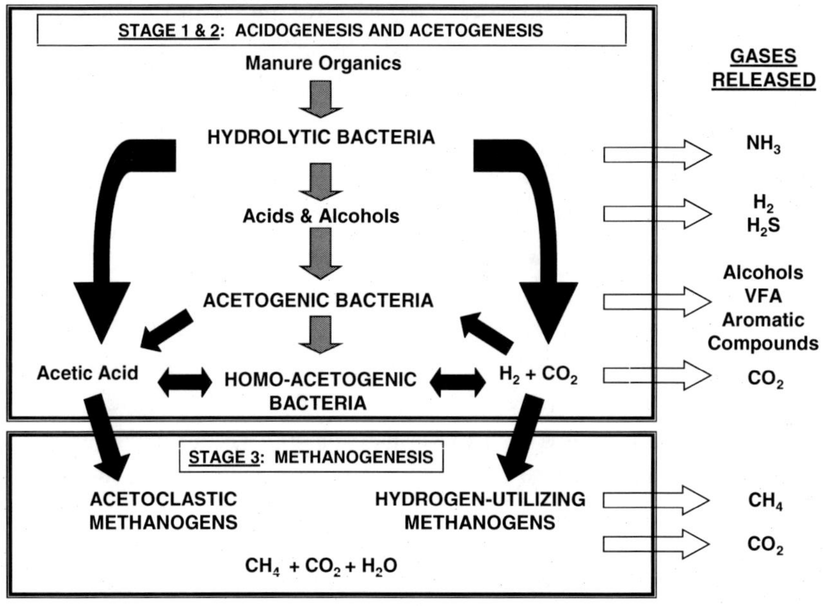

Figure 7.1. Anaerobic fermentation pathways (Watts et al., 1994).

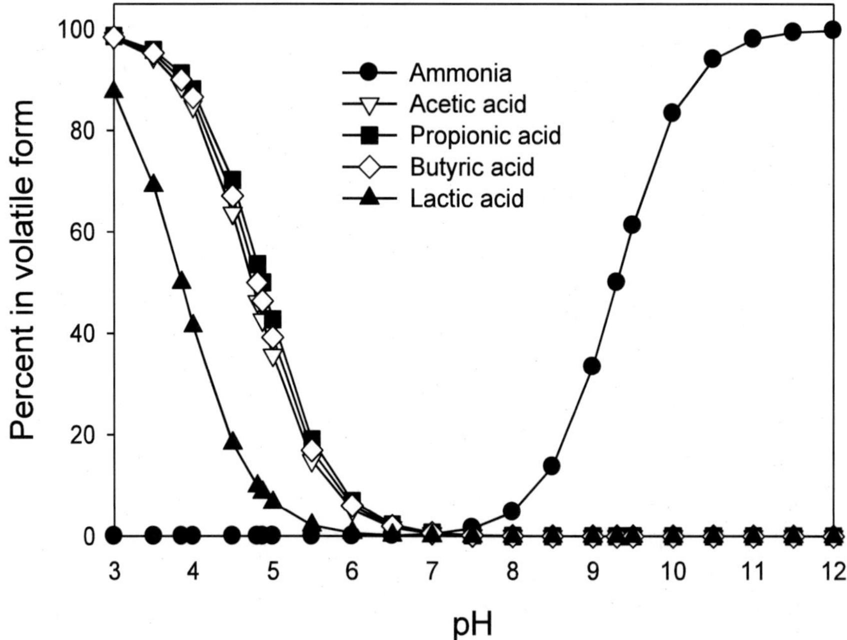

Figure 7.2. Dependence of volatile manure components on pH. Curves were developed based upon the relationship between pH and the acid disassociation constants of the various compounds.

7.4.4. Sulfides

Reduced sulfur compounds, such as H_2S and methylsulfides, have very low odor detection thresholds (ppb) and are characteristic of a 'rotten eggs' smell. These reduced sulfur compounds can be formed in two ways, i) by assimilatory and dissimilatory sulfate reduction or ii) through amino acid fermentation. In sulfate reduction, microorganisms utilize exogenous sulfate either for energy production as the anaerobic electron acceptor (dissimilatory) or as a source of sulfur for synthesis of the reduced sulfur-containing amino acids methionine and cysteine (assimilatory). The amino acid fermentation mechanism for generating malodorous sulfide compounds is similar to the mechanism for the formation of aromatic odor compounds—deamination of the amino acids methionine and cysteine, which already contain reduced sulfur in their side chains, yields H_2S and methylsulfide. Longer chain methylsulfides, such as dimethyl trisulfide, are also emitted from manures.

7.4.5. Amines/ammonia

The most abundant nitrogen-containing odor compound associated with animal manures is ammonia, but a number of other amines (organic compounds containing reduced nitrogen groups) have also been detected. Under normal manure and soil pHs, ammonia is principally in the ammonium (NH_4^+) form. It is formed through the action of urease, which converts urea in the urine into CO_2 and NH_3. Ureases are produced by a wide variety of manure and soil microorganisms and are widely distributed in manure-impacted environments. In fact, urea is rarely detected in these environments, and only when conditions are very cold or dry. Amines, such as methylamine, have also been detected in manure-impacted environments, and these amines are formed during amino acid fermentation via a decarboxylation reaction.

7.4.6. Other

A number of other odor compounds (ketones, esters, ethers, and aldehydes) have been detected at very low concentrations in manure or in the atmospheres of animal production sites. Some of these compounds, such as ketones, are undoubtedly produced through fermentation by specific microbial communities, but other products are not easily identifiable as fermentation products. The most probable explanation is that these odor compounds are secondary products that have formed via reaction with oxygen and other odor compounds in the manure. One example is the formation of esters from volatile fatty acids and alcohols under acidic conditions. Another suite of odorous compounds are also produced when reactive sulfides oxidize in the atmosphere. A third source contributing to the mix of odor compounds at animal production sites are cleaning compounds utilized during animal production, specifically pinenes. Together, these fermentation products, chemicals in cleaning products, and secondary products generated in manures contribute to the overall manure odor of animal production sites.

From the very initial chemical analyses of manure odor compounds, considerable effort has been made to identify which compound(s) are to blame for 'manure odor'. Although H_2S and NH_3 correlate with odor measurements from human panels, a large portion of the variation is not explained by either of these gases. Further research into volatile organic compounds, specifically aromatic compounds like skatole, expanded the search for a master odorant correlating with odor. However, as manure odor field and laboratory research progress, the scientific consensus has coalesced around the idea that no single compound is responsible for manure odor, rather the mixture of several odorant molecules produces in humans the perception of manure odor (Spoelstra, 1980; Yasuhara, 1980). Current research shows that mixtures of aromatic compounds and VFAs most closely correlate to odor (Powers et al., 1999; Zahn et al., 1997; Zahn et al., 2001; Zhu et al., 1999). Recognizing that multiple compounds are responsible for manure odor has led to the formulation of artificial manure odor mixtures containing VFAs and aromatic compounds, which are useful for developing more controlled emission experiments and for training human odor panelists (Qu and Feddes, 2007). An important aspect of odor emission to note is that the chemical composition of an odor plume changes with distance from a manure site. Recent research indicates that aromatic compounds, specifically *p*-cresol, may be more important at longer distances from a site compared to VFAs, which are important near manure storage/accumulation sites (Koziel et al., 2006).

It is important to recognize that the composition and relative abundance of odor compounds within the atmospheres of animal production sites will not exactly mirror the composition and relative abundance of odor compounds within the manures. A number of physical and chemical factors will influence the emission of odor compounds. One of the more obvious chemical factors is pH (Figure 7. 2). Based upon the pH of the medium, the fraction of volatile fatty acid and ammonia available for emission will change—a property of the acid/base disassociation constant (Dewes et al., 1990; Miller and Berry, 2005). The normal pH of manures ranges from 6 to 8.5, but atypical results outside this neutral range may be observed. Another very important chemical attribute of odor compounds is that their vapor pressures vary over a wide range, which affects how easily an odor compound in the liquid phase will go into the gas phase. Two important physical factors affecting all manure emissions are temperature and air speed above the manure surface (Zhu et al., 2000). Increasing either one will enhance the emission of odor compounds by providing energy needed to vaporize the odor compounds or by reducing diffusive barriers (or partial pressure of particular odor compounds) at the manure surface. Since some physical parameters change through time, either naturally or purposely through management, the emission of odor compounds will be dynamic.

7.5. BIOCHEMICAL ORIGINS OF ODOR COMPOUNDS

It is important to recognize that manure odor formation is a two-phase process. In the first phase, odor compounds are produced within the digestive tracts of animals where an anaerobic microbial community utilizes substrates flowing through the gut. In cattle, the benefit of this microbial relationship is very strong in that cattle rely upon microbes in the rumen and large intestine to convert complex carbohydrates (cellulose and starch) into VFAs that will be adsorbed and utilized for energy. In swine, the animal is not as overtly dependent upon microbes to provide the basics for life, but swine likely benefit from microbial byproducts in the lower intestine, during which odor compounds are formed. The second phase of odor compound formation occurs outside the animal where conditions may or may not be favorable for complete anaerobic fermentation of substrates in the accumulating manure. The proportion of odor compounds formed during the first (internal) and secondary (external) odor production phases and their relative contribution to air quality has not been fully described. Under optimum laboratory conditions, the content of odor compounds in incubated fresh manures will increase several fold. Whether or not these 'optimum' conditions are prevalent in manure-impacted environments is debatable, and hence this is an excellent area for future odor research.

As described in a previous section of this chapter, microorganisms can metabolize a variety of potential substrates to produce odorous compounds. The range of available substrates in the manure includes starch, proteins, lipids, and nonstarch carbohydrates (like cellulose and hemicellulose). The availability of each type of substrate differs; protein and starch are much more water soluble than lipids or cellulose and hence, more available to microorganisms. The relative amounts of substrates in manures differ among animal species, which is the result of not only different diets fed to the animals but more importantly a function of differences in their respective digestive strategies, i.e., rumen versus monogastric. Feedlot cattle are typically fed a diet rich in starch energy and based upon locally produced grains or forage component. The grain is usually processed (dry rolled/cracked, steam flaked, high moisture harvested) to make the starches more available to digestion in order to optimize the feed conversion of the animal. Over-processing of grains, poor bunk management, or other factors may shift ruminal fermentation towards increased production of lactic acid which can potentially induce lactic acidosis in the animal, which decreases animal production efficiency. Swine and poultry are typically fed a diet that is higher in protein, and more finely ground. Aside from this major difference, the diets fed to each animal species will differ across regions because available grains and forages vary from region to region. Finally, individual animals will usually be fed a range of diets during their production cycle in order to transition between pasture to feedlot or as their nutritional requirements change during maturation.

Fresh manure composition has typically focused on elemental analysis and gross measures of organic matter content. While this information is valuable for nutrient management plans and for wastewater treatment, it lacks the detail needed to understand which components of the organic matter pool are fueling odor production during manure accumulation or storage. Two lines of evidence—loss of substrate through time and production of protein-specific fermentation products—have been used to establish the relationships between odor compound production and the manure substrates used to fuel odor compound production.

Direct measures of substrates (carbohydrates and proteins) in manures and soils are complicated by the diverse matrix of interfering substances, but methods modified from techniques used to analyze fiber content in animal feeds have been developed which offer insights into manure substrate composition (Miller et al., 2006). These protocols utilize gravimetric and enzymatic methods to separate substrates based upon their solubility in hot detergents. For starch, dried, ground manure samples are heated to dissolve starch (and some other components). Solubilized starch is then specifically hydrolyzed using amyloglucosidase, and the resultant free dissolved glucose can then be quantified in the solution phase. Cellulose and other non-soluble organic molecules can be quantified by gravimetric loss after filtration, drying, and ignition. Measuring the protein content of manures is also done through a roundabout method, since some manure constituents interfere with direct colorimetric protein methods, such as the Lowry and Bradford protein assay. Another technique borrowed from nutritionists is the crude protein method where total nitrogen is measured in a combusted sample and then multiplied by 6.25, a number that factors in the amount of N in protein. Since manures can have large quantities of ammonia, the amount of crude protein would be overestimated. To avoid this pitfall, manures can be made alkali and then dried to drive off ammonia prior to combustion analysis. Nitrate can also interfere with protein estimations in cattle feedlot manure/soil mixtures. In this case, it is better to divide a sample, extract and quantify inorganic nitrogen compounds in one subsample, and measure total N (inorganic and organic) in the other subsample.

Relying solely on changes in substrate concentrations during manure incubations may not be enough to determine which substrates are fueling odor compound production. Unique insights into the process can be made by examining the odor compounds in the fresh manure sample and during anaerobic incubation. The accumulation of branched-chain VFAs and aromatic compounds provides a clear indication that protein is available and being utilized by the resident microbial community, since these specific products are only formed during protein fermentation (Le et al., 2005a; Mackie et al., 1998; Rappert and Müller, 2005). However, not all protein is fermented to branched-chain and aromatic amino acids. Studies of mixed culture microbial fermentations that rely solely on protein for energy production and cell biosynthesis found that 16 to 21% of the total VFAs was branched-chain VFAs (Macfarlane et al., 1992; Smith and Macfarlane, 1998). Using this molar ratio of branched-chain VFAs to total VFAs can provide insights into the importance of protein fermentation relative to carbohydrate fermentation. For instance, if 8% of the total VFAs were branched-chain VFAs, one could estimate that 40 to 50% of the substrate fermented was protein.

Laboratory incubations of manure slurries that monitor specific substrates have identified clear patterns in substrate utilization and specific odor compounds indicating which substrates are utilized and the extent of utilization (Archibeque et al., 2007; Miller and Varel, 2001; Miller and Varel, 2002; Miller and Varel, 2003; Miller et al., 2006). In beef cattle manures from animal fed a dry-rolled corn (DRC) diet, starch seems to be preferred even when other substrates are available; addition of starch to beef cattle manures prevented protein fermentation, and no malodorous branched-chain VFAs or aromatic compounds accumulated (Miller and Varel, 2001; Miller and Varel, 2002). Odor compound accumulation in cattle manure appears to be self-limiting due to lactate accumulation and pH-associated VFA toxicity. Even in the aged manure from pens where beef cattle were fed DRC, starch was still utilized, and only a small amount of protein fermentation occurred in manure slurries. Furthermore, adding starch to aged, starch-depleted DRC-fed cattle manures temporarily circumvented the low rate of protein fermentation. For the microbial communities in DRC manure, starch is clearly the king.

Because beef cattle transitioning from pasture to feedlot are unable to go immediately to a high-grain finishing diet, they may be fed a wide variety of forage-based diets including silages and hays. Laboratory incubation studies have examined manures from beef cattle fed either corn silage or brome grass hay diets prior to a DRC-based finishing diet. These studies determined that lactic acid and VFAs accumulated in the DRC and corn silage manures at the expense of starch, based on starch loss and the production of exclusively straight-chain VFAs (Miller et al., 2006). In the brome grass manure, however, accumulation of branched-chain VFAs and aromatic compounds and the low starch availability indicated that protein in the feces was the primary source for odor compound production. Substrate additions of starch, protein, and cellulose in additional manure slurries confirmed that starch availability was the primary factor determining accumulation and composition of malodorous fermentation products, but when starch was unavailable, as in the brome grass diet, protein was utilized by manure microorganisms.

A major shift in beef cattle nutrition has occurred as ethanol production has claimed a larger and larger portion of grain production. Most cattle feedlot operations in the northern Great Plains and many in the southern Great Plains are replacing a portion of the processed grains with the by-products of ethanol production, namely distillers' grains. Because of this shift in ingredients, finishing diets contain less starch and more protein than the finishing diets fed ten years ago (Vasconcelos and Galyean, 2007). As expected, these changes in diet have also impacted odor compound composition in the fresh and in incubated manures (Hao et al., 2009; Spiehs and Varel, 2009; Varel et al., 2008). Although microbial substrates were not measured in these studies, the composition of fermentation products describes a clear picture of protein fermentation playing a larger role as the percentage of distillers' byproducts increased in the diet. Fresh manure from animals fed high concentrations of distillers' byproducts (35% or greater) contained more branched-chain VFAs than animals fed lower concentrations of distillers' byproducts, and when the manure was slurried and incubated, the proportion of branched-chain VFAs and aromatic compounds increased proportionate to the amount of distillers' byproducts in the diet. As less starch (and more protein) was fed to the animal in the distiller's diet, products of protein fermentation accumulated.

Understanding fermentation reactions in beef cattle manure slurries is an important step toward understanding how manure odors are produced in feedlots, but slurry conditions are rarely observed in the feedlot. Incubation studies of manure mixed into soils provides a better approximation of real world conditions. Fortunately, soil incubation studies support the findings of the slurry studies (Berry and Miller, 2005; Miller and Berry, 2005). There are clear losses of starch when the soils were completely saturated, but no change was observed in organic nitrogen content.

Comparing fresh and fermented beef cattle manure with fresh and fermented swine manure illustrates the differences that digestion and substrate availability have on odor compounds (Table 7.1). In strong contrast to beef cattle manure from animals fed DRC diets, starch was not the predominant substrate in fresh swine manure. Initial starch content in cattle feedlot manure from cattle fed DRC-based diets was nearly six-fold greater than the starch content of swine manure (Miller and Varel, 2003). For organic nitrogen (protein), the situation was reversed, with initial protein content in the swine manure twice as high as in the cattle feedlot manure. Results of swine manure incubations consistently indicate that a very different process is responsible for the accumulation of odor compounds in swine manures (Figure 7.3). Examining the trends in substrate consumption during slurry incubations indicate that both protein and starch are utilized by microorganisms in the swine slurries, whereas protein concentrations actually increased in the cattle manure slurry incubations due to the large loss of other organic nutrients such as starch. The final composition of fermentation products (Table 7.1) confirms the importance of protein fermentation during swine manure incubation—both branched-chain VFAs and aromatic compounds are enriched relative to straight-chain VFAs (acetate, propionate, and butyrate). Some other interesting features are the higher concentrations of lactate, alcohols, and butyrate in the cattle manure slurries compared to the swine, which may be indicative of starch as the original substrate.

High concentrations of lactate in the cattle manure slurries would also act to limit further production of odor compounds by contributing to VFA toxicity, where a drastic increase in VFAs concentration leads to a drastic drop in pH (Switzenbaum et al., 1990). These differences in butyrate, branched-chain VFAs, and aromatic compound content are likely to impact differences in perceived odor, and these differences are directly attributable to the action of microorganisms on available substrate in the manure. Thus, substrate availability will likely 1) influence the diversity and function of the manure microbial community and 2) have a direct impact on the composition of odorous fermentation products produced in the manure.

Substrate composition and availability in the manure likely select for certain fermentative bacteria capable of utilizing available substrates. *Clostridium, Lactobacillus, Bacillus,* and *Streptococcus* microorganisms are four of the dominant groups of microorganisms in cattle and swine manures (Cotta et al., 2003; Leung and Topp, 2001; Ouwerkerk and Klieve, 2001; Whitehead and Cotta, 2001). Bacteria in these groups are widely known for their ability to utilize polysaccharides and proteins in anaerobic environments. Furthermore, some of these microorganisms have the ability to form spores, which gives them an additional advantage in feedlot soil environments that experience widely fluctuating moisture, temperature, and oxidative regimes. It is likely that differences in substrate availability between cattle and swine manure select for different bacterial species within the *Clostridium, Lactobacillus,* and *Bacillus* groups. The swine manure microbial community appears to be receptive to a wider variety of substrates, whereas the cattle manure microbial community specializes in starch fermentation to the exclusion of other substrates. Thus, the diet and digestive processes of swine and cattle control manure substrate composition and potential for post excretion odor production.

Table 7.1. Initial manure and final fermentation product composition in fresh swine.

Item	Swine manure(corn/soy bean diet)	Cattle manure (dry rolled corn diet)
Initial manure composition (% of DM)		
Starch	4.8 ± 0.3	28.0 ± 0.5
Organic N	3.3 ± 0.1	1.6 ± 0.1
Final fermentation products (μmoles/g DM)		
Lactate	23 ± 2	524 ± 141
Total alcohols	29 ± 11	221 ± 46
Total VFAs	1018 ± 65	1800 ± 194
Acetate	412 ± 39	470 ± 89
Propionate	227 ± 11	317 ± 13
Butyrate	169 ± 7	1001 ± 259
Total branched-chain VFAs	97.1 ± 8.5	5.5 ± 0.4
Total aromatics	33.3 ± 1.8	20.3 ± 1.3

[†] Fermentation product composition from the final collection time determined from samples collected on d 36 in fresh cattle manure slurries (Miller and Varel, 2001) and d 37 in fresh swine manure slurries. Reported values are the average of five samples ± 1 SE.

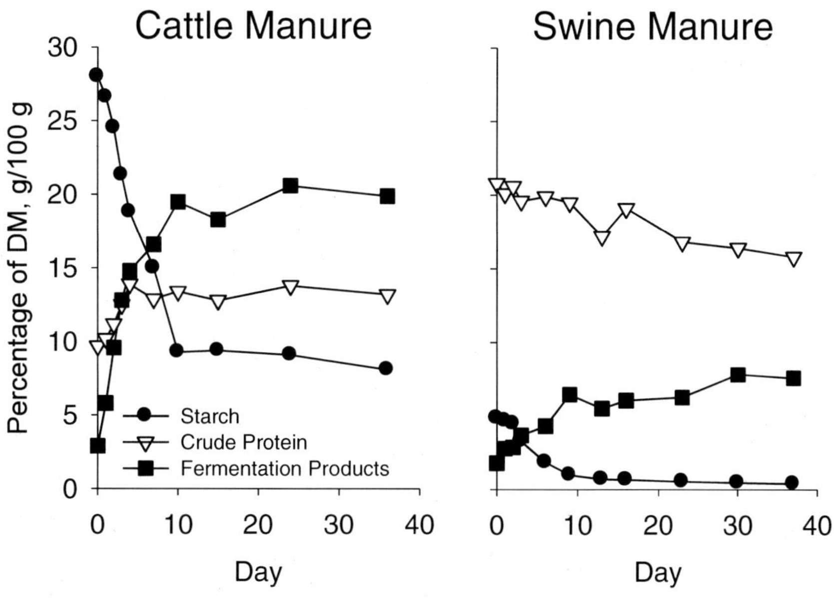

Figure 7.3. Accumulation of fermentation products and consumption of substrates in fresh cattle and swine manure during five weeks of incubation (Miller and Varel, 2002; Miller and Varel, 2003).

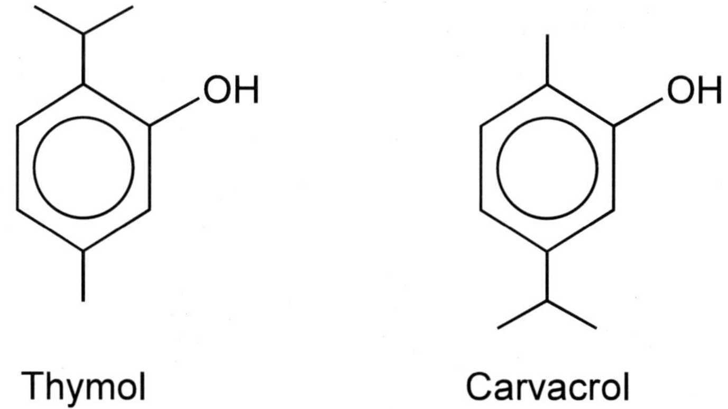

Figure 7.4. Structure of thymol and carvacrol.

7.6. MITIGATING MANURE ENVIRONMENTAL ISSUES

7.6.1. Diet Manipulation

Ask any layman to judge the odors from swine or beef cattle production, and they are more likely to find odors from swine production to be more objectionable than odors from cattle production. Why is that so? Manure storage, handling, and treatment practices are very different between (and within) swine production and cattle feedlot systems and are likely to account for a portion of the differences in perceived odor. However, most of the perceived odor difference between feedlot cattle and swine is likely attributable to differences in odor compound composition with the key difference being the greater branched-chain VFAs and aromatic compound content in swine manure compared to beef cattle manure. As mentioned earlier in this chapter, each odor compound is the product of a specific fermentation pathway, which in turn, is strongly influenced by substrate availability and/or content in swine and cattle feedlot manures. One approach to control manure odors involves diet manipulation to decrease the substrates during digestion and in the manure that are fermented to odor compounds.

Because starch and protein are the chief substrates for odor compound production, reducing their abundance in manure is the logical starting point. However, this approach is not as simple as it may seem, since decreasing either starch or protein may drastically decrease the growth and health of the animal, particularly in swine. Studies described earlier in this chapter show that beef cattle can use a wide variety of feedstuffs, some with high starch content and some with low starch content, for growth. Archibeque et al. (2006) fed cattle either a high moisture corn-based diet (HMC) or a DRC-based diet. Feeding the HMC-based diet decreased fecal starch excretion by 25%, and decreased VFAs accumulation during incubation by 50% compared to the DRC-based diet. Manure protein content was not affected. However, branched-chain VFAs accumulation increased by 50% during the incubation, and protein concentrations decreased in step with branched-chain VFAs accumulation. One explanation for these results is that the fecal microbial community changed from a predominately starch fermenting community to a more metabolically diverse community capable of utilizing both starch and protein. A similar result has been observed in numerous distillers' grain studies where starch content was low in the diet, very little starch was excreted, and protein fermentation products were more abundant than with high-grain diets (Hao et al., 2009; Spiehs and Varel, 2009; Varel et al., 2008). The microbial community changes to utilize whatever substrates are available. In these cases there is a move from starch to protein utilization.

In swine, decreasing ammonia emissions has been a major research focus with odor reduction being a secondary objective. Hence, a large number of diet studies have been undertaken to develop diets that decrease ammonia emissions by limiting the amount of protein fed. The general strategy employed is to decrease protein intake without affecting animal performance, which in turn reduces nitrogen excretion. Limiting protein intake will not only decrease nitrogen excretion, but should also decrease protein substrate in the lower gastrointestinal tract fueling branched-chain and aromatic compound production. Any amino acid deficiencies could be met by supplementing specific amino acids in the diet, which may substantially increase the cost of the diet. Excellent review articles have summarized research on the effect of low-protein diets on odorous compound production (Le et al., 2005b; Sutton et al., 1999). Readers should consult those papers for a more complete description of specific research studies. In general, decreased crude protein intake (with amino acid supplementation) reduced ammonia emissions by 30% or more and also reduced the concentration of protein-derived odorous compounds (aromatic compounds, branched-chain VFAs, and sulfides) in fresh swine manures. However, not all studies have supported this conclusion, and the effect of decreased protein in manures on odorous compound production was inconsistent in stored manures. Carbohydrate supplementation is an alternative path that could be used to alter odorous compound accumulation in swine manure. Recent research utilizing resistant starches (gelatinized) to shift fermentation in the pig colon away from amino acids and towards carbohydrates indicate that both branched-chain VFAs and indole compounds were reduced when resistant starch was included in the diet (Willig et al., 2005). A study examining two levels of cellulose and protein also showed that carbohydrate supplementation had an effect on the concentration of specific VFAs and aromatic compounds in both fresh feces and stored manure, but a general consensus on the effect of cellulose on odor was not clear with some key aromatic compounds increasing (Kerr et al., 2006).

7.6.2. Microbial Metabolic Inhibitors

A critical point in controlling odor emissions is regulating the volatilization rate. There are numerous factors which influence the volatilization rate of odorous compounds: source concentration; surface area; net radiation; air temperature; surface temperature; wind velocity; manure slurry pH, turbulence within the manure slurry, air turbulence above the manure slurry, and relative humidity. By using a cover over a lagoon or waste storage basin the volatilization rate can be reduced by decreasing the solar radiation and direct wind velocity stripping off the VOC. Various other engineered processes can be used to affect these and other factors controlling volatilization rate (Zahn et al., 1997).

If the microbial formation of the VOC can be controlled, the volatilization rate can be reduced. In most livestock production facilities it is not possible to control the environment in which complete degradation of the waste to methane and carbon dioxide occurs. Conventional anaerobic digesters for production of methane were popular during the 1970s and 1980s; however, economics and the technical expertise required to operate these digesters have diminished their popularity (Morse et al., 1996). Similarly, aerobic treatment is not economically feasible, and it does little to conserve nutrients. Therefore, inhibiting the microbial fermentation of organic matter in waste before manure microorganisms metabolize the substrate in fresh manures (Figure 7. 1) should prevent formation of additional malodorous VOC and reduce potential odor. Specific antimicrobial chemicals inhibit fermentation, and their application in manure-impacted environments and storage systems should reduce odor, retain nutrients, help to eliminate certain pathogens, and reduce the emission of greenhouse gases.

The requirements of chemicals that might be used to control the microbial fermentation of manure should include additives which are safe in the environment, inexpensive, and easy to apply. One group of chemical additives that fits these requirements is the naturally occurring antimicrobial plant-derived oils (Beuchat, 1994; Helander et al., 1998; Ultee et al., 1999). Essential oils are secondary metabolites of plants, they are volatile consisting mostly of terpenes and oxygenated derivatives, and are used for flavor, fragrances, spices, antiseptic and preservative action. Oils with the most effective antimicrobial activity in descending order are phenolic compounds, alcohols, aldehydes, ketones, ethers, and hydrocarbons (Charai et al., 1996; Dorman and Deans, 2000). Two extensively studied phenolic compounds are thymol (5-methyl-2-isopropylphenol) and carvacrol (5-isopropyl-2-methylphenol) (Figure 7. 4). In practice, carvacrol is added to different products such as baked goods (16 ppm), nonalcoholic beverages (28 ppm/0.18 mM), and chewing gum (8 ppm) (Ultee et al., 1999). Thymol is a component in many different products including soaps, toothpastes, shampoos, deodorants, and mouthwashes (Manou et al., 1998; Shapiro et al., 1994). These chemicals, like most plant essential oils, are generally recognized as safe (GRAS). Their antimicrobial mode of action consists of interactions with cell membranes which changes the permeability for cations like H^+ and K^+ (Ultee et al., 1999). The dissipation of ion gradients leads to loss of turgor pressure, inhibition of DNA synthesis, enzyme activity, overall metabolic activities, and finally to cell death.

Various chemicals and a number of plant essential oils have been evaluated for their ability to control the production of short-chain fatty acids, L-lactate and gas, and to reduce total anaerobic bacteria and fecal coliforms in stored waste (Varel and Miller, 2000; Varel and Miller, 2001b). Some of these chemical additives, α-pinene, limonene, camphor, borneol, fenchol, eugenol, geraniol, 2-bromoethanesulfonic acid, anthraquinone, monensin, N,N[1]-dicylohexylcarbodiimide, and methylglyoxal, had little inhibitory effect on the fermentation of cattle waste. Others, such as chlorhexidine diacetate, iodoacetic acid, and diphenyliodonium, were inhibitory to cattle but not swine waste (Varel, 2002).

Studies suggest that thymol and carvacrol are the most effective in controlling the parameters listed above. Data indicate that a combination of carvacrol and thymol, each at 1000 mg/L and 500 mg/L pinene, will inhibit any new production of VFAs in cattle waste for 23 days (Varel and Miller, 2001a). Data in this study also indicated that a combination of carvacrol, thymol, and pinene, at both concentrations evaluated, 750 or 1000 mg/L each of carvacrol and thymol, reduced the number of viable anaerobic bacteria within 2 days in the waste when compared with the controls. A complete bactericidal effect was not observed after 14 days, even at the higher concentration of carvacrol and thymol (1000 mg/L); however, the number of organisms remained low similar to the 2 day population. Previous studies have suggested that a combination of carvacrol and thymol would provide better antimicrobial action, rather than a higher concentration of one alone (Manou et al., 1998; Paster et al., 1995). However, further studies indicated that thymol or carvacrol can be used individually, and each is as effective as the combination as long as the total concentration added is the same (Varel, 2002; Varel and Miller, 2001b; Varel and Wells, 2007). Based on cost, thymol would be the oil of choice (Varel and Wells, 2007).

Obviously, the residual thymol in the waste must be considered. Field studies with thymol in cattle feedlot pens indicated that the concentration of topically applied thymol decreases within days after application under aerobic conditions (Varel et al., 2006). There are several pathways for phenolic compound degradation (Fang et al., 2006). Soil microorganisms are known to degrade some of the monoterpenoid plant essential oils (van der Werf et al., 1999) and some are degraded under anaerobic conditions (Harder and Probian, 1995). Indeed, essential oils are used as a carbon and energy source by ubiquitously occurring soil microorganisms, and they would not accumulate in soil if environmental conditions favor growth of these organisms (Vokou and Liotiri, 1999). These chemicals are also volatile, thus a fraction of the thymol would be lost once the waste is transferred from a pit to cropland. However, under anaerobic conditions, similar to those used in our laboratory studies and field studies, we have found thymol and carvacrol are not degraded for over 60 days (Varel, 2002; Varel and Wells, 2007). In livestock production facilities this appears to be ideal, since odor (VFAs) is produced under anaerobic conditions, during which time it is desirable that these antimicrobial oils not be degraded. However, at the same time it is desirable to have these plant oils degraded at some point to avoid a build-up in the soil environment. Therefore, it is speculated that once the livestock wastes are spread on the soil surface (aerobic conditions); any essential oil would be broken down and metabolized by the soil microbes.

Another additive that may be effective in reducing emissions from livestock facilities is amending the waste with iron. Laboratory studies have demonstrated that malodorous compounds, primarily VFAs, associated with swine waste can be removed by amending the waste with alternate electron acceptors. In laboratory incubations Fe(III) and a novel dissimilatory Fe(III)-reducing organism reduced total VFAs (Coates et al., 2005). Another laboratory study added nitrate, oxidized iron, oxidized manganese, or sulfate to manure slurries and found that nitrate and oxidized iron reduced maximum VFAs concentrations and shortened the period of time that VFAs persisted in the bottles (Miller, 2001). Field applications have not been evaluated. Addition of specific microorganisms or enzymes to an open ecosystem such as a manure storage basin is unlikely to mitigate odor emissions since naturally selected microorganisms have adapted to these ecosystems and are not easily displaced by additions of a few select microorganisms without some other selective pressure imposed.

Most of the ammonia emissions from livestock wastes originate from hydrolysis of urea (Bierman et al., 1999; Van Horn et al., 1996; Varel et al., 1999). Beef cattle raised in a CAFO and fed a high protein diet excrete approximately 60% of their nitrogen in urine and the remainder in feces. Up to 97% of urinary nitrogen is in the form of urea, which is readily hydrolyzed by microbial urease to ammonia shortly after excretion. Ammonia emissions in Europe have increased by more than 50% during the past 30 years with livestock production identified as the primary contributor (Bouwman et al., 1997; McCrory and Hobbs, 2001; Pain et al., 1998). In the United States, EPA predictions also indicate an increase in ammonia emission associated with increase animal production through 2030 (US EPA, 1995).

Urease inhibitors have been used successfully, on a short-term basis (2 wk), to prevent ammonia emissions from urea-based fertilizers when they are applied to soil (Byrnes and Freney, 1995). Similarly with cattle feedlot manure, the urease inhibitor, N-(n-butyl) thiophosphoric triamide (NBPT), has the potential to prevent ammonia emission and retain urea nitrogen in manure (Parker et al., 2005; Shi et al., 2001; Varel et al., 1999). Urease inhibitors bind to the active nickel site of the enzyme and prevent urea hydrolysis to ammonia and carbon dioxide (Mobley et al., 1995). Field studies with NBPT have demonstrated that urea nitrogen can be retained in beef cattle feedlot waste if it is applied once per week to the feedlot surface (Varel et al., 1999; Parker et al., 2005). However, once applications of NBPT to the pen surface were discontinued, all urea was eventually hydrolyzed and much of the ammonia nitrogen was emitted to the air indicating that urease inhibitors were either quickly broken down or inhibited. Repeated reapplication makes the cost of NBPT an economic consideration (McCrory and Hobbs, 2001), but with decreasing NBPT costs and new research directed specifically to develop products that can prolong the effectiveness of urease inhibitors in manure environments, the economic costs of this approach become more feasible (Varel and Wells, 2007).

Efforts to control ammonia emissions from CAFO should be complementary to controlling odor emissions. Combining an antimicrobial additive with NBPT may prolong the inhibitory period of the urease inhibitor. Results from a recent study indicated that thymol had a supplemental effect to the urease inhibitor, NBPT, by prolonging the time before urea in cattle waste slurries was hydrolyzed by microbial urease (Varel et al., 2007). Increasing the concentration of NBPT to 80 mg/kg (from 20 mg/kg in other studies) prevented most of the urea from being hydrolyzed up to 14 d, but then a gradual hydrolysis began and continued to d 56. Thus, the antimicrobial chemical thymol may offer a solution to extending the activity of NBPT. Unfortunately in a more recent field study with swine waste, a combination of NBPT and thymol only produced a short-term response (6 to 10 d) in conserving urea in pits with swine manure (Varel and Wells, 2007). Weekly additions of NBPT may be needed to overcome urease inactivation, similar to what is observed with cattle waste (Varel et al., 1999). Treating manure slurries with 160 mg NBPT/kg had little effect on the rate at which urea was hydrolyzed (Varel et al., 2007). This observation supports an earlier study finding that microorganisms degrade NBPT (Byrnes and Freney, 1995). Whether an antimicrobial chemical can be specifically added to control urease production is still unknown. Further efforts are needed in this area to maximize the return of nitrogen to agronomic crops, while decreasing losses to the atmosphere (Galloway et al., 2003).

7.7. CONCLUSION

Manure odors are complex mixtures of individual chemical compounds that change from site to site and with distance from a particular site. Odor compounds originate in manure, both fresh and stored, and are the result of the fermentative decomposition of substrates in the manure. A diverse group of microorganisms are involved and are in part selected by the availability of easily degraded substrates. Chief among the substrates identified are protein and starch. Fermentation of proteins result in specific suites of odor compounds, branched-chain VFAs and aromatic compounds, that, when detected, are indicators of protein fermentation. In beef cattle manures and soils, starch is the dominant substrate, but in swine manures protein is as important as starch for odorous compound production. With these insights, studies to reduce odor through diet modification have been (and continue to be) conducted indicating the preeminent role of diet in manure odor. In the future, effective management of manure emissions will likely involve both diet manipulation and a complementary strategy of microbial inhibition of both odor compound formation and urea hydrolysis, but continued research is needed to identify the best practices for all manure-impacted environments.

REFERENCES

Archibeque, S.L., D.N. Miller, H.C. Freetly, And C.L. Ferrell. 2006. Feeding High-Moisture Corn Instead Of Dry-Rolled Corn Reduces Odorous Compound Production In Manure Of Finishing Beef Cattle Without Decreasing Performance. *J. Anim. Sci.* 84:1767-1777.

Archibeque, S.L., D.N. Miller, H.C. Freetly, E.D. Berry, And C.L. Ferrell. 2007. The Influence Of Oscillating Dietary Protein Concentrations On Finishing Cattle. I. Feedlot Performance And Odorous Compound Production. *J. Anim. Sci.* 85:1487-1495.

Berry, E.D., And D.N. Miller. 2005. Cattle Feedlot Soil Moisture And Manure Content: .II. Impact On *Escherichia Coli* O157. *J. Environ. Qual.* 34:656-663.

Beuchat, L.R. 1994. Antimicrobial Properties Of Spices And Their Essential Oils, P. 167-179, *In* V. M. Dillon And R. G. Board, Eds. *Natural Antimicrobial Systems And Food Preservation.* Cab International, Wallingford.

Bierman, S., G.E. Erickson, T.J. Klopfenstein, R.A. Stock, And D.H. Shain. 1999. Evaluation Of Nitrogen And Organic Matter Balance In The Feedlot As Affected By Level And Source Of Dietary Fiber. *J. Anim. Sci.* 77:1645-1653.

Bouwman, A.F., D.S. Lee, W.A.H. Asman, F.J. Dentener, K.W. Van Der Hoek, And J.G.J. Olivier. 1997. A Global High-Resolution Emission Inventory For Ammonia. *Global Biogeochem. Cycles* 11:561-587.

Bulliner IV, E.A., J.A. Koziel, L. Cai, And D. Wright. 2006. Characterization Of Livestock Odors Using Steel Plates, Solid-Phase Microextraction, And Multidimensional Gas Chromatography-Mass Spectrometry-Olfactometry. *J. Air Waste Manage. Assoc.* 56:1391-1403.

Byrnes, B.H., And J.R. Freney. 1995. Recent Developments On The Use Of Urease Inhibitors In The Tropics. *Fert. Res.* 42:251-259.

Charai, M., M. Mosaddak, And M. Faid. 1996. Chemical Composition And Antimicrobial Activities Of Two Aromatic Plants: *Origanum Majorana L.* And *O. Compactum* Benth. *J. Essential Oil Res.* 8:657-664.

Coates, J.D., K.A. Cole, U. Michaelidou, J. Patrick, M.J. Mcinerney, And L.A. Achenbach. 2005. Biological Control Of Hog Waste Odor Through Stimulated Microbial Fe(Iii) Reduction. *Appl. Environ. Microbiol.* 71:4728-4735.

Cotta, M.A., T.R. Whitehead, And R.L. Zeltwanger. 2003. Isolation, Characterization And Comparison Of Bacteria From Swine Faeces And Manure Storage Pits. *Environ. Microbiol.* 5:737-745.

Dewes, T., L. Schmitt, U. Valentin, And E. Ahrens. 1990. Nitrogen Losses During The Storage Of Liquid Livestock Manures. *Biol. Wastes* 31:241-250.

Dorman, H.J.D., And S.G. Deans. 2000. Antimicrobial Agents From Plants: Antibacterial Activity Of Plant Volatile Oils. *J. Appl. Microbiol.* 88:308-316.

Fang, H.H.P., D.W. Liang, T. Zhang, And Y. Liu. 2006. Anaerobic Treatment Of Phenol In Wastewater Under Thermophilic Condition. *Water Res.* 40:427-434.

Galloway, J.N., J.D. Aber, J.W. Erisman, S.P. Seitzinger, R.W. Howarth, E.B. Cowling, And B.J. Cosby. 2003. *The Nitrogen Cascade. Bioscience* 53:341-356.

Gralapp, A.K., W.J. Powers, And D.S. Bundy. 2001. Comparison Of Olfactometry, Gas Chromatography, And Electronic Nose Technology For Measurement Of Indoor Air From Swine Facilities. *Trans ASAE* 44:1283-1290.

Hao, X., M.B. Benke, D.J. Gibb, A. Stronks, G. Travis, And T.A. Mcallister. 2009. Effects Of Dried Distillers' Grains With Solubles (Wheat-Based) In Feedlot Cattle Diets On Feces And Manure Composition. *J. Environ. Qual.* 38:1709-1718.

Harder, J., And C. Probian. 1995. Microbial Degradation Of Monoterpenes In The Absence Of Molecular Oxygen. *Appl. Environ. Microbiol.* 61:3804-3808.

Hartung, J., And V.R. Phillips. 1994. Control Of Gaseous Emissions From Livestock Buildings And Manure Stores. *J. Agric. Eng. Res.* 57:173-189.

Helander, I.M., H.L. Alakomi, K. Latva-Kala, T. Mattila-Sandholm, I. Pol, E.J. Smid, L.G.M. Gorris, And A. Von Wright. 1998. Characterization Of The Action Of Selected Essential Oil Components On Gram-Negative Bacteria. *J. Agric. Food Chem.* 46:3590-3595.

Keener, K.M., J. Zhang, R.W. Bottcher, And R.D. Munilla. 2002. Evaluation Of Thermal Desorption For The Measurement Of Artificial Swine Odorants in the Vapor Phase. *Trans. ASAE* 45:1579-1584.

Kerr, B.J., C.J. Ziemer, S.L. Trabue, J.D. Crouse, And T.B. Parkin. 2006. Manure Composition Of Swine As Affected By Dietary Protein And Cellulose Concentrations. *J. Anim. Sci.* 84:1584-.

Koziel, J.A., L. Cai, D.W. Wright, And S.J. Hoff. 2006. Solid-Phase Microextraction As A Novel Air Sampling Technology For Improved, Gc-Olfactometry-Based Assessment Of Livestock Odors. *J. Chromatogr. Sci.* 44:451-457.

Le, P.D., A.J.A. Aarnink, N.W.M. Ogink, P.M. Becker, And M.W.A. Verstegen. 2005a. Odour From Animal Production Facilities: Its Relationship To Diet. *Nutr. Res. Rev.* 18:3-30.

Leung, K., And E. Topp. 2001. Bacterial Community Dynamics In Liquid Swine Manure During Storage: Molecular Analysis Using Dgge/Pcr Of 16s Rdna. Fems Microbiol. *Ecol.* 38:169-177.

Lo, Y.C.M., J.A. Koziel, L. Cai, S.J. Hoff, W.S. Jenks, And H. Xin. 2008. Simultaneous Chemical And Sensory Characterization Of Volatile Organic Compounds And Semi-Volatile Organic Compounds Emitted From Swine Manure Using Solid Phase Microextraction And Multidimensional Gas Chromatography-Mass Spectrometry-Olfactometry. J. *Environ. Qual.* 37:521-534.

Loughrin, J.H., N. Lovanh, And R. Mahmood. 2008. Equilibrium Sampling Used To Monitor Malodors In A Swine Waste Lagoon. *J. Environ. Qual.* 37:1-6.

Macfarlane, G.T., G.R. Gibson, E. Beatty, And J.H. Cummings. 1992. Estimation Of Short-Chain Fatty Acid Production From Protein By Human Intestinal Bacteria Based On Branched-Chain Fatty Acid Measurements. *Fems Microbiol. Ecol.* 101:81-88.

Mackie, R.I., P.G. Stroot, And V.H. Varel. 1998. Biochemical Identification And Biological Origin Of Key Odor Components In Livestock Waste. *J. Anim. Sci.* 76:1331-1342.

Manou, I., L. Bouillard, M.J. Devleeschouwer, And A.O. Barel. 1998. Evaluation Of The Preservative Properties Of *Thymus Vulgaris* Essential Oil In Topically Applied Formulations Under A Challenge Test. *J. Appl. Microbiol.* 84:368-376.

McCrory, D.F., And P.J. Hobbs. 2001. Additives To Reduce Ammonia And Odor Emissions From Livestock Wastes: A Review. *J. Environ. Qual.* 30:345-355.

Miller, D.N. 2001. Accumulation And Consumption Of Odorous Compounds In Feedlot Soils Under Aerobic, Fermentative, And Anaerobic Respiratory Conditions. *J. Anim. Sci.* 79:2503-2512.

Miller, D.N., And V.H. Varel. 2001. In Vitro Study Of The Biochemical Origin And Production Limits Of Odorous Compounds In Cattle Feedlots. *J. Anim. Sci.* 79:2949-2956.

Miller, D.N., And V.H. Varel. 2002. An In Vitro Study Of Manure Composition On The Biochemical Origins, Composition, And Accumulation Of Odorous Compounds In Cattle Feedlots. *J. Anim. Sci.* 80:2214-2222.

Miller, D.N., And V.H. Varel. 2003. Swine Manure Composition Affects The Biochemical Origins, Composition, And Accumulation Of Odorous Compounds. *J. Anim. Sci.* 81:2131-2138.

Miller, D.N., And E.D. Berry. 2005. Cattle Feedlot Soil Moisture And Manure Content: I. Impacts On Greenhouse Gases, Odor Compounds, Nitrogen Losses, And Dust. *J. Environ. Qual.* 34:644-655.

Miller, D.N., And B.L. Woodbury. 2006. A Solid-Phase Microextraction Chamber Method For Analysis Of Manure Volatiles. *J. Environ. Qual.* 35:2383-2394.

Miller, D.N., E.D. Berry, J.E. Wells, C.L. Ferrell, S.L. Archibeque, And H.C. Freetly. 2006. Influence Of Genotype And Diet On Steer Performance, Manure Odor, And Carriage Of Pathogenic And Other Fecal Bacteria. Iii. Odorous Compound Production. *J. Anim. Sci.* 84:2433-2445.

Mobley, H.L.T., M.D. Island, And R.P. Hausinger. 1995. Molecular Biology Of Microbial Ureases. *Microbiol. Rev.* 59:451-480.

Morse, D., J.C. Guthrie, And R. Mutters. 1996. Anaerobic Digester Survey Of California Dairy Producers. *J. Dairy Sci.* 79:149-153.

O'neill, D.H., And V.R. Phillips. 1992. A Review Of The Control Of Odour Nuisance From Livestock Buildings: Part 3. Properties Of The Odorous Substances Which Have Been Identified In Livestock Wastes Or In Air Around Them. *J. Agric. Eng. Res.* 53:23-50.

Ouwerkerk, D., And A.V. Klieve. 2001. Bacterial Diversity Within Feedlot Manure. *Anaerobe* 7:59-66.

Pain, B.F., T.J. Van Der Weerden, B.J. Chambers, V.R. Phillips, And S.C. Jarvis. 1998. A New Inventory For Ammonia Emissions From U.K. Agriculture. *Atmos. Environ.* 32:309-313.

Parker, D.B., S. Pandrangi, L.K. Almas, M.B. Rhoades, N.A. Cole, L.W. Greene, And J.A. Koziel. 2005. Rate And Frequency Of Urease Inhibitor Application For Minimizing Ammonia Emissions From Beef Cattle Feedyards. *Trans. ASAE* 48:787-793.

Paster, N., M. Menasherov, U. Ravid, And B. Juven. 1995. Antifungal Activity Of Oregano And Thyme Essential Oils Applied As Fumigants Against Fungi Attacking Stored Grain. *J. Food Prot.* 58:81-85.

Powers, W.J., H.H. Van Horn, A.C. Wilkie, C.J. Wilcox, And R.A. Nordstedt. 1999. Effects Of Anaerobic Digestion And Additives To Effluent Or Cattle Feed On Odor And Odorant Concentrations. *J. Anim. Sci.* 77:1412-1421.

Qu, G., M.M. Omotoso, M.G. El-Din, And J.J.R. Feddes. 2008. Development Of An Integrated Sensor To Measure Odors. *Environ. Monit. Assess.* 144:277-283.

Qu, G., And J.J.R. Feddes. 2007. Development Of A Reference Artificial Swine Odor. *Trans. ASABE* 50:1789-1793.

Rappert, S., And R. Müller. 2005. Odor Compounds In Waste Gas Emissions From Agricultural Operations And Food Industries. *Waste Manage.* 25:887-907.

Schiffman, S.S., J.L. Bennett, And J.H. Raymer. 2001. Quantification Of Odors And Odorants From Swine Operations In North Carolina. Agric. *Forest Meterol* 108:213-240.

Shabtay, A., U. Ravid, A. Brosh, R. Baybikov, H. Eitam, And Y. Laor. 2009. Dynamics Of Offensive Gas-Phase Odorants In Fresh And Aged Feces Throughout The Development Of Beef Cattle. *J. Anim. Sci.* 87:1835-1848.

Shapiro, S., A. Meier, And B. Guggenheim. 1994. The Antimicrobial Activity Of Essential Oils And Essential Oil Components Towards Oral Bacteria. *Oral Microbiol. Immun.* 9:202-208.

Shi, Y., D.B. Parker, N.A. Cole, B.W. Auvermann, And J.E. Mehlhorn. 2001. Surface Amendments To Minimize Ammonia Emissions From Beef Cattle Feedlots. *Trans. ASAE* 44:677-682.

Smith, E.A., And G.T. Macfarlane. 1998. Enumeration Of Amino Acid Fermenting Bacteria In The Human Large Intestine: Effects Of Ph And Starch On Peptide Metabolism And Dissimilation Of Amino Acids. *Fems. Microbiol. Ecol.* 25:355-368.

Spiehs, M.J., And V.H. Varel. 2009. Nutrient Excretion And Odorant Production In Manure From Cattle Fed Corn Wet Distillers Grains With Solubles. *J. Anim. Sci.* 87:2977-2984.

Spoelstra, S.F. 1980. Origin Of Objectionable Odorous Components In Piggery Wastes And The Possibility Of Applying Indicator Components For Studying Odour Development. A*gric. Environ.* 5:241-260.

Sutton, A.L., K.B. Kephart, M.W.A. Verstegen, T.T. Canh, And P.J. Hobbs. 1999. Potential For Reduction Of Odorous Compounds In Swine Manure Through Diet Modification. *J. Anim. Sci.* 77:430-439.

Switzenbaum, M.S., E. Giraldo-Gomez, And R.F. Hickey. 1990. Monitoring Of The Anaerobic Methane Fermentation Process. *Enzyme Microb. Technol.* 12:722-730.

Trabue, S.L., J.C. Anhalt, And J.A. Zahn. 2006. Bias Of Tedlar Bags In The Measurement Of Agricultural Odorants. *J. Environ. Qual.* 35:1668-1677.

Ultee, A., E.P.W. Kets, And E.J. Smid. 1999. Mechanisms Of Action Of Carvacrol On The Food-Borne Pathogen. *Appl. Environ. Microbiol.* 65:4606-4610.

US EPA. 1995. Ap-42 Compilation Of Air Pollutant Emission Factors, Volume I: Stationary Point Sources. Us Environmental Protection Agency, Office Of Air Quality Planning And Standards. Research Triangle Park, Nc. Http://Www.Epa.Gov/Ttn/Chief/Ap42/Toc_Kwrd.Pdf.

Van Der Werf, M.J., H.J. Swarts, And J.A.M. De Bont. 1999. *Rhodococcus Erythropolis* Dcl14 Contains A Novel Degradation Pathway For Limonene. *Appl. Environ. Microbiol.* 65:2092-2102.

Van Horn, H.H., G.L. Newton, And W.E. Kunkle. 1996. Ruminant Nutrition From An Environmental Perspective: Factors Affecting Whole-Farm Nutrient Balance. *J. Anim. Sci.* 74:3082-3102.

Varel, V.H. 2002. Carvacrol And Thymol Reduce Swine Waste Odor And Pathogens: Stability Of Oils. *Curr. Microbiol.* 44:38-43.

Varel, V.H., And D.N. Miller. 2000. Effect Of Antimicrobial Agents On Livestock Waste Emissions. *Curr. Microbiol.* 40:392-397.

Varel, V.H., And D.N. Miller. 2001a. Effect Of Carvacrol And Thymol On Odor Emissions From Livestock Wastes. *Water Sci. Technol.* 44:143-148.

Varel, V.H., And D.N. Miller. 2001b. Plant-Derived Oils Reduce Pathogens And Gaseous Emissions From Stored Cattle Waste. *Appl. Environ. Microbiol.* 67:1366-1370.

Varel, V.H., And J.E. Wells. 2007. Influence Of Thymol And A Urease Inhibitor On Coliform Bacteria, Odor, Urea, And Methane From A Swine Production Manure Pit. *J. Environ. Qual.* 36:773-779.

Varel, V.H., J.A. Neinaber, And H.C. Freetly. 1999. Conservation Of Nitrogen In Cattle Feedlot Waste With Urease Inhibitors. *J. Anim. Sci.* 77:1162-1168.

Varel, V.H., D.N. Miller, And E.D. Berry. 2006. Incorporation Of Thymol Into Corncob Granules For Reduction Of Odor And Pathogens In Feedlot Cattle Waste. *J. Anim. Sci.* 84:481-487.

Varel, V.H., J.E. Wells, And D.N. Miller. 2007. Combination Of A Urease Inhibitor And A Plant Essential Oil To Control Coliform Bacteria, Odour Production, And Ammonia Loss From Cattle Waste. *J. Appl. Microbiol.* 102:472-477.

Varel, V.H., J.E. Wells, E.D. Berry, M.J. Spiehs, D.N. Miller, C.L. Ferrell, S.D. Shackelford, And M. Koohmaraie. 2008. Odorant Production And Persistence Of Escherichia Coli In

Manure Slurries From Cattle Fed Zero, Twenty, Forty, Or Sixty Percent Wet Distillers Grains With Solubles. *J. Anim. Sci.* 86:3617-3627.

Vasconcelos, J.T., And M.L. Galyean. 2007. Nutritional Recommendations Of Feedlot Consulting Nutritionsists: The 2007 Texas Tech University Survey. *J Anim. Sci.* 85:2772-2781.

Vokou, D., And S. Liotiri. 1999. Stimulation Of Soil Microbial Activity By Essential Oils. *Chemoecology* 9:41-45.

Watts, P.J., M. Jones, S.C. Lott, R.W. Tucker, And R.J. Smith. 1994. Feedlot Odor Emissions Following Heavy Rainfall. *Trans. ASAE* 37:629-636.

Whitehead, T.R., And M.A. Cotta. 2001. Characterisation And Comparison Of Microbial Populations In Swine Faeces And Manure Storage Pits By 16s Rdna Gene Sequence Analyses. *Anaerobe* 7:181-187.

Willig, S., D. Losel, And R. Claus. 2005. Effects Of Resistant Potato Starch On Odor Emission From Feces In Swine Production Units. *J. Agric. Food Chem.* 53:1173-1178.

Wright, D.W., D.K. Eaton, L.T. Nielsen, F.W. Kuhrt, J.A. Koziel, J.P. Spinhirne, And D.B. Parker. 2005. Multidimensional Gas Chromatography-Olfactometry For The Identification And Prioritization Of Malodors From Confined Animal Feeding Operations. *J. Agric. Food Chem.* 53:8663-8672.

Yasuhara, A. 1980. Relation Between Odor And Odorous Components In Solid Swine Manure. Chemosphere 9:587-592.

Zahn, J.A., J.L. Hatfield, D.A. Laird, R.L. Pfeiffer, Y.S. Do, And A.A. Dispirito. 1997. Characterization Of Volatile Organic Emissions And Wastes From A Swine Production Facility. *J. Environ. Qual.* 26:1687-1696.

Zahn, J.A., A.A. Dispirito, Y.S. Do, B.E. Brooks, E.E. Cooper, And J.L. Hatfield. 2001. Correlation Of Human Olfactory Responses To Airborne Concentrations Of Malodorous Volatile Organic Compounds Emitted From Swine Effluent. *J. Environ. Qual.* 30:624-634.

Zhu, J., L. Jacobson, D. Schmidt, And R. Nicolai. 2000. Daily Variation In Odor And Gas Emissions From Animal Facilities. *Appl. Eng. Agric.* 16:153-158.

Zhu, J., G.L. Riskowski, And M. Torremorell. 1999. Volatile Fatty Acids As Odor Indicators In Swine Manure - A Critical Review. *Trans. ASAE* 42:175-182.

In: Environmental Chemistry of Animal Manure
Editor: Zhongqi He

ISBN: 978-1-62808-641-6
© 2013 Nova Science Publishers, Inc.

Chapter 8

MANURE AMINO ACID COMPOUNDS AND THEIR BIOAVAILABILITY

Zhongqi He[1], and Daniel C. Olk[2]*

8.1. INTRODUCTION

Historically, microbial nitrogen (N) mineralization in soil has been viewed as the most critical aspect of the N cycle. In the traditional view, only N that is mineralized by microbes in excess of their own demand and released back into soil (i. e. net N mineralization) is available to plants (Kahmen et al., 2009). Amino acids (AA) are widely presumed to be the primary pool of organic N in soil. Thus, identification and quantification of these AA species have obvious significance in understanding N mineralization and cycling (Olk, 2008). However, our growing recognition of the importance of AA is not matched by knowledge of the amounts and types of AA in the soil (Warren, 2008).

During the last two decades, a new paradigm of terrestrial N cycling has begun to emerge that recognizes the potential for plants to acquire N from organic sources (Chapin et al., 1993; Rothstein, 2009). Weigelt et al. (2005) selected five grass species that were grown in a gradient from fertilized, productive pastures to extensive, low productive pastures to test their capability to take up inorganic N and a range of AA. All five grass species were able to take up directly a diversity of soil amino acids of varying complexity. Werdin-Pfisterer et al. (2009) examined water extractable AA composition of soils across a boreal forest in interior Alaska, USA. They found that the amino acid pool was dominated by six AAs: glutamine acid, glutamine, aspartic acid, asparagine, alanine, and histidine, which accounted for approximately 80% of the total water extractable soil amino acid pool. Furthermore, the AA concentrations were an order of magnitude higher in coniferous-dominated late successional stages than in early deciduous-dominated stages. The authors (Weigelt et al., 2005) concluded

---

* Corresponding author: Zhongqi.He@ars.usda.gov
[1]USDA-ARS, New England Plant, Soil, and Water Laboratory, Orono, ME 04469, USA
[2]USDA-ARS, National Laboratory for Agriculture and the Environment, Ames, IA 50011, USA

that AA compounds are important constituents of the biogeochemical diverse soil N pool in the boreal forest of interior Alaska.

In addition, L-tryptophan is a precursor of the growth hormone indole-3-acetic acid and is known to stimulate plant growth at extremely low concentrations (Arshad et al., 1995). Ahmad et al. (2008) found that enrichment of composted organic wastes with N and L-tryptophan can change them into a value-added organic product that could be used as a soil amendment at rates as low as 300 kg ha^{-1} to increase maize crop production on a sustainable basis. Wang et al. (2007) proposed that asparagine and glutamine may be used to partially replace nitrate-N nutrition in pak-choi (*Brassica chinensis* L.) growth to improve pak-choi shoot quality.

Amino acids have been long known to be present in animal manure and other wastes or composts (Hacking et al., 1977; Hirai et al., 1983). Earlier research on manure AA distribution was based on the potential re-use of the manure AA as a feed ingredient for animals (Blair, 1974). Recently, Larkin et al. (2006) investigated soil microbial community characteristics impacted by manure application, concluding that dairy manure had greater effects than swine manure on soil microbial utilization of carbohydrates and AA. Bol et al. (2008) examined the long-term dynamics of AA from a bare fallow soil experiment in unamended control plots and plots treated with ammonium sulphate, ammonium nitrate, sodium nitrate or with animal manure. Their data show that, with time, soil N, C and AA content were increased only in the plots with animal manure. Scheller and Raupp (2005) investigated AA and soil organic matter contents of topsoil in a long term trial with farmyard manure and mineral fertilizers. Higher AA contents in manure fertilized plots were observed even at the lowest rate of fertilizer application. This observation indicates that differences between the treatments not only depend on the AA supply from manure, but are also influenced by an altered AA metabolism in the soil.

Although some information on manure AA contents and their bioavailability has been accumulated over years, a review of the relevant literature has not yet been published. Therefore, the objectives of this chapter were to (1) compile and summarize the AA concentrations and proportions in various animal manure samples, and (2) review and discuss soil AA distribution impacted by manure application. We hope this review is helpful to identify gaps in knowledge and propose new research approaches in studying the dynamics of manure AA for sustainable and environment-friendly agricultural production.

8.2. PROTEINACEOUS AMINO ACIDS AND THEIR MEASUREMENTS

Amino acids are the monomeric units of proteins. Analysis of a vast number of proteins from almost every conceivable sources has shown that all proteins are composed primarily of 20 "standard" AA (Table 8.1) (Voet and Voet, 1990). The 20 AA compounds can be classified to three groups: (1) nine compounds with nonpolar side chains; (2) six compounds with uncharged polar side chains; and (3) five compounds with charged polar side chains. It is worth noting that the dimeric compound cysteine is composed of two molecules of the amino acid cystine, although the literature has sometimes overlooked this distinction. One should be aware that the 20 AA compounds are not the only AA occurring in biological systems. Many other AA compounds, common or rare, have been found in nature (He and Spain, 1999;

Potter et al., 2001; Voet and Voet, 1990). For example, poly-γ-glutamic acid, a swine manure-based fermentation product, could be used as environmental-friendly fertilizer synergists to increase nutrient utilization and to enhance the yield and quality of crops (Chen et al., 2005). 2,6-Diaminopimelic acid (DAPA) has been used as a biomarker for bacteria in animal ruminal digesta and feces (Martin et al., 1994; Karr-Lilienthal et al., 2004).

To measure AA composition, a soil or manure sample must be hydrolyzed first with HCl, HF-HCl or other hydrolyzing reagents (Liang et al., 1996; Martens and Loeffelmann, 2003; Scheller and Raupp, 2005). Due to the acid hydrolysis process, asparagine and glutamine are converted to aspartic acid and glutamic acid, respectively, and in an HCl extraction tryptophan is also lost. For this reason, only 17 proteinaceous standard AAs are commonly detected in the HCl hydrolysate of a sample.

High-performance liquid chromatography (HPLC), gas chromatography (GC) and gas chromatography–mass spectrometry (GC-MS) are traditional methods used to quantify individual AA compounds (Warren, 2008). To improve the resolution and sensitivity, these chromatographic techniques can be carried out by pre- or post-column derivatization (Bruckner et al., 1995; Ilisz et al., 2008; Zahradnickova et al., 2008). For example, o-phthalaldehyde (OPA) is a pre-column derivatization reagent in reversed phase HPLC (Jarrett et al., 1986) and ninhydrin (triketohydrindene hydrate) is the post- column derivatization reagent in cation exchange chromatography (Olk, 2008). Two recently developed methods used in soil and relevant samples are capillary electrophoresis with laser-induced fluorescence (CE-LIF) (Warren, 2008) and anion chromatography–pulsed amperometry (Martens and Loeffelmann, 2003).

Capillary electrophoresis (CE) is a relatively new technique that is increasingly being used in separation and characterization of components of complex mixtures in soil research (He et al., 2010; 2008b). Whereas CE with ultraviolet-visible (UV-vis) detection has been used to quantify AA and amino sugars in plant leaves (Warren and Adams, 2000), the detection limit with UV-vis detection is typically one or two orders of magnitude lower that for GC and HPLC (Warren, 2008) . Thus, Warren (2008) further tested the CE method with laser-induced fluorescence (LIF). Because most AA compounds do not possess fluorophores, they were derived with 3-(2-furoyl)quinoline-2-carboxyaldehyde to form intensely fluorescent isoindoles. The CE-LIF method separated the 17 common AA within 12 min, with detection limits between 7 and 250 nM. Using this method, this research group has analyzed AA in 1 M KCl extracts from grassland and forest soils (Warren, 2008; 2009).

Alternatively, Martens and Loeffelmann (2003) reported the use of ion chromatography coupled with a new method of amperometric detection to determine AA compounds in soil samples. The method involved acid digestion with a nonoxidizing acid, 4 M methanesulfonic acid (MSA), neutralization of excess acidity, and dilution as the only steps of sample preparation. Although Martens and Loeffelmann (2003) claimed that 4 M MSA does not destroy tryptophan, serine, and threonine, a tryptophan peak is not apparent in the chromatographs of their MSA extracts. The use of amperometric detection gives the added advantage of detecting both AA compounds and amino sugars in a single chromatographic analysis (Figure 8.1). This method has been used to quantify AA compounds in soil, plant and manure samples (He et al., 2009a; Martens et al., 2006; Olk et al., 2008). In addition, this method has been used to monitor soil mineralization of 11 added AA compounds in laboratory incubations to investigate the contribution of soil AA mineralization to N_2O flux from semi-arid soils (McLain and Marten, 2005). Based on Olk (2008), the pulsed amperometry analysis for amino compounds may be incrementally better than the conventional approach by HCl extraction, cation exchange chromatography, ninhydrin derivatization, and detection by visible light absorption, even though both amino acids and sugars (i. e. galactosamine and glucosamine) could also be detected by the conventional approach.

8.3. AMINO ACID DISTRIBUTION IN ANIMAL MANURE

In 1974, Blair (1974) reported the chemical composition of poultry litter. The main features of those poultry litter samples were a high content of nitrogen (43.2 g kg^{-1} of air dried matter), of which 39% (16.7 g kg^{-1} of air dried matter) was protein- or AA-N. Based on the distribution of AA in these poultry litter samples (Figure 8.2), Blair (1974) claimed that the proteins and AA in poultry litter are broadly equivalent to those of a cereal such as barley (*Hordeum vulgare* L.). Blair (1974) further hypothesized that this protein or AA fraction is composed of undigested feed protein, endogenous fecal and urinary compounds and internal secretions, spilled feed, and probably products of bacterial fermentation both in the lower gut of the bird and in the accumulated manure in the poultry house. This argument seems reasonable, as the order of the relative abundance of these AA compounds is basically consistent with the average occurrence of these AA compounds in protein (Table 8.1).

Table 8.1. Formulae and molecular mass of the "standard" amino acids of proteins[a].

Name	Symbol	Formula (group)[b]	Molecular mass	Average occurrence in proteins (%)
Alanine	Ala	$CH_3CH(NH_2)COOH$ (n)	89.1	9.0
Arginine	Arg	$(NH_2)_2CNH(CH_2)_3CH(NH_2)COOH$ (c)	175.2	4.7
Asparagine	Asn	$NH_2COCH_2CH(NH_2)COOH$ (u)	132.2	4.4
Aspartic acid	Asp	$HOOCCH_2CH(NH_2)COOH$ (c)	133.2	5.5
Cysteine	Cys	$HSCH_2CH(NH_2)COOH$ (u)	121.2	2.8
Glutamic acid	Glu	$HOOC(CH_2)_2CH(NH_2)COOH$ (c)	147.2	6.2
Glutamine	Gln	$NH_2CO(CH_2)_2CH(NH_2)COOH$ (u)	146.2	3.9
Glycine	Gly	$CH_2(NH_2)COOH$ (n)	75.1	7.5
Histidine	His	$(C_3N_2H_3)CH_2CH(NH_2)COOH$ (c)	155.2	2.1
Isoleucine	Ile	$CH_3CH_2CH(CH_3)CH(NH_2)COOH$ (n)	131.2	4.6
Leucine	Leu	$(CH3)_2CHCH_2CH(NH_2)COOH$ (n)	131.2	7.5
Lysine	Lys	$NH_2(CH_2)_4CH(NH_2)COOH$ (c)	147.2	7.0
Methionine	Met	$CH_3S(CH_2)_2CH(NH_2)COOH$ (n)	149.2	1.7
Phenylalanine	Phe	$(C_6H_5)CH_2CH(NH_2)COOH$ (n)	165.2	3.5
Proline	Pro	$(NHCH_2CH_2CH_2)COOH$ (n)	115.1	4.6
Serine	Ser	$HOCH_2CH(NH_2)COOH$ (u)	105.1	7.1
Threonine	Thr	$CH_3CH(OH)CH(NH_2)COOH$ (u)	119.2	6.0
Tryptophan	Trp	$(C_8H_6N)CH_2CH(NH_2)COOH$ (n)	204.2	1.1
Tyrosine	Tyr	$HO(C_6H_4)CH_2CH(NH_2)COOH$ (u)	181.2	3.5
Valine	Val	$(CH_3)_2CHCH(NH_2)COOH$ (n)	117.1	6.9

[a] Based on Voet and Voet (1990).

[b] (n)-with nonpolar side chains; (u) with uncharged polar side chains; and (c) with charged polar side chains.

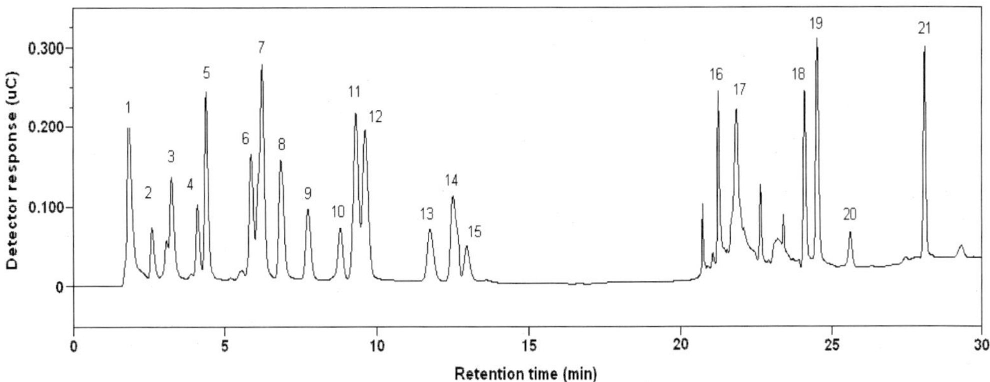

Figure 8.1. A representative chromatogram of amino acids and amino sugars obtained from poultry manure by extraction with methanesulfonic acid, followed by anion chromatographic separation and pulsed amperometric detection. Peaks are numbered to represent the following compounds: (1) arginine, (2) ornithine, (3) lysine, (4) galactosamine, (5) glucosamine, (6) alanine, (7) threonine, (8) glycine, (9) valine, (10) hydroxyproline, (11) serine, (12) proline, (13) isoleucine, (14) leucine, (15) methionine, (16) histidine, (17) phenylalanine, (18) glutamic acid plus glutamine, (19) aspartic acid plus asparagine, (20) cystine, and (21) tyrosine. The detector response is in units of charge.

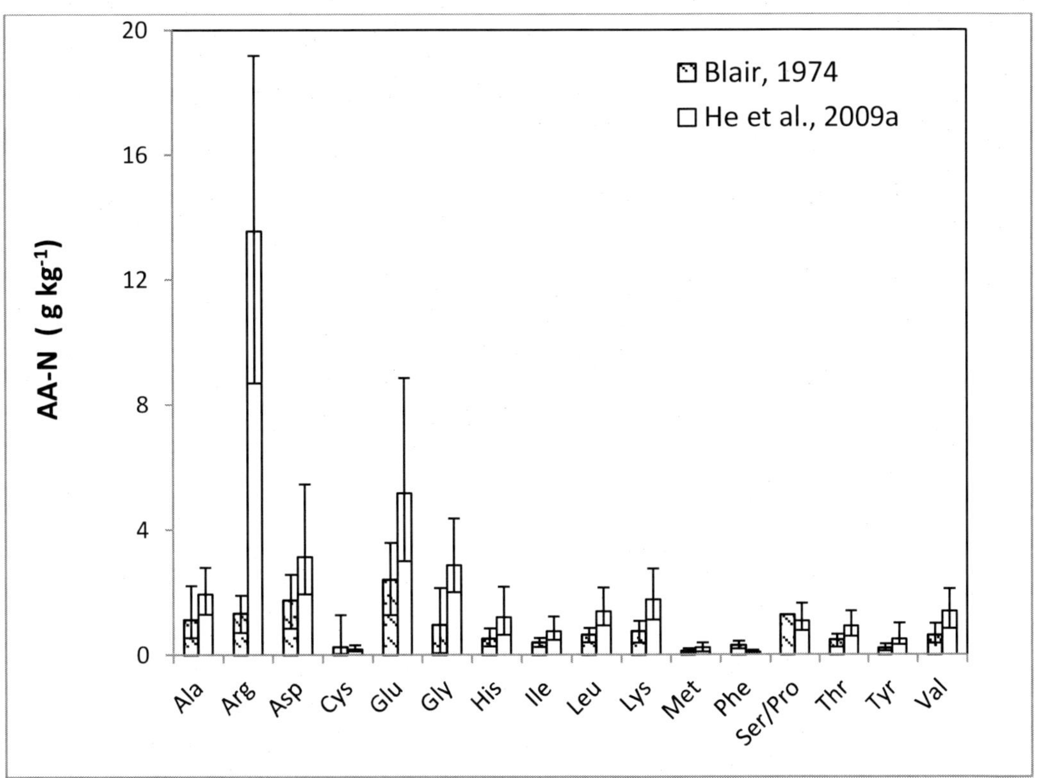

Figure 8.2. Amino acid (AA)-N contents in poultry litter for Blair (1974) and He et al. (2009a). Data are presented as means (g kg^{-1} of air dry matter) with the range of variation among the samples. Data from He et al. (2009a) are means of 23 poultry litter samples. Blair (1974) did not report the number of samples. Amino acid abbreviations are explained in Table 8.1.

Table 8.2. Correlation coefficients (r) and probability of statistical significance (p in parentheses) between either amino-N input or total N input and soil total N content in 10 long-term experiments[a].

Name	Period	Treatments	r and p between AA-N input and soil total N content	r and p between total N input and soil total N content
IOSDV, Berlin-Dahlem	1984-2003 (20 years)	Unfertilized; min; FYM; FYM + min; crop residues; crop residues + min	0.30 (p = 0.20)	0.07 (p = 0.42)
LTE, Darmstadt	1981-1995 (15 years)	Min; FYM; FYM + bd preparations; each type at 3 rates	0.77 (p = 0.002)	0.42 (p = 0.061)
DOK trial, Therwil	1978-1998 (21 years)	5 cropping systems fertilized with composted FYM + bd preparations; rotted FYM; FYM + min.; min solely; unfertilized	0.67 (p = 0.05)	0.00 (p = 0.50)
K trial, Järna	1958-1985 (28 years)	Unfertilized; min at 3 rates; comp. FYM; the same plus bd spray preparations; fresh FYM; fresh FYM + min	0.31 (p = 0.14)	0.35 (p = 0.11)
Static Fertilization Experiment , Bad Lauchstädt plus Legumes	1926-1982 (57 years)	Unfertilized; legumes; min; min + legumes; FYM at 2 rates; FYM at 2 rates + min; FYM at 2 rates + legumes; FYM at 2 rates + min + legumes	0.75 (p = 0.0003)	0.63 (p = 0.002)
Static Fertilization Experiment, Bad Lauchstädt	1903-1930 (28 years)	Unfertilized; min; FYM at 2 rates; FYM at 2 rates + min	0.75 (p = 0.018)	0.73 (p = 0.019)
IOSDV, Puch	1984-2004 (21 years)	Unfertilized; min; FYM; straw; straw + legumes; slurry; straw + slurry; all the organic treatments were also combined with min	0.54 (p = 0.007)	0.38 (p = 0.04)
Fertilizer Combination Trial, Seehausen	1967-1996 (30 years)	Unfertilized; min; FYM + min; high rate of FYM; high rate of FYM + min	0.89 (p = 0.014)	0.80 (p = 0.025)
IOSDV, Speyer	1983-2003 (21 years)	Unfertilized; min at 2 rates; FYM; FYM + 2 rates of min; crop residues; crop residues + 2 rates of min	0.58 (p = 0.015)	0.30 (p = 0.13)
Static Soil Fertility Trial, Thyrow	1937-1978 (42 years)	Mineral fertilizer at 2 rates; FYM at 2 rates; FYM at 2 rates + min; green manure; green manure + min; straw; straw + min	0.80 (p = 0.0006)	0.31(p = 0.10)

[a] Adapted from Schuler et al. (2008). IOSDV = international organic nitrogen long-term trial, min = mineral fertilizer, FYM = farmyard manure, bd = biodynamic.

Recently, He et al. (2009a) reported the AA composition in 23 poultry litter samples (Figure 8.2). The contents of the 17 AA compounds varied among the 23 samples. The contents of most AA compounds in the 23 samples were greater than those of Blair's manure samples (1974). Correlation analysis reveals a correlation coefficient of 0.556, which is statistically significant at $P<0.05$, between the two sets of data. In other words, there was some similarity in the pattern of AA contents between these recent samples and those manures analyzed decades ago; however, the absolute amounts of AA contents were greater in the latest manure samples. Only the contents of cystine, phenylalanine and serine/proline in the 23 poultry litter samples were roughly equivalent to, or lower than those in Blair's work (1974). He et al. (2009a) reported that AA-N represented 75% of total N in the 23 samples, compared to 39% in Blair's work. Several possible causes could explain the systematically higher AA contents in the 23 poultry samples, including different procedures for drying and preparing manure samples, different analytical methods, and increasing protein/AA supplies in poultry feedstuff.

An AA profile of solids recovered from hog manure was reported by Iniguez-Covarrubias et al. (1986). The authors reported grams of each AA per 100 g of protein, rather than the AA or AA-N contents per kg of dry matter. For this reason, we are not able to discuss and compare these data with the AA contents of other manures reported in literature. However, it is worth mentioning that the AA profile of the hog manure solids was comparable to that of commercial cattle feed (Iniguez-Covarrubias et al., 1986). For example, the most abundant AA in both materials was glutamic acid, with 20.5and 22.6 g per 100 g of protein, respectively. Methionine was the least abundant AA with 2.2 and 1.1 g per 100 g of protein, respectively. These data suggest the nutritional potential of hog manure as cattle feed.

Liang et al. (1996) reported AA contents in water extracts of stockpiled and composted dairy manures. Total amounts of AA-N in the water extracts of the stockpiled manures were five times greater than those extracted from the composted manure, with 7.16 and 1.37 g AA-N kg^{-1} of dry matter, respectively. The most abundant AA compounds were neutral and acidic compounds (Figure 8.3). The water extract of the composted dairy manure was relatively richer in aspartic acid, glutamic acid, lysine, proline, glycine, alanine, isoleucine and leucine, but depleted in serine and valine compared to the water extracts of the stockpiled manure. Moreover, arginine, threonine, tyrosine, phenylalanine, and methionine were not present in the water extract of the composted manure. These changes in AA-N contents in the water extracts of stockpiled and composted manures reflect decomposition of proteinaceous materials during composting of the dairy manure (Liang et al., 1996).

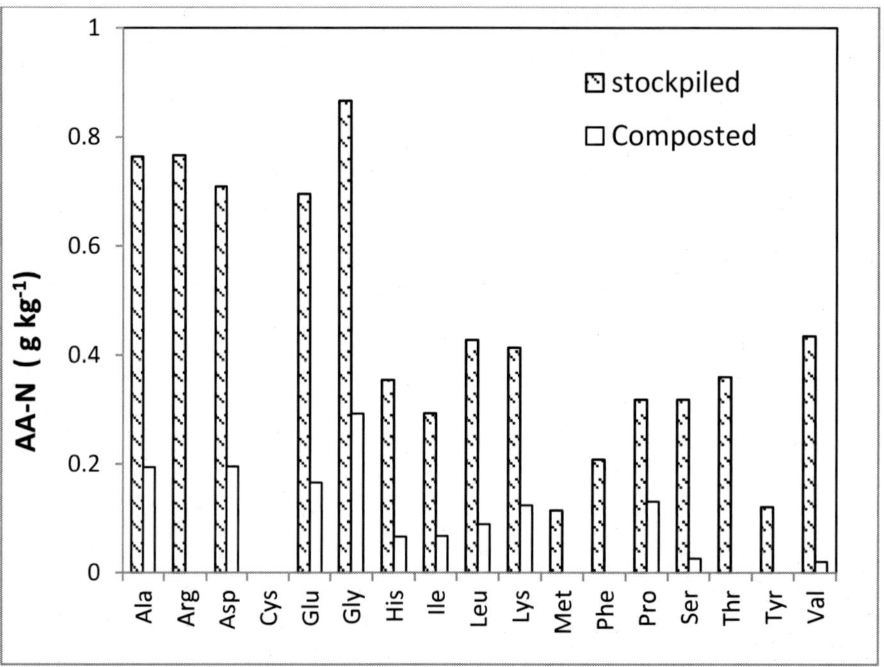

Figure 8.3. Amino acid (AA)-N contents in water extracts of stockpiled and composted dairy manures. Amino acid abbreviations are explained in Table 8.1. Data are adapted from Liang et al. (1996).

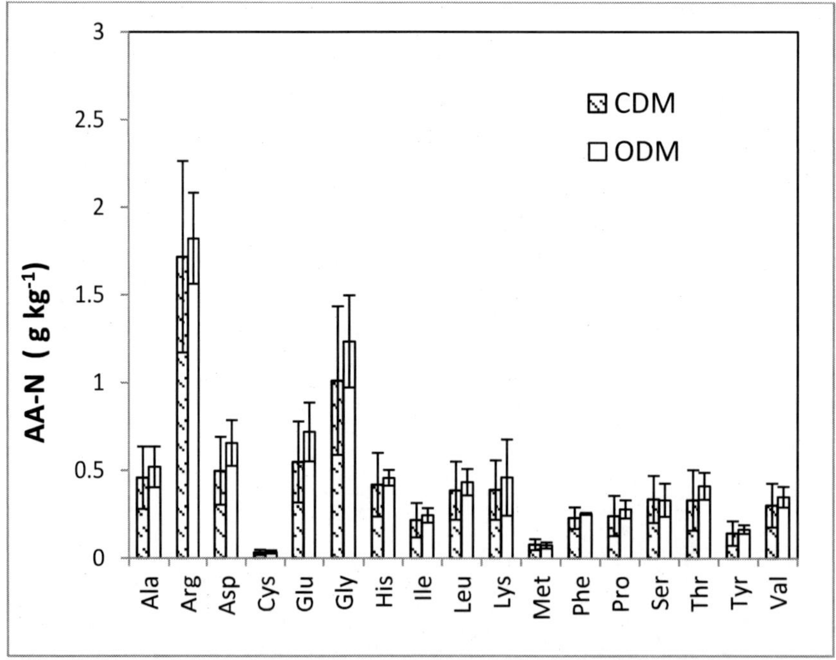

Figure 8.4. Amino acid (AA)-N contents in conventional dairy manure (CDM) and organic dairy manure (ODM). Data (He et al., unpublished) are presented (g kg^{-1} of air dry matter) with standard deviation bars and are based on four samples for CDM and 14 samples for ODM. Amino acid abbreviations are explained in Table 8.1.

Organic dairy production has increased rapidly in recent years. For example, the number of organic dairy farms in Maine, USA, increased from less than 5 small to medium-size farms in 1998 to more than 60 in 2005 (He et al., 2006). Organic dairy farms have significant differences from their conventional counterparts, including fewer imports of protein and energy feeds, a higher proportion of forage crops in the ration, and increased reliance on manure and compost as nutrient sources, rather than chemical fertilizers. These differences may significantly impact the availability, utilization, and cycling of manure nutrients (He et al., 2009c). Thus, we (He et al. unpublished data) compared the AA distributions in conventional and organic dairy manures (Figure 8.4). Individual AA levels in the dairy manure are generally only 10 to 40% of those in poultry litter, except for phenylalanine which is about three times higher in the dairy manure (Figure 8.2). However, these individual AA concentrations are about two times higher than the concentrations of AA compounds in water extracts of dairy manure reported in the literature (Figure 8.3) (Liang et al., 1996). The distribution of AA in these dairy manures is more similar to that of whole poultry litter than to that of water extracts of dairy manure. This observation implies that not all manure-bound AA compounds have the same solubility. Previously, Hawkins et al. (2006) determined dissolved free and combined amino acids in surface runoff and drainage waters from drained and undrained grassland under different fertilizer management. They found both had larger concentrations in the waters exported from undrained soils than from drained soils, and attributed the difference in part to the assumption that less proteinaceous material may have been transferred down the soil profile in drained soil compared with material moving laterally in undrained soil. Whereas the AA distribution patterns in the two types of dairy manures are similar, the concentration of each AA compound is always equal or greater (up to 22% higher with glycine) in organic dairy manure than in conventional dairy manure (Figure 8.4). This observation implies either a protein-rich diet of cows on organic farms or less digestibility of AA or proteins in organic dairy feedings compared to those in conventional feeding stuffs. If the second possibility is further confirmed, it would suggest that best management practices of organic dairy manure farms should include measurement of AA digestibility of organic dairy feeding stuffs.

8.4. SELECTIVE AMINO COMPOUNDS IN ANIMAL MANURE

Only a couple of publications have evaluated the contents of selected manure AA compounds, such as tryptophan which was not detectable in the routine AA analysis. Hacking et al. (1977) collected excreta from laying hens at daily and weekly intervals and determined their concentrations of available methionine, leucine, and tryptophan using a microbiological method (Ford, 1964). Their data indicated that one-week storage of the excreta resulted in a decrease in N content from 59.5 g kg^{-1} to 54.1 g kg^{-1} and in the moisture content from 108.5 g kg^{-1} to 83.3 g kg^{-1}. However, the mean values of available methionine, tryptophan and leucine were affected little by storage and were 0.34, 0.56 and 1.04 g per 16 g of manure N, respectively. Given the interest in re-using animal manures as livestock feed, these data provided some nutrient information on poultry manure as a potential dietary ingredient for both ruminants and poultry.

Arkhipchenko et al. (2006) determined the concentration of tryptophan in manure using a high-performance liquid chromatography method. The samples included (a) poultry litter dung, (b) pig farm activated sludge, (c) sediment from pig farm liquid waste, (d) liquid dung from poultry plants, and (e) three fertilizers produced from these manure materials via aerobic or anaerobic transformation and drying. They found the poultry litter dung contained the highest level of tryptophan (460 mg kg^{-1}), followed by activated sludge (363 mg kg^{-1}), liquid dung (226 mg kg^{-1}), and wastewater sediment (142 mg kg^{-1}). The tryptophan concentration in fertilizers produced from these manure varied considerably depending on the type of fertilizer. The largest total concentration of tryptophan, 388.5 mg kg^{-1} (0.04% of the sample weight), was observed for a fertilizer sample, produced from a mixture of activated sludge and pig farm wastewater sediment. The amount of this amino acid in this fertilizer was more than 10 times greater than those of the two fertilizers produced from poultry dung, in which only 5.2% and 7.7% of the initial tryptophan concentration were retained after aerobic and anaerobic processings, respectively. Soil tryptophan levels have been reported to vary from 0.08 to 5.8 mg kg^{-1} (Lebuhn and Hartmann, 1993; Lebuhn et al., 1994). In case of lower soil tryptophan levels, application of these organic fertilizers at the recommended rate of 1–5 g per kg of soil would enrich the soil with 0.02–1.9 mg exogenous tryptophan per kg soil (Arkhipchenko et al., 2006). As a physiological stimulation by exogenous tryptophan is expressed within a range of 0.01–3 mg kg^{-1} of soil (Ahmad et al., 2008; Arshad et al., 1995), the application of the organic fertilizers has the potential of raising the soil tryptophan concentration to physiologically relevant levels (Arkhipchenko et al., 2006).

8.5. AMINO ACIDS IN SOILS WITH LONG-TERM MANURE APPLICATION

Beavis and Mott (1996) investigated the effects of land use on the amino acid composition of soils. AA concentrations were measured in hydrolysates of manured and unmanured soils collected in 1881 and 1980 from the Broadbalk continuous wheat experiment at Rothamsted (England). The authors selected 15 well resolved and reliably identified AA (i. e. Lys, His, Arg, Asp, Thr, Ser, Glu, Gly, Ala, Val, Met, Ile, Leu, Tyr, and Phe) to represent the AA composition. They found that all the measured AAs except Ile, Met and Tyr were of higher concentration in the 1980 farmyard manure treated soil than in the 1980 untreated soil, and all the measured AAs except His, Arg, Met, Tyr and Phe were of higher concentration in the farmyard manure treated soil of 1980 than in the farmyard manure treated soil of 1881. These results show that there is a measurable distinction in AA distribution pattern between a soil treated with farm yard manure and an untreated soil. The manure-treated soil appears to be stable in AA composition over the sampling interval, but the authors were not able to rule out the possibility that the composition of the farm yard manure alone was responsible for this constancy, or indeed for the difference between the treated and untreated soils. Furthermore, the authors concluded that a treatment effect can be measured in the relative proportions (i. e. percent of individual AA of total molar AA concentration) after removal of the absolute concentration trends, thus showing the AA distribution does not simply reflect the changing concentration of soil organic matter. Discriminant analysis of the treatment groups, after the effect of absolute AA concentration had been removed, suggests that it may be possible to identify an AA "fingerprint" for classifying soils to reflect land use.

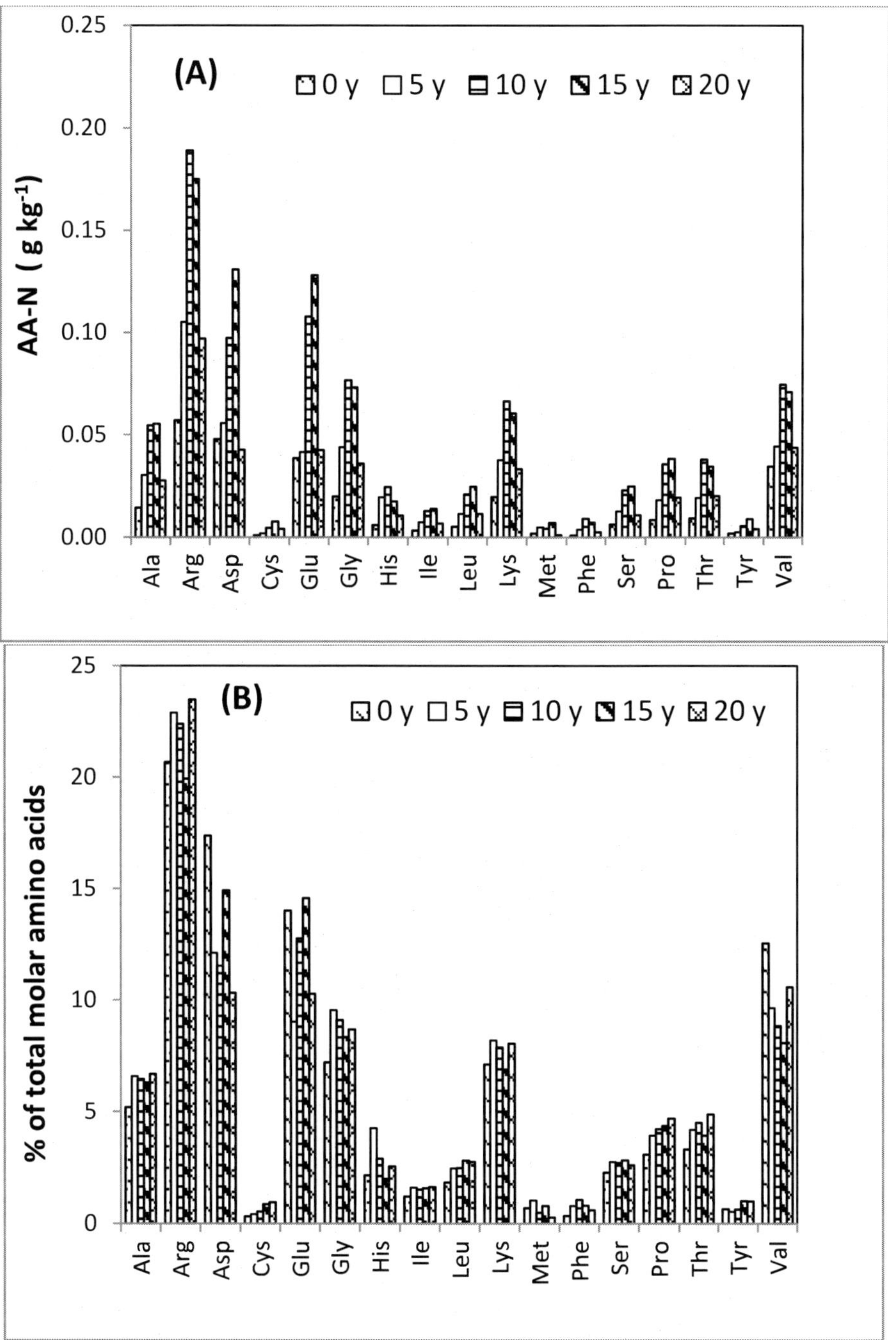

Figure 8.5. Amino acid (AA) contents in Alabama pasture soils with 0, 5, 10, 15, and 20 years of poultry litter application. (A) AA-N in mg kg^{-1} of soil; (B) Percent of each AA of total molar AA concentration for each soil. Amino acid abbreviations are explained in Table 8.1. Data are adapted from He et al. (2009a).

Scheller and Raupp (2005) evaluated soil AA and organic matter contents of topsoils in a long-term (15-year) experiment that had been amended with either mineral fertilizer (MIN), composted cattle manure (CM), or composted cattle farmyard manure with the addition of biodynamic compost and field preparations (CMBD) (Heinze, 2010). The authors measured total 16 hydrolyzable AA compounds without cysteine and tryptophan. With MIN, the lowest contents of total soil C and N that were measured in the top 25 cm soils were 0.80% C and 0.069% N, with CM they were 0.90% C and 0.080% N, and with CMBD they were 1.08% C and 0.094% N. The soil AA compounds showed the same trend among treatments as did organic C and total N contents: MIN<CM<CMBD. The 16 measured AA compounds comprised about 43 to 54% of total N in these soils. Apart from the least abundant AA's (histidine and methionine), the difference in AA contents between the mineral treatment and the two manure treatments were significant (P=0.05). Except for arginine and proline, there were also significant differences between MIN and CM and between CM and CMBD. However, increasing application rates of either mineral fertilizer or composted manure had no clear influence on levels of the 16 AA (Scheller and Raupp, 2005).

A long-term field experiment of land applied poultry litter to pasture soils had been maintained for two decades in the Sand Mountain region of north Alabama, USA (He et al., 2008a; 2009b). Soil samples in five fields had received poultry litter continuously for the past 0 (control), 5, 10, 15, and 20 years at rates of 0, 2.27, 2.27, 3.63, and 1.36 Mg ha^{-1} yr^{-1}, respectively. These fields were hayed and fertilized with litter only and were being managed for pasture without supplemental inorganic N, P or K fertilizers. All 17 AA compounds were detected in the pasture soils with or without poultry litter application (Figure 8.5A). Their concentrations were about 2-3 magnitudes lower than those in poultry litter (Figure 8.2). However, their relative abundances were similar to those of poultry litter. That is, arginine, aspartic acid, and glutamic acid were the most abundant AAs in these pasture soils. is the concentrations of the 16 amino acids generally varied in the sequence 0 y < 5 y ≈ 20 y < 10 y ≈ 15 y. This observation suggests that AA compounds from poultry litter had accumulated in these pasture soils. Previously, He et al. (2009b) demonstrated that the levels of total P, Mehlich-3 extractable P, and total N in these soils were related to cumulative amounts of poultry litter applied, with an order of 0 y < 5 y < 10 y < 20 y□ < 15 y. However, the build-up pattern of AA in these soils was not consistent with the order of annual or cumulative rates of poultry litter application, implying that AA compounds from poultry litter in the five time treatments were differentially mineralized or lost from the soil by runoff or other pathways. Furthermore, the percentage of AA-N in total soil N decreased with the year of poultry litter application as AA-N accounted for 86% of total N in the control pasture soil, but only 77%, 71%, 69% and 40% in the soils with 5-, 10-, 15- and 20-year poultry litter application, respectively. This observation indicated that repeated poultry litter application not only contributed AA-N, but also accelerated AA-N transformation in soil. However, when each AA is expressed as molar percent of total amino acids (Figure 8.5B), there is much less difference in distribution pattern between these soils although the relative abundances of these AA compounds are similar to those expressed in mg AA-N kg^{-1} of soil.

Schuler et al. (2008) tested the role of AA-N input to the soil in humus formation using data from10 long-term experiments with different fertilization and crop rotation treatments (Table 8.2). In 7 of the 10 experiments, the AA-N input was significantly ($P \leq 0.05$) correlated with the N content of the soil, with the correlation coefficients between 0.54 and

0.89 (Table 8.2). On the other hand, in 9 of 10 cases the correlation between total N input and soil N was weaker and in many cases statistically not significant. From these data, Schuler et al. (2008) concluded that the AA-N supply was more important for accumulation of soil N than was the total amount of N applied by various organic fertilizers, including manure. This conclusion suggests greater lability of inorganic N (i. e. NH_4^+ and NO_3^-), thus causing it to have less effect on the accumulation of soil N, although Schuler et al. (2008) do not state this suggestion. Furthermore, the linear regression between amino-N input and the N surplus in fertilized compared to unfertilized treatments for all organic treatments showed a much lower probability of statistical significance (p=0.158) than does the regression for manure treatments (p=0.071), implying that AA-N from farmyard manure more effectively builds stable soil N than does AA-N from other organic fertilizers (Table 8.2). Schuler et al. (2008) proposed that the role of AA-N and its fate during composting (and other processes of organic matter decomposition) should be investigated in more detail in order to better understand the importance of AA-N in soil N accumulation and synthesis of humic substances in agricultural soils.

An analysis of the relative $^{15}N/^{14}N$ ratio of an analyte to the relative $^{15}N/^{14}N$ ratio of a standard (expressed as $\delta^{15}N$ value in $^0/_{00}$) has been widely used to study processes and systems involving amino acids (Bol et al., 2004; Sacks and Brenna, 2005). Simpson et al. (1999) reported $\delta^{15}N$ values for AA in cattle manure and topsoils (15-cm depth) from manured and unmanured grassland on an experimental farm in Northumberland, England. Their data demonstrated that a distinction between soil organic matter derived from either manure or vegetation cover can be based on $\delta^{15}N$ values of some hydrophobic soil AA compounds (i. e. AA with nonpolar side chains in Table 8.1). Specifically, valine, alanine, leucine and isoleucine were isotopically heavier in manured grassland than in unmanured grassland by between 1.8 and 7.3 $^0/_{00}$. In contrast, hydrophilic AA compounds provided no distinction in land management practices.

To elucidate those soil AA-N dynamics that are affected by annual N fertilizer and manure applications, Bol et al. (2008) examined the long-term dynamics of AA compounds in a bare fallow soil experiment established in 1928 at Versailles, France. The management practices were unamended control and addition of ammonium sulphate, ammonium nitrate, sodium nitrate or manure (a horse manure and wheat straw mix). Topsoil (0-25 cm) from 1929, 1963, and 1997 was analyzed for total C, N and ^{15}N contents and distribution of 18 AA compounds. From 1929 to 1997 , soil N, C and AA contents were reduced in the control and inorganic fertilizer treatments, but increased in manured treatments. However, the absolute N loss during this time was 3-11 times greater with manure than other four treatments, due to the much higher N annual inputs applied to the manure treatment. These observations are similar to those observed by others (Schuler et al., 2008; Wander et al., 2007).

Bol et al. (2008) obtained further insights of soil AA changes with this manure application by $\delta^{15}N$ isotope analysis of individual AA compounds. For example, $\delta^{15}N$ of leucine in manured samples became more enriched from 1929 to 1997 than in the other treatments. The variation of alanine $\delta^{15}N$ values among treatments was related to source input in the sequence control<N fertilizers<manure, supporting the observation by Simpson et al. (1999) that alanine $\delta^{15}N$ value in manured soil was ca. $2^0/_{00}$ higher than that in unmanured soils. Manure application also tended to increase histidine $\delta^{15}N$ values compared to the

other four treatments. The authors attributed this increase to the presence of relatively undecomposed, but enriched $\delta^{15}N$ plant-derived material in the manure.

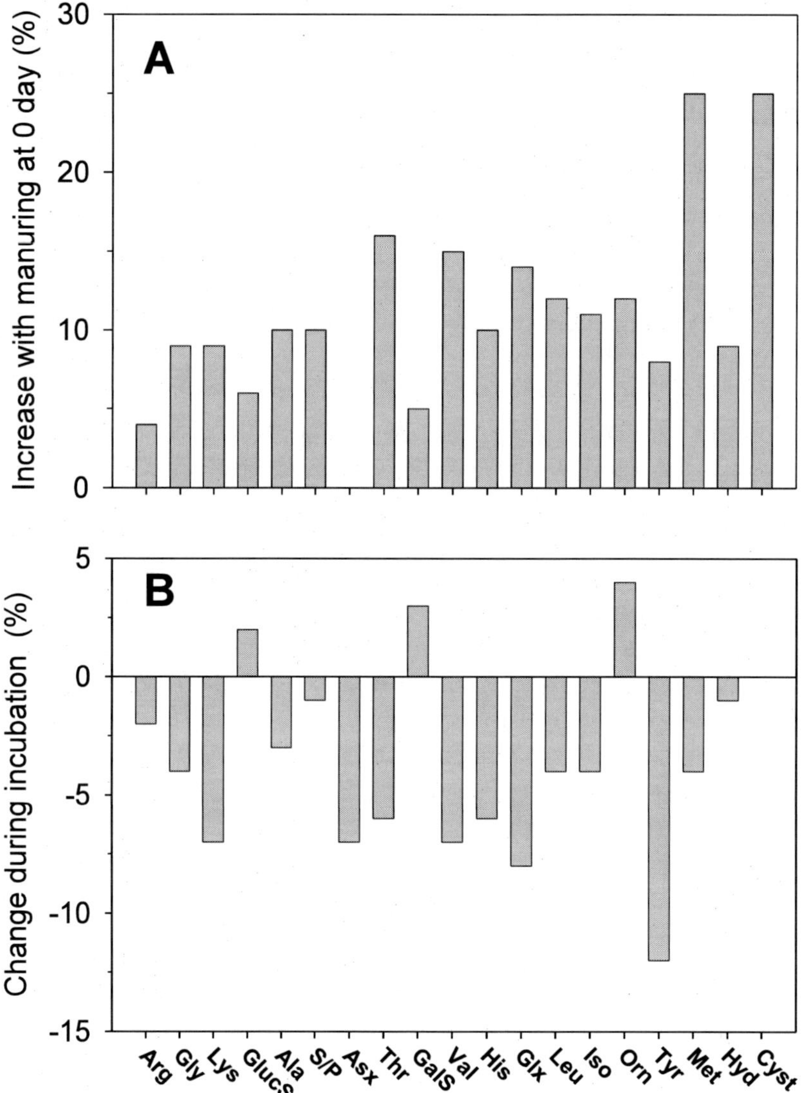

Figure 8.6. Mean percent change for nine soils in concentration of amino acids and amino sugars (a) with addition of dairy slurry manure, and (b) during a subsequent 28-day incubation. Amino acid abbreviations are explained in Table 8.1. Figure is modified from Olk et al. (2008) with permission of the Soil Science Society of America.

8.6. SHORT-TERM IMPACTS OF MANURE APPLICATION ON SOIL AMINO ACIDS

In a controlled incubation, dairy slurry manure was applied to nine soils from six U.S. states (Olk et al., 2008). Immediately after manure application, soil concentrations were found to numerically increase with manuring for 16 of 17 amino acids (Figure 8.6A). After a 28-day aerobic incubation, soil levels decreased for 16 of the 17 amino compounds by on average 11% (Figure 8.6B), suggesting either modest mineralization of the amino acids or their chemical stabilization onto the soil beyond extraction. The remaining amino acid (ornithine) along with two amino sugars (glucosamine and galactosamine) increased during the incubation by on average 2%, and they are also the compounds of solely microbial origin. These results suggest some microbially driven decomposition of the amended manure during the 28-day incubation.

Olk et al. (2008) found soil type affected several properties involving soil manure, including percent of soil N that was extractable as AA-N plus amino sugar-N (amino-N), the increase in extractable amino-N within minutes of manure addition, and the amount of N extracted as each amino compound with or without manuring. No single soil property appeared to be primarily responsible for these relationships. The extraction appeared to be more efficient for coarser soils (sandy) than finer soils (silty and clayey). The authors viewed soil type as a more important factor of extractable amino-N after the 28-d incubation than was the degree of N mineralization during the incubation. Across all nine soils they identified 41-43% of total soil as amino-N by extracting with methanesulfonic acid and separating with anion exchange chromatography, which is only marginally better than previous estimates in the literature using HCl extraction and cation exchange chromatography.

To study the impacts of animal manure on soil amino acids during slightly longer durations, we (He et al. unpublished data) evaluated AA-N distribution and changes in rhizospheric soils amended with two levels of poultry manure. A soil (sandy loam, no established soil series designation; coarse-loamy, mixed, frigid, Typic Haplorthod) was collected from a USDA-ARS research site at Newport, Maine, USA (Waldrip et al., 2009). This soil was mixed with sand (3:1) and with no poultry manure added (Control), soil with poultry manure added to meet crop P demands (Low PM, 3.8 g poultry manure kg^{-1} of soil), and soil with poultry manure added 5 times greater than crop P demand (High PM, 19.0 g poultry manure kg^{-1} of soil). Rye grass was grown in greenhouse in pots holding this soil mixture for up to 16 weeks (Waldrip-Dail et al., unpublished data). At 4 weeks and 16 weeks after ryegrass planting, the tightly adhering rhizospheric soil was gently brushed off roots of sampled plants with a paintbrush and collected for AA-N analysis. At the 4-week sampling of the rhizospheric control soil lacking manure input, arginine was the most abundant AA, and cysteine and methionine were the least abundant AA (Figure 8.7), as observed in other soils. The relative abundances of other AA compounds in the rhizospheric control soil were not always in the same order of AA distribution as with the pasture Alabama soil (Figure 8.5), but the general trend was similar. Another interesting observation is that the measured AA levels in both low and high manure-amended soils were lower than the theoretically calculated AA contents based on the control soils and contribution of manure AAs (Figure 8.7). The only exceptions were tyrosine in low manure-amended soil and valine in high manure-amended soil. Indeed, in some cases (such as aspartic acid, glutamic acid, and histidine), the levels of AAs were lower in the manure-amended soils than in the control soil. In those cases, manure application might have increased soil microbial or root-releasing enzyme activities to decompose or utilize these AA compounds, or these changes might represent random error in measurement.

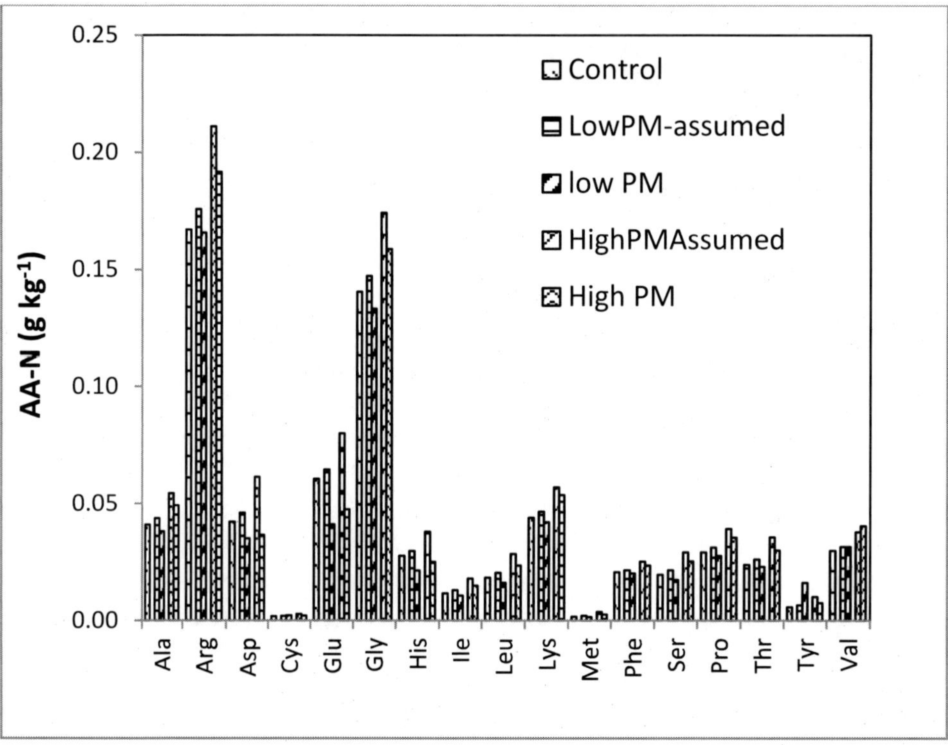

Figure 8.7. Amino acid (AA)-N contents in rhizospheric soils from a poultry manure/ryegrass study four weeks after planting of ryegrass. Data are from He et al. (unpublished data). Treatments include soil with no poultry manure added (Control); soil with poultry manure added to meet crop P demands (Low PM, 3.8 g poultry manure kg^{-1} of soil), and soil with poultry manure added 5 times greater than crop P demand (High PM, 19.0 g poultry manure kg^{-1} of soil). Low PM-assumed and High PM-assumed are calculated based on the AA contents of the control soil and the AA contents of applied manure. Amino acid abbreviations are explained in Table 8.1.

We further compared the changes of AA contents between 4 weeks and 16 weeks after planting of the ryegrass (Figure 8.8). For the unamended control, the levels of 15 of 17 AA compounds (except for arginine and phenylalanine) decreased during this growing period, indicating consumption of these AA compounds by microorganisms or plants. On the other hand, in the low poultry manure-amended soil, the levels of 16 of 17 AA compounds (except for tyrosine) increased in the same period. The increase probably reflects the release of amino acids from amended manure through stimulated soil microbial activity. Further increase in the rate of manure application from 3.8 g kg^{-1} of soil to 19.0 g kg^{-1} of soil did not maintain widespread increases in AA levels, however, as only five AA compounds (arginine, aspartic acid, cysteine, glutamic acid, and histidine) were more abundant 16 week after rye grass planting than at 4 weeks. These observations indicate that the impacts of poultry litter application on soil AA levels are quite complicated, especially in the presence of growing plants. The analysis of AA levels in the bulk soils and in soils without plant grown under the same green house conditions are under way. Those data may be useful in discriminating the roles of plant root-secreted and microbial activities in manure AA decomposition and utilization in an agricultural soil.

8.7. Conclusion

Amino acids have been long known to be present in animal manure. Amino acid compounds are widely presumed to be the primary pool of organic nitrogen (N) in soil, which provides nutrition for plant growth through N mineralization. This chapter reviews existing knowledge on AA distributions in animal manures. Although there are 20 standard protein AA compounds, only 17 standard AA compounds are normally reported in most studies, due to the lability of asparagine, glutamine, and tryptophan under assay conditions. Published data demonstrate that manure AA distribution profiles are more or less similar to those of proteins. However, management practices (such as organic feeding or manure composting) did impact the levels of AA compounds in manures. Several publications have demonstrated that repeated application of animal manure impacted soil AA distribution. The impact is more complex than simple build-up of manure AA compounds in soils, as microbial activities affect the cycling of manure AA compounds. As there have been only limited reports on distribution of AA compounds in animal manure and manure-amended soils, we hope this chapter is helpful in identifying gaps in knowledge and developing new research activities for better utilizing manure AA for sustainable and environment-friendly agricultural production.

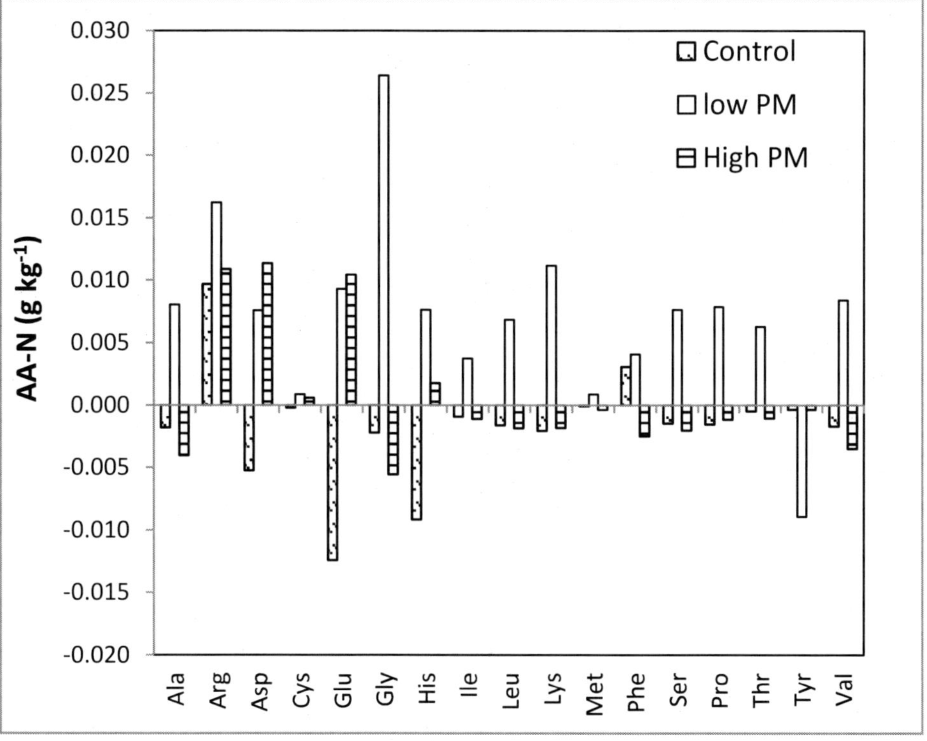

Figure 8.8. Changes in amino acid (AA)-N contents of rhizospheric soils from a poultry manure/ryegrass study between 4 weeks and 16 weeks after planting of ryegrass. Data are from He et al. (unpublished data). Treatments include soil with no poultry manure added (Control); soil with poultry manure added to meet crop P demands (Low PM, 3.8 g poultry manure kg⁻¹ of soil), and soil with poultry manure added 5 times greater than crop P demand (High PM, 19.0 g poultry manure kg⁻¹ of soil). Amino acid abbreviations are explained in Table 8.1.

REFERENCES

Ahmad, R., A. Khalid, M. Arshad, Z.A. Zahir, and T. Mahmood. 2008. Effect of compost enriched with N and L-tryptophan on soil and maize. *Agron. Sustain. Dev.* 28:299-305.

Arkhipchenko, I.A., A.I. Shaposhnikov, and L.V. Kravchenko. 2006. Tryptophan concentration of animal wastes and organic fertilizers. *Appl. Soil Ecol.* 34:62-64.

Arshad, M., A. Hussain, and A. Shakoor. 1995. Effect of soil applied L-tryptophan on growth and chemical composition of cotton. *J. Plant Nutr.* 18:317-329.

Beavis, J., and C.J.B. Mott. 1996. Effects of land use on the amino acid composition of soils. 1. Manured and unmanured soils from the Broadbalk continuous wheat experiment, Rothamsted, England. *Geoderma* 72:259-270.

Blair, B. 1974. Evaluation of dehydrated poultry waste as a feed ingredient for poultry. *Fed. Proc. Fed. Am. Soc. Exp. Biol.* 33:1934-1936.

Bol, R., N.J. Ostle, K.J. Petzke, C. Chenu, and J. Balesdent. 2008. Amino acid N-15 in long-term bare fallow soils: influence of annual N fertilizer and manure applications. *Eur. J. Soil Sci.*:617-629.

Bol, R., N.J. Ostle, C. Chenu, K.J. Petzke, R.A. Werner, and J. Balesdent. 2004. Long term changes in the distribution and d^{15}N values of individual soil amino acids in the absences of vegetation and fertilizer. *Isotopes in Environ. and Health Studies.* 40:243-256.

Bruckner, H., T. Westhauser, and H. Godel. 1995. Liquid-chromatographic determination of D-amino-acids and L-amino-acids by derivatization with O-phthaldialdehyde and N-isobutyryl-L-cysteine-- Applications with reference to the analysis of peptidic antibiotics, toxins, drugs and pharmaceutically used amino-acids. *J. Chromatogr.* A 711:201-215.

Chapin, F.S., L. Moilanen, and L. Kielland. 1993. Preferential use of organic nitrogen for growth by a nonmycorrhizal arctic sedge. *Nature* 361:150-153.

Chen, X.N., S.W. Chen, M. Sun, and Z.N. Yu. 2005. High yield of poly-gamma-glutamic acid from *Bacillus subtilis* by solid-state fermentation using swine manure as the basis of a solid substrate. *Bioresour. Technol.* 96:1872-1879.

Ford, J.E. 1964. A microbiological method for assessing the nutritional value of proteins. 3. Further studies on the measurement of available amino acids. *Br. J. Nutr.* 18:449-460.

Hacking, A., M.T. Dervish, and W.R. Rosser. 1977. Available amino acid content and microbiological condition of dried poultry litter. *Br. Poult. Sci.* 18:443-448.

Hawkins, J.M.B., D. Scholefield, and J. Braven. 2006. Dissolved free and combined amino acids in surface runoff and drainage waters from drained and undrained grassland under different fertilizer management. *Environ. Sci. Technol.* 40:4887-4893.

He, Z., and J.C. Spain. 1999. Preparation of 2-aminomuconate from 2-aminophenol by coupled enzymatic dioxygenation and dehydrogenation reactions. *J. Ind. Microbiol. Biotechnol.* 23:138-142.

He, Z., T.S. Griffin, and C.W. Honeycutt. 2006. Soil phosphorus dynamics in response to dairy manure and inorganic fertilizer applications. *Soil Sci.* 171:598-609.

He, Z., I.A. Tazisong, Z.N. Senwo, and D. Zhang. 2008a. Soil properties and macro cations status impacted by long-term applied poultry litter. *Commun. Soil Sci. Plant Anal.* 39:858-872.

He, Z., Z.N. Senwo, I.A. Tazisong, and D.A. Martens. 2009a. Amino compounds in poultry litter, litter-amended soil and plants. *The Proceedings of the International Plant Nutrition*

Colloquium XVI., Vol. http://repositories.cdlib.org/ipnc/xvi/1048. On-line publication, Sacramento, CA,.

He, Z., T. Ohno, D.C. Olk, and F. Wu. 2010. Capillary electrophoresis profiles and fluorophore components of humic acids in Nebraska corn and Philippine rice soils. *Geoderma* 156:143-151.

He, Z., I.A. Tazisong, Z.N. Senwo, C.W. Honeycutt, and D. Zhang. 2009b. Nitrogen and phosphorus accumulation in pasture soil from repeated poultry litter application. Commun. *Soil Sci. Plant Anal.* 40:587-599.

He, Z., T. Ohno, F.C. Wu, D.C. Olk, C.W. Honeycutt, and M. Olanya. 2008b. Capillary electrophoresis and fluorescence excitation-emission matrix spectroscopy for characterization of humic substances. *Soil Sci. Soc. Am. J.* 72:1248-1255.

He, Z., C.W. Honeycutt, T.S. Griffin, B.J. Cade-Menun, P.J. Pellechia, and Z. Dou. 2009c. Phosphorus forms in conventional and organic dairy manure identified by solution and solid state P-31 NMR spectroscopy. *J. Environ. Qual.* 38:1909-1918.

Heinze, S., J. Raupp, and R.G. Joergensen. 2010. Effects of fertilizer and spatial heterogeneity in soil pH on microbial biomass indices in a long-term field trial of organic agriculture. *Plant Soil* 328:203-215.

Hirai, M.F., V. Chanyasak, and H. Kubota. 1983. A standard measurement for compost maturity. *BioCycle* 24(6):54-56.

Ilisz, I., R. Berkecz, and A. Peter. 2008. Application of chiral derivatizing agents in the high-performance liquid chromatographic separation of amino acid enantiomers: A review. *J. Pharm. Biomed. Anal.* 47:1-15.

Iniguez-Covarrubias, G., M.D.J. Franco-Gomez, M. Pena-Romero, and A. Ciurlizza-Guizar. 1986. Evaluation of the protein quality of solids recovered from hog manure slurry. *Agric. Wastes.* 16:113-120.

Jarrett, H. W., K. D. Cooksy, B. Ellis and J. M. Anderson. 1986. The separation of o-phthalaldehyde derivatives of amino acids by reversed-phase chromatography on octylsilica columns. *Anal. Biochem.* 153:189–198.

Kahmen, A., S.J. Livesley, and S.K. Arndt. 2009. High potential, but low actual, glycine uptake of dominant plant species in three Australian land-use types with intermediate N availability. *Plant Soil* 325:109-121.

Karr-Lilienthal, L.K., C.M. Grieshop, J.K. Spears, A.R. Patil, G.L. Czarnecki-Maulden, N.R. Merchen, and G.C. Fahey. 2004. Estimation of the proportion of bacterial nitrogen in canine feces using diaminopimelic acid as an internal bacterial marker. *J. Anim. Sci.* 82:1707-1712.

Larkin, R.P., C.W. Honeycutt, and T.S. Griffin. 2006. Effect of swine and dairy manure amendments on microbial communities in three soils as influenced by environmental conditions. *Biol. Fertil. Soils.* 43:51-61.

Lebuhn, M., and A. Hartmann. 1993. Method for the determination of indole-3-acetic acid and related compounds of L-tryptophan catabolism in soils. *J. Chromatogr.* A 629:255-266.

Lebuhn, M., B. Heilmann, and A. Hartmann. 1994. Effects of drying/rewetting stress on microbial auxin production and L-tryptophan catabolism in soils. *Biol. Fertil. Soils.* 18:302-310.

Liang, B.C., E.C. Gregorich, M. Schnitzer, and H. Schulten. 1996. Characterization of water extracts of two manures and their adsorption on soils. *Soil Sci. Soc. Am. J.* 60:1758-1763.

Martens, D.A., and K.L. Loeffelmann. 2003. Soil amino acid composition quantified by acid hydrolysis and anion chromatography-pulsed amperometry. *J. Agric. Food Chem.* 51:6521-6529.

Martens, D.A., D.B. Jaynes, T.S. Colvin, T.C. Kaspar, and D.L. Karlen. 2006. Soil organic nitrogen enrichment following soybean in an Iowa corn-soybean rotation. *Soil Sci. Soc. Am. J.* 70:382-392.

Martin, C., A.G. Williams, and B. Michaletdoreau. 1994. Isolation and characteristics of the protozoal and bacterial fractions from bovine ruminal contents. *J. Anim. Sci.* 72:2962-2968.

McLain, J.E.T., and D.A. Martens. 2005. Nitrous oxide flux from soil amino acid mineralization. *Soil Biol. Biochem.* 37:289-299.

Olk, D.C., A. Fortuna, and C.W. Honeycutt. 2008. Using anion chromatography-pulsed amperometry to measure amino compounds in dairy manure-amended soils. *Soil Sci. Soc. Am. J.* 72:1711-1720.

Potter, M., F.B. Oppermann-Sanio, and A. Steinbuchel. 2001. Cultivation of bacteria producing polyamino acids with liquid manure as carbon and nitrogen source. *Appl. Environ. Microbiol.* 67:617-622.

Rothstein, D.E. 2009. Soil amino-acid availability across a temperate-forest fertility gradient. *Biogeochemistry* 92:201-215.

Sacks, G.L., and J.T. Brenna. 2005. N-15/N-14 position-specific isotopic analyses of polynitrogenous amino acids. *Anal. Chem.* 77:1013-1019.

Scheller, E., and J. Raupp. 2005. Amino acid and soil organic matter content of topsoil in a long term trial with farmyard manure and mineral fertilizers. *Biol. Agric. Hortic.* 22:379-397.

Schuler, C., J. Raupp, and E. George. 2008. *The importance of amino-N for humus formation studied by comparing amino-N input to the soil and soil total nitrogen content in long-term experiments.*, http://orgprints.org/view/projects/conference.htm 6th IFOAM Organic World Congress, Modena, Italy,.

Simpson, I.A., R. Bol, I.A. Bull, R.P. Evershed, and K.J. Petzke. 1999. Compound specific stable isotope signals in anthropogenic soils as indicators of early land management. *Rapid Comm. Mass Spectrom.* 13:1315-1319.

Voet, D., and J. Voet. 1990. Chapter 4. Amino acids, p. 59-74 *Biochemistry*. John Wiley & Son, Inc., New York, N.Y.

Waldrip-Dail, H., Z. He, M.S. Erich, and C.W. Honeycutt. 2009. Soil phosphorus dynamics in response to poultry manure amendment. *Soil Sci.* 174:195-201.

Wander, M.M., W. Yun, W.A. Goldstein, S. Aref, and S.A. Khan. 2007. Organic N and particulate organic matter fractions in organic and conventional farming systems with a history of manure application. *Plant Soil* 291:311-321.

Wang, H.J., L.H. Wu, M.Y. Wang, Y.H. Zhu, Q.N. Tao, and F.S. Zhang. 2007. Effects of amino acids replacing nitrate on growth, nitrate accumulation, and macroelement concentrations in pak-choi (Brassica chinensis L.). *Pedosphere* 17:595-600.

Warren, C.R. 2008. Rapid and sensitive quantification of amino acids in soil extracts by capillary electrophoresis with laser-induced fluorescence. *Soil Biol. Biochem.* 40:916-923.

Warren, C.R. 2009. Uptake of inorganic and amino acid nitrogen from soil by Eucalyptus regnans and Eucalyptus pauciflora seedlings. *Tree Physiol.* 29:401-409.

Warren, C.R., and A. M. Adams. 2000. Capillary electrophoresis for the determination of major amino acids and sugars in foliage: application to the nitrogen nutrition of sclerphyllous species. *J. Experi. Bot.* 51:1147-1157.

Weigelt, A., R. Bol, and R.D. Bardgett. 2005. Preferential uptake of soil nitrogen forms by grassland plant species. *Oecologia.* 142:627-635.

Werdin-Pfisterer, N.R., K. Kielland, and R.D. Boone. 2009. Soil amino acid composition across a boreal forest successional sequence. *Soil Biol. Biochem.*41:1210-1220.

Zahradnickova, H., P. Hartvich, P. Simek, and P. Husek. 2008. Gas chromatographic analysis of amino acid enantiomers in Carbetocin peptide hydrolysates after fast derivatization with pentafluoropropyl chloroformate. *Amino Acids* 35:445-450.

In: Environmental Chemistry of Animal Manure
Editor: Zhongqi He

ISBN: 978-1-62808-641-6
© 2013 Nova Science Publishers, Inc.

Chapter 9

DETERMINANTS AND PROCESSES OF MANURE NITROGEN AVAILABILITY

C. Wayne Honeycutt[1,], James F. Hunt[2], Timothy S. Griffin[3], Zhongqi He[1] and Robert P. Larkin[1]*

9.1. INTRODUCTION

The value of animal production in the United States well exceeds $100 billion annually (USDA Census of Agriculture, 2007). Manure generated from these industries exceeds 834,000 Mg of dry matter per day (Honeycutt et al., 2005b), with nitrogen (N) concentrations ranging from approximately 15-55 kg N Mg^{-1} dry matter (Griffin et al., 2003). Effectively managing this N to optimize its recycling to crops, while also minimizing adverse environmental consequences of manure application to cropland, requires thorough knowledge of the nutrient transformation processes and controlling factors involved when manure is added to soil. For another recent review of this area, readers are referred to Beegle et al. (2008).

9.2. NITROGEN CONTENT OF ANIMAL MANURE

The content of various N fractions in manure is strongly dependent upon animal species and age, diet (Broderick, 2003), manure condition (i.e., fresh, solid, slurry, bedding, etc.), collection procedure and storage method (Beauchamp, 1983). The term "farmyard manure" refers to manure that contains plant fibers such as straw and other bedding materials, while the term "slurry" refers to a liquid mixture of feces and urine usually collected from systems

---

[*] Corresponding Author – Wayne.Honeycutt@wdc.usda.gov
[1]USDA-Agricultural Research Service; New England Plant, Soil & Water Laboratory; Orono, ME 04469 USA
[2]Department of Plant, Soil & Environmental Sciences; University of Maine; Orono, ME 04469 USA
[3]Friedman School of Nutrition Science and Policy; Tufts University; Boston, MA 02111 USA

where slats or concrete floors are used. Most urine N is present as urea (McCrory and Hobbs, 2001), which can be quickly hydrolyzed into NH_4^+ by urease, an extracellular enzyme found in feces and soil. In contrast, while some of the organic N contained in feces is present as labile proteins, peptides and amino acids; most is bound in organic compounds of greater microbial stability (Beauchamp, 1983).

9.2.1. Dairy Manure

Assuming N excretion in milk averages 25% of dietary N intake in dairy cows, the remaining 75% of N intake is excreted in feces and urine (Sorensen, 2004). Although estimates vary, between 10 and 20% of the initial total N in cattle/dairy manure (as excreted) is present as easily mineralizable urea (Van Kessel and Reeves, 2002; Kirchmann, 1991). In a survey of 29 dairy manure sludges, solids and composts from Central Valley, CA, Pettygrove et al. (2009) reported a median NH_4-N content of 6% of total N. Van Kessel and Reeves (2002) found widely varying amounts of both NH_4-N $(0.3 - 4.7$ kg m^{-3}) and organic N $(0.4 - 5.6$ kg m^{-3}) in their characterization of 100 dairy manures. In an analysis of dairy manure slurry from five farms in southern Minnesota, Russelle et al. (2007) found initial total N content varied between 1.9-4.3 kg m^{-3}. These authors reported that initial NH_4-N comprised 41-81% of the slurry total N content.

9.2.2. Swine manure. Total N in fresh swine manure has been estimated to consist of approximately 56% NH_4-N and 44% organic N (Chastain et al., 2001). As with all stored manures, the amount of readily available N in these forms is highly dependent upon storage time and storage conditions (Koelsch and Shapiro, 2006). Storage in ponds leads to N stratification, with potentially volatile NH_4^+ present primarily at the surface and greater amounts of organic N associated with solids at depth. The bottom solids, typically referred to as sludges, are often stored for over a decade before being removed (Moore et al., 2005). The chemical composition of swine sludge from nine different anaerobic lagoons in North Carolina was characterized by Moore et al. (2005). They reported initial mean Kjeldahl N and NH_4-N of the sludge as 5.0 and 1.3 kg m^{-3}, respectively. Russelle et al. (2007) compared swine manure slurry applied to nine farms in southern Minnesota between 2004 and 2006. They reported initial total N content to vary from 4.1 to 8.9 kg m^{-3}, with NH_4-N comprising between 72 and 89% of total N. These studies underscore the importance of sampling technique and timely analysis when quantifying manure N content.

9.2.3. Poultry manure. Poultry manure N is a solid mixture of organic N containing feces and easily mineralizable uric acid (Hester et al., 1940; Qafoku et al., 2001). Poultry manure is often applied as litter, which is a mixture of manure, feathers, bedding, unused feed and other materials that are collected when poultry houses are cleaned (Moore, 1998). Fresh poultry manure N is comprised of about 70% uric acid and about 30% undigested protein (Huff et al., 1984). In aerobic, moist environments, uric acid N can be quickly hydrolyzed by the enzyme uricase to the more easily hydrolyzable urea (Nahm, 2003). Nearly all of the NH_4^+ found in poultry manure is the result of this conversion, which generally occurs during storage. Poultry manure typically contains the highest initial amount of N of any manure source, with about 1 to 6% N by weight and C:N ratios varying from 1 to 30, depending on whether it is fresh or composted (Stephenson et al., 1990; Nahm, 2003). Because of its high contents of N, P, and soluble salt; proper handling, storage and application of poultry manure are critical to avoid crop "burning" and excessive nutrient losses to the environment (Karlen et al., 1998; Nahm, 2003).

Some studies have found the organic N fraction in poultry manure to be more available (60-100%) within the first 6 months of application than the organic N fraction of mammalian manures (Bitzer and Sims, 1988; Van Faassen and Van Dijk, 1987). Other studies have found little to no mineralization from poultry manure when incubated with soil (Preusch et al., 2002; Kirchmann, 1991; Burger and Venturea, 2008). Such conflicting results may possibly reflect storage impacts on uric acid content.

9.3. TRANSFORMATION PROCESSES OF MANURE NITROGEN

Although a portion of manure organic N may remain unavailable for long periods of time, all animal-excreted N is potentially vulnerable to microbial and chemical transformations. Rates of N mineralization, immobilization, nitrification, volatilization, and leaching are all influenced to varying degrees by manure type, soil texture and condition (Schjonning et al., 1999), crop type, and by the method and timing of manure storage, treatment, and application (Kruse et al., 2004; Griffin, 2008). In short, the characteristics of both the manure and the soil environment to which it is applied all influence turnover and availability of manure N. It is little wonder that determining the quantities and rates of N turnover for different manure types has been a goal of soil and agricultural scientists for decades (Schimel and Bennett, 2004).

9.3.1. Mineralization and Immobilization

Many studies have reported immediate decline in soil inorganic N following manure addition. Researchers have often attributed these losses to microbial immobilization, although NH_3 volatilization and denitrification may also play important roles. In some cases direct evidence of the fate of manure N has been measured (Calderon et al., 2005; Burger and Ventura, 2008), while in other cases the evidence is inferred. For example, in some studies microbial immobilization of inorganic N has been shown, as reflected by decreased soil inorganic N concentrations immediately following manure addition (Sorensen and Amato, 2002; Calderon et al., 2005). Microbial assimilation of C requires consumption of N in an amount consistent with the internal C:N ratio of the microbe (Alexander, 1977). If insufficient N is present in the decomposing manure, then net N immobilization occurs (Kirchmann, 1991).

9.3.1.1. The Effect of C:N Ratio

The critical C:N ratio of manure mineralizing microbes has been reported to vary widely, from about 15 to 40 (Gilmour and Skinner, 1999; Van Kessel et al., 2000; Cabrera et al., 2005). Specific requirements of the assimilating microbial biomass, the stability of manure N compounds added, and the ever-changing characteristics of the soil environment all influence whether mineralization or immobilization predominates following manure addition. Manures with higher organic C content often have been observed to produce greater apparent immobilization (Chadwick et al., 2000; Van Kessel and Reeves, 2002). Incubation studies have shown that the type of organic C initially present in manure also strongly influences the timing and extent of N mineralization. For example, initially high amounts of water soluble C and N have been well correlated with N mineralization within the first weeks of incubation (Jensen et al., 2005). Thus, one simple and common method for estimating potentially available N (PAN) in animal manure is based on the amount of NH_4^+ and urea/uric acid initially present. Manure NH_4^+ content can be easily determined using near-infrared spectroscopy (DeFerrari et al., 2007). The amount of uric acid or urea in manures can be determined through reverse phase HPLC (Eiteman et al., 1994) and various colorimetric analyses (Nahm, 2003).

Volatile fatty acids are soluble intermediate products of anaerobic fermentation with high C:N ratios, and their decomposition by microbes leads to N immobilization (Sorensen, 1998; Van Kessel et al., 2000) or denitrification (Paul and Beauchamp, 1989). Bechini and Marino (2009) conducted a 181-day incubation study of five different liquid dairy manures in soils of varying clay content but similar pH and C:N ratio. In addition to concluding that the N fertilizer replacement value of liquid dairy manures was due to their NH_4^+ content, the authors also identified three general categories of N response based upon initial C content. The first category was characterized by a large initial labile C pool (23-28% of organic C) and high C:organic N ratio (~18) conducive to a relatively long-lasting phase of immobilization. The second category of liquid manure was characterized by a small labile C pool (12% of organic C) and a high C:N ratio (~23). These dairy manures underwent initial mineralization, followed by a period of little change, which in turn was followed by a second period of re-mineralization. Finally, the third N-response category in the study consisted of manures having a very large amount of soluble C (47% of organic C) but an initially low C:N ratio (10). Manures belonging to this last group responded to incubation with a short initial period of immobilization, followed by a long and continuous period of mineralization. The researchers concluded that initial labile C content was an important predictor of available N only in manures with low initial C:organic N ratios. Conversely, in manures with initially high C:organic N ratios, C:N ratio was a better gauge of overall mineralization than soluble C content. In a review of 100 ^{15}N dilution studies of forest, shrubland, grassland and agricultural soils, Booth et al. (2005) found strong inverse correlations between soil N mineralization rates and soil C:N ratio. In their data set, these authors also found significant correlations between N mineralization rate and both C and N concentrations, reflecting the importance of both heterotrophic microbial biomass and available substrate on these processes.

In a study of 107 incubated dairy manures in a fine sandy loam soil, Calderon et al. (2004) found that net N mineralization was generally observed when initial manure C:N was < 16, while net immobilization generally occurred with initial manure C:N ratios > 19. A follow-up 70 day litterbag study involving four types of solid cattle manure reached somewhat different conclusions regarding the N transforming effect of soluble C and low C:N ratios (Calderon et al., 2005). In that study, direct measurements were made of microbial biomass N and denitrification rate so that N losses following manure application were not assumed to only be due to immobilization. It was found that rather than undergoing rapid immobilization, solid manures having high soluble C content were far more vulnerable to denitrification. In that study denitrification rates ranged between 15 and 39% of added N, with the greatest losses found in manures having high NH_4^+ content. The authors proposed that manures having large amounts of labile C and low C:N ratio (< 15), provided significant substrate for high microbial activity that in turn led to oxygen depleted conditions in soil microsites, thus favoring denitrification. On the other hand, manures with C:N ratios greater than 15 generally resulted in immobilization.

9.3.1.2. Predicting Potential N Mineralization

A 121-day incubation study (at 25 $^{\circ}$C) of broiler chicken manure in sandy loam soils was used to validate estimates of potentially mineralizable N (PMN) as predicted from both near-infrared spectroscopy (NIRS) and water soluble organic N (WSON) (Qafoku et al., 2001). In that study, strong correlations were found between PMN and both WSON (r^2=0.87) and NIRS-predicted N (r^2=0.82). An investigation of the PMN of chicken and turkey manure in loam and clay loam Iowa soils was made by Diaz et al. (2008) using aerobic incubation. These workers found very close agreement between the amount of NO_3^- produced after 14 days and the initial total water soluble N (TWSN) content of manure. Because NO_3^- in that study did not increase after 14 days, PMN release was assumed to be completed by that time. Initial mean TWSN concentrations were found to be higher than initial NH_4^+ and uric acid alone, indicating an additional labile organic N source in these manures. The researchers concluded that TWSN concentration may be a better overall field test of PMN for poultry manure than the more traditional NH_4^+ plus uric acid determination.

Long-term N mineralization has been correlated with either total initial N content and C:N ratio (Nett et al., 2010), or non-detergent extractable N (Jensen et al., 2005; Griffin et al., 2005). Van Kessel and Reeves (2000) found weak correlation (r=0.35) between N mineralization and the ratio of initial acid detergent fiber (ADF) C to total initial manure N in dairy manure. In a 176-day aerobic incubation study in sandy loam and silt loam soils, Griffin et al. (2005) examined the relationship between recalcitrant fibrous C content and N availability from nine dairy manures. They reported mineralization rates between 0.08 and 0.12 mg N kg^{-1} soil d^{-1}, or about 10% of N added in manure. They also found that the combination of initial organic N and NH_4^+ were good predictors of PMN, while both non-detergent fiber (NDF) C and total C content had strong negative correlations with mineralization and nitrification. In their study of 47 animal wastes encompassing a large compositional range, Morvan et al. (2006) found net mineralization in 70% of the wastes by the end of the 224-day incubation period, with slurries and farmyard manures generally showing initial immobilization followed by net mineralization. By the end of the incubation, the amount of organic N mineralized from the wastes was generally less than 20%. Due to the wide variability in the wastes, the authors could find no better overall predictor for PMN than N content and C:N ratio.

Many other chemical indices have been developed to predict available N from manure. Some of these methods include pepsin-extractable N (Castellanos and Pratt, 1981) and thermal fractionation (Velthof, et al. 1998). In a study of sheep, poultry, cow and un-composted pig manures by Antil et al. (2009), thermal fractionation was found to be a poor predictor of plant-available N. In contrast, simple C:N$_{(organic)}$ ratios were found to be at least as good a predictor of organic N availability as was pepsin extraction (Antil et al., 2009).

9.3.2. Nitrification and Denitrification

Nitrification and denitrification can occur simultaneously under many circumstances, but in general, the different redox environments required for their major pathways favor one process over the other. Coarse-textured, dry soils with high aeration generally favor nitrification, while finer textured, wetter soils are more conducive to denitrification (Alexander, 1977). Production of N_2O can occur through either denitrification or direct NH_4^+ oxidation by chemoautotrophic nitrifying microbes (Blackmer et al., 1980). Thus, N_2O emissions can occur under either oxic or anoxic conditions, although the latter environments generally produce the greatest amounts of N_2O (Hayakawa et al., 2009; Ding et al., 2007). Fertilizer and manure additions to soils increase the rates of both transformation processes because they supply more N and C substrate, thereby changing the chemical and physical properties of the soil (Firestone et al., 1980; Mosier et al., 1998; Kroeze et al., 1999). Lessard et al. (1996) examined N_2O fluxes from dairy cattle-manured soils under maize from April to October. Maximum fluxes of N_2O-N emission were 0.3, 7.6 and 21.9 g m^{-2} for manure application rates of 0, 170 and 339 kg N ha^{-1}, respectively, and coincided with high levels of NO_3^- and soil water content. For the two manured soils, 67% of the total N_2O flux occurred within the first 49 days and amounted to approximately 1% of the organic N added to soil. Ginting et al. (2003) found no differences in N_2O flux from manured, composted or inorganically fertilized soils 4 years after their most recent amendment.

The easy transport of nitrate to ground and surface waters makes it a serious pollutant in some areas. It has been estimated that for the 15 years between 1982 and 1997, crop or pasture N needs were exceeded in approximately 75% of all U.S. counties where manure-producing farms were located (Gollehon and Caswell, 2000). An 8-year field study by Goulding et al. (2000) of long-term, manure-amended soils found weather patterns to be the dominant factor associated with NO_3^- loss. A comparison of study results at the historic Broadbalk experimental site in Rothamsted, England from 1878-1883 and 1990-1998, during which times essentially identical manure treatments were applied, revealed 74% greater rainfall-adjusted mean NO_3^- losses for the earlier period.

The quantity and quality of soil organic matter plays an important role in regulating nitrification rates by influencing competition between heterotrophic bacteria and autotrophic nitrifiers. A delay between NH_4^+ production/addition and the onset of nitrification may reflect the growth rate of nitrifying autotrophs (Meyer et al., 2002). Autotrophic nitrifiers utilize N for both energy production and growth, while in heterotrophic microorganisms N usually is utilized for growth alone (Alexander, 1977). High SOM C:N ratios appear to encourage the proliferation of heterotrophic microbes, while simultaneously discouraging nitrification by these same organisms and limiting autotrophic nitrification to isolated C-depleted microsites (Chen and Stark, 2000). In their extensive isotopic review study, Booth et al. (2005) observed that the relationship between nitrification and N mineralization rates can be described by a log-linear curve, characterized by only minute increases in nitrification with greatly increasing soil mineralization rates. In the early part of the 20th century it was thought that nitrification rates were severely impeded, if not halted, below pH 4.5, although an exact limiting pH varied with soil (Myrold, 1998). Today's view has changed with heterotrophic nitrification being acknowledged as a contributor to nitrate production, although autotrophic (chemolithotrophic) bacteria still appear to be the dominant nitrifying agent in acidic agricultural soils (DeBoer and Kowalchuk, 2001). Although capable of utilizing inorganic N sources, most heterotrophic microorganisms preferentially use organic N sources for nitrification (Muller et al., 2003).

In a study of N transformation rates in liquid/solid cattle, liquid hog and turkey manure, Burger and Ventura (2008) reported that more than 99% of the inorganic N in manure amended soils was present as NO_3^- after 180 days. In that study, soil water contents were kept intentionally low, with water filled pore space (WFPS) less than 60%, while slurry applications were uniform in order to prevent the formation of O_2-depleted "hotspots". As a result, minimal denitrification rates were measured (< 1.5 g N_2O-N kg^{-1} day^{-1}). The amount of NO_3^- produced by each manure type was directly related to the initial amount of NH_4^+ present. Similarly, both composted dairy and liquid cattle manures increased nitrifier activity in a silt loam by 300% (Habteselassie et al., 2006b) over a 5-year period. A 176-day aerobic incubation study of the N transformations in 9 dairy manures (Griffin et al., 2005) found that the majority of manures behaved in a manner consistent with an exponential, single pool model, with an early rapid NO_3^- build up (3-5 weeks) followed by slower accumulation as NH_4^+. Estimated growing season mineralization of the different manures averaged about 10% of initial organic N, and the same general nitrification pattern was observed in both a sandy loam and a silt loam soil. Some studies have shown slower rates of nitrification in finer textured soils, presumably due to physical protection of the substrate and lower aeration (Sørensen et al., 1994; Chadwick et al., 2000).

Denitrification is a bacterially mediated process whereby certain oxidized forms of N such as NO_3^- and NO_2^- can be utilized as alternative electron acceptors by certain microbial species, and reduced to the volatile gases N_2, N_2O and NO. In 2007, agricultural sources reportedly accounted for approximately 72% of all nitrous oxide emissions in the U.S., with manure contributing approximately 5% (US EPA, 2009). Denitrification not only reduces the efficiency of manure-derived N by plants, but atmospheric nitrous oxide is also an important greenhouse gas (Lowe et al., 2007). Due to variability in soils, environmental conditions and manure characteristics, estimates of manure N loss from denitrification differ widely, from less than 1% (Paul et al., 1993) to more than 39% (Calderon et al., 2005).

While soil temperature (Dobbie et al., 1999) and water (Skiba and Smith, 2003) strongly impact denitrification, manure characteristics also influence denitrification. Velthof et al. (2003) found approximately 0.5-1.9% of applied N to be lost as N_2O for poultry manure and 1.8-3.0% of applied N lost as N_2O for cattle slurry. The highest N_2O emissions were from liquid pig manure, which ranged between 7.3 and 13.9% of applied N. These results compare with approximately 2-4% of N lost as N_2O when applied as inorganic fertilizer. Velthof et al. (2003) found positive correlations of high N_2O emissions with manure inorganic N and easily mineralizable N and C. High N_2O emissions from liquid pig manure were attributed to the large amounts of soluble C, particularly the volatile fatty acid (VFA) component. Pig manure slurries have been found to contain greater than 30% of their total C as VFAs (Paul and Beauchamp, 1989; Kirchmann and Lundvall, 1993).

Jones et al. (2007) studied N_2O emission from a temperate grassland having poorly drained clay loam soils that received both mineral fertilizer and poultry/cattle slurry manure additions (300 kg N ha^{-1} y^{-1}) for two consecutive years. Manure treatments resulted in much higher and more sustained rates of N_2O production than did NH_4NO_3 treatments, without increasing grass productivity. It was also found that surface application of the manures led to a concentrated buildup of total and mineral N near the surface where it could be most vulnerable to denitrification processes. The authors suggested that deep ploughing of manure might help conserve more of the added N. In contrast, a 6 week dairy manure-incubation study by Calderon et al. (2004) found average denitrification losses of about 5% for added manure N and about 30% for added NH_4-N. Water filled pore space (WFPS) in the incubated soils averaged 23-26%, which is much lower than the WFPS (> 70%) generally associated with denitrification.

In regions where animal manures are in short supply, pelletized manure is a popular fertilizer resource due to reduced transportation, storage and handling costs (Hara et al., 2003). One drawback of these manures may be an increase in N_2O emission relative to other N sources under certain circumstances (Jones et al., 2007). This concern was addressed in a recent study by Hayakawa et al. (2009), where N_2O and NO emissions were measured from an Andisol treated for 2 years with 120 kg N ha^{-1} as pelletized poultry manure, poultry manure and inorganic fertilizer. They found generally higher N_2O emissions from the manured soils, with highest emissions from the pelletized manure treatment. In addition, they found higher conversion rates of N_2O to NO during wetter conditions. The authors reasoned that conditions inside the pellets were more conducive to denitrification following rainfall and the heterotrophic microbial activity that followed.

Short of reducing denitrification loss from manured soils, some researchers have studied ways of lowering the mole fraction of N_2O emitted relative to N_2. Firestone and Davidson (1989) found that N_2O formation was favored by manures having high initial NO_3^-, lower relative pH, and a high percentage of soluble C in their organic C pool. These authors concluded that liquid manures should act to favor N_2 production during denitrification because they have low initial NO_3^- contents and high pH. In contrast, Meijide et al. (2007) examined denitrification of digested and untreated pig slurry. Production of N_2O through chemoautotrophic nitrification was prevented through the use of dicyandiamide (DCD). The authors concluded that the high C content of the untreated slurry led to greater production of N_2 relative to N_2O. Digestion and composting processes, along with nitrification inhibitors, have been recommended for reducing N emission losses from animal manures (Vallejo et al., 2006).

9.3.3. Ammonia Volatilization

Burger and Ventura (2008) studied the loss of ^{15}N-labeled NH_4^+ from incubated liquid dairy manure and found direct evidence for its microbial utilization. Volatilization losses in that study were < 1%, while gross N immobilization was found to be 13% of the total initial N, or 32% of initial NH_4^+. Generally, volatilization loss is relatively small once manure is applied to soil. However, N losses from manure through NH_3 volatilization can be substantial, especially when manure is stored in anaerobic lagoons (Karlen et al., 1998). Information on the mechanisms of ammonia volatilization and its associated mitigation techniques is provided in Chapter 6.

9.4. MEASURING MANURE NITROGEN AVAILABILITY

9.4.1. Laboratory Determinations

Understanding manure N availability to crops is key to ensuring adequate nutrient supply to optimize yield while avoiding over-application. Potentially mineralizable N in manure has traditionally been estimated by incubating a known amount of manure in a soil for a certain period of time and then comparing the N content present at the conclusion with some aspect of the amount present initially. Such biological incubations are most easily conducted in the laboratory where a number of confounding environmental variables, such as temperature and water, are more easily controlled. However, field incubations are also possible (Honeycutt et al., 2005b). Incubation studies are generally labor-intensive, and require long time periods to allow for the natural progression of N turnover to occur (Cusick et al., 2006). As such, many attempts have been made to use chemical methods to identify initial manure characteristics that closely correlate with N availability for a particular manure type (Castellanos and Pratt, 1981). Examples of some of these potential mineralization indices include C:N ratio (Morvan et al., 2006; Nett et al., 2010), NH_4-N content (Burger and Ventura, 2008), water soluble N (Qafoku et al., 2001; Diaz et al., 2008), total soil N (Cabrera and Kissel, 1988), pepsin extracted N (Castellanos and Pratt, 1981), thermal fractionated N (Velthof, et al., 1998; Antil et al., 2009), non-detergent extractable N (Jensen et al., 2005), total urea (Eiteman et al., 1994), total organic N (Griffin et al., 2005), acid-detergent fiber C (Van Kessel and Reeves, 2002, N predicted from near-infrared reflectance spectroscopy (Qafoku et al., 2001), and N related to UV absorbance at various wavelengths following $NaHCO_3$ extraction (Sharifi et al., 2008). Biological and chemical methods have often been used in conjunction with one another (Qafoku et al., 2001).

The most common manure incubation method measures the overall net change in soil inorganic N over a given time interval. Net N transformation calculations cannot separately quantify N productive processes such as mineralization, or N consumptive processes, such as denitrification, immobilization and volatilization (Habteselassie et al., 2006a). As a result, high rates of a productive or a consumptive process may be masked by even higher rates of an opposing process. Thus, significant N turnover in the soil may be occurring even when incubation or field results indicate little net change in available N concentration (Jansson and Persson, 1982). Measurements of the gross rates of individual N transformation processes can be made through the use of isotope labeling; a technique that enables researchers to follow ^{15}N-labeled manure components through a series of time-dependent transformations. For example, a laboratory ^{15}N tracer study on N transformation rates in a sandy clay loam soil, found that gross mineralization rates following addition of cattle slurry were increased by a factor of two over an NH_4Cl/KNO_3 control, despite little differences in net mineralization rates (Muller et al., 2003). Similarly, Griffin (2007) used ^{15}N isotope pool dilution techniques to compare net and gross N transformations occurring in a sandy loam soil following application of four different dairy manures. He found gross mineralization rates (1.21 mg N kg^{-1} soil day^{-1}) were 2-9 times faster than net mineralization rates of those same manures. The difference between gross and net mineralization rate narrowed over the 56-58 day study, but was still significant. Gross immobilization rates were faster in the four manured soils than in the control soil after 56 days, and were found to be well correlated with initial C content.

Again, measurement of net N mineralization alone would not have revealed the extent of N turnover following manure addition.

9.4.2. Effects of Soil Properties and Environmental Variables on N Availability

Developing management practices that optimize recycling of manure-derived N to crops, while minimizing adverse environmental consequences of manure application to cropland, is hindered by the often dramatic impacts exerted by soil properties, manure composition and climate on manure N availability.

Particle size and structure are two soil properties that exert significant influence on microbially-mediated processes such as mineralization (Paul & Clark, 1989; Van Veen & Kuikman, 1990). Primary mechanisms for this influence include physical protection by formation of clay-organic material complexes, physical separation, such as organic material entrapment in soil aggregates, and through the influence of soil texture/structure on microbial growth by affecting water relations (VanVeen & Kuikman, 1990). Sorensen *et al.* (1994) examined the influence of soil texture on plant-availability of ^{15}N labelled sheep feces for three soils of varying texture but identical clay mineralogy. Nitrogen immobilization was greatest in soil with the highest clay content. In a similar study, Sorensen & Jensen (1995) felt that plant-availability of manure N may be higher in coarse-textured than in fine-textured soils. Similarly, more sheep urine-derived N was immobilized in a sandy loam soil (36%) than in a sandy soil (13%) (Sorensen & Jensen, 1996). Soil aggregation can have a modifying effect on the influence of soil texture on decomposition and mineralization processes (Haynes & Beare, 1997).

The importance of environmental conditions, including temperature, water and aeration, on microbial activity is well documented (Atlas, 1988; Honeycutt *et al.*, 1988; Paul & Clark, 1989). Soil thermal units (degree days) were recently shown to be useful for describing the influence of temperature on N availability from dairy (*Bos taurus*), swine (*Sus scrofa*), and poultry (*Gallus gallus*) manures (Griffin & Honeycutt, 2000). They found NO_3^- formation from dairy, poultry and swine manures could be described with the same thermal unit relationship.

Water content can strongly influence decomposition and nutrient turnover in soils (Stanford & Epstein, 1974; Sommers *et al.*, 1981; Doel *et al.*, 1990). Maximum aerobic microbial activity occurs when 60% of the pore space is filled with water for a wide range of soils (Linn & Doran, 1984). Interactive effects of soil water potential and soil temperature on decomposition have also been observed in some studies (Wildung *et al.*, 1975; Clark & Gilmour, 1983).

Most laboratory studies on manure N availability have been conducted at constant water content. Field soils, however, are subjected to cycles of wetting and drying. Wetting/drying cycles can have significant impact on N transformation processes. In manure-amended soils, highest N_2O fluxes were observed following rainfall (Lessard *et al.*, 1996). Rewetting of dried soil results in a flush of microbial activity, possibly owing to both chemical and physical disturbances of the soil organic matter (Kieft *et al.*, 1987; van Gestel *et al.*, 1991). Previously protected organic matter within aggregates may become exposed to microbial attack following rewetting events (Beare *et al.*, 1994; Franzluebbers & Arshad, 1997). Greater N_2O

losses were measured in a field study with poultry manure-amended soils than in unamended soils following rainfall (Coyne *et al.*, 1995). However, N_2O losses were less than 1% of the total N added in the manure.

Cabrera (1993) felt that drying and rewetting increased mineralization rate of the more stable soil organic matter N pool. In contrast, repeated wetting and drying appeared to increase the resistance of cowpea [*Vigna unguiculata* (L.) Walp.] plant N compounds to microbial decomposition (Franzluebbers *et al.*, 1994). De Bruin *et al.* (1989) observed both net N mineralization and net nitrification to be higher under wetting/drying conditions. Appel (1998) reported greater N mineralization following drying-rewetting than for undried soil in eight out of ten soils. In two soils with approximately 70% sand, N mineralization was greatest in the undried treatment.

Flush of CO_2 following rewetting of dry soil is often attributed to mineralization of microbial cells that had perished from desiccation (Jenkinson & Powlson, 1976; Bottner, 1985). However, Kieft *et al.* (1987) demonstrated that rapid water potential increase is a major perturbation, resulting in plasmoptysis of sensitive microbial populations that may be accompanied by release of N (amino acids; quaternary ammonium compounds) and P. Kieft *et al.* (1987) felt that differences in microbial response to water potential increase may be indicative of the microbial population's ability to accumulate organic intracellular solutes and rapidly modulate their internal water potential.

Table 9.1. Studies of manure nitrogen mineralization by nationally coordinated United States Department of Agriculture-Agricultural Research Service (USDA-ARS) laboratories.

Location	Mean annual temperature (°C)	Mean annual precipitation (mm)	Manure	Soil great group	Associated Reference
Orono, ME	5.2	993	Dairy, Swine	Haplorthod	Griffin et al., 2002 (L)[b];
Mississippi State, MS	17.7	1453	Poultry	Paleudult	Honeycutt et al., 2005a (L)
Auburn, AL	16.7	1397	Poultry, dairy	Paleudult	Sistani et al., 2008 (L&F)
Madison, WI	7.5	927	Dairy	Argiudoll, Hapludalf, Glossudalf, Glossoboralf	Watts et al., 2007 (L); 2010 (F) Wu et al., 2007 (F)
Lincoln, NE	10.9	767	Beef, swine	Argiudoll, Ustipsamment	Wienhold et al., 2009 (F)
Pendleton, OR	10.2	439	Beef	Haploxeroll, Haplocalcid	-
Tifton, GA	15.2	1202	Poultry	Kandiudult	Hubbard et al., 2008 (F)
Champaign, IL[a]	10.9	1041	All	Argiudoll	Griffin et al., 2008 (L)

[a] Soil from this location was used as a reference across all participating laboratories.
[b] L=laboratory study; F=field study.

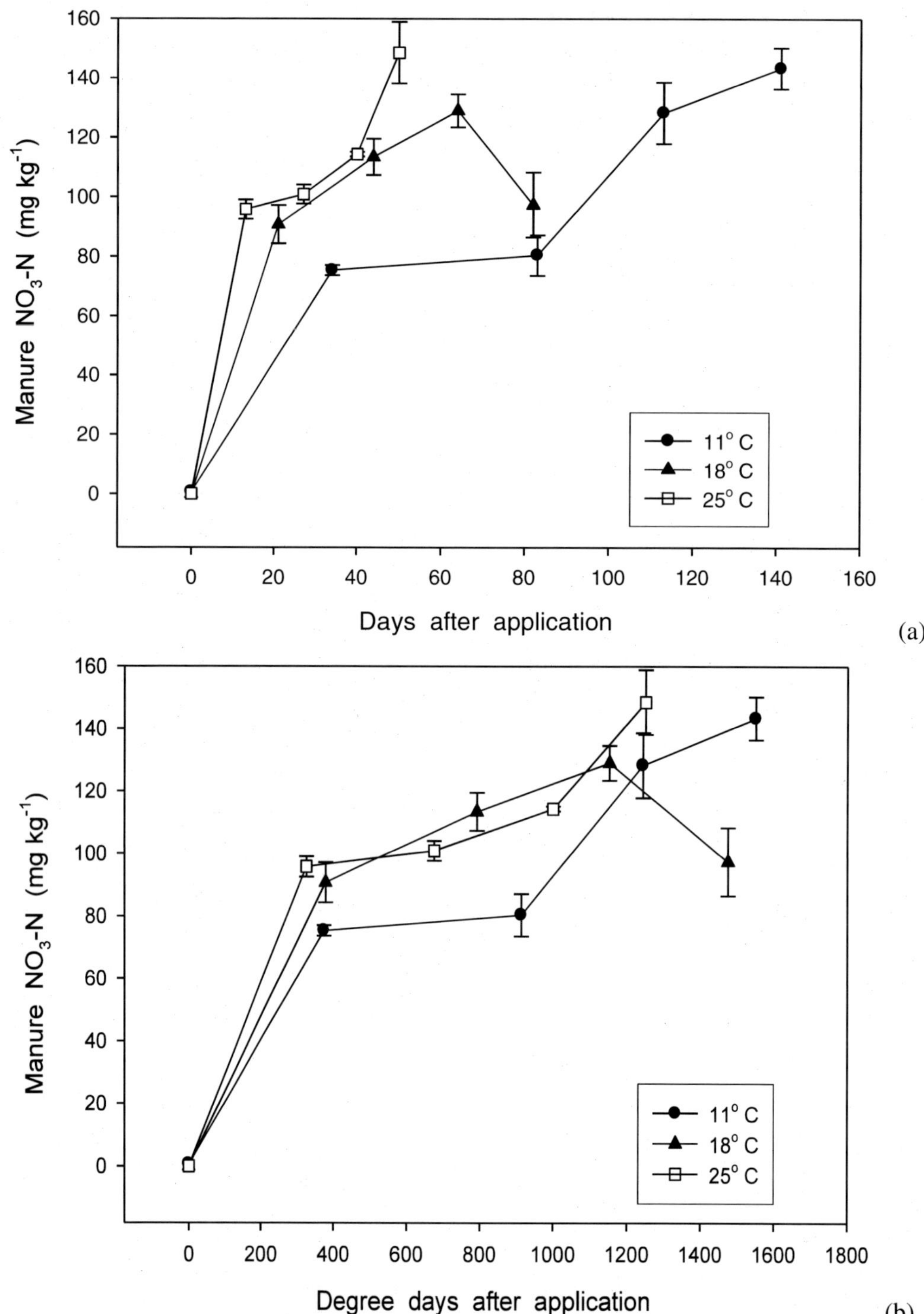

Figure 9.1. Manure-derived nitrate in the dairy manure amended Caribou soil incubated at constant water content and three temperatures in relation to (a) days after application and (b) degree days after application. (After Honeycutt et al., 2005a)

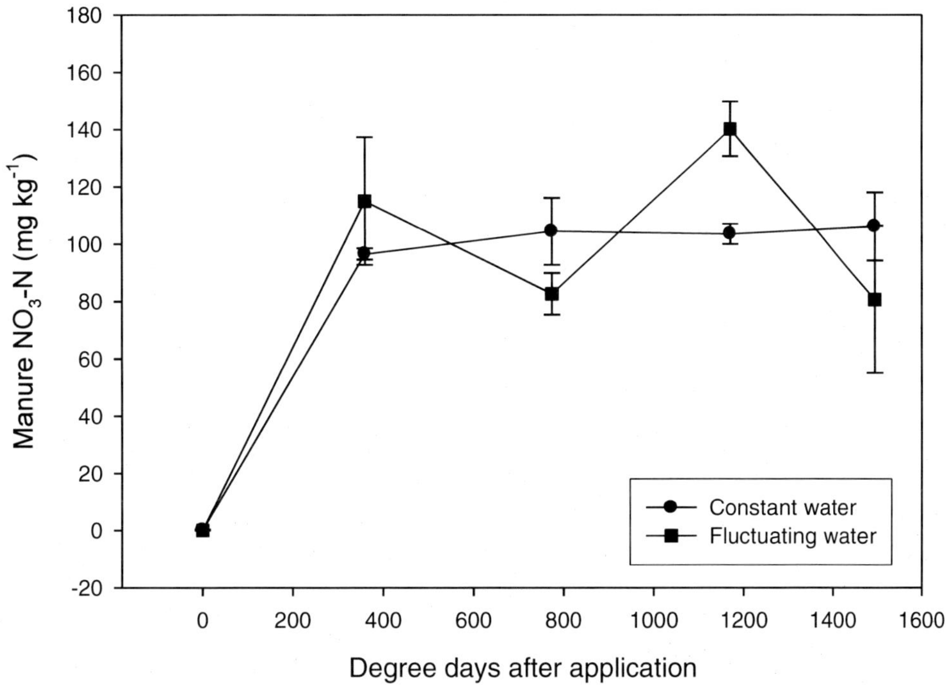

Figure 9.2. Manure-derived nitrate in relation to degree days after application for the Newport soil incubated at 18°C under fluctuating and constant water regimes. (After Honeycutt et al., 2005a)

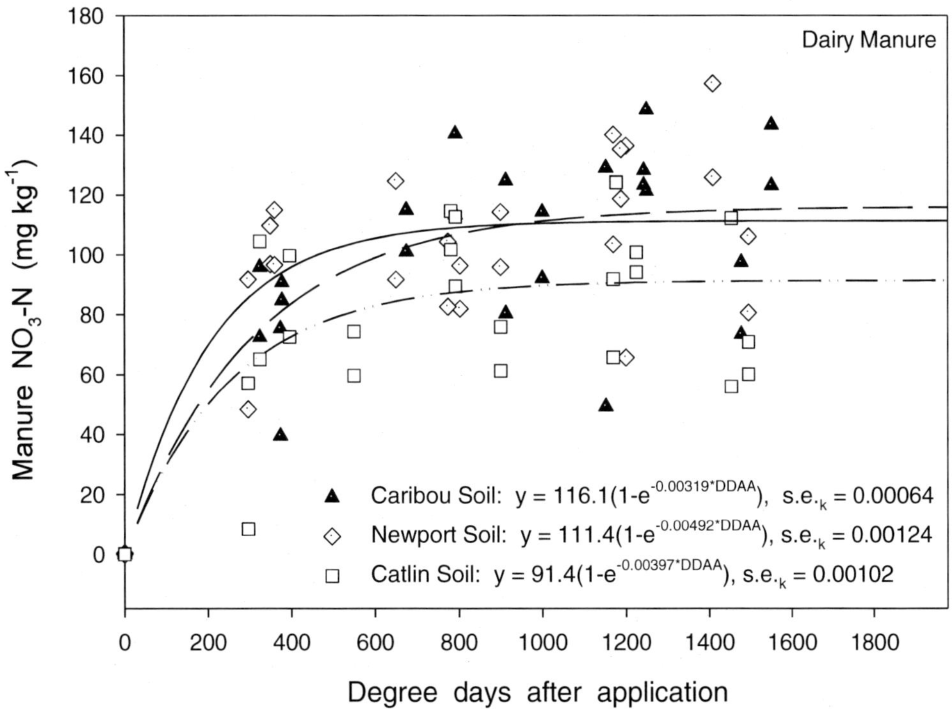

Figure 9.3. Manure-derived nitrate in relation to degree days after application for the Caribou, Newport, and Illinois (Catlin) soils incubated at three temperatures and two water regimes.

These studies emphasize the importance of soil properties and environmental variables in regulating N availability from soil-applied manures. Important soil physical properties include soil texture and aggregation. Key environmental variables include soil temperature and soil water content. Although water content *per se* is important, flushes in microbial activity and N transformations may by associated with wetting/drying cycles that occur under field conditions.

9.4.3. Case Study

Nationally Coordinated Research on Manure Nitrogen Mineralization by USDA-ARS. The National Program focus of USDA-ARS provides an opportunity to coordinate research on problems of national and global significance. A team of USDA-ARS scientists conducted a nationally coordinated research project to develop predictive relationships that quantify the impacts of key soil, environmental, and soil X environmental factors on manure N mineralization (Table 9.1). Experimental design and research protocols were developed and used in common across all participating locations. Laboratory incubations were conducted at each location with a minimum of three soils, three temperatures, two wetting/drying regimes, and two manure treatments (with and without), as described by Honeycutt et al. (2005b).

A soil from the central U.S. (Catlin silt loam; fine-silty, mixed, superactive, mesic, Oxyaquic Arguidoll) was used as an internal reference across all locations.

Griffin et al. (2008) used this opportunity to evaluate the feasibility of conducting coordinated laboratory incubations by comparing soil organic N mineralization from the Catlin silt loam across six locations. After excluding time=0 data, N mineralization rate did not differ for five of the six locations evaluated, averaging 0.0216 mg N kg^{-1} (degree day)$^{-1}$. This demonstrated the feasibility of employing standardized incubation study protocols across laboratories.

Honeycutt et al. (2005a) used this protocol (Honeycutt et al., 2005b) to evaluate the impacts of temperature, water regime, and soil on N availability from a liquid dairy manure. Two Maine soils [Caribou sandy loam (fine-loamy, mixed, frigid Typic Haplorthod) and 'Newport' loam (unnamed variant of a Bangor silt loam; coarse-loamy, mixed, frigid Typic Haplorthod)] and the Catlin silt loam from Illinois were incubated at three temperatures and two water regimes (constant and fluctuating). Temperature strongly influenced NO_3^- concentration over time, but this effect could be described with the thermal unit (degree day) concept (Figure 9.1). The fluctuating water regime had no significant impact on manure-derived NO_3^- (Figure 9.2). Changes in manure NO_3^- over thermal time could be described with a single exponential equation. However, less manure NO_3^- was observed in an Illinois soil than in two Maine soils (Figure 9.3).

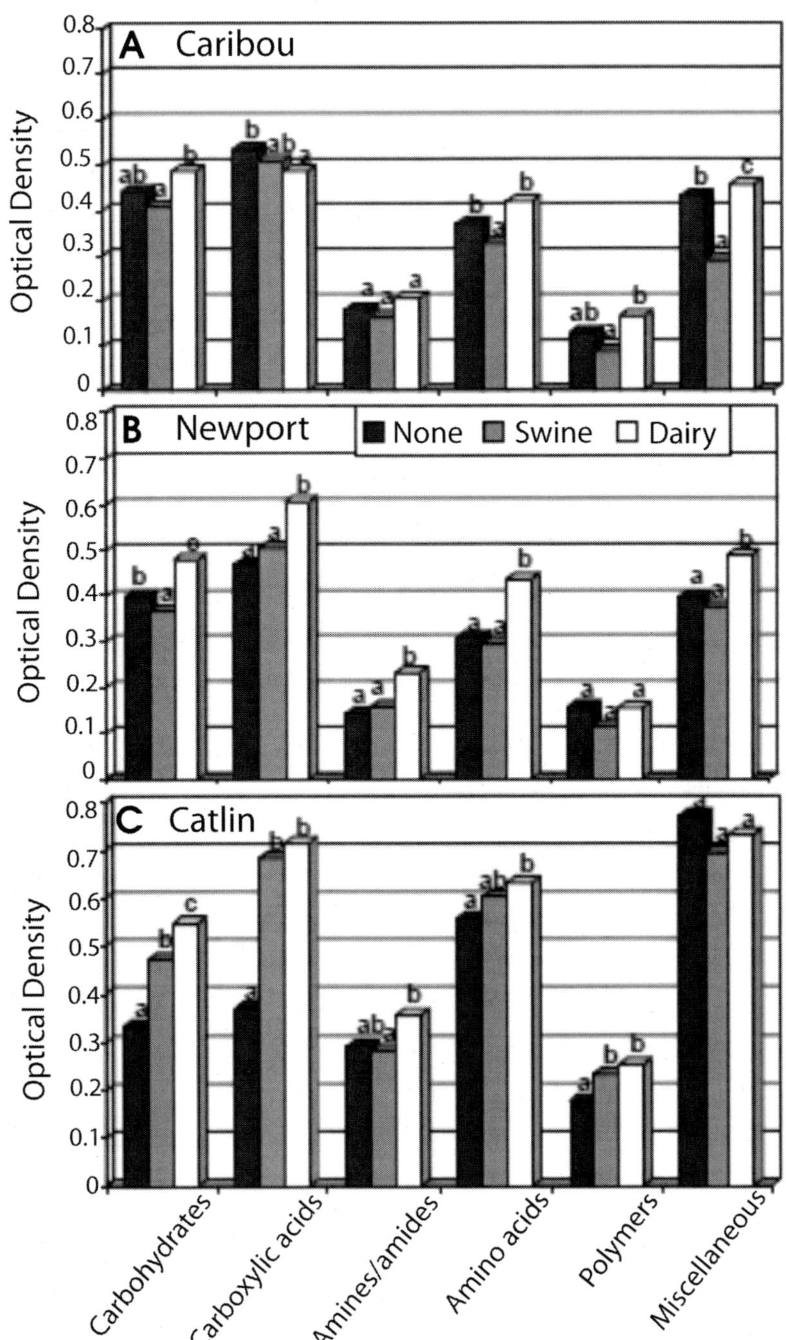

Figure 9.4. Effects of manure amendments on utilization of different substrate guilds (as represented by optical density). (a) Caribou, (b) Newport, and (c) Catlin soils. Guilds are represented as carbohydrates, carboxylic acids, amines/amides, amino acids, polymers, and miscellaneous. Amendment treatments were none (no amendment), dairy manure, and swine manure. Different letters above columns in the same substrate guild for each soil type indicate significantly different according to Fisher's protected LSD at P=0.05. [After Larkin et al., 2006].

Upon evaluating these same soils, Larkin et al. (2006) found that dairy manure amendment resulted in increased utilization of carbohydrates and amino acids in all three soils, increased polymer use in Caribou and Catlin soils, increased carboxylic acid and amine use in Newport and Catlin soils, but decreased carboxylic acid use in Caribou soil (Figure 9.4). Overall, Larkin et al. (2006) observed that the Catlin soil had higher bacterial counts, greater utilization of most substrates and substrate groups, and greater abundance of soil microbial biomarkers than the two Maine soils. These characteristics indicated a greater potential for N immobilization in Catlin soil than in the other soils, which is consistent with the lower final NO_3^- levels observed in this soil after manure addition (Griffin et al. 2002; Honeycutt et al., 2005a).

Field studies analyzing *in-situ* soil cores following manure application have sometimes been used to validate or compare with laboratory results (Honeycutt, 1999). Thus, Honeycutt et al. (2005b) also offered a field study procedure to complement laboratory studies by using the same soils and manures at a given location. Wienhold (2007) concluded that *in situ* field techniques provide reasonable estimates of N mineralization. Honeycutt (1999) found N mineralization from soil organic matter under field conditions was overestimated by laboratory predictions; however, field measurements of N mineralization from both ground and unground crop residue were closely predicted by a complementary laboratory study.

9.5. CONCLUSION

Increased animal production to feed a growing human population, concomitant with increasing loss of farm land available for manure application, make it vitally important that practices be developed for optimizing nutrient recycling from manure to crops. To assist in reaching this goal, methods for measuring N turnover and availability from manure are fairly well established. However, complexity of the many transformation processes such as mineralization, immobilization, nitrification, denitrification, and ammonia volatilization, along with the modifying and interacting impacts of soil physical, chemical, and biological properties, animal species and diets, manure storage and handling techniques, and climatic variables such as temperature and water dynamics, combine to present significant challenges for developing cost-effective and environmentally sound manure N management and utilization practices. Only with a thorough understanding of the independent and interactive components and drivers of N availability will we be able to reach this goal. This chapter is offered as a summary of recent research contributing to our current state of knowledge in this important area.

REFERENCES

Alexander, M. 1977. *Introduction to soil microbiology.* John Wiley and Sons, New York.
Antil, R.S., B.H. Janssen, and A. Lantinga. 2009. Laboratory and greenhouse assessment of plant availability of organic N in animal manure. *Nutr. Cycl. Agroecosyst.* 85:95–106.

Appel, T. 1998. Non-biomass soil organic N - the substrate for N mineralization flushes following soil drying-rewetting and for organic N rendered $CaCl_2$-extractable upon soil drying. *Soil Biol. Biochem.* 30:1145-1156.

Atlas, R.M. 1988. *Microbiology: Fundamentals and applications.* 2nd ed. Macmillan, New York, NY, USA.

Beauchamp, E.G. 1983. Response of corn to nitrogen in preplant and side-dress applications of liquid dairy cattle manure. *Can. J. Soil Sci.* 63:377–386.

Beare, M.H., P.F. Hendrix, and D.C. Coleman. 1994. Aggregate-protected and unprotected organic matter pools in conventional- and no-tillage soils. *Soil Sci. Soc. Am. J.* 58:787-795.

Bechini, L., and P. Marino. 2009. Short-term nitrogen fertilizing value of liquid dairy manures is mainly due to ammonium. *Soil Sci. Soc. Am. J.* 73:2159–2169.

Beegle, D. B., K.A. Kelling, and M. A. Schmitt. 2008. Nitrogen from Animal Manures. pp 823-881. *In* J. Schepers and W.R. Raun (ed.) Nitrogen in Agricultural Systems. *Agron. Monogr.* 49. ASA, Madison, WI.

Bitzer, C.C., and J.T. Sims. 1988. Estimating the availability of nitrogen in poultry manure through laboratory and field studies. *J. Environ. Qual.* 17:47–54.

Blackmer, A.M., J.M. Bremner, and E.L. Schmidt. 1980. Production of nitrous oxide by ammonia-oxidizing chemoautotrophic microorganisms in soil. *Appl. Environ. Microbiol.* 40:1060–1066.

Booth, M., J.M. Stark, and E. Rastetter. 2005. Controls on nitrogen cycling in terrestrial ecosystems: a synthetic analysis of literature data. *Ecol. Monogr.* 75:139–157.

Bottner, P. 1985. Response of microbial biomass to alternate moist and dry conditions in a soil incubated with ^{14}C and ^{15}N-labelled plant material. *Soil Biol. Biochem.* 17:329-337.

Broderick, G.A. 2003. Effects of varying dietary protein and energy levels on the production of lactating dairy cows. *J. Dairy Sci.* 86:1370–1381.

Burger, M., and R.T. Venturea. 2008. Nitrogen immobilization and mineralization kinetics of cattle, hog, and turkey manure applied to soil. *Soil Sci. Soc. Am. J.* 72:1570–1579.

Cabrera, M.L. 1993. Modeling the flush of nitrogen mineralization caused by drying and rewetting soils. *Soil Sci. Soc. Am. J.* 57:63-66.

Cabrera, M.L., and D.E. Kissel. 1988. Potentially mineralizable nitrogen in disturbed and undisturbed soil samples. *Soil Sci. Soc. Am. J.* 52:1010–1015.

Cabrera, M.L., D.E. Kissel, and M.F. Vigil. 2005. Nitrogen mineralization from organic residues: research opportunities. *J. Environ. Qual.* 34:75–79.

Calderon, F.J., G.W. McCarty, J.S. Van Kessel, and J.B. Reeves III. 2004. Carbon and nitrogen dynamics during incubation of manured soil. *Soil Sci. Soc. Am. J.* 68:1592–1599.

Calderon, F.J., G.W. McCarty, and J.B. Reeves. 2005. Nitrapyrin delays denitrification on manured soils. *Soil Sci.* 170:350–359.

Castellanos, J.Z., and P.F. Pratt. 1981. Mineralization of manure nitrogen—correlation with laboratory indexes. *Soil Sci. Soc. Am. J.* 45:354–357.

Chadwick, D.R., F. John, B.F. Pain, B.J. Chambers, and J. Williams. 2000. Plant uptake of nitrogen from the organic nitrogen fraction in animal manures: a laboratory experiment. *J. Agric. Sci.* 134:159–168.

Chastain, J.P., J.J. Camberato, and J.E. Albrecht. 2001. Nutrient content of livestock and poultry manure. Clemson University. *http://www.clemson.edu/cafls/departments/ agbioeng/faculty_staff/research/chastain_nutrient_content_manure.pdf*

Chen, J., and J.M. Stark. 2000. Plant species effects and carbon and nitrogen cycling in a sagebrush–crested wheatgrass soil. *Soil Biol. Biochem.* 32:47–57.

Clark, M.D., and J.T. Gilmour. 1983. The effect of temperature on decomposition at optimum and saturated soil water contents. *Soil Sci. Soc. Am. J.* 47:927-929.

Coyne, M.S., A. Villalba, and R.L. Blevins. 1995. Nitrous oxide loss from poultry manure-amended soil after rain. *J. Environ. Qual.* 24:1091-1096.

Cusick, P.R., K.A. Kelling, J.M. Powell, and G.R. Munoz. 2006. Estimates of residual dairy manure nitrogen availability using various techniques. *J. Environ. Qual.* 35:2170–2177.

DeBoer, W., and G. A. Kowalchuk. 2001. Nitrification in acid soils: micro-organisms and mechanisms. *Soil Biol. Biochem.* 33:853–866.

De Bruin, B., F.W.T. Penning De Vries, L.W. Van Broekhoven, N. Vertregt, and S.C. Van De Geijn. 1989. Net nitrogen mineralization, nitrification and CO_2 production in alternating moisture conditions in an unfertilized low-humus sandy soil from the Sahel. *Plant Soil* 113:69-78.

De Ferrari, G., P. Marino Gallina, G. Cabassi, L. Bechini, and T. Maggiore. 2007. Near infrared spectral analysis of cattle slurries from Lombardy (Northern Italy) breeding farms. p. 630–637. *In* G.R. Burling-Claridge et al. (ed.) NIR 2005– NIR in action. Making a difference. Near Infrared Spectroscopy. *Proc. Int. Conf.,* Auckland, New Zealand. 9–15 April 2005. New Zealand Near Infrared Spectroscopy Soc., Hamilton, New Zealand.

Deenik, J. 2006. Nitrogen mineralization potential in important agricultural soils of Hawaii. *Coop. Ext. Serv.* Pub., CTAHR, SCM-15.

Diaz, D.A.R., J.E. Sawyer, and A.P. Mallarino. 2008. Poultry manure supply of potentially available nitrogen with soil incubation. *Agron. J.* 100:1310–1317.

Ding, W., L. Meng, Z. Cat, and F. Han. 2007. Effects of long- term amendment of organic manure and nitrogen fertilizer on nitrous oxide emission in a sandy loam soil. J. *Environ. Sci.* 19:185–193.

Dobbie, K. E., I. P. McTaggart, and K. A. Smith. 1999. Nitrous oxide emissions from intensive agricultural systems: Variations between crops and seasons, key driving variables, and mean emission factors. *J. Geophys. Res.*104:26,891– 26,899.

Dobbie, K.E. and K.A. Smith. 2003. Nitrous oxide emission factors for agricultural soil in Great Britain: the impact of soil water-filled pore space and other controlling variables. *Global Change Biol.* 9:204–218.

Doel, D.S., C.W. Honeycutt, and W.A. Halteman. 1990. Soil water effects on the use of heat units to predict crop residue carbon and nitrogen mineralization. *Biol. Fertil. Soils* 10:102-106.

Egelkraut, T.M., D.E. Kissel, and M.L. Cabrera. 2000. Effect of soil texture on nitrogen mineralized from cotton residues and compost. *J. Environ. Qual.* 29:1518–1522.

Eiteman, M.A., R.M. Gordillo, and M.L. Cabrera. 1994. Analysis of oxonic acid, uric acid, creatine, allantoin, xanthine and hypoxanthine in poultry litter by reverse phase HPLC. Fresenius Z. *Anal. Chem.* 348:680–683.

Firestone, M.K., and E.A. Davidson. 1989. Microbiological basis of NO and N_2O production and consumption in Soil. p. 7–21. *In* M.O. Andreae and D.S. Schimel (ed.) *Exchange of trace gases between terrestrial ecosystems and the atmosphere.* John Wiley & Sons, Chichester, UK.

Firestone, M.K., R.B. Firestone, and J.M. Tiedje. 1980. Nitrous oxide from soil denitrification: factors controlling its biological production. *Science* 208:749–751.

Franzluebbers, A.J., and M.A. Arshad. 1997. Soil microbial biomass and mineralizable carbon of water-stable aggregates. *Soil Sci. Soc. Am. J.* 61:1090-1097.

Franzluebbers, K., R.W. Weaver, A.S.R. Juo, and A.J. Franzluebbers. 1994. Carbon and nitrogen mineralization from cowpea plants part decomposing in moist and in repeatedly dried and wetted soil. *Soil Biol. Biochem.* 26:1379-1387.

Gilmour, J.T., and V. Skinner. 1999. Predicting plant available nitrogen in land-applied biosolids. *J. Environ. Qual.* 28:1122–1126.

Ginting, D., A. Kessavalou, B. Eghball, and J.W. Doran. 2003. Greenhouse gas emissions and soil indicators four years after manure and compost applications. *J. Environ. Qual.* 32:23–32.

Gollehon, N., and M. Caswell. 2000. Confined animal production poses manure management problems. *Agric. Outlook* 274:12–18.

Goulding, K.W.T., P.R. Poulton, C.P. Webster, and T. Howe. 2000. Nitrate leaching from the Broadbalk Wheat Experiment, Rothamsted, UK, as influenced by fertilizer and manure inputs and the weather. *Soil Use Manage.* 16:244–250.

Griffin, T.S. 2007. Estimates of gross transformation rates of dairy manure N using ^{15}N pool dilution. *Commun. Soil Sci. Plant Anal.* 38:1451–1465.

Griffin, T.S., C.W. Honeycutt, S.L. Albrecht, K.R. Sistani, H.A. Torbert, B.J. Wienhold, B.L. Woodbury, R.K. Hubbard, and J.M. Powell. 2008. Nationally coordinated evaluation of soil nitrogen mineralization rate using a standardized aerobic incubation protocol. Commun. *Soil Sci. Plant Anal.* 39:257-258.

Griffin, T.S., Z. He, and C.W. Honeycutt. 2005. Manure composition affects net transformation of nitrogen from dairy manures. *Plant Soil* 273:29–38.

Griffin, T.S. and C.W. Honeycutt. 2000. Using growing degree days to predict nitrogen availability from livestock manures. *Soil Sci. Soc. Am.* J. 64:1876-1882.

Griffin, T.S., C.W. Honeycutt, and Z. He. 2002. Effects of temperature, soil water status, and soil type on swine slurry nitrogen transformations. *Biol. Fertil. Soils* 36:442–446.

Griffin, T.S., C.W. Honeycutt, and Z. He. 2003. Changes in soil phosphorus from manure application. S*oil Sci. Soc. Am. J.* 67:645-653.

Habteselassie, M.Y., J.M. Stark, B.E. Miller, S.G. Thacker, and J.M. Norton. 2006a. Gross nitrogen transformations in an agricultural soil after repeated dairy-waste application. *Soil Sci. Soc. Am. J.* 70:1338–1348.

Habteselassie, M.Y., B.E. Miller, S.G. Thacker, J.M. Stark, and J.N. Norton. 2006b. Soil nitrogen and nutrient dynamics after repeated application of treated dairy-waste. *Soil Sci. Soc. Am. J.* 70:1328–1337.

Hara, M., H. Ishikawa, and Y. Furuichi. 2003. Manufacturing conditions and handling improvement effects of pelletized compost from swine manure using a uniaxial extruder. (In Japanese with English abstract.) Japan. *J. Soil Sci. Plant Nutr.* 74:1–7.

Hartung, J., and V.R. Phillips. 1994. Control of gaseous emissions from livestock buildings and manure stores. *J. Agric. Eng. Res.* 57:173–189.

Hassink, J. 1992. Effects of soil texture and structure on carbon and nitrogen mineralization in grassland soils. *Biol. Fertil. Soils* 14:126–134.

Hayakawa, A., H. Akiyama, S. Sudo, and K. Yagi. 2009. N_2O and NO emissions from an Andisol field as influenced by pelleted poultry manure. *Soil Biol. Biochem.* 41:521–529.

Haynes, R.J., and M.H. Beare. 1997. Influence of six crop species on aggregate stability and some labile organic matter fractions. *Soil Biol. Biochem.* 29:1647-1653.

Hester, H.R., H.E. Essex, and F.C. Mann. 1940. Secretion of urine in the chicken (*Gallus domesticus*). *Am. J. Physiol.* 128:592–602.

Honeycutt, C.W. 1999. Nitrogen mineralization from soil organic matter and crop residues: Field validation of laboratory predictions. *Soil Sci. Soc. Am. J.* 63:134-141.

Honeycutt, C. W., T.S. Griffin and Z. He. 2005a. Manure nitrogen availability: dairy manure in northeast and central U.S. soils. *Biol. Agric. Hort.* 23:199–214.

Honeycutt, C.W., T.S. Griffin, B.J. Weinhold, B. Eghball, S.L. Albrecht, J.M. Powell, B.L. Woodbury, K.R. Sistani, P.K. Hubbard, H.A. Torbert, R.A. Eigenberg, R.J. Wright, and M.D. Jawson. 2005b. Protocols from nationally coordinated laboratory and field research on manure nitrogen mineralization. Commun. *Soil Sci. Plant Anal.* 36:2807–2822.

Honeycutt, C.W., L.M. Zibilske, and W.M. Clapham. 1988. Heat units for describing carbon mineralization and predicting net nitrogen mineralization. *Soil Sci. Soc. Am. J.* 52:1346-1350.

Hubbard, R.K., D.D. Bosch, L.K. Marshall, T.C. Strickland, D. Rowland, T.S. Griffin, C.W. Honeycutt, S.L. Albrecht, K.R. Sistani, H.A. Torbert, B.J. Wienhold, B.L. Woodbury, and J.M. Powell. 2008. Nitrogen mineralization from broiler litter applied to southeastern Coastal Plain soils. *J. Soil Water Conserv.* 63:182–192.

Huff , W.E., G.W. Malone, and G.W. Chaloupka. 1984. Effect of litter treatment on broiler performance and certain litter quality parameters. *Poult. Sci.* 63:2167–2171.

Jansson, S.L., and J. Persson. 1982. Mineralization and immobilization of soil nitrogen. p. 229–252 *In* F.J. Stevenson (ed.) *Nitrogen in Agricultural Soils. Agron. Monogr.* 22. ASA, Madison, WI.

Jenkinson, D.S., and D.S. Powlson. 1976. The effects of biocidal treatments on metabolism in soil - V. A method for measuring soil biomass. *Soil Biol. Biochem.* 8:209-213.

Jensen, L.S., T. Salo, F. Palmason, T.A. Breland, T.M. Henriksen, B. Stenberg, A. Pedersen, C. Lundstrom, and M. Esala, M. 2005. Influence of biochemical quality on C and N mineralization from a broad variety of plant materials in soil. *Plant Soil* 273:307–326.

Jones, S.K., R.M. Rees, U.M. Skiba, and B.C. Ball. 2007. Influence of organic and mineral N fertiliser on N_2O fluxes from a temperate grassland. *Agric. Ecosyst. Environ.* 121:74–83.

Karlen, D.L., J.R. Russell, and A.P. Mallarino. 1998. *Animal Waste Utilization: Effective Use of Manure as a Soil Resource.* p. 287–290. *In* J.L. Hatfield and B.A. Stewart (ed.) Ann Arbor Press, Chicago.

Kieft, L.T., E. Soroker, and M.K. Firestone. 1987. Microbial biomass response to a rapid increase in water potential when dry soil is wetted. *Soil Biol. Biochem.* 19:119-126.

Kirchmann, H. 1991. Carbon and nitrogen mineralization in fresh and anaerobic animal manures during incubation with soils. *Swed. J. Agric. Res.* 21:165–173.

Kirchmann, H., and A. Lundvall. 1993. Relationship between N immobilization and volatile fatty acids in soil. *Biol. Fertil. Soils* 15:161–164.

Knudsen, M.T., I.B.S. Kristensen, J. Berntsen, B.M. Petersen, and E.S. Kristensen. 2006. Estimated N leaching losses for organic and conventional farming in Denmark. *J. Agric. Sci.* 144:135–149.

Koelsch, R., and C. Shapiro. 2006. Determining crop available nutrients from manure. *Neb. Ext. Guide 494*. Univ. Nebraska, Lincoln.

Kohler, K., W.H.M. Duynisveld, and J. Bottcher. 2006. Nitrogen fertilization and nitrate leaching into groundwater on arable sandy soils. *J. Plant Nutr. Soil Sci.* 169:185–195.

Kroeze, C., A. Mosier, and L. Bouwman. 1999. Closing the global N_2O budget: a retrospective analysis 1500–1994. *Glob. Biogeochem. Cycles* 13:1–8.

Kruse, J., D.E. Kissel, and M.L. Cabrera. 2004. Effects of drying and rewetting on carbon and nitrogen mineralization in soils and incorporated residues. *Nutr. Cycling Agroecosyst.* 69:247–256.

Larkin, R.P., C.W. Honeycutt, and T.S. Griffin. 2006. Effect of swine and dairy manure amendments on microbial communities in three soils as influenced by environmental conditions. *Biol. Fertil. Soils.* 43:51-61.

Lessard, R., P. Rochette, E.G. Gregorich, E. Pattey, and R.L. Desjardins. 1996. Nitrous oxide fluxes from manure-amended soil under maize. *J. Environ. Qual.* 25:1371–1377.

Linn, D.M., and J.W. Doran. 1984. Effect of water-filled pore space on carbon dioxide and nitrous oxide production in tilled and nontilled soils. *Soil Sci. Soc. Am. J.* 48:1267-1272.

Lowe, G. Myhre, J. Nganga, R. Prinn, G. Raga, M. Schulz and R. Van Dorland. 2007. Changes in Atmospheric Constituents and in Radiative Forcing. p. *In* S.D. Solomon et al. (ed.) *Climate Change 2007: The Physical Science Basis.* Contribution of Working Group I to the Fourth Assessment Report of the Intergovernmental Panel on Climate Change. Cambridge University Press, Cambridge, U.K.

McCrory, D.F., and P.J. Hobbs. 2001. Additives to reduce ammonia and odor emissions from livestock wastes: a review. J. *Environ. Qual.* 30:345–355.

Meijide, A., J.A. Diez, L. Sanchez-Martin, S. Lopez-Fernandez, and A. Vallejo. 2007.

Nitrogen oxide emissions from an irrigated maize crop amended with treated pig slurries and composts in a Mediterranean climate. *Ag. Ecosys. Environ.* 121:383–394.

Meyer, R.L., T. Kjaer, and P.R. Neils. 2002. Nitrification and denitrification near a soil–manure interface studied with a nitrate-nitrite biosensor. *Soil Sci. Soc. Am.* J. 66:498–506.

Moore, A.D., D.W. Israel, and R.L. Mikkelsen. 2005. Nitrogen availability of anaerobic swine lagoon sludge: sludge source effects. *Biores. Tech.* 96:323–329.

Moore, P.A. 1998. Best management practices for poultry manure utilization that enhance agricultural productivity and reduce pollution. p. 92. *In* J.L. Hatfield and B.A. Stewart (ed.) *Animal Waste Utilization: Effective Use of Manure as a Soil Resource.* Ann Arbor Press, Chicago.

Morvan, T., B. Nicolardot, and L. Pean. 2006. Biochemical composition and kinetics of C and N mineralization of animal wastes: A typological approach. *Biol. Fertil. Soils.* 42:513–522.

Mosier, A., C. Kroeze, C. Nevison, O. Oenema, S. Seitzinger, and O. Van Cleemput. 1998. Closing the global N_2O budget: nitrous oxide emissions through the agricultural nitrogen cycle. *Nutr. Cycl. Agroecosyst.* 52:225–248.

Muller, C., R.L. Stevens, and R.J. Laughlin. 2003. Evidence of carbon stimulated N transformations in grassland soil after slurry application. *Soil Biol. Biochem.* 35:285–293.

Myrold, D. D. 1998. Transformations of nitrogen. p. 259–294 *In* D. M. Sylvia et al. (ed.) Principles and applications of soil microbiology. Prentice Hall, Upper Saddle River, NJ.

Nahm, K.H. 2003. Evaluation of the nitrogen content in poultry manure. *World Poult. Sci. J.* 59:77–88.

Nett, L., S. Averesch, S. Ruppel, J. Ruhlmann, C. Feller, E. George, and M. Fink. 2010. Does long-term farmyard manure fertilization affect short-term nitrogen mineralization from farmyard manure? *Biol. Fertil. Soils* 46:159–167.

Nordmeyer, H., and J. Richter. 1985. Incubation experiments on nitrogen mineralization in loess and sandy soils. *Plant Soil* 83:433–445.

Oorts, K., H. Bossuyt, J. Labreuche, R. Merckx, B. Nicolardot. 2007. Carbon and nitrogen stocks in relation to organic matter fractions, aggregation and pore size distribution in no-tillage and conventional tillage in northern France. *Eur. J. Soil Sci.* 58:248–259.

Paul, J.W., and E.G. Beauchamp.1989. Effect of carbon constituents in manure on denitrification in soil. *Can. J. Soil Sci.* 69:49–61.

Paul, E.A., and F.E. Clark. 1989. *Soil Microbiology and Biochemistry*. Academic Press. New York, NY, U.S.A.

Paul, J.W., E.G. Beauchamp, and X. Zhang. 1993. Nitrous and nitric-oxide emissions during nitrification and denitrification from manure-amended soil in the laboratory. *Can. J. Soil Sci. 73*:539–553.

Pettygrove, G.S., A.L. Heinrich, and D.M. Crohn. 2009. Manure nitrogen mineralization. *University of California Cooperative Extension Manure Technical Bulletin Series.* Available at http://manuremanagement.ucdavis.edu

Powell, J.M., M.A. Wattiaux, G.A. Broderick, V.R. Moreira, and M.D. Casler. 2006. Dairy diet impacts on fecal chemical properties and nitrogen cycling in soils. *Soil Sci. Soc. Am. J. 70*:786–794.

Preusch, P.L., P.R. Adler, L.J. Sikora, and T.J. Tworkoski. 2002. Nitrogen and phosphorus availability in composted and uncomposted poultry litter. *J. Environ. Qual.* 31:205 – 207.

Qafoku, O.S., M.L. Cabrera, W.R. Windham, and N.S. Hill. 2001. Rapid methods to determine potentially mineralizable nitrogen in broiler litter. *J. Environ. Qual.* 30:217–221.

Russelle, M., K. Blanchet, G. Randall, and L. Everett. 2007. Nitrogen availability from liquid swine and dairy manure: Results of on-farm trials in Minnesota. Coop. Ext. Rep. Univ. of Minnesota, St. Paul. Available at http://www.extension.umn.edu/distribution/livestocksystems/DI8583.html

Schimel, J. P., and J. Bennett. 2004. Nitrogen mineralization: challenges of a changing paradigm. *Ecology.* 85:591–602.

Schjonning, P., I.K. Thomsen, J.P. Mobert, H. de Jonge, K. Kreisensen, and B.T. Christensen. 1999. Turnover of organic matter in differently textured soils: I. Physical characteristics of structurally disturbed and intact soils. *Geoderma* 89:177–198.

Sharifi, M., B.J. Zebarth, D.L. Burton, C.A. Grant, and G.A. Porter. 2008. Organic amendment history and crop rotation effects on soil nitrogen mineralization potential and soil nitrogen supply in a potato cropping system. *Agron. J.* 100:1562–1572.

Sistani, K.R., A. Adeli, S.L. McGowen, H. Tewolde, and G.E. Brink. 2008. Laboratory and field evaluation of broiler litter nitrogen mineralization. *Bioresource Tech.* 99:2603-2611.

Skiba, U., and K.A. Smith. 2000. The control of nitrous oxide emissions from agricultural and natural soils. Chemosphere. *Global Change Science* 2:379–386.

Sommers, L.E., Gilmour, C.M., Wildung, R.E., and S.M. Beck. 1981. The effect of water potential on decomposition processes in soil. In *Water Potential Relations in Soil Microbiology*, (J.F. Parr, W.R. Gardner, and L.F. Elliott, eds.), pp. 97-119. *Soil Sci. Soc. Am.*; Madison, WI, U.S.A.

Sorensen, P. 1998. Carbon mineralization, nitrogen immobilization and pH change in soil after adding volatile fatty acids. *Eur. J. Soil Sci.* 49:457–462.

Sørensen, P. 2004. Immobilisation, remineralisation and residual effects in subsequent crops of dairy cattle slurry nitrogen compared to mineral fertiliser nitrogen. *Plant Soil* 267:285–296.

Sorensen, P., and M. Amato. 2002. Remineralisation and residual effects of N after application of pig slurry to soil. *Eur. J. Agric.* 16:81–95.

Sorensen, P., and E.S. Jensen. 1995. Mineralization of carbon and nitrogen from fresh and anaerobically stored sheep manure in soil of different texture. *Biol. Fertil. soils* 19:29–35.

Sorensen, P., and E.S. Jensen. 1996. The fate of fresh and stored ^{15}N-labelled sheep urine and urea applied to a sandy and a sandy loam soil using different application strategies. *Plant Soil 183*:213-220.

Sørensen, P., E.S. Jensen, and N.E. Nielsen. 1994. The fate of ^{15}N-labeled organic nitrogen in sheep manure applied to soils of different texture under field conditions. *Plant Soil* 162:39–47.

Stanford, G., and E. Epstein. 1974. Nitrogen mineralization-water relations in soils. *Soil Sci. Soc. Amer. Proc.* 38:103-107.

USDA Census of Agriculture. 2007. *http://www.agcensus.usda.gov/Publications/2007*.

US EPA. 2009. U.S. emissions inventory 2009: Inventory of U.S. greenhouse gas emissions and sinks: 1990-2007. *http://www.epa.gov/nitrousoxide/sources.html*.

Vallejo, A., L. Garci'a-Torres, A. Arce, S. Lopez-Fernandez, and L. Sanchez-Martin. 2006. Nitrogen oxides emission from soils bearing a potato crop as influenced by fertilization with treated pig slurries and composts. *Soil Biol. Biochem.* 38:2782–2793.

Van Faassen, H.G., and H.Van Dijk. 1987. Manure as a source of nitrogen and phosphorous in soils. p. 27–45. *In* H.G. van de Meer et al. (ed.) Animal Manure on Grassland and Fodder Crops. *Proc. Symp. Eur. Grassl.* Fed. Martinus Nijhoff Publishers, Boston, MA.

Van Gestel, M., J.N. Ladd, and M. Amato. 1991. Carbon and nitrogen mineralization from two soils of contrasting texture and microaggregate stability: Influence of sequential fumigation, drying and storage. *Soil Biol. Biochem.* 23:313-322.

Van Kessel, J.S., and J.B. Reeves III. 2002. Nitrogen mineralization potential of dairy manures and its relationship to composition. *Biol. Fertil. Soils* 36:118–123.

Van Kessel, J.S., J.B. Reeves III, and J.J. Meisinger. 2000. Nitrogen and carbon mineralization of potential manure components. *J. Environ. Qual.* 29:1669–1677.

Van Veen, J.A., and P.J. Kuikman. 1990. Soil structure aspects of decomposition of organic matter by micro-organisms. *Biogeochem.* 11:213-233.

Velthof, G.L., P.J. Kuikman, and O. Oenema. 2003. Nitrous oxide emission from animal manures applied to soil under controlled conditions. *Biol. Fertil. Soils* 37:221–230.

Velthof, G.W., M.L. Van Beusichem, W.M.F. Raijmakers, and B.H. Janssen. 1998. Relationship between availability indices and plant uptake of nitrogen and phosphorus from organic products. *Plant Soil* 200:215–226.

Vinther, F. P., E.M. Hansen, and J. Eriksen. 2006. Leaching of soil organic carbon and nitrogen in sandy soils after cultivating grass-clover swards. *Biol. Fertil. Soils* 43:12–19.

Watts, D.B., H.A. Torbert, and S.A. Prior. 2007. Mineralization of nitrogen in soils amended with dairy manure as affected by wetting/drying cycles. *Commun. Soil Sci. Plant Anal.* 38:2103–2116.

Watts, D.B., H.A. Torbert, and S.A. Prior. 2010. Soil property and landscape position effects on seasonal nitrogen mineralization of composted dairy manure. *Soil Sci.* 175:27–35.

Wienhold, B. J. 2007. Comparison of laboratory methods and an *in situ* method for estimating nitrogen mineralization in an irrigated silt-loam soil. *Commun. Soil Sci. Plant Anal.* 38:1721–1732.

Wienhold, B.J., G.E. Varvel, and W.W. Wilhelm. 2009. Container and installation time effects on soil Moisture, temperature, and inorganic nitrogen retention for an in situ nitrogen mineralization method. *Commun. Soil Sci. Plant Anal.* 40:2044-2057.

Wildung, R.E., T.R. Garland, and R.L. Buschbom. 1975. The interdependent effects of soil temperature and water content on respiration rate and plant root decomposition in arid and grassland soils. *Soil Biol. Biochem.* 7:372-378.

Wu, Z., and J.M. Powell. 2007. Dairy manure type, application rate, and frequency impact plants and soils. *Soil Sci. Soc. Am. J.* 71:1306–1313.

PART III. PHOSPHORUS FORMS AND LABILITY

In: Environmental Chemistry of Animal Manure   ISBN: 978-1-62808-641-6
Editor: Zhongqi He   © 2013 Nova Science Publishers, Inc.

Chapter 10

SOLUBILITY OF MANURE PHOSPHORUS CHARACTERIZED BY SELECTIVE AND SEQUENTIAL EXTRACTIONS

John D. Toth[1,], Zhengxia Dou[1] and Zhongqi He[2]*

10.1. INTRODUCTION

Phosphorus (P) availability is governed by the P in soil or water that is made available by desorption and dissolution processes for uptake by plants in terrestrial and aquatic ecosystems (Sharpley, 2000). Solubility of manure P is a critical factor in evaluating manure's nutrient value in agriculture and its role in eutrophication of surface waters. Kuo (1996) pointed out that the quantity of labile P, the concentration of P in soil solution, as well as P buffering capacity affecting the distribution of P between the solution and solid phases, are the primary factors characterizing soil P availability. To evaluate P availability in soils, numerous soil tests have been developed which extract varying amounts of P, depending on the types of extractants used. The extractants include, but are not limited to, water or buffered salt solutions, anion exchange resins, and diluted acids or buffered alkaline solutions with or without a complexing agent (Kuo, 1996). Whereas the principles of these soil P extractions could be applied to manure P research, the different physico-chemical properties of animal manure should be recognized (He et al., 2003). For example, He et al. (2006a) observed that the sequentially-extracted HCl fraction of animal manure, especially poultry litter/manure, contained a large portion of organic P (P_o) which would not have been measured by a soil-based protocol.

This chapter reviews solubility of P from food animal manures (in general, swine, dairy, beef and poultry) and manure products (those undergoing further processing such as storage, composting, pelletizing, etc.) determined through individual and sequential extractions with

---

[*] Corresponding author – jdtoth@vet.upenn.edu
[1] University of Pennsylvania, School of Veterinary Medicine, Kennett Square, PA 19348
[2] USDA-ARS, New England Plant, Soil and Water Laboratory, Orono, ME 04469, USA

the emphasis on methodological variations. Three broad classes of extraction methods, water, other simple individual extractants, and sequential fractionation, are covered, followed by examples of research on characterization of P forms in different types of animal manures, effects of manure handling, processing, use of chemical amendments to bind P, and dietary manipulation to reduce soluble P excretion.

10.2. WATER-EXTRACTABLE PHOSPHORUS

Water-extractable soil P has been used for several decades as a predictor of the potential for losses of P in runoff from surface soil horizons (Pote et al., 1996; 1999) and in leachate from soils enriched in P (Koopmans et al., 2006). Pote et al. (1996; 1999) found a high degree of correlation between water-extractable P (WEP) and the biologically active P fractions (r^2=0.82 to 0.93) extracted from the 0-2 cm soil depth in four Ultisols. Water-extractable P was as good a predictor of dissolved reactive P and bioavailable P as other standard soil test P methods and iron oxide paper strips. More recently, WEP has become a critical part of Phosphorus Indices. The concept was proposed by Lemunyon and Gilbert (1993) as a phosphorus management tool to identify areas and practices enhancing the risk of water pollution due to P movement off-farm. The P Index uses field, soil type, topographic, background soil P level, and P application rate and method parameters divided into "source" and "transport" categories to rank a given field for P loss vulnerability. The P Index has been developed into a flexible and widely applicable tool for nutrient management planning, and has now been adapted for use in some form by most states in the US (Sharpley et al., 2003). As a component of the P Index, WEP as a function of manure or manure product type, management and application method is currently being employed to develop P source coefficients (Elliott et al., 2006; Shober and Sims, 2007), which are weighting factors related to potential for P loss in runoff and may differ depending on locally- or regionally-specific manure types, application methods and farm management decisions.

10.2.1. Methodology

A factor complicating WEP use in nutrient management planning is the diversity of laboratory handling and analytical methods and the potential difficulty of comparing results from different studies (Vadas and Kleinman, 2006). Across-study variability is found in use of fresh or dried manure material, sample:extractant ratio, length of extraction (shaking), filtration method, and choice of analytical instruments (Table 10.1). Self-Davis and Miller (2000) proposed as a standard method a 20:200 ratio between the poultry manure (g wet basis:mL) in their study and the water extractant, with 2 hr shaking, while Sharpley and Moyer (2000), based on analyses of manures and manure composts suggested a 1:200 g:mL sample dry weight:extractant ratio and 1 hr shaking. Several recent studies have examined WEP methodology issues with the goal of a widely-applicable standard procedure. Increasing the sample:extractant ratio in nearly all manure types increased WEP release. In poultry litter, Vadas and Kleinman (2006) found a non-linear increase in WEP at decreasing sample:extractant ratios, so 1:10<1:50<1:150 but 1:200=1:250, while Haggard et al. (2005)

noted 5 to 13 times more WEP in 1:200 compared to 1:10 ratio poultry litters and pelletted litter products. Similarly, Mamo et al. (2007) saw more WEP released from fresh, refrigerated feedlot beef manure at higher extractant ratios. Kleinman et al. (2002) found relatively little increase in WEP in dairy, poultry and swine manure extracts as the ratio was increased from 1:10 to 1:40, but substantially more WEP at 1:200 than at the 1:40 ratio, and suggested dissolution of sparingly-soluble Ca phosphates as a reason. There was less difference between the 1:200 and 1:40 g:mL extractant ratios in dairy than swine or poultry manures. The highest correlation of WEP with runoff P was at a sample:extractant ratio of 1:100 from a range of manures, poultry litters and biosolids (Kleinman et al., 2007).

Drying method of manure samples has given equivocal results for the amount of WEP released relative to fresh manure. Air-drying or oven-drying tended to increase WEP release for swine and some poultry manures (Vadas and Kleinman, 2006), but decrease WEP for dairy manure (Vadas and Kleinman, 2006; Chapuis-Lardy et al., 2003). McDowell et al. (2008) noted a decline of up to 61% in WEP released from air-dried manure compared to fresh dairy manure. In broiler manure, Sistani et al. (2001) found no significant difference in WEP between fresh and oven-dried (105°C) but less WEP released from freeze-dried, air-dried or oven-dried at 65°C.

Increasing shaking time increased WEP release, but often by a small amount. Dou et al. (2000) found 9% more inorganic P (P_i) released in water extracts of layer poultry manure and 11% more from dairy manure as shaking time increased from 1 to 16 hr. Mamo et al. (2007) noted significantly greater WEP release at shaking times of 0.5 to 2 hr from beef manure at 1:200 and 1:100 g:mL sample:extractant ratios, but no difference at the 1:10 ratio. A logarithmic relationship between shaking times up to 24 hr and WEP release from dairy, layer poultry and swine manures was found by Kleinman et al. (2002), with 70% of total WEP released into solution in the first hr.

As with shaking time, method of filtration had relatively little effect on measurement of WEP. Kleinman et al. (2002) found up to 10% more WEP in dairy and poultry manure extracts filtered through coarse paper (Whatman 1) compared to 0.45 μm membranes, and suggested colloidal P not retained by the coarse filter paper could be hydrolyzed by acidic colorimetric reagents or inductively coupled plasma spectroscopy (ICP). No differences as a function of filtration type were noted in their swine slurry sample, nor by Mamo et al. (2007), using coarse or fine filter paper or 0.45 μm membranes on beef manure extracts. Pore size of the filtering medium is an important component of Haygarth and Sharpley's proposed classes of WEP forms (2000), though many studies of WEP do not deal specifically with those functional classes. Toor et al. (2006), in a review of the literature, noted that if centrifugation and decanting is used without filtration, WEP may be substantially overestimated, although Wolf et al. (2005) found no significant differences in WEP analyzed by ICP between Whatman 40 paper filtration and centrifugation followed by decanting.

Vadas and Kleinman (2006) found acid digestion of extracts of dairy, poultry and swine manure yielded on average 22% greater total WEP than inorganic WEP measured by colorimetry; however, Wolf et al. (2005) reported 7% lower WEP from ICP analysis compared to Murphy-Riley colorimetric analysis, and suggested interference from extraction of colored compounds by the water extractant may be the cause.

Table 10.1. Details of the various methods used to determine WEP in manures.

Sample conditions	Sample:extractant ratio (DM basis unless noted)	Extraction time	Filtration	Analytical methods†	Reference
Freeze-dried	2:25	1 hr	0.45 μm membrane	Not specified	Angel et al., 2005
Fresh	1:60 to 1:250	1 hr, 1 hr repeated	Whatman 2	Ascorbic acid method (P_i), ICP (P_t)	Chapuis-Lardy et al., 2003
Fresh	2:98 wet basis	1 hr	Whatman 42	Ascorbic acid method (P_i), ICP (P_t)	Chapuis-Lardy et al., 2004
Air-dried	1:10	1 hr	Whatman 42	Ascorbic acid method (P_i)	Codling et al., 2000
Fresh	1:10 wet basis	2 hr	Centrifuged, 0.45 μm	Ascorbic acid method (P_i)	Do et al., 2005
Fresh	1:10 to 1:200 wet basis	2 hr	Centrifuged, 0.45 μm	Ascorbic acid method (P_i), ICP (P_t)	Haggard et al., 2005
Fresh	1:10 to 1:200	1 min to 24 hr	Centrifuged, Whatman 1 or 0.45 μm	Ascorbic acid method (P_i)	Kleinman et al., 2002
Fresh	1:200	1 hr	Centrifuge and decant	ICP (P_t)	Kleinman et al., 2005
Dried	1:100	1 hr	0.45 μm membrane	ICP (P_t)	Leytem and Thacker, 2008
Oven-dried	1:10	1 hr	0.45 μm membrane	ICP (P_t)	Leytem et al., 2004
Fresh	1:10	1 hr	Whatman 40	ICP (P_t)	Maguire et al., 2006
Fresh	1:10	Not specified	Centrifuged, Whatman 40	ICP (P_t)	Maguire et al., 2006
Fresh	1:10 to 1:200	0.5 to 2 hr	0.45 μm, Whatman 40 or 42	Ascorbic acid method (P_i)	Mamo et al., 2007
Fresh or oven-dried	1:333 or 2:98 wet basis	1 hr	Centrifuged, Whatman 42	Colorimetric method (P_i), ICP (P_t)	McDowell et al., 2008
Oven-dried	1:15	30 min	Whatman 2V	ICP (P_t)	Miles et al., 2003
Fresh	1:10	2 hr	Centrifuged, 0.45 μm	Ascorbic acid method (P_i)	Moore and Miller, 1994
Fresh	1:10 wet basis	2 hr	Centrifuged, 0.45 μm membrane	Ascorbic acid method (P_i)	Self-Davis and Moore, 2000
Fresh	1:10 wet basis	1 hr	0.45 μm membrane	ICP (P_t)	Sims and Luka-McCafferty, 2002
Fresh, oven-, air-, or freeze-drying	1:15	30 min	Whatman 2V	Ascorbic acid method (P_i)	Sistani et al., 2001

Sample conditions	Sample:extractant ratio (DM basis unless noted)	Extraction time	Filtration	Analytical methods[†]	Reference
Fresh	1:10 to 1:250	1 hr	Centrifuged, 0.45 μm or Whatman 40	Ascorbic acid method (P_i), ICP (P_t)	Toor et al., 2007
Fresh, oven- or air-dried	1:10 to 1:250	1 hr, 1 hr repeated	0.45 μm membrane	Ascorbic acid method (P_i), acid digestion (P_t)	Vadas and Kleinman, 2006
Fresh	1:200	Not specified	0.45 μm membrane	Ascorbic acid method (P_i), acid digestion (P_t)	Vadas et al., 2007
Fresh	1:200	1 hr	Whatman 40 or centrifuge and decant	Ascorbic acid method (P_i), ICP (P_t)	Wolf et al., 2005

[†] "Ascorbic acid method" refers here to a variety of colorimetric procedures, including the Murphy-Riley method (1962) and He and Honeycutt's modified Mo blue method (2005), in which ascorbate serves as reducing agent for the quantitative reaction of P with the color reagent. Details of methodological approaches can be found in the individual citations. "Colorimetric method" is used when further details are not provided by the authors.

Taking into account correlation with runoff P measurements, interlaboratory repeatability and ease and feasibility of laboratory procedures, Kleinman et al. (2007) proposed the following as a standard protocol for WEP measurement for manures: 1:100 sample:extractant ratio, 1 hr shaking, centrifugation followed if necessary by filtration through Whatman 1 paper, and analysis by ICP or colorimetry.

10.2.2. Variability in WEP Across Manure Types

Water-extractable P from a range of manures from food animal species and storage and handling procedures (Table 10.2) were reported by Kleinman et al. (2005) and Sharpley and Moyer (2000). Based on analyses of 24 fresh manure samples of each manure type WEP ranged from 1.63 to 2.49 g P kg^{-1} for dairy, 2.85 to 5.10 g kg^{-1} for poultry litter, 6.08 to 8.52 g P kg^{-1} for poultry manure, and 5.01 to 7.38 g kg^{-1} for swine slurry (Sharpley and Moyer, 2000). As a proportion of manure total P, WEP was lowest in beef manure (17%), broiler litter and layer manure (12 and 20%, respectively), intermediate in swine manure and turkey litter (28 and 34%), and at 70%, greatest in dairy manure (Kleinman et al., 2005).

10.2.3. Effects of Manure Handling, Composting and Addition of P-binding Chemicals

With composting of manure, Sharpley and Moyer (2000) found little difference in WEP in their dairy samples (mean of 2.09 g P kg^{-1} in fresh manure compared to 2.39 g kg^{-1} after composting), but a significant decrease in WEP in the poultry manure composted with wood chips and other plant material (mean of 7.30 g kg^{-1} in fresh manure compared to 2.11 g kg^{-1} after composting).

Table 10.2. Water-extractable P results from selected studies on animal manures and composts.

Animal species	Manure sample handling or management	Reference
Broiler poultry, turkey, swine	Conventional, reduced mineral P or phytase-containing diets	Angel et al., 2005
Broiler poultry	Conventional, low-phytate corn or phytase-containing diets	Applegate et al., 2003
Dairy	Slurry with or without commercial chalk-clay amendment	Chapuis-Lardy et al., 2003
Dairy	Diets with a range of P concentrations	Chapuis-Lardy ct al., 2004
Broiler poultry	Litter amended with Al- and Fe-based water treatment residuals	Codling et al., 2000.
Dairy	Amended with polymers, Al or Fe-based chemicals or coal combustion byproduct	Dao and Daniel, 2002
Beef	Fresh or composted, receiving high-pH amendments	Dao, 1999
Broiler poultry	Litter receiving Al, Ca, K or Fe-based chemical amendments	Do et al., 2005
Dairy	Different aerobic and anaerobic digestion systems	Güngör and Karthikeyan, 2005, 2006
Broiler poultry	Laboratory sample:extractant ratios	Haggard et al., 2005
Dairy, layer poultry, swine	Laboratory sample handling and analytical methods	Kleinman et al., 2002
Dairy, beef, broiler and layer poultry, turkey, swine	Fresh manure and manure under different storage conditions	Kleinman et al., 2005
Dairy, beef, broiler and layer poultry, swine	Laboratory sample handling and analytical methods	Kleinman et al., 2006
Dairy, broiler and layer poultry, turkey, swine	Fresh or with different manure storage conditions, laboratory sample handling and analytical methods	Kleinman et al., 2007
Swine	Different grain-based diets	Leytem and Thacker, 2008
Dairy, beef, swine, poultry	Fresh, stored slurry, composted	Leytem et al., 2004
Broiler poultry	Breeder manure, with or without phytase, under different storage conditions	Maguire et al., 2004
Broiler poultry	Litter amended with Ca-based liming materials	Maguire et al., 2006
Beef, layer poultry, swine	Fresh or stored, sample:extractant ratios, shaking time, type of filter paper	Mamo et al., 2007
Dairy	Comparison of 4 dairy feeding systems, confinement vs. grazing	McDowell et al., 2008

Animal species	Manure sample handling or management	Reference
Broiler poultry	Conventional, low-phytate corn or phytase-containing diets	Miles et al., 2003
Broiler poultry	Receiving Al, Ca or Fe-based chemical amendments	Moore and Miller, 1994
Broiler poultry	Bark or straw-based litter, dried at different temperatures	Robinson and Sharpley, 1995
Broiler poultry	Litter repeatedly amended with aluminum sulfate	Sims and Luka-McCafferty, 2002
Broiler poultry	Fresh or dried by different methods	Sistani et al., 2001
Broiler poultry	Fresh or granulated with chemical amendments	Toor et al., 2007
Dairy, poultry, swine	Presence/absence of bedding materials, different drying methods	Vadas and Kleinman, 2006
Dairy, broiler poultry	Manure on porous sheet on field soil, interacting with natural rainfall	Vadas et al., 2007
Dairy, broiler and layer poultry, swine	Laboratory sample handling and analytical methods	Wolf et al., 2005

Vadas et al. (2007) and Robinson and Sharpley (1995) examined changes in WEP in manures and litters exposed to actual or simulated variable field conditions. Vadas et al. (2007) spread dairy manure and poultry litter on porous fabric sheets on soil outdoors and monitored WEP trends over up to 18 months. Water-extractable P decreased over the first 2 months of the study and then stabilized at 10 to 20% of the initial WEP concentrations. In the Robinson and Sharpley (1995) study, dried, ground poultry litter was spread across a leachate collection apparatus, and simulated rainfall events were alternated with drying at different temperatures. Forty percent of the total amount of P_i leached from the litter was lost in the first simulated rainfall, and 80 to 95% after 5 rainfalls.

Both high and low pH amendments have been used to reduce WEP concentrations in animal manures, generally poultry litter and manure. Moore and Miller (1994) treated poultry litter with a range of Al-, Ca- or Fe-containing chemicals and incubated the samples for 1 wk at ambient temperature. After incubation, WEP was determined on 1:10 dry matter (DM):extractant ratio samples. Selected high- and low-pH amendments lowered WEP in the litters from >2000 to <1 mg P kg^{-1}. Among the most effective amendments were alum, quicklime, slaked lime, and the Fe-containing chemicals. Gypsum and sodium aluminate reduced WEP by 50 to 60%. Maguire et al. (2006) treated layer manure and broiler litter with calcium oxide and calcium hydroxide at rates of addition of 2.5 to 15% wet weight and incubated samples 1 d. Water-extractable P was measured at 1:10 manure:extractant ratio by ICP. Water-extractable P was reduced compared to unamended controls by >90% with amendment rates of at least 10% calcium oxide. High-pH, Al- and Fe-rich water treatment plant residuals were used by Codling et al. (2000) to control solubility of P in poultry litter. Both Al- and Fe-rich residuals applied at rates of 25 to 100 g kg^{-1} litter reduced WEP by 39 to 88% following 7 wk incubation. Sims and Luka-McCafferty (2002) expanded chemical amendment studies to the farm scale; 97 poultry houses had alum applied to the litter on the house floor at 0.09 to 0.135 kg bird^{-1} seven times over the 16 mo study. An equivalent number of non-amended poultry houses served as controls. At the end of the study, WEP in the litter was determined on 1:10 litter:extractant samples and measured by both ICP and colorimetrically. After alum treatment, soluble P declined compared to controls by 67 and 73% (ICP and colorimetric determination, respectively).

These and other laboratory- and field-based studies demonstrate that a variety of chemicals and water-treatment byproducts can effectively tie up soluble P in poultry litters and manures. Other treatments, including industrial byproducts, and their effects on P solubility in other animal manure types are discussed in later sections.

10.2.4. Dietary Manipulation Effects

Modifying the constituents of food animal diets has been proposed as a simple, at-the-source means to reduce manure P concentrations (Valk et al., 2000; Maguire et al., 2005). In dairy cattle, excess P in the diet is largely contributed by mineral P supplementation, in the misperception that high-P diets improve milk production and reproductive success (Wu et al., 2000). In a study of P feeding levels and manure P concentration on commercial dairies, Chapuis-Lardy et al. (2004) found WEP increased from 2.25 g kg^{-1} fecal DM in a herd with 0.39% diet P to 6.35 g P kg^{-1} in a herd fed 0.60% P. Mixed stepwise regression of feed and fecal parameters' influence on fecal P excretion showed that diet P was the dominant factor, although fecal Ca, pH and stage of lactation also were significant contributors.

Table 10.3. Other single-extraction solutions used on animal manures and byproducts.

Animal species	Manure sample handling or management	Reference
Poultry	Three fresh samples extracted with sodium acetate buffer (pH 5.0)	Dail et al., 2007
Dairy	Manure extracted with dilute HCl solution	Dou et al., 2007, 2010
Dairy, poultry	Mixed dairy and poultry manures, palletized, extracted with NaHCO$_3$	Hadas et al., 1990
Dairy	Fresh from 13 farms, extracted with sodium acetate buffer (pH 5.0)	He et al., 2004b
Broiler poultry	Litter extracted with range of dilute HCl solutions	Tasistro et al., 2004
Broiler poultry	Litter extracted with range of dilute HCl or buffer solutions (pH 6.0)	Tasistro et al., 2007

In swine and poultry diets, phytase is now generally used as a diet additive that can release P from phytate, which is a relatively unavailable P source for monogastric animals. Phytase supplementation allows the reduction or elimination of mineral P in the diet, otherwise necessary for proper P nutrition in swine and poultry. There has been some concern that the P released by action of phytase may have higher solubility and pose more of a risk for P losses in runoff when manure is land-applied (Angel et al., 2005; Applegate et al., 2003). Angel et al. (2005) added antimicrobial agents to poultry and swine diets with added phytase and concluded that post-excretion increases in WEP were due to microbial activity and not to inherently higher WEP in phytase-supplemented diets.

10.3. SINGLE EXTRACTIONS IN DILUTE ACID AND OTHER SOLUTIONS

In addition to water as a single extractant for animal manures and manure products, other solutions have been used to evaluate P solubility to fulfill different research objectives (Table 10.3). In a study of the relative P availability in the soil from amendment with pelletized manure with or without fertilizer P enrichment, Hadas et al. (1990) extracted ground, mixed poultry and cattle manure formed into 4-6 mm pellets with 0.5 mol L^{-1} sodium bicarbonate solution, noting that soil test P-Olsen P is extracted under similar conditions (Kuo, 1996). Phosphorus release was greatest from ground pellets followed by broken and whole pellets. Inhibition of P release was thought to be related to high concentrations of ammonium and high pH in the intact pellets. Similarly, sodium acetate buffer (100 mmol L^{-1}, pH 5.0) has been tested as a single extractant to determine plant-available P in dairy manures because extraction conditions are close to that for Morgan (1.24 mol L^{-1} sodium acetate buffer, pH 4.8) or modified Morgan P (0.62 mol L^{-1} NH_4OH + 1.25 mol L^{-1} acetic acid, pH 4.8) solutions, which have been used for soil P testing (He et al., 2004b). In the 13 dairy manures tested, total P extracted by the acetate buffer had an average of 6221 mg P kg^{-1} DM with a standard deviation (SD) of 1811. Total P in both H_2O and $NaHCO_3$ fractions had an average of 6104 mg P kg^{-1} DM with SD of 1668. The average of total P in all three fractions was 6669 mg P kg kg^{-1} DM with SD of 1701. These data indicate that the amount of P extracted by the single sodium acetate buffer solution (100 mmol L^{-1}, pH 5.0) from dairy manure was equal to the summed amount of P extracted by the series of H_2O, $NaHCO_3$, and NaOH solutions frequently used in sequential fractionation methods (described in detail in later sections). However, dairy manure contained too little HCl-extracted P to test the correlation with the portion of sequentially extracted HCl-P. Poultry manure contained small quantities of NaOH-extractable P and a relatively large amount of HCl-extractable P, thus providing an opportunity to verify the correlation between acetate buffer-extracted P and sequentially-extracted P (Dail et al., 2007). In one sample, acetate-extractable P was twice that extracted by H_2O, $NaHCO_3$, and NaOH together, but less than the total extracted by H_2O, $NaHCO_3$, NaOH, and HCl. In a second sample, acetate-extractable P was about 22 % higher than that in the first three fractions but less than the total extracted in the four fractions, suggesting that the NaOH-P fraction of animal manure was extractable by sodium acetate buffer. These observations imply that, whereas sequentially extracted NaOH- and HCl-extractable P in soil are considered less plant-available than H_2O and $NaHCO_3$ fractions, the counterpart P in animal manure should not be assumed to be so.

Tasistro et al. (2004) argued that the normally alkaline pH of poultry litter limits the solubility of P forms; however, low soil pH could increase P solubility from poultry litter after field application. Therefore, they assumed that the use of WEP concentrations measured at the original litter pH might lead to an underestimation of the risk of P contamination of runoff water. They measured WEP in broiler and breeder poultry litter (1:200 manure:extractant ratio) at original pH values of 7.6 to 8.5, and at pH of 6 and 7 after acidification with HCl. Their results show that WEP measured at lower pH increased by 24 to 69% compared to that of WEP extracted at unmodified pH. In a second experiment, Tasistro et al. (2007) compared soluble P concentrations extracted from poultry wastes at three pHs: 1) at natural pH, using deionized water (DIw); 2) after titrating DIw suspensions with 0.5 mol L^{-1} hydrochloric acid (HCl) to pH end-points 3.0, 4.0, and 6.0; and 3) at pH 6.0 with a variety of buffering compounds. Extracting solutions adjusted to pH 3 to 4 extracted approximately 20% more molybdate-reactive P (MRP) from low-P layer manures than solutions at pH 6, and up to 40% more MRP from high-P layer manures than at pH 6. Compared to WEP, total soluble P increased by 60 to 140% in the acidified extractant at pH 6. Buffers extracted more soluble minerals than suspensions acidified with HCl, probably because of their complexation ability. The 2-(*N*-morpholino)ethanesulfonic acid (MES) buffer showed minimal metal complexation, suggesting that it was the most suitable buffer compound tested for extracting P at a stable pH value of 6. These results highlight the importance of measuring WEP under conditions similar to those encountered in the soil after litter application. Tasistro et al. (2007) argued that soluble P from poultry manure measured at pH 6.0 rather than at unadjusted original pH would be a more correct input to simulation models as the extraction would be done at a pH representative of the environment from which runoff is most likely to occur. However, to date there are no modeling data to support their claims.

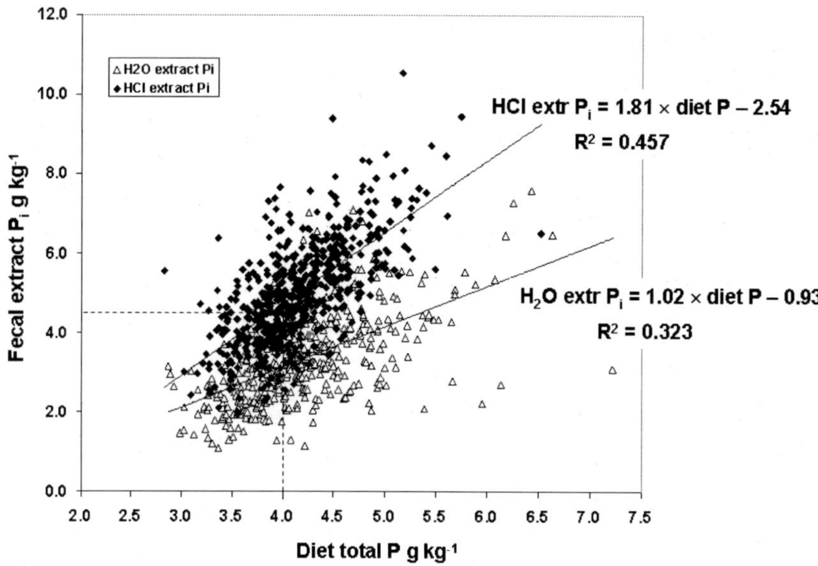

Figure 10.1. Provisional indicator of overfeeding P in dairy cattle, based on 0.1% HCl extraction of fecal samples. Inorganic P in extracts of 4.75 g kg^{-1} manure DM corresponds to a diet P concentration of 0.40%. Adapted from Dou et al. (2010).

Although a one-hr water extract of dairy manure seemed initially to be a good choice for an indicator of P overfeeding (Dou et al., 2002), additional research (Chapuis-Lardy et al., 2004) suggested that other factors such as fecal Ca concentration, sample handling and pH also had a significant impact on the relationship between diet P and fecal P. To overcome the confounding effects of Ca concentration, etc., a series of dilute hydrochloric, acetic, and citric acid solutions were chosen as extractants to determine which, if any, could improve the correlation between dairy diet P and fecal P (Dou et al., 2007). Manure samples collected from 25 commercial dairies were extracted 1 hr in acid solutions at a 2:98 sample wet weight:extractant ratio. Inorganic P released from dairy manure in 0.1% HCl solution was closely correlated with diet P concentration (R^2=0.69 compared to R^2=0.33 for water as extractant). A fecal P overfeeding indicator was proposed using the 0.1% HCl extracting solution and tested on a large set of fecal and feed samples collected from commercial dairy farms in the Northeast and Mid-Atlantic states (Dou et al., 2010). Based on results from controlled feeding trials, and corresponding to a diet total P level of 0.40% (4.0 g kg^{-1}), Dou and colleagues proposed a provisional benchmark HCl fecal extract concentration of 4.75 g P kg^{-1} manure DM. Comparing herd average HCl fecal extract P (>90 farms, 5 to 17 cows per herd with samples collected quarterly from 2002 through 2004) with total mixed ration diet P concentration, the coefficient of determination of the extract P to diet P relationship was 0.46 (Figure 10.1). Results indicated that 316 of 575, or 55%, of the herd average datapoints may have been associated with P overfeeding. Using the provisional benchmark HCl extract fecal P concentration of 4.75 g kg^{-1} and the regression parameters derived from the 575 sample dataset (Figure 10.1), 75% of diet P concentrations were predicted correctly as being above or below 4.0 g P kg^{-1}.

10.4. P FRACTIONATION BY SEQUENTIAL EXTRACTIONS

Expanding on early twentieth-century research into P forms in animal manures, McAuliffe and Peech (1949) proposed an extraction sequence that separated manure P into phospholipid, protein-bound, acid-soluble and P_i fractions. Phospholipids were extracted from oven-dried sheep and cow feces with a 3:1 alcohol-ether mixture. Insoluble residues were shaken with 5% trichloroacetic acid (TCA), and the resulting soluble fraction contained what were identified as P_i and acid-soluble P_o. The acid-soluble P_o fraction consisted largely of phytate or phytic acid. Residues from the acid extraction were considered to contain protein-bound P. In the sheep feces there was an average of 10.3 g total P kg^{-1}, of which 82% were P_i, 0.3% phospholipid P, 12% protein-bound P and 3% acid-soluble P_o; in the cow feces the comparable fractions of the 4.7 g total P kg^{-1} were 50% P_i, 0.7% phospholipid P, 33% protein-bound P, and 9.4% acid soluble P.

Fletcher (1955) reported that in the early twentieth century, some farmers in the eastern US added phosphoric acid to barnyard manures, believing them to be deficient in P needed for crop growth. Due to dietary and other management practices that have changed since then, Barnett (1994a, b) noted that his manure samples tended to have higher total P concentrations than those reported in Peperzak et al. (1959) and earlier studies. With a modification of the McAuliffe and Peech procedure, Barnett (1994a, b) reported substantial differences in relative proportions of P fractions between the ruminants (beef and dairy cattle) and monogastric species (swine and layer and broiler poultry). Total P was much lower in the cattle manures, 7 to 9 g P kg^{-1}, than in swine (29 g P kg^{-1}) and poultry (18 to 24 g P kg^{-1}). While the relative proportion of total P as P_i and phospholipid was similar across species, acid soluble P, including phytic acid, was three to five times higher in manures from swine and poultry. Monogastric animals such as swine and poultry have a lower ability to assimilate phytate P than ruminants. Correspondingly, residual P, derived from nucleic acids, was greater in cattle than swine and poultry manures. In the last decade or so, a modified Hedley sequential fractionation has been more widely used to characterize manure P (He and Dou, 2010) and results using this method will be the subject of the next section.

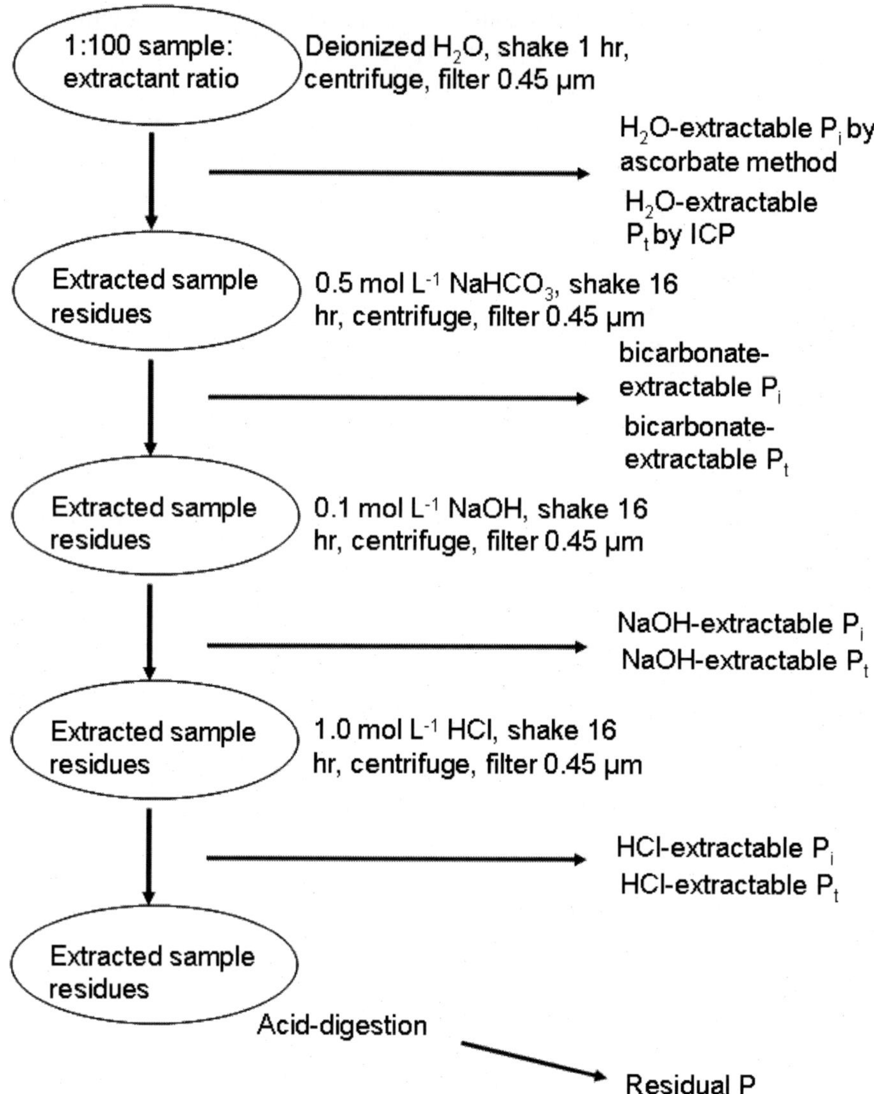

Figure 10.2. A representative diagram of the modified Hedley sequential P fractionation method used in manure P characterization. Refer to the main text for discussion of the variations in methods used by different researchers.

Table 10.4. Modifications to the Hedley sequential fractionation procedure as used in manure fractionation analyses.

Drying method	Sample:extractant ratio	Extraction time	Analytical methods[†]	Reference
Fresh, oven-dried	1:100	16 hr	Ascorbic acid method (P_i), acid digestion (P_t)	Ajiboye et al., 2004
Oven-dried	1:100	16 hr	Colorimetric (P_i), acid digestion (P_t)	Codling, 2006

Drying method	Sample:extractant ratio	Extraction time	Analytical methods[†]	Reference
Oven-, air-, or freeze-dried	1:100	2 hr water, 16 hr other	Ascorbic acid method (P_i), acid digestion (P_t)	Dail et al., 2007
Oven-dried	1:100	1 hr, repeated	Ascorbic acid method (P_i), ICP (P_t)	Dou et al., 2000
Oven-dried	1:100	1 hr, repeated	Ascorbic acid method (P_i), ICP (P_t)	Dou et al., 2003
Freeze-dried	1:50	16 hr	Ascorbic acid method (P_i), acid digestion (P_t)	He et al., 2003
Freeze- or air-dried	1:100	2 hr water, 16 hr other	Ascorbic acid method (P_i), acid digestion (P_t)	He et al., 2004a, b; 2006b
Oven-dried	1:200	1 hr water, 16 hr other	ICP (P_t)	Maguire et al., 2004
Fresh, air-dried	1:100	1 hr, repeated	Colorimetric (P_i), acid digestion (P_t)	McDowell and Stewart, 2005
Oven-dried	1:200	1 hr water, 16 hr other	Ascorbic acid method (P_i), ICP (P_t)	McGrath et al., 2005
Fresh	1:200	1 hr water, 16 hr other	Ascorbic acid method (P_i), acid digestion (P_t)	Sharpley and Moyer, 2000
Oven-dried	1:200	1 hr water, 16 hr other	ICP (P_t)	Toor et al., 2005
Freeze-dried	1:60	1 hr	ICP (P_t)	Turner and Leytem, 2004
Oven-dried	1:200	16 hr	Ascorbic acid method (P_i), ICP (P_t)	Warren et al., 2008
Air-dried	1:100	1 hr, repeated	Ascorbic acid method (P_i), acid digestion (P_t)	Wienhold and Miller, 2004.
Air-dried	1:60	4 hr water, 16 hr other	Ascorbic acid method (P_i), acid digestion (P_t)	Ylivainio et al., 2008

[†] Refer to footnote, Table 10.1.

10.4.1. Adaptations of the Hedley Sequential Procedure in Manure P Research

Hedley and colleagues (1982a, b) developed a soil sequential fractionation scheme with quantitative fractions of P forms separated using a series of acidic or basic solutions with 16 hr shaking: an anion exchange resin followed by 0.5 mol L^{-1} $NaHCO_3$, 0.1 mol L^{-1} NaOH, 1.0 mol L^{-1} HCl, and acid digestion of the residues. Products of the extraction series were defined as labile water-soluble P, labile adsorbed P (in bicarbonate extracts), labile P associated with

Al and Fe (in NaOH), Ca-bound P (in HCl), and recalcitrant insoluble P in the residues, respectively. These definitions of P chemical groups are generally accepted, although not on an exclusive basis (He et al., 2003), and by itself the Hedley procedure cannot identify specific P-containing inorganic or organic compounds, which require other analytical tools. Comprehensive reviews of the Hedley sequential extraction procedure for soils are given in Cross and Schlesinger (1995) and more recently in Negassa and Leinweber (2009).

For manure P research, a wide variety of methodological variations have been applied to the basic framework of the Hedley fractionation procedure, including choice of extractants, sample drying, sample-to-extractant ratios, extraction and agitation time, and analytical methods (Table 10.4). One adaptation of the Hedley sequential fractionation for animal manures is presented in Figure 10.2. In addition to the replacement of the anion exchange resin with water extractant in the first step, another noteworthy change is inclusion of P_o in the HCl fraction. In the original procedure, no P_o was included in the HCl fractions because the amount of organic phosphorus in the HCl fractions of their samples was negligible based on preliminary measurement (Hedley et al., 1982b). Others have found considerable P_o in the HCl fraction with some manure samples containing more P_o than P_i (He et al., 2006a). Apparently, it should be determined experimentally whether the HCl fraction of a sample contains P_i only or both P_i and P_o (He et al., 2006a; 2008). Readers should be aware that some of the papers reviewed in this chapter have not determined if any P_o was present in the HCl fractions of their manure samples; thus, these data might be incomplete or less accurate.

While Leinweber et al. (1997) retained the anion exchange resin in the first extraction step, Sui et al. (1997) suggested that a simple one-step water extraction was more representative of bioavailable labile P than resin-exchanged, and nearly all subsequent studies have used water as the initial extractant. Ajiboye et al. (2004) and Dail et al. (2007) examined the effect of drying method on relative distribution of P among the sequential fractions. Ajiboye and colleagues tested oven-dried compared to fresh dairy and swine manures, and found water-extractable P_o was converted to water-extractable P_i in swine manure, with little change in the other fractions, while in their dairy samples bicarbonate-extractable P was hydrolyzed to water-extractable P_i. Dail et al. (2007) noted that all drying methods tested (air-, oven-, and freeze-drying) increased water-soluble P_i and decreased bicarbonate-extractable P_i and NaOH-extractable P_o in layer poultry manures. Although use of dried samples has an advantage in convenience over moist or wet samples, both groups of researchers caution that sample handling and drying must be taken into account when comparing results of different studies.

Table 10.5. Sequential fractionation methods in animal manure studies

Animal species	Manure sample handling or management	Reference
	Alcohol-ether and TCA fractionation	
Dairy, beef, poultry, swine	Fresh	Barnett, 1994 a, b
Cattle, sheep	Fresh or aged	McAuliffe and Peech, 1949
Dairy, beef, poultry, swine	Fresh or aged	Peperzak et al., 1959
	Modified Hedley fractionation	
Beef, dairy, swine	Fresh or oven-dried	Ajiboye et al., 2004

Animal species	Manure sample handling or management	Reference
Dairy, beef, swine, poultry	Fresh	Ajiboye et al., 2007
Poultry	Litter, litter ash	Codling, 2006
Poultry	Method of drying manure	Dail et al., 2007
Dairy, poultry	Individual or repeated sequential extractions	Dou et al., 2000
Dairy	3 diet P concentrations	Dou et al., 2002
Dairy, broiler poultry, swine	Addition of aluminum sulfate or coal combustion byproducts	Dou et al., 2003b
Poultry	Manure with litter	He et al., 2006b
Swine, cattle	Freeze-dried	He et al., 2003; 2004a
Dairy	Fresh from 13 herds	He et al., 2004b
Poultry	Air-dried	He et al., 2006b, 2008
Poultry litter, biosolids	Stored at ambient temperature or frozen, granulated	He et al., 2010
Swine slurry, poultry	Fresh	Leinweber et al., 1997
Broiler poultry, turkey	Litter from animals fed low or high non-phytate P diets, or diets with phytase	Maguire et al., 2004
Cattle, sheep, deer	Fresh or air-dried	McDowell and Stewart, 2005
Broiler poultry	Litter from animals fed low non-phytate P diets, or diets with phytase	McGrath et al., 2005
Dairy, poultry, swine	Manures, litter, composts	Sharpley and Moyer, 2000
Broiler poultry, turkey	Conventional or low-phytate corn diets	Toor et al., 2005
Dairy	Fresh or stored slurry, fed low or high P diets	Toor et al., 2005
Beef, broiler poultry, swine	Fresh	Turner and Leytem, 2004
Dairy, fox, composted fox and mink manure	Fresh	Uusitalo et al. 2006
Broiler poultry	Addition of aluminum sulfate	Warren et al., 2008
Swine	Conventional or low-phytate corn diets	Wienhold and Miller, 2004
Dairy, fox	Fresh manure, composted or composted and pelletized (fox only)	Ylivainio et al., 2008

Sample-to-extractant ratio is generally 1:100 or 200, although higher proportions of sample to extractant are sometimes employed. Duration of agitation and extraction in the original Hedley procedure was 16 hr. Dou et al. (2000) found small differences in P extracted with shaking times ranging from 1 to 16 hours. Most of the P, 68% of the extractable total P (P_t) in water extracts of dairy manure and 75% in layer poultry manure, was released in the first hour of agitation. Based on these results they proposed using repeated, sequential 1 hr extractions with the number of extractions ranging from seven for water and $NaHCO_3$ to a maximum of three for NaOH and HCl. Most other modified Hedley sequential extraction schemes either retained the 16 hr agitation time, or reduced the time to 1 to 4 hr for water extraction only (Table 10.4).

Total P in extracts is determined by one of several versions of acid digestion to release organic P, followed by colorimetry, or by ICP. A frequently-used analytical method for determination of orthophosphate is a colorimetric procedure developed by Murphy and Riley (1962). This method uses ascorbic acid reduction in which ammonium molybdate reacts with phosphate ions to form a reduced Mo-phosphoric acid complex that produces a blue color having an intensity that is proportional to the concentration of orthophosphate in the original sample. However, the assumption that molybdate-reactive P is exclusively derived from orthophosphate in the original samples is not strictly correct. The strongly-acid Murphy-Riley color reagent may also hydrolyze poly-, pyro- and other condensed phosphates (Dick and Tabatabai, 1977), and labile P_o (McKelvie et al., 1995), leading to overestimation of orthophosphate. A modification of Dick and Tabatabai's procedure was tested by He and Honeycutt (2005), with an increase in sensitivity compared to previous methods. Wolf et al. (2005) also noted that interference with colored compounds in extracts may overestimate P_i, but suggested that the difference compared to extract P_t measured by ICP was relatively small ($\leq 10\%$).

Given the differences in chemical forms of P in soils compared to animal manures and the difficulty in assigning binding mechanisms applicable in the soil-based Hedley methods to manure P, simple colorimetric and ICP analyses of manure extracts have largely been replaced by more sophisticated methodology. Developed in the past decade, these newer methods can provide more detailed information on specific inorganic and organic P chemical species in extracts and/or residues from the sequential fractionation procedure. These additional methods include solution ^{31}P nuclear magnetic resonance (^{31}P NMR) spectroscopy (Ajiboye et al., 2007; He et al., 2008; Turner and Leytem, 2004), x-ray absorption near-edge structure (XANES) spectroscopy (Ajiboye et al., 2007; Seiter et al., 2008), and enzymatic hydrolysis (He and Honeycutt, 2001; He et al., 2004a, b; 2006b). However, colorimetric and ICP analyses of sequential fractionation extracts may continue to have applicability in studies of manure P dynamics in the agroecosystem.

10.4.2. Variations in P Fractions among Manure Types

The modified Hedley sequential method has been used to identify shifts in manure P fractions in manures from different animal species, as a result of modifying animal diets, and manure management and handling procedures such as composting, drying, and amendment with P-binding chemicals (Table 10.5). Phosphorus distribution patterns in six types of manures are shown in Figure 10.3. More than 70% of total P in the dairy and swine manures were in the labile H_2O and $NaHCO_3$ fractions; in contrast, about 70% of total P was in more stable NaOH, HCl, and residual fractions of the poultry litter and fox manure. The patterns of P_o distribution differed from those of P_i in these manures. A large proportion of total P was in organic forms in the NaOH and HCl fractions of dairy manure, swine manure, and poultry litter; with 54, 43 and 90% of P in the NaOH extracts as P_o in the three manures, respectively, and 40 and 60% of total P as P_o in the HCl fractions of swine manure and poultry litter. Total extracted P in fox manure had a pattern similar to poultry litter. Unlike poultry litter, however, there was little or no P_o in the NaOH and HCl fractions of fox manure. This may be because foxes are fed bone meal which has a high level of sparingly-soluble P_i (Ylivainio et al., 2008). McDowell and Stewart (2005) reported 72% of total P in the H_2O and $NaHCO_3$ fractions of fresh dung of grazing dairy cattle, similar to the P distribution of dairy manure in Figure 10.3. In both deer and sheep dung samples, McDowell and Stewart (2005) reported about 60% of total P in the labile H_2O and $NaHCO_3$ fractions (Figure 10.3), which were lower than those in dairy manure. Unfortunately, they did not measure the total P (thus, P_o) in the HCl fractions, and total manure P was the sum of the five fractions, rather than an independent measurement of acid-digested manure. Thus, the assumption that the more aggressively-extracted $HCl-P_o$ was included in residual P (Huang et al., 2008) may not be applicable for the two manures. For this reason, the P partitioning pattern suggested by the authors may not be correct if indeed a significant amount of P_o was present in the HCl fractions of the two samples.

In an investigation of P fractions in a dairy and a layer poultry manure sample, Dou et al. (2000) found the proportions of total P released from dairy manure were 70% for the H_2O extractant, 14% for $NaHCO_3$, 6% for NaOH, and 5% for HCl. In the layer poultry sample the results were 49% for the H_2O extractant, 19% for $NaHCO_3$, 5% for NaOH, and 25%, for HCl. Inorganic P predominated in the H_2O, $NaHCO_3$, and HCl extracts while most of the P in the NaOH extracts was organic. The authors noted that the high degree of solubility of the manure P in water suggests a pronounced risk of loss from the soil under conditions favoring runoff.

Sharpley and Moyer (2000) used a different modification of the Hedley fractionation to examine P forms in dairy and poultry manure, swine slurry, poultry litter and dairy and poultry composts. The dairy manure samples differed from the other manures in the study in that P_i as a fraction of total P was lower and proportion of P_o and residual P fractions were higher. Compared to the dairy manure, the other manures and composts had a greater percentage of total P in the P_i fraction extracted by bicarbonate or HCl. Water-extractable P could thus serve as a predictor of runoff P loss. Turner and Leytem (2004) sequentially fractionated broiler litter, swine and beef manures and noted substantially more P in the bicarbonate fraction (43% of total extractable P) in the beef manure than the litter or swine manure samples (5 and 23%, respectively), and more total extractable P in the HCl fraction of the broiler litter (48%) compared to the beef and swine manures (6 and 9%).

10.4.3. Effects of Manure Processing and Composting and Amendment with P-binding Chemicals

Sharpley and Moyer (2000) examined differences between dairy and poultry manures and their composts using sequential fractionation. In the dairy compost, P_i was largely shifted from water-soluble to bicarbonate- and HCl-soluble fractions. However, in the compost prepared from layer poultry manure, very little difference was seen in fraction distribution. McGrath et al. (2005) found that broiler litter stored dry for 440 days had P fraction distributions little changed from initial conditions, but noted substantial increases in P_i in litter stored wet.

Codling (2006) investigated the changes in P fractions in poultry litter after combustion, which is one of a number of manure processing options for excess litter. Combustion to ash reduced P_i from 55% in the water-extractable fraction in the fresh litter to as little as 1.5% in the ash, while the less-soluble HCl fraction increased from 34 to 82% in the ash. Processing litter by burning to provide a component in fertilizer mixtures may be a feasible way to limit the loss of soluble P when land-applied.

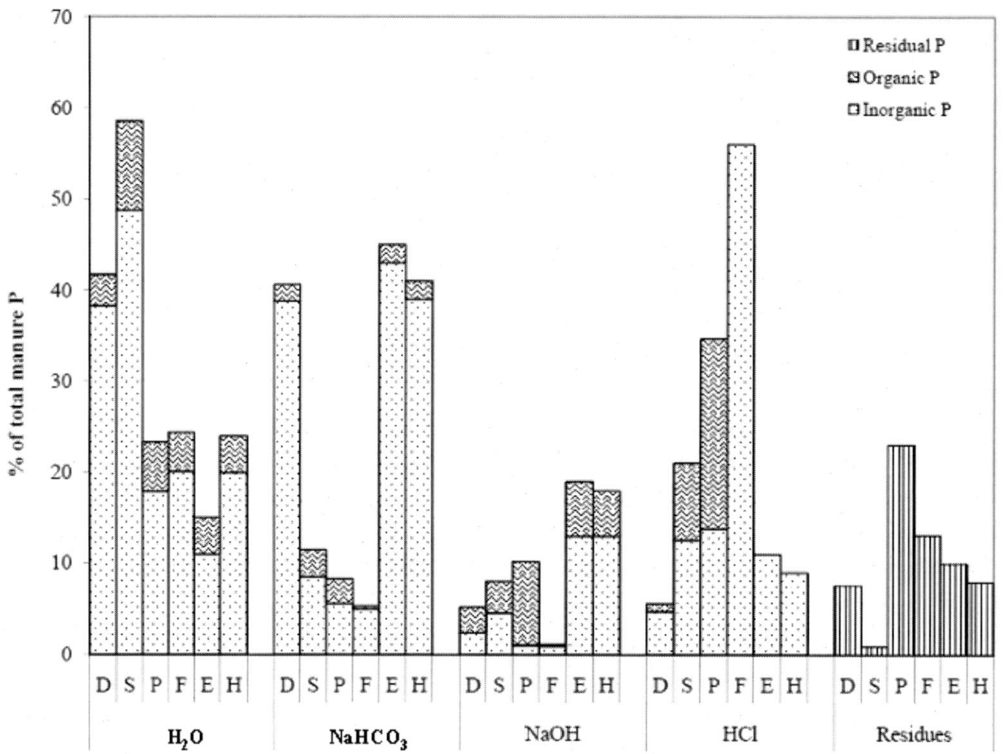

Figure 10.3. Phosphorus distribution patterns in sequentially-extracted H$_2$O, 0.5 mol L^{-1} NaHCO$_3$ (pH 8.5), 0.1 mol L^{-1} NaOH, 1 mol L^{-1} HCl and residual fractions of animal manure. Data are from He et al. (2004b) for dairy manure (D) with 9.2 g P kg^{-1} DM; Wienhold and Miller (2004) for swine manure (S) with 25.8 g P kg^{-1} DM; He et al. (2006b) for poultry litter (P) with 22.3 g P kg^{-1} DM; Uusitalo et al. (2007) for fox manure (F) with 34.3 g P kg^{-1} DM, and McDowell and Stewart (2005) for fresh deer (E) and sheep (H) dung samples. Note that McDowell and Stewart (2005) did not measure organic P content in the HCl fractions, and the total P was the sum of the five fractions with 7.4 and 8.0 g P kg^{-1} DM for deer and sheep dung samples, respectively. Figure adapted from He and Dou (2010).

Coal combustion flyash products and alum were used by Dou et al. (2003b) to treat dairy, swine and broiler poultry litter. Water-soluble P was reduced by all chemicals in the study, with the exception of anthracite refuse ash, and moved into the bicarbonate-soluble fraction (coal flyash treatments) or NaOH fraction (alum). Warren et al. (2008) treated broiler litter with alum and incubated the samples for up to 93 days. There were immediate and significant shifts away from P in the water-soluble fraction and into the organic portion of the NaOH and HCl fractions. The authors suggested that the alum effect may be more complex than simple formation of Al-P associations.

10.4.4. Dietary Manipulation Effects

Manipulating animal diets in order to reduce P excretion would be expected to alter the distribution of P in the Hedley sequential fractions. In a study of diet P levels in dairy cows, Dou et al. (2002) performed sequential fractionation on feces from cows in controlled feeding trials receiving 3.1 to 6.7 g P kg^{-1} in their diets. Almost all differences in P fractions across the feeding groups were found in the water extracts. Water-soluble P concentrations in the 3.4, 5.1 and 6.7 g P kg^{-1} diets were 2.91, 7.13, and 10.46 g P kg^{-1}, corresponding to 56, 77 and 83%, respectively, of total P in the manures. Bicarbonate-, NaOH- and HCl-extracted P were unaffected by increase in diet P levels, indicating P in excess of animal metabolic needs was excreted in soluble form in the feces. He et al. (2004b) examined variability in P fractions across 13 dairy herds with fecal total P ranging from 4.1 to 18.3 g kg^{-1}. The majority of P was extracted by water (12 to 44% as P$_i$ and 2 to 23% P$_o$), and most of the remainder was bicarbonate-extractable (4 to 44% as P$_i$ and 2 to 27% P$_o$).

In swine and poultry, modifying diets to manage P losses involves feeding low-phytate grains and forage crops and supplementation with phytase enzyme. A comparison of P fractions in swine fed traditional corn versus low-phytate corn (Wienhold and Miller, 2004) showed that while less P was found in the water extracts of manure from swine fed low-phytate corn (9.7 to 10.2 g P kg^{-1} over 2 years) than the traditional diet (13.6 to 17.0 g kg^{-1}), the relative distribution of P across the water, bicarbonate, NaOH and HCl extracts did not differ between the diets. Most of the total P was in the water-extractable fraction (60%). In a study of poultry fed low or high non-phytase P with and without phytase supplementation, Maguire et al. (2004) found that birds fed low-P diets plus phytase compared to high-P industry-recommended diets had reductions in water-soluble P of 38% and 55% in turkey and broiler poultry, respectively, and 24 and 21% in the bicarbonate-soluble fraction from turkey and broilers.

10.5. Conclusions

The increasing awareness of the severity of the problem of P derived from agricultural production moving off-farm and threatening water quality has led to the search for methods to characterize the forms and potential solubilities of P in food animal manures and manure products. A number of manure handling, processing, and management options, such as composting, drying, dietary manipulation, and amendment with P-binding chemicals may have substantial effects on P dynamics in manure and the soil to which it is applied. Choice of which soluble P method a researcher will use depends largely on the questions to be answered. Water-extractable P tests, which are closely correlated with P concentrations in runoff and leaching, are useful for characterization of manures for potential loss in the agroecosystem. With the advent of the P Index concept, incorporating topographic, soil, field management, manure type, and application rate parameters, among others, manures and manure products can be assigned a P source coefficient, a weighting factor based on potential for off-field P movement. While methodological issues are still in a state of flux, Kleinman et al. (2006; 2007) have proposed standardization of analytical procedures and this should prove very useful for coordinating soluble P results across the multitude of studies that have been undertaken. Other single-extraction methods have been used to solve particular issues in manure P management, including changes in poultry manure pH after field application and effect on P solubility. Dilute acid extraction of manure has been proposed as a test for P overfeeding in dairy cows. Finally, sequential extractions can delineate classes of manure P chemical species and their potential for lability in the environment. Development of this range of tools for agricultural scientists and food animal producers has a great potential to assist in farm and nutrient management decision-making for agricultural sustainability, farm profitability and protection of the environment.

REFERENCES

Ajiboye, B., O.O. Akinremi, and G.J. Racz. 2004. Laboratory characterization of phosphorus in fresh and oven-dried organic amendments. *J. Environ. Qual.* 33:1062-1069.

Ajiboye, B., O.O. Akinremi, Y. Hu, and D.N. Flaten. 2007. Phosphorus speciation and sequential extracts of organic amendments using nuclear magnetic resonance and X-ray absorption near-edge structure spectroscopies. *J. Environ. Qual.* 36:1563-1576.

Angel, C.R., W.J. Powers, T.J. Applegate, N.M. Tamim, and M.C. Christian. 2005. Influence of phytase on water-soluble phosphorus in poultry and swine manure. *J. Environ. Qual.* 34:563-571.

Applegate, T.J., B.C. Joern, D.L. Nussbaum-Wagler, and R. Angel. 2003. Water-soluble phosphorus in fresh broiler litter is dependent upon phosphorus concentration fed but not on fungal phytase supplementation. *Poultry Sci.* 82:1024-1029.

Barnett, G.M. 1994a. Phosphorus forms in manure. *Bioresour. Technol.* 49:139-147.

Barnett, G.M. 1994b. Manure fractionation. *Bioresour. Technol.* 49:149-155.

Chapuis-Lardy, L., E.J.M. Temminghoff, and R.G.M. De Goede. 2003. Effects of different treatments of cattle slurry manure on water-extractable phosphorus. *Neth. J. Agric. Sci.* 51:91-102.

Chapuis-Lardy, L., J. Fiorini, J. Toth, and Z. Dou. 2004. Phosphorus concentration and solubility in dairy feces: Variability and affecting factors. *J. Dairy Sci.* 87:4334-4341.

Codling, E.E., R.L. Chaney, and C.L. Mulchi. 2000. Use of aluminum- and iron-rich residues to immobilize phosphorus in poultry litter and litter-amended soils. *J. Environ. Qual.* 29:1924-1931.

Codling, E.E. 2006. Laboratory characterization of extractable phosphorus in poultry litter and poultry litter ash. *Soil Sci.* 171:858-864.

Cross, A.F. and W.H. Schlesinger. 1995. A literature review and evaluation of the Hedley fractionation: Applications to the biogeochemical cycle of soil phosphorus in natural ecosystems. *Geoderma* 64:197-214.

Dail, H.W., Z. He, M.S. Erich, and C.W. Honeycutt. 2007. Effect of drying on phosphorus distribution in poultry manure. *Commun. Soil Sci. Plant Anal.* 38:1879-1895.

Dao, T.H. 1999. Coamendments to modify phosphorus extractability and nitrogen/phosphorus ratio in feedlot manure and composted manure. *J. Environ. Qual.* 28:1114-1121.

Dao, T.H. and T.C. Daniel. 2002. Particulate and dissolved phosphorus chemical separation and phosphorus release from treated dairy manure. *J. Environ. Qual.* 31:1388-1398.

Dick, W.A. and M.A. Tabatabai. 1977. Determination of orthophosphate in aqueous solutions containing labile organic and inorganic phosphorus compounds. *J. Environ. Qual.* 6:82-85.

Do, J.C., I.H. Choi, and K.H. Nahm. 2005. Effects of chemically amended litter on broiler performances, atmospheric ammonia concentration, and phosphorus solubility in litter. *Poultry Sci.* 84:679-686.

Dou, Z., J.D. Toth, D.T. Galligan, C.F. Ramberg, Jr. and J.D. Ferguson. 2000. Laboratory procedures for characterizing manure phosphorus. *J. Environ. Qual.* 29:508-514.

Dou, Z., K.F. Knowlton, R.A. Kohn, Z. Wu, L.D. Satter, G. Zhang, J.D. Toth, and J.D. Ferguson. 2002. Phosphorus characteristics of dairy feces affected by diets. *J. Environ. Qual. 31*:2058-2065.

Dou, Z., J.D. Ferguson, J. Fiorini, J.D. Toth, S.M. Alexander, L.E. Chase, C.M. Ryan, K.F. Knowlton, R.A. Kohn, A.B. Peterson, J.T. Sims, and Z. Wu. 2003a. Phosphorus feeding levels and critical control points on dairy farms. *J. Dairy Sci.* 86:3787-3795.

Dou, Z., G.Y. Zhang, W.L. Stout, J.D. Toth, and J.D. Ferguson. 2003b. Efficacy of alum and coal combustion by-products in stabilizing manure phosphorus. *J. Environ. Qual.* 32:1490-1497.

Dou, Z., C.F. Ramberg, Jr., L. Chapuis-Lardy, J.D. Toth, Y. Wang, R.J. Munson, Z. Wu, L.E. Chase, R.A. Kohn, K.F. Knowlton, and J.D. Ferguson. 2007. A novel test for measuring and managing potential phosphorus loss from dairy cattle feces. *Environ. Sci. Technol.* 41:4361-4366.

Dou, Z., C.F. Ramberg, Jr., L. Chapuis-Lardy, J.D. Toth, Z. Wu, L.E. Chase, R.A. Kohn, K.F. Knowlton, and J.D. Ferguson. 2010. A fecal test for assessing phosphorus overfeeding on dairy farms: Evaluation using extensive farm data. *J. Dairy Sci.* 93:830-839.

Elliott, H.A., R.C. Brandt, P.J.A. Kleinman, A.N. Sharpley, and D.B. Beegle. 2006. Estimating source coefficients for phosphorus source indices. *J. Environ. Qual.* 35:2195-2201.

Fletcher, S.W. 1955. *Pennsylvania agriculture and country life,* 1840-1940. Pennsylvania Historical and Museum Commission, Harrisburg, PA.

Güngör, K. and K.G. Karthikeyan. 2005. Influence of anaerobic digestion on dairy manure phosphorus extractability. *Trans. ASAE* 48:1497-1507.

Güngör, K. and K.G. Karthikeyan. 2006. Phosphorus forms and extractability in dairy manure: A case study for Wisconsin on-farm aerobic digesters. *Bioresour. Technol.* 99:425-436.

Hadas, A., B. Bar Yosef, and R. Portnoy. 1990. Extractability of phosphorus in manure pellets enriched with fertilizer phosphorus. *Soil Sci. Soc. Am. J.* 54:443-448.

Haggard, B.E., P.A. Vadas, D.R. Smith, P.B. DeLaune, and P.A. Moore, Jr. 2005. Effect of poultry litter to water ratios on extractable phosphorus content and its relation to runoff phosphorus concentrations. *Biosyst. Eng.* 92:409-417.

Haygarth, P.M. and A.N. Sharpley. 2000. Terminology for phosphorus transfer. *J. Environ. Qual.* 29:10-15.

He, Z., and Z. Dou. 2010. Phosphorus forms in animal manure and the impact on soil P status. *In* C.S. Dellaguardia (ed.) *Manure: Management, uses and environmental impacts.* Nova Science Publishers, Inc., Hauppauge, NY.

He, Z. and C.W. Honeycutt. 2001. Enzymatic characterization of organic phosphorus in animal manure. *J. Environ. Qual.* 30:1685-1692.

He, Z. and C.W. Honeycutt. 2005. A modified molybdenum blue method for orthophosphate determination suitable for investigating enzymatic hydrolysis of organic phosphates. *Commun. Soil Sci. Plant Anal.* 36:1373-1383.

He, Z., C.W. Honeycutt, and T.S. Griffin. 2003. Comparative investigation of sequentially extracted phosphorus fractions in a sandy loam soil and a swine manure. *Commun. Soil Sci. Plant Anal.* 34:1729-1742.

He, Z., T.S. Griffin, and C.W. Honeycutt. 2004a. Enzymatic hydrolysis of organic phosphorus in swine manure and soil. *J. Environ. Qual.* 33:367-372.

He, Z., T.S. Griffin, and C.W. Honeycutt. 2004b. Phosphorus distribution in dairy manure. *J. Environ. Qual.* 33:1528-1534.

He, Z., A. Fortuna, Z.N. Senwo, I.A. Tazisong, C.W. Honeycutt, and T.S. Griffin. 2006a. Hydrochloric fractions in Hedley fractionation may contain inorganic and organic phosphates. *Soil Sci. Soc. Am. J.* 70:893-899.

He, Z., Z.N. Senwo, R.N. Mankolo, and C.W. Honeycutt. 2006b. Phosphorus fractions in poultry litter characterized by sequential fractionation coupled with phosphatase hydrolysis. *J. Food Agri. Environ.* 4:304-312.

He, Z., C.W. Honeycutt, B.J. Cade-Menun, Z.N. Senwo, and I.A. Tazisong. 2008. Phosphorus in poultry litter and soil: Enzymatic and nuclear magnetic resonance characterization. *Soil Sci. Soc. Am. J.* 72:1425-1433.

He, Z., H. Zhang, G.S. Toor, Z. Dou, C.W. Honeycutt, B.E. Haggard, and M.S. Reiter. 2010. Phosphorus distribution in sequentially extracted fractions of biosolids, poultry litter, and granulated products. *Soil Sci.* 175:154-161.

Hedley, M.J., J.W.B. Stewart, and B.S. Chauhan. 1982a. Changes in inorganic and organic soil phosphorus fractions induced by cultivation practices and by laboratory incubations. *Soil Sci. Soc. Am. J.* 54:970-976.

Hedley, M.J., R.E. White, and P.H. Nye. 1982b. Plant-induced changes in the rhizosphere of rape (*Brassica napus* var. Emerald) seedlings. III. Changes in *L* value, soil phosphate fractions and phosphatase activity. *New Phytol.* 91:45-56.

Huang, X.-L., Y. Chen, and M. Shenker. 2008. Chemical fractionation of phosphorus in stabilized biosolids. *J. Environ. Qual.* 37:1949-1958.

Kleinman, P.J.A., A.N. Sharpley, A.M. Wolf, D.B. Beegle, and P.A. Moore, Jr. 2002. Measuring water-extractable phosphorus in manure as an indicator of phosphorus in runoff. *Soil Sci. Soc. Am. J.* 66:2009-2015.

Kleinman, P.J.A., A.M. Wolf, A.N. Sharpley, D.B. Beegle, and L.S. Saporito. 2005. Survey of water-extractable phosphorus in livestock manures. *Soil Sci. Soc. Am. J.* 69:701-708.

Kleinman, P.J.A., A.N. Sharpley, A.M. Wolf, D.B. Beegle, H.A. Elliott, J.L. Weld, and R. Brandt. 2006. Developing an environmental manure test for the phosphorus index. *Commun. Soil Sci. Plant Anal.* 37:2137-2155.

Kleinman, P., D. Sullivan, A. Wolf, R. Brandt, Z. Dou, H. Elliot, J. Kovar, A. Leytem, R. Maguire, P. Moore, L. Saporito, A. Sharpley, A. Shober, T. Sims, J. Toth, G. Toor, H. Zhang, and T. Zhang. 2007. Selection of a water-extractable phosphorus test for manures and biosolids as an indicator of runoff loss potential. *J. Environ. Qual.* 36:1357-1367.

Koopmans, G.F., W.J. Chardon, P.H.M. Dekker, P.F.A.M. Römkens, and O.F. Schoumans. 2006. Comparing different extraction methods for estimating phosphorus solubility in various soil types. *Soil Sci.* 171:103-116.

Kuo, S. 1996. Phosphorus. p. 869-919. *In* D.L. Sparks (ed.) *Methods of soil analysis. Part 3, chemical methods.* SSSA and ASA, Madison, WI.

Leinweber, P., L. Haumaier and W. Zech. 1997. Sequential extractions and ^{31}P-NMR spectroscopy of phosphorus forms in animal manure, whole soils and particle-size separates from a densely populated livestock area in northwest Germany. *Biol. Fert. Soils* 25:89-94.

Lemunyon, J.L. and R.G. Gilbert. 1993. The concept and need for a phosphorus assessment tool. *J. Prod. Agric.* 6:483-486.

Leytem, A.B. and P.A. Thacker. 2008. Fecal phosphorus excretion and characterization from swine fed diets containing a variety of cereal grains. *J. Anim. Vet. Adv.* 7:113-120.

Leytem, A.B., B.L. Turner, and P.A. Thacker. 2004. Phosphorus composition of manure from swine fed low-phytate grains: Evidence for hydrolysis in the animal. *J. Environ. Qual.* 33:2380-2383.

Maguire, R.O., J.T. Sims, W.W. Saylor, B.L. Turner, R. Angel, and T.J. Applegate. 2004. Influence of phytase addition to poultry diets on phosphorus forms and solubility in litters and amended soils. *J. Environ. Qual.* 33:2306-2316.

Maguire, R.O., Z. Dou, J.T. Sims, J. Brake, and B.C. Joern. 2005. Dietary strategies for reduced phosphorus excretion and improved water quality. *J. Environ. Qual.* 34:2093-2103.

Maguire, R.O., D. Hesterberg, A. Gernat, K. Anderson, M. Wineland, and J. Grimes. 2006. Liming poultry manures to decrease soluble phosphorus and suppress the bacteria population. *J. Environ. Qual.* 35:849-857.

Mamo, M., C. Wortmann, and C. Brubaker. 2007. Manure phosphorus fractions: Development of analytical methods and variation with manure types. *Commun. Soil Sci. Plant Anal.* 38:935-947.

McAuliffe, C. and M. Peech. 1949. Utilization by plants of phosphorus in farm manure: I. Labeling of phosphorus with [32]P. *Soil Sci.* 68:179-184.

McDowell, R.W. and I. Stewart. 2005. Phosphorus in fresh and dry dung of grazing dairy cattle, deer, and sheep: Sequential fraction and phosphorus-31 nuclear magnetic resonance analyses. *J. Environ. Qual.* 34:598-607.

McDowell, R.W., Z. Dou, J.D. Toth, B.J. Cade-Menun, P.J.A. Kleinman, K. Soder, and L. Saporito. 2008. A comparison of phosphorus speciation and potential bioavailability in feed and feces of different dairy herds using 31P nuclear magnetic resonance spectroscopy. *J. Environ. Qual.* 37:741-752.

McGrath, J.M., J.T. Sims, R.O. Maguire, W.W. Saylor, C.R. Angel, and B.L. Turner. 2005. Broiler diet modification and litter storage: Impacts on phosphorus in litters, soils, and runoff. *J. Environ. Qual.* 34:1896-1909.

McKelvie, I.D., D.M.W. Peat, and P.J. Worsfold. 1995. Techniques for the quantification and speciation of phosphorus in natural waters. *Anal. Proc. Incl. Anal. Commun.* 32:437-445.

Miles, D.M., P.A. Moore, Jr., D.R. Smith, D.W. Rice, H.L. Stilborn, D.R. Rowe, B.D. Lott, S.L. Branton, and J.D. Simmons. 2003. Total and water-soluble phosphorus in broiler litter over three flocks with alum litter treatment and dietary inclusion of high available phosphorus corn and phytase supplementation. *Poultry Sci.* 82:1544-1549.

Moore, P.A., Jr. and D.M. Miller. 1994. Decreasing phosphorus solubility in poultry litter with aluminum, calcium, and iron amendments. *J. Environ. Qual.* 23:325-330.

Murphy, J. and J.P. Riley. 1962. A modified single solution method for the determination of phosphate in natural waters. *Anal. Chim. Acta* 27:31-36.

Negassa, W. and P. Leinweber. 2009. How does the Hedley sequential phosphorus fractionation reflect impacts of land use and management on soil phosphorus: A review. *J. Plant Nutr. Soil Sci.* 172:305-325.

Peperzak, P., A.G. Caldwell, R.R. Hunziger, and C.A. Black. 1959. Phosphorus fractions in manures. *Soil Sci.* 87:293-302.

Pote, D.H., T.C. Daniel, A.N. Sharpley, P.A. Moore, Jr., D.R. Edwards, and D.J. Nichols. 1996. Relating extractable soil phosphorus to phosphorus losses in runoff. *Soil Sci. Soc. Am. J.* 60:855-859.

Pote, D.H., T.C. Daniel, D.J. Nichols, A.N. Sharpley, P.A. Moore, Jr., D.M. Miller, and D.R. Edwards. 1999. Relationship between phosphorus levels in three Ultisols and phosphorus concentrations in runoff. *J. Environ. Qual.* 28:170-175.

Robinson, J.S. and A.N. Sharpley. 1995. Release of nitrogen and phosphorus from poultry litter. *J. Environ. Qual.* 24:62-67.

Seiter, J.M., K.E. Staats-Borda, M. Ginder-Vogel, and D.L. Sparks. 2008. XANES spectroscopic analysis of phosphorus speciation in alum-amended poultry litter. *J. Environ. Qual.* 37:477-485.

Self-Davis, M.L. and P.A. Moore, Jr. 2000. Determining water-soluble phosphorus in animal manure. p. 74-76. *In* G.M. Pierzynski (ed.) *Methods of phosphorus analysis for soils, sediments, residuals, and waters. Southern Coop. Ser. Bull. No. 396.* SERA-IEG. N.C. State Univ., Raleigh, NC.

Sharpley, A.N. 2000. Phosphorus availability. p. D18-D38. *In* M.E. Sumner (ed.) *Handbook of soil science.* CRC Press, Boca Raton, FL.

Sharpley, A.N. and B. Moyer. 2000. Phosphorus forms in manure and compost and their release during simulated rainfall. *J. Environ. Qual.* 29:1462-1469.

Sharpley, A.N., J.L. Weld, D.B. Beegle, P.J.A. Kleinman, W.J. Gburek, P.A. Moore, Jr., and G. Mullins. 2003. Development of phosphorus indices for nutrient management planning strategies in the United States. *J. Soil Water Conserv.* 58:137-152.

Shober, A.L. and J.T. Sims. 2007. Integrating phosphorus source and soil properties into risk assessments for phosphorus loss. *Soil Sci. Soc. Am. J.* 71:551-560.

Sims, J.T. and N.J. Luka-McCafferty. 2002. On-farm evaluation of aluminum sulfate (alum) as a poultry litter amendment: Effects on litter properties. *J. Environ. Qual.* 31:2066-2073.

Sistani, K.R., D.M. Miles, D.E. Rowe, G.E. Brink, and S.L. McGowen. 2001. Impact of drying method, dietary phosphorus levels, and methodology on phosphorus chemistry of broiler manure. *Commun. Soil Sci. Plant Anal.* 32:2783-2793.

Sui, Y., M.L. Thompson, and C. Shang. 1999. Fractionation of phosphorus in a Mollisol amended with biosolids. *Soil Sci. Soc. Am. J.* 63:1174-1180.

Tasistro, A.S., M.L. Cabrera, and D.E. Kissel. 2004. Water soluble phosphorus released by poultry litter: Effect of extraction pH and time after application. *Nutr. Cycl. Agroecosyst.* 68:223-234.

Tasistro, A.S., M.L. Cabrera, Y.B. Zhao, D.E. Kissel, K. Xia, and D.H. Franklin. 2007. Soluble phosphorus released by poultry wastes in acidified aqueous extracts. *Commun. Soil Sci. Plant Anal.* 37:1395-1410.

Toor, G.S., J.D. Peak, and J.T. Sims. 2005. Phosphorus speciation in broiler litter and turkey manure produced from modified diets. *J. Environ. Qual.* 34:687-697.

Toor, G.S., S. Hunger, J.D. Peak, J.T. Sims, and D.L. Sparks. 2006. Advances in the characterization of phosphorus in organic wastes: Environmental and agronomic applications. *Adv. Agron.* 89:1-72.

Toor, G.S., B.E. Haggard, M.S. Reiter, T.C. Daniel, and A.M. Donoghue. 2007. Phosphorus solubility in poultry litters and granulates: Influence of litter treatments and extraction ratios. *Trans. ASABE* 50:533-542.

Turner, B.L. and A.B. Leytem. 2004. Phosphorus compounds in sequential extracts of animal manures: Chemical speciation and a novel fractionation procedure. *Environ. Sci. Technol.* 38:6101-6108.

Uusitalo, R., K. Ylivainio, E. Turtola, and A. Kangas. 2007. Accumulation and translocation of sparingly soluble manure phosphorus in different types of soils after long-term excessive inputs. *Agric. Food Sci.* 16:317-331.

Vadas, P.A. and P.J.A. Kleinman. 2006. Effect of methodology in estimating and interpreting water-extractable phosphorus in animal manures. *J. Environ. Qual.* 35:1151-1159.

Vadas, P.A., R.D. Harmel, and P.J.A. Kleinman. 2007. Transformations of soil and manure phosphorus after surface application to field plots. *Nutr. Cycl. Agroecosyst.* 77:83-99.

Valk, J., J.A. Metcalf, and P.J.A. Withers. 2000. Prospects for minimizing phosphorus excretion in ruminants by dietary manipulation. *J. Environ. Qual.* 29:28-36.

Warren, J.G., C.J. Penn, J.M. McGrath, and K. Sistani. 2008. The impact of alum addition on organic P transformations in poultry litter and litter-amended soil. *J. Environ. Qual.* 37:469-476.

Wienhold, B.J. and P.S. Miller. 2004. Phosphorus fractionation in manure from swine fed traditional and low-phytate corn diets. *J. Environ. Qual.* 33:389-393.

Wolf, A.M., P.J.A. Kleinman, A.N. Sharpley, and D.B. Beegle. 2005. Development of a water-extractable phosphorus test for manure: An interlaboratory study. *Soil Sci. Soc. Am. J.* 69:695-700.

Wu, Z., L.D. Satter, and R. Sojo. 2000. Milk production, reproductive performance and fecal excretion of phosphorus by dairy cows fed three amounts of phosphorus. *J. Dairy Sci.* 83:1028-1041.

Ylivainio, K., R. Uusitalo, and E. Turtola. 2008. Meat bone meal and fox manure as P sources for ryegrass (*Lolium multiflorum*) grown on a limed soil. *Nutr. Cycl. Agroecosyst.* 81:267-278.

In: Environmental Chemistry of Animal Manure
Editor: Zhongqi He

ISBN: 978-1-62808-641-6
© 2013 Nova Science Publishers, Inc.

Chapter 11

ENZYMATIC HYDROLYSIS
OF ORGANIC PHOSPHORUS

Zhongqi He[*] and C. Wayne Honeycutt

11.1. INTRODUCTION

Phosphorus occurs in both inorganic and organic forms. Most of organic P (P_o) is present in a certain form of phosphoric esters. In nature, phosphatases catalyze chemical reactions releasing orthophosphate from various P_o compounds (Webb, 1992). Because most, if not all P_o, must be hydrolyzed to inorganic P (P_i) prior to uptake by plants and microorganisms, enzymatic hydrolysis provides an estimate of hydrolyzable, and thus the bioavailable, P_o in environmental samples (He et al., 2006f; Herbes et al., 1975; Sannigrahi et al., 2006; Seeling and Jungk, 1996). As early as 1975, Herbe et al. (1975) enzymatically characterized soluble P_o in lake water samples. Fox and Comerford (1992) investigated the bioavailability of P_o in forested spodosols with wheat germ acid phosphatase and found that 20-30% of the water-extracted P_o was enzymatically hydrolyzable. Pant et al. (1994b) reported that phytase released nearly twice (55% of P_o) the quantity of P as did acid and alkaline phosphatases (20-28% of P_o) in water extracts of 4 soils. In 2001, He and Honeycutt (2001) first examined the release of P_o in the sequentially-extracted fractions of swine manure and cattle manure by enzymatic hydrolysis. Subsequently, enzymatic hydrolysis has been used in characterizing numerous animal manures and manure–amended soils under various management practices by this group and others, which are reviewed in this chapter.

---

[*] Corresponding author: Zhongqi.He@ars.usda.gov USDA-ARS, Orono, ME 04469, USA

11.2. PHOSPHORUS FORMS AND PHOSPHATASES IN NATURE

11.2.1. Forms of P_o in Nature

Phosphorus is an essential nutrient for all organisms. It is involved in biosynthesis of diverse cellular components, including nucleic acids, proteins, lipids, and sugars. Phosphorus in animal manure originates from the undigested/unabsorbed part of animal feed materials and from animal metabolites or associated microflora. Microbial activities in the animal's intestinal systems, as well as in excreted manure, can produce biomass containing P_o. Therefore, knowledge of the various P forms present in nature may help elucidate P_o identities in animal manure.

Various forms of P_o exist in nature (Cheshire and Anderson, 1975; Dalal, 1977; Jokanovic, 2001). A large portion of P_o is present as phosphoesters in which the the P atom is connected with the organic moiety through P-O-C bonds. These phosphate esters can be classified as phosphomonoester, phosphodiester, and phosphotriester. Common phosphomonoesters include sugar phosphates, mononucleotides, inositol phosphates, and protein phosphates. Phytate [IP6, *Myo*-inositol (1,2,3,4,5,6) hexa*kis*phosphate] is the primary storage compound of P in seeds accounting for up to 80% of the total seed P and contributing as much as 1.5% to the seed dry weight (Bohn et al., 2008). It contains a 6-C ring with 1 H and 1 phosphate attached to each C. Each of the 6 phosphate groups is attached to the inositol ring by an ester linkage and retains two replaceable hydrogens (He et al., 2006e). The negatively charged phosphate in IP6 can strongly bind to Al, Ca, Cu, Fe, K, Mg, Mn and Zn cations making them insoluble and the major P_o form in the environment (Bohn et al., 2008; He et al., 2006e).

Nucleic acids, phospholipids, and teichoic acids are three major groups of phosphodiesters, however, naturally occurring phosphotriesters are rare. In fact, some synthetic phosphotriesters such as the pesticide paraoxon (diethyl *p*-nitrophenyl phosphate), and insecticide metabolite coroxon (3-chloro-7-hydroxy-4-methylcoumarin diethyl phosphate) are toxic. Organic polyphosphates (P-O-P) are formed when P-O bonds of two or more esters are condensed to anhydride bond(s). Compounds, such as, nicotinamide adenine dinucleotide (NAD), adenosine diphosphate (ADP), and adenosine triphosphate (ATP), belong to this category. When the atom P forms bonds with N, S, and C atoms in addition to P-O-C bonds, these chemicals are frequently referred to as organophosphoric compounds. Organophosphonates with a direct C-P bond have been found in both eukaryotic and prokaryotic organisms (Cheshire and Anderson, 1975; Dalal, 1977; Jokanovic, 2001; Wanner, 1994). Due to their structural similarity to phosphate esters, synthetic organophosphoric compounds often act as inhibitors of enzymes that involve phosphoryl transfer reactions (Jokanvic, 2001; Wanner, 1994). Thus, many of these naturally-occurring or synthetic compounds are used as antibacterial, antiviral, and antitumor agents, as well as herbicides and insecticides, such as O,O-diethyl-O-(3-chloro-4-methylumbelliferone)thiophosphate (coumaphos) and N-(phosphonomethyl)glycine (glyphosate). Turner et al. (2004) compiled a list of naturally occurring and synthetic P compounds frequently found in the environment and/or used in P research. For more information, one can refer to a comprehensive reference book written by Corbridge (2000) which gives a unique coverage of the whole field of phosphorus chemistry.

11.2.2. Phosphate-releasing Enzymes

Enzymes are proteins which catalyze reactions to cleave specific chemical bonds. In nature, phosphatases catalyze the hydrolysis of certain P_o compounds to produce inorganic orthophosphate (Webb, 1992). There are four major groups of phosphatase enzymes, that include (a) phosphoric monoester hydrolases (EC 3.1.3), (b) phosphoric diester hydrolases (EC 3.1.4), (c) phosphoric triester hydrolases (EC 3.1.8), and (d) polyphosphate hydrolases (EC 3.6.1).

Biological hydrolysis of phosphomonoesters is accomplished by a diverse group of phosphoric monoester hydrolases. There are five well-characterized classes of phosphoric monoester hydrolases: (i) alkaline phosphatases, (ii) purple acid phosphatases, (iii) low molecular weight acid phosphatases, (iv) high molecular weight acid phosphatases, and (v) protein phosphatases (Vincent et al., 1992). Phytases are phosphomonoester hydrolases which can catalyze the hydrolysis of phytate (Webb, 1992). Phytases have been found in diverse organisms (Bohn et al., 2008) including bacteria from manure-relevant soils (Hill et al., 2007). The monoester bond in diphosphoric and triphosphoric compounds can also be cleaved by diphosphoric monoester hydrolase (EC 3.1.7) and triphosphoric monoester hydrolase (EC 3.1.5) to produce corresponding organic compounds and di- or triphosphate. For example, dGTP can be hydrolyzed to deoxyguanosine and triphosphate.

Phosphodiesterases catalyze the reactions to cleave the P-O bond in phosphoric diesters (Webb, 1992). Phospholipases (EC 3.1.4), exonulceases (EC 3.1.11-16), endonucleases (EC 3.1.21-31), and teichoicases (Kusser and Fiedler, 1983) are categories in the diester class. Phosphodiesterases generally hydrolyze only one P-O bond in a diester compound. Therefore, a monoester hydrolase is required to release P_i from derived P monoesters. In contrast to a large number of mono- and diester phosphatases, only a couple of phosphoric triester hydrolases have been reported (Devorshak and Roe, 2001). This may reflect natural selection, because few phosphoric triesters are found in nature. In addition to cleaving the P-O bond, triesterases are capable of hydrolyzing a variety of organophosphoric compounds with P-C, P-N, P-F, and P-S bonds (Lai et al., 1995). These enzymes are also called organophosphorus hydrolases and are important in degrading (detoxification) artificial organophosphorus residues in the environment (Lai et al., 1995).

The anhydride (P-O-P) bond in pyro- and polyphosphates is hydrolyzed by a group of polyphosphatases in EC 3.6.1 (Webb, 1992). Nucleoside di- and triphosphates are hydrolyzed by relevant enzymes, such as ATPase and GDPase. Longer chain phosphate polyesters can be hydrolyzed by either endo- or exopolyphosphatases (Sethuraman et al., 2001). Phosphonatase (EC 3.11.1) catalyzes the hydrolysis of the P-C bond in phosphonates (Quinn et al., 2007). Phosphoamidase (EC 3.9.1.1) acts on the P-N bond in N-phosphocreatine and other phosphoamides. However, the phosphoamidase in rat renal microsome is actually an alkaline phosphatase (EC 3.1.3.1) with this additional activity (Nishino et al., 1994).

11.3. SOLUBLE INORGANIC PHOSPHORUS DETERMINATION

Enzymatically hydrolyzable P_o is determined by the difference in soluble P_i in incubated samples in the presence and absence (control) of enzymes. Thus, reliable measurement of soluble P_i is a critical factor to determine the accuracy of enzymatically hydrolyzable P_o. Perhaps the most commonly used method to determine soluble P_i is the ascorbic acid–molybdenum blue colorimetric method (Murphy and Riley, 1962). Over the years, modifications have been made to improve the robustness of this method. For example, Dick and Tabatabai (1977) reported reduction of errors due to labile P hydrolysis by complexation of excess molybdate ions to prevent further formation of the blue chromophore from orthophosphate derived from acid labile P hydrolysis. He and Honeycutt (2005) further investigated this method, and reported a higher absorption peak at 850 nm in addition to the reported peak at 700 nm. The absorption coefficient at 850 nm is 45-49% higher than that at 700 nm, and linear up to, at least, 80 nmol P_i in 1 mL assay solution. Therefore, adaption of the readings at 850 nm improved the sensitivities of P_i determination by about 45%.

Another concern in determining hydrolyzable P_o is the interference by protein precipitates under the acidic conditions for P_i determination (He and Honeycutt, 2001; Pant et al., 1994b; Shand and Smith, 1997). An approach to reduce protein interference is to keep the enzyme concentrations low (Bunemann, 2008; He and Honeycutt, 2001; Turner et al., 2002). However, reducing enzyme concentrations may lead to incomplete hydrolysis of relevant hydrolyzable P_o compounds, especially in the case of high P_i concentrations which can inhibit phosphatase activities (Juma and Tabatabai, 1978). This enzyme precipitation can be prevented by the addition of 2% SDS (sodium dodecyl sulfate) (He and Honeycutt, 2005) or dimethyl sulfoxide (Shand and Smith, 1997). On the other hand, Dao (2003) used high-performance anion chromatography to improve the accuracy of soluble P_i determination in the presence of high concentrations of IP6 which caused the colorimetric method to overestimate the soluble P_i concentrations. The author (Dao, 2003) claimed that concentrations of soluble P_i determined by anion chromatography in the 0.016 to 0.16 mM range were unaffected by the presence of IP6. However, one should be cautious as anion chromatography could not separate myo-inositol mono-,di- and triphosphates from orthophosphate (i. e. soluble P_i).

In some literature (Bunemann, 2008; He et al., 2009d; Ron Vaz et al., 1993), the terms "molybdate-reactive P" (MRP) and "molybdate-unreactive P" (MUP) have been used to describe the soluble P determined in an ascorbic acid–molybdenum blue method and difference between total P and molybdate-reactive P, respectively. With reduced interferences, however, the measurement of free phosphate by methods in Dick and Tabatabai (1977), He and Honeycutt (2005), and Dao (2003) should be considered more soluble P_i than MUP. In this chapter, no effort has been made to distinguish P_i and P_o from MRP and MUP in order to facilitate data comparison.

11.4. Identification and Quantification of Organic Phosphorus Forms by Enzymatic Hydrolysis

11.4.1. Method Development

He and Honeycutt (2001) proposed to characterize P_o in animal manure by phosphatase hydrolysis about a decade ago. Their first effort was focused on developing an appropriate approach for evaluating soluble P_o in animal manure. One of the important factors to determine the accuracy of classification of soluble P_o is the substrate specificity of phosphatases to be used. The substrate specificity of selected commercially-available enzymes on 14 P compounds is shown in Table 11.1. All enzymes show rather high activity on various phosphomonoesters, but little or no activity on organic pyrophosphate NAD and polyphosphodiesters (RNA and DNA). Another striking point was the inability of alkaline phosphatase to release phytate P. It is should pointed out that the high substrate concentration (10 mM for most substrates) and short incubation time (30 min) used in the substrate specificity experiment were for showing the maximum substrate difference among the four enzymes, rather than for showing the completeness of P_o hydrolysis. Similarly, Hayes et al. (2000) investigated substrate specificities, rather than the completeness of hydrolysis, of commercial preparations of wheat germ acid phosphatase and fungal phytase on seven P compounds as the former is more critical for selecting an enzyme. Completeness can be more easily reached by increasing the enzyme amount and incubation time, diluting the P_o concentration in samples, and their combinations. Based on the different substrate specificities, He and Honeycutt (2001) proposed that P release by alkaline phosphatase may reflect the content of most simple phosphmonoesters and the difference between P released by wheat phytase and alkaline phosphatase may reflect the content of phytate P. Furthermore, nuclease P1 from *Penicilium citrinum* cleaves the endonucleotic bonds of RNA and DNA. Nucleotide pyrophosphatase from *Crotalus adamanteus* venom cleaves the dinucleotide bond of NAD. The P in the monoesters produced by these two enzymes is then released by a phosphatase. Based on the differential substrate specificity of these enzymes, He and Honeycutt (2001) categorized five types of P_o forms in animal manure: (1) simple monoester P is given by the difference in P contents determined in the presence and absence of alkaline phosphatase hydrolysis; (2) phytate-like P is given by the difference in P contents determined with phytase and alkaline phosphatase hydrolysis; (3) DNA-like P is the difference of P contents due to the hydrolysis with or without nuclease P1 preincubation; (4) organic pyrophosphate is given by the difference in P contents due to the hydrolysis with or without nucleotide pyrophosphatase preincubation; and (5) the difference in P of total P_o and the sum of the enzyme-hydrolyzed P_o is the hydrolyzable P_o.

With this procedure, He and Honeycutt (2001) examined the release of P_o in sequentially-extracted fractions of swine manure and cattle manure. About 49% of P in swine manure and 44% in cattle manure were present in organic forms, mainly distributed in H_2O, $NaHCO_3$, and NaOH fractions. The enzymatic hydrolysis treatment provided more information on P forms found in the sequential fractions than the conventionally determined values of P_i and P_o (Table 11.2). The four designated P_o forms were distributed unequally in the sequential extractions. Phytate-like P was most prevalent in the H_2O fraction and accounted for 39% of P_o in swine manure and 17% of P_o in cattle manure. This finding is consistent with the lack of

 Zhongqi He and C. Wayne Honeycutt

phytase enzymes in swine digestion systems (Wodzinski and Ullah, 1996). Simple orthophosphate monoesters comprised the major P_o form identified in the NaOH fractions. A small percentage of DNA-like P was found in each fraction (1.1-12%). Organic pyrophosphate was found only in the H_2O extracts.

Table 11.1. Relative activity of four commercial enzyme preparations on 14 phosphorus compounds.

Enzyme	Acid phosphatase	Alkaline phosphatase	Fungal phytase	Wheat phytase
Source	Type 1, wheat germ	Type VII-S, Bovine intestinal mucosa	*Aspergillus ficuum*	Wheat
EC number	3.1.3.2	3.1.3.1	3.1.3.8	3.1.3.26
Function	Release of P_i from $R\text{-}PO_4$ (monoester)	Release of P_i from $R\text{-}PO_4$ (monoester)	Release of P_i from phytate	Release of P_i from phytate
Optimal conditions:				
pH	4.8	9.8	2.5	5.15
Temperature (^{0}C)	37	37	37	55
Actual assay conditions:				
pH	5.0	5.0	2.5	5.0
Temperature (^{0}C)[a]	22	22	22	22
Substrate:				
Phytate	5.9[b]	0.5	29.7	37.8
p-Nitrophenol	70.7	100	97.5	100
Glucose 6-phosphate	24.7	29.3	79.6	49.7
Glucose 1-phosphate	3.1	47.3	94.3	29.2
Fructose 6-phosphate	8.5	20.9	79.6	26.3
Ribose-5-phosphate	15.8	33.4	70.1	31.1
AMP	7.5	48.2	39.9	25.3
Glycerophosphate	23.7	45.0	72.9	43.2
NAD	4.1	1.5	2.5	0.3
Pyrophosphate	100	70.3	100	76.5
ADP	55.9	50.9	63.2	46.7
ATP	44.3	64.9	47.3	78.8
RNA	1.5	9.7	5.0	6.4
DNA	0.0	0.0	0.0	0.0

Adapted from He and Honeycutt (2001).

[a] All incubations were carried out at room temperature (22^0C) for 30 min in order to keep the data more comparable. All substrates but RNA and DNA contained 10 mM total P. The concentrations of RNA from Bakers Yeast and DNA from salmon testes were 3 mg substrate per 1 ml reaction mixture.

[b] % of model P compound hydrolyzed.

Table 11.2. Phosphorus forms in sequentially extracted fractions of swine and cattle manures revealed by enzymatic treatments.

P_o form	Swine manure fraction				Cattle manure fraction			
	H_2O	NaHCO$_3$	NaOH	HCl	H_2O	NaHCO$_3$	NaOH	HCl
				% P in each fraction				
Inorganic P	61.1	38.9	18.5	88.3	57.8	52.5	35.0	92.4
Simple monoester	2.8	2.0	35.0	1.7	0.3	0.5	9.8	1.3
Phytate-like	15.2	2.0	4.2	2.5	7.2	0	0	0
DNA-like	2.1	0.7	1.5	-[a]	5.1	4.2	2.1	-
Pyrophosphate	1.6	0	0.2	-	6.8	0	0	-
Nonhydrolyzable	18.6	56.6	40.8	7.7	23.2	42.3	54.6	6.1

Adapted from He and Dou (2010).
[a] Not determined.

The enzymatic approach reported by He and Honeycutt (2001) provides a baseline for classifying P_o in animal manure. However, a drawback of the approach is that the incubation conditions (such as, cofactors, buffer media and pH) for alkaline phosphatase differ from those of other P_i-releasing enzymes. The requirements of different incubation conditions not only makepreparing the reaction mixtures inconvenient, but it may also introduce errors due to different rates of spontaneous P_o hydrolysis and interferences of the two reaction media during the P-assay (He and Honeycutt, 2001; Pant et al., 1994a; 1994b). However, a single set of incubation conditions would likely reduce these systematic errors. For this purpose, He et al. (2004a) further evaluated the substrate specificity of potato acid phosphatase (Table 11.3). This enzyme has not been previously used to investigate P_o hydrolysis in either soils or animal manure, and it shows optimal activity at pH 4.8 and 37 ^{0}C , close to the conditions for other phosphatases tested previously in He and Honeycutt (2001). Potato acid phosphatase is active in the hydrolysis of the P compounds with the same characteristics to those for alkaline phosphatase (Table 11.3). Therefore, P_i released by potato acid phosphatase may reflect the content of simple phosphomonoester as previously reported for alkaline phosphatase. Substitution of potato acid phosphatase would make it possible to apply the same incubation conditions (pH 5.0, 37 ^{0}C) for all enzymes. Therefore He et al. (2004a) modified the approach by using potato acid phosphatase, and acid phosphatases from potato and wheat germ, and both phosphatases plus nuclease P1 to identify and quantify simple monoester P, phytate-like P, and polynucleotide-like P, respectively, in 100 mM Na acetate (pH 5.0). Wheat germ phosphatase preparation could be substituted by commercial fungal phytase preparation dependent on its availability (He et al., 2008b). Whereas the method was originally designed to characterize P_o forms in sequentially-extracted fractions of manure and soil samples (He et al., 2004a), it has also been used in manure and humic samples from other matrices (He et al., 2009a; 2009b; 2007). Recently, this method has been adapted to a micro-plate reader format that allows researchers to process many samples quickly, affordably and with relative ease (Johnson and Hill, 2010).

Table 11.3. Hydrolysis of model P compounds.

Compound[a]	PP	PP/GP	PP/GP/NP
		% of P hydrolyzed	
Phytate	-0.1	91.9	-[b]
p-Nitrophenyl phosphate	101.5	100.1	-
Glucose 6-phosphate	96.1	95.8	-
Glucose 1-phosphate	11.2	18.3	-
Fructose 6 –phosphate	91.7	97.1	-
Ribose 5-phosphate	89.8	91.2	-
AMP	59.0	88.8	-
Glycerophosphate	89.0	103.5	-
NAD	8.2	53.8	-
Pyrophosphate	101.0	100.3	-
ADP	97.8	100.3	-
ATP	100.8	101.0	-
RNA	15.9	45.8	95.2
DNA	2.9	16.7	96.0

Adapted from He et al. (2004a).

[a] 0.1 mM except 0.064 mM for RNA and 0.070 mM for DNA) by incubation at 37°C for 1 h with potato acid phosphatase (PP, 0.25 U mL^{-1}), the combination of PP and wheat germ acid phosphatase (PP/GP, 0.25 U mL^{-1} each) and the combination plus nuclease P1 (2 U mL^{-1}) (PP/GP/NP) in 100 mM sodium acetate (pH 5.0), respectively.

[b] not tested.

Table 11.4. Effects of storage on inorganic P (P_i), enzymatically hydrolyzable organic P (P_e), and non-hydrolyzable organic P (P_{ue}) in manure extracts and residual resuspensions.

Manure	Prior to storage[a]			After storage[b]		
	% P_t[c]			% P_t		
	P_i	P_e	P_{ue}	P_i	P_e	P_{ue}
			Swine manure			
Extract:	52.6	10.1	3.2	60.0	5.8	23.1
Residual:						
pH 5.0	2.9	3.4	1.4	3.4	2.4	3.6
pH 9.0	2.0	0.5	2.0	2.3	0.4	3.6
pH 2.5	0.8	0.5	0.3	0.9	1.1	1.8
			Cattle manure			
Extract:	58.6	5.2	15.4	55.2	2.6	27.3
Residual:						
pH 5.0	3.1	0.2	2.5	1.9	-[d]	2.5
pH 9.0	2.5	0.2	1.6	2.8	-	3.3
pH 2.5	0.7	0.5	2.2	0.4	0.0	4.1

Adapted from(He et al. (2003).

[a] Fresh manures treated and stored at -20 °C.

[b] Swine manure at 4 °C, and cattle manure at 22 °C for a year prior to sample treatments.

[c] Total P was 4800, and 5400 mg kg^{-1} dry matter for swine and cattle manures, separately.

[d] Negative values were observed.

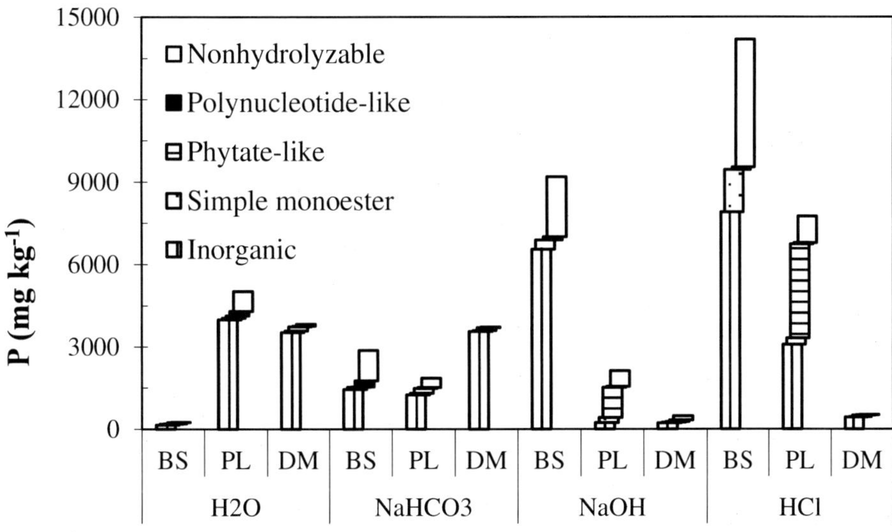

Figure 11.1. Phosphorus distribution patterns in sequentially extracted H_2O, 0.5 M $NaHCO_3$ (pH8.5), 0.1 M NaOH, and 1 M HCl fractions of biosolids (BS), dairy manure (DM), and poultry litter (PL). Data are adapted from He et al. (2004b; 2006c; 2010).

11.4.2. Organic P in Dairy Manure, Poultry Litter and Biosolids

With the modified procedure, He and his collaborators (He et al., 2004b; 2006c; 2010) further investigated P distribution in dairy manure, poultry litter and biosolids (Figure 11.1). The distribution patterns of P among the three types of wastes are different. If both P in H_2O and $NaHCO_3$ fractions are labile P, >85 % of diary manure P were in these two labile fractions, however, only about 50% of poultry litter P are in the labile fractions. More than 85% of P in biosolids is in the stable NaOH and HCl fractions. Furthermore, enzymatic hydrolysis demonstrated that dairy manure does not contain much P_o. The higher P levels in water and sodium hydroxide fractions of poultry litter was due to the portion of P_o. Another significant difference is that, in hydroxide and acid fractions of poultry litter, 50 to 90% of P was present in organic forms, and most of them are phytate P and non-hydrolyzable P_o (Figure 11.1). Biosolids contained moderate amounts of P_o. But unlike poultry litter, the majority of P_o in biosolids was non-hydrolyzable by the three enzymes used. These results in Table 11.2 and Figure 11.1 demonstrated that the patterns of P distribution are different among swine manure, cattle dairy manure, poultry litter, and biosolids. Therefore, for the best P management strategy we should consider the particular P chemistry in each type of manure or biosolids.

11.5. BIOAVAILABILITY OF ORGANIC PHOSPHORUS CHARACTERIZED BY ENZYMATIC HYDROLYSIS

11.5.1. Effects of Storage on Bioavailability of Manure P

Organic phosphates are considered bioavailable if they can be hydrolyzed to P_i (George et al., 2007; Yadav and Tarafdar, 2003). Thus, the degree of P_o hydrolysis is an important characteristic of P_o in terms of plant nutrition. The bioavailability of P_o may be assessed by estimating the P_o that can be hydrolyzed to P_i with an excess amount of phosphate-releasing enzymes (phosphatases). Numerous commercially available phosphatase enzymes have been used for this purpose (Bunemann, 2008). Sequential fractionation coupled with phosphatase hydrolysis provides useful information on P forms and their distribution. However, it is known that a portion of labile P_o can be hydrolyzed under strong alkaline or acidic conditions (He et al., 2006d; Leinweber et al., 1997; Pant et al., 1994b). To reduce such concerns, He et al. (2003; 2006d) designed alternative schemes to directly determine hydrolyzable P_o in animal manure extracts and resuspensions under mild conditions. These schemes have been used to investigate the effects of storage on bioavailability of manure P in swine and cattle manures, which had been separately stored at -20, 4 or 22 °C for a year. Due to the broad substrate specificity of the three phosphatases used and the mild extraction and resuspension conditions, He et al. (2003) assumed that all P_i and P_e determined in the sequential incubation scheme were labile or bioavailable, and the unaccounted P portion (*i. e.* P_t- P_i- P_e −P_{ue}) is recalcitrant P, where P_e = enzymatically hydrolyzable P_o, and P_{ue} = non-hydrolyszble P_o. Assuming that the P species in manure stored at −20 °C were the same as those present in fresh manure, changes in P distribution following storage were obvious (Table 11.4). In fresh manures, as represented by the -20 °C samples, the percentage of bioavailable P was the same for swine manure (72.8□%) and cattle manure (71.6%), although less recalcitrant P was present in cattle manure (7.8□%) than in swine manure (21.3□%). Comparison of these data with bioavailable P after storage, from the 4 °C swine sample (76.3%) and the 20 °C cattle sample (62.9%), revealed that storage did not markedly change the level of bioavailable P in either manure. However, the soluble but enzymatically unhydrolyzable organic P (P_{ue}) in extract fractions of both manure samples increased significantly following storage (from 3.2 to 23.1% of P_t for swine manure and from 15.4 to 27.3% of P_t for cattle manure) (Table 11.4). These data indicate that the major change during the animal manure stored for a year resulted in an increase in P_{ue}, suggesting that manure P solubility may increase with storage, however, the increase would not produce more bioavailable P in the manure.

11.5.2. Fe(III)-associated Labile P

The reducing agent dithionite buffered with bicarbonate has been used frequently to promote the reduction of Fe(III) to Fe (II), thus releasing Fe-related P_i (Ruiz et al., 1997; Scalenghe et al., 2002; Uusitalo and Turtola, 2003). He et al. (2006a) applied this reducing agent to investigate the Fe-labile P_o in animal manure. After removal of soluble phosphate, poultry litter and dairy manure solids were incubated in sodium acetate buffer (pH 5.0) with the reducing agent sodium dithionite and/or fungal phytase to identify insoluble manure P species. The results indicated that poultry litter contained most labile P (9.7% of insoluble P), reducible P_i (3.2% of insoluble P), non-Fe-related (reducible-irrelevant) P_o (5.7% of insoluble P), and Fe-related (reducible) P_o (16.2% of insoluble P). In dairy manure, 51.5% of insoluble P was most labile P, 28.1% non-Fe-related P_o, and 20.4% Fe-related P_o, but little or no reducible P_i. Previous research has shown that as soils become reduced, soluble P increased (Vadas and Sims, 1998). The portion of the Fe-labile P_i and P_o in animal manure identified by the dithionite/phytase incubation should have, at least partially, contributed to the soluble P increase in the poultry litter-amended soils under reduction conditions observed by Vadas and Sims (1998). Recently, solution ^{31}P NMR Spectroscopy identified more polyphosphate and pyrophosphate in sodium acetate buffer (pH 5.0)-dithionite extracts than in NaOH-EDTA extracts of dairy manure (He et al., 2009e). This observation further confirms that acetate-dithionite is a milder extractant than NaOH-EDTA for labile P species from animal manure.

11.6. Ligand- and Phytase-Labile Organic Phosphorus in Manure

11.6.1. Effects of Metal ions on Hydrolysis of IP6 by Fungal Phytase

Special attention has been paid to the hydrolysis of phytate P in animal manure as phytate is the major P_o form and has high affinity to metal cations (Dao, 2003; He et al., 2006a; 2006e). He et al. (2006a) reported that the inhibitory effect of Fe(III) on the enzymatic hydrolysis of IP6 increased with increased concentrations of Fe(III). Whereas almost all IP6 P (90%) was released by commercial fungal phytase in a 30-h incubation period in the absence of Fe(III), only about 50% of IP6 P was hydrolyzed when the ratio of IP6 P:Fe was 6:2. The P_o hydrolyzed was a minor part (<3%) of total IP6 when the ratio was 6:6 or 6:12. Meanwhile, He et al. (2006a) demonstrated that dithionite could reduce the inhibition of Fe(III) on enzymatic hydrolysis of IP6 by fungal phytase through conversion of Fe(III) to Fe(II) which is more soluble. Dao (2003) observed that the hydrolysis of IP6 by fungal phytase decreased by 40-50% as IP6 P:Ca ratio reached 6:6. This decrease in hydrolysis was 27 and 32% at cation to IP6 P ratios of 1:6 for Al and Fe, respectively, while reaching more than 99% at a mole ratio of 6:6. Dao (2004a) then tested various organic ligands to disassociate Ca, Al or Fe cations from IP6, and their impact on promoting hydrolysis of IP6 in dairy wastewater by commercial fungal phytase preparation. The most efficient organic ligands were 1,2-cyclohexane diaminotetraacetate (CDTA) and ethylenediamine tetraacetate (EDTA). Inclusion of a ligand such as EDTA made it possible to directly measure P_o in dairy manure hydrolyzed by fungal phytase without extraction (Dao, 2003; 2004a).

11.6.2. Fractionation of Manure Bioactive P Pools

Based on the role of ligands in mobilizing enzyme-labile P, Dao (2004a) proposed an *in situ* fractionation of manure bioactive P pools. With this procedure, four P fractions were differentiated. They included water-extractable phosphate-P (WEP), ligand-exchangeable inorganic phosphate-P (EEP_i), organic EDTA-exchangeable and phytase-labile P (EDTA-PHP), and the all-included total bioactive P (WEP + EEPi + EDTA-PHP).

Dao et al. (2006) then evaluated water soluble- and labile complexed (i.e. metal-associated) P and the mechanisms controlling P solubilization in dairy manure. Among the 107 manures collected across five US, northeastern states, WEP averaged 16% of the total P with a standard deviation of 14.8%. The addition of EDTA released EEP_i (15.3 $\pm$ 8.3% of total P) primarily associated with Ca and Mg and additional P_o fraction (35.9 $\pm$ 15.6% of total P) was hydrolyzable by fungal phytase. Including EDTA in enzymatic incubation only slightly increased hydrolyzable P_o fraction by an average 3.7% of the total P. Furthermore, the total EDTA-PHP was related to EDTA-extractable Ca and Mg contents of the manure, indicating that a fraction of this P pool was derived from complexed Ca and Mg IP6 and other inositol P (Dao et al., 2006). Recent solution and solid state ^{31}P NMR spectroscopies have confirmed that the majority of P in dairy manure are Ca and Mg P_i and P_o with varying solubilities (He et al., 2009e).

In addition to the commercial fungal phytase preparation, Dao and Hoang (2008) tested fungal phytases prepared from the cultures of five strains of *Aspergillus*. Indeed, the K_m value (0.21 mM) of the in-house prepared *Aspergillus ficumm* phytases is greater than that (0.098 mM) of the commercial *Aspergillus ficumm* enzyme, indicating a lower affinity of the in-house purified preparation for substrate IP6 (Dao and Hoang, 2008). The higher affinity by the commercially prepared enzyme could be due to the fact that it contains a considerable amount of contaminant phosphatase activities (Hayes et al., 2000; Tazisong et al., 2008). However, the multiple phosphatase activities should be considered an advantage for total hydrolyzable P_o determination as the enzymatic hydrolysis of P_o was more complete under such circumstances. Although both enzyme preparations effected the complete hydrolysis of phytate under a long term (overnight) incubation, this in-house preparation was used to evaluate the hydrolyzable P_o in 71 PL samples collected across poultry producing regions of Arkansas, Maryland and Oklahoma of the US (Dao and Hoang, 2008). Overall, hydrolyzable P_o quantified by the enzyme preparation averaged 54% of the total P with a standard deviation of 14%. These data on total hydrolyzable P_o in PL are consistent with those measured and reported under similar conditions using a single commercial fungal phytase (He et al., 2009b).

11.7. BIOAVAILABILITY OF MANURE P IN SOILS EVALUATED BY ENZYMATIC HYDROLYSIS

11.7.1. Aerobic Incubation

Enzymatic hydrolysis could also be used to evaluate the fate of manure P after application to soils coupled with laboratory incubations and pot experiments to determine its short-term fate. He's group (He et al., 2004c; 2006b; Waldrip-Dail et al., 2009) conducted laboratory aerobic incubation experiments to assess the P dynamics of two Maine soils in response to dairy and poultry manure application. Since the mechanism of interconverting P species can be complex, He et al. (2004c) assumed that interconverting P species among the various fractions are mainly a physicochemical process while transformation of the various P species within a given fraction are mainly mediated by biological and biochemical processes. They measured P_i, P_{ho}, and P_{no} (enzymatically hydrolyzable and nonhydrolyzable P_o, respectively) in the sequentially-extracted fractions of manure-amended soils at various incubation times and analyzed their patterns of change over time. The P species in the H_2O, 0.5 M $NaHCO_3$ (pH 8.5) and 0.1 M NaOH soil extracts were quantified over the 108-day incubation period. Phosphorus in 1.0 M HCl was not determined since these dairy manures contained only a very small amount of HCl-extracted P (He and Honeycutt, 2001; He et al., 2004b); therefore, this fraction should not significantly impact soil P dynamics. Similar patterns of P dynamics were observed in the two soils. Added P_i from either chemical fertilizer or animal manure was found mainly in the $NaHCO_3$ and NaOH fractions (Figure 11.2). In the H_2O fraction, P_i attained a low and stable level after an initial rapid decrease. Concentrations of other P fractions (species) fluctuated during the incubation period. Furthermore, the fluctuations were observed in complementary patterns between P_i in the $NaHCO_3$ and NaOH fractions, as well as between labile P (inorganic and hydrolysable organic) and P_{no} in the NaOH fraction (Figure 11.2). These complementary fluctuations implied an interchange of P species during incubation. The interchange among P fractions is considered to reflect how P dynamics maintain a balance between labile and immobile P in soils. Furthermore, He et al. (2006b) observed that bioavailable P (P_i and P_{ho}), rather than total extracted P (P_i and total P_o), in water and bicarbonate fractions were strongly related to applied manure P. In other words, the portion of P_{no} in soils is not immediately available for plant growth although it may be potentially bioavailable in the long term. This indicates the importance of distinguishing the P_{ho} portions from total P_o for better nutrient management. Thus, when quantifying P_o portions in fractions for a short-term nutrient management (such as for a growth season), He et al. (2006b) suggest evaluating their bioavailability by enzymatic hydrolysis.

A significant portion of P in poultry manure/litter is HCl-extracted P, differs from the relative P fraction ratios of dairy manure (Figure 11.1). However, in some previous research, the portion of HCl-extracted P has not been measured, thus its fate was not accounted for upon being applied to soil (He et al., 2006g). Thus, Waldrip-Dail et al. (2009) applied aerobic incubation to derive information on the fate of stable HCl-extracted P in poultry manure (PM) upon application to soils (Waldrip-Dail et al., 2009). Changes in P forms in PM-amended soils during the incubation followed the same patterns as control soils, particularly in $NaHCO_3$- and NaOH-extractable P. However, unlike the dairy manure amendment (He and Honeycutt, 2001; He et al., 2004b), applied PM, P did not additively increase extractable P concentrations. With up to 58% of extractable P in the HCl fraction of the poultry manure samples, which had relatively equal amounts of P_i and P_o (Dail et al., 2007), the P_i level in the HCl fraction of PM-amended soils remained relatively stable over the course of the incubation. In contrast, the P_{no} fraction rapidly became enzymatically hydrolyzable, with a portion remaining in the HCl fractions and some transformed to other fractions. This observation indicates that the stable HCl-extractable P in PM has rapidly degraded to more labile P forms in these amended soils. Whereas sequentially extracted HCl P in soil is considered less plant available (Cross and Schlesinger, 1995), the incubation data from this experiment confirmed that this may not be true for the HCl-extracted P in animal manure (Dail et al., 2007). Therefore, soil amendments with moderate levels of poultry manure does not appear to cause accumulation of stable P forms, and that the large HCl-P fraction (31.1% of total P) in PM could act as a source of plant-available P over a short term (e. g. a growing season) (Waldrip-Dail et al., 2009). P transformations to more bioavailable forms is likely microbially-mediated (Hill et al., 2007)

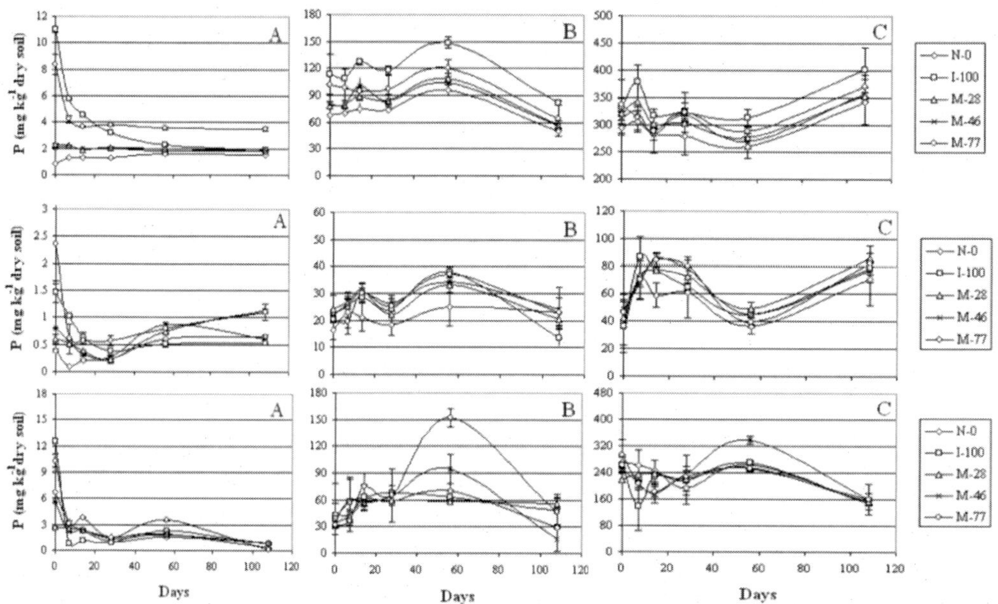

Figure 11.2. Phosphorus dynamics in relation to incubation time for H_2O (A), $NaHCO_3$ (B), and NaOH (C) fractions extracted from Newport (Maine) sandy loam soil with no P added (N-0), 100 mg inorganic P kg^{-1} dry soil (I-100), 28 mg dairy manure P kg^{-1} dry soil (M-28), 46 mg dairy manure P kg^{-1} dry soil (M-46), and 77 mg dairy manure P kg^{-1} dry soil (M-77). Upper row, inorganic P (P_i); Middle row, enzymatically hydrolyzable organic P (P_{ho}); and Lower row, enzymatically nonhydrolyzable organic P (P_{no}). Data are adapted from He et al. (2004c).

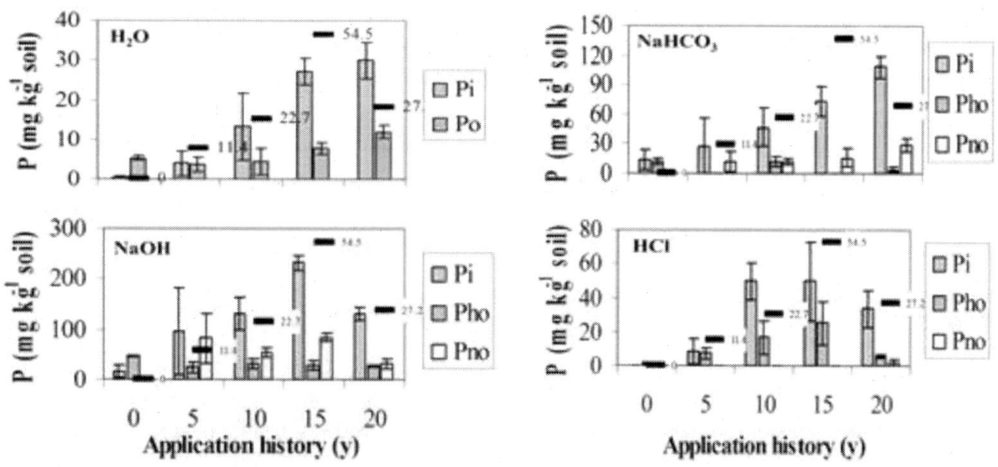

Figure 11.3. Concentrations of inorganic (P_i), organic (P_o) or enzymatically hydrolyzable organic (P_{ho}) , and enzymatically nonhydrolyzable organic (P_{no}) phosphorus in sequentially extracted H_2O, 0.5 M $NaHCO_3$ (pH 8.5), 0.1 M NaOH, and 1 M HCl fractions of five pasture soils with 0, 5, 10, 15, 20 year of poultry litter application, respectively (He et al., unpublished data). Values represent the cumulative amount of poultry litter applied (Mg/ha). Vertical bars indicate standard deviations (n=3).

11.7.2. Pot Experiments

Rao and Dao (2008) conducted a potted plant growth experiment to characterize cattle manure P mineralization as modified by iron amendments and uptake by pigeon pea and soybean. Triple superphosphate, untreated, or manure amended with Fe at 1:1 or 1:3 molar ratio of manure P:Fe, was applied to Dale silt loam (fine-silty, mixed, superactive, thermic Pachic Haplustolls) at the rate of 20 mg kg^{-1}. The *in situ* fractionation of soil bioactive P pools was conducted using the procedure developed by Dao (2003). Rao and Dao (2008) observed temporal changes in ligand-exchangeable fractions, i.e., EEPi and EDTA-PHP, which suggested that plant P was taken up from the inorganic (WEP +EEPi) P pools. However, non-cropped soils corroborated the observations that enzyme-mediated hydrolysis of P_o pools generated biologically derived orthophosphate and replenished the water-soluble and insoluble (WEP + EEPi) fractions to account for the P uptake by grain legume crops. Thus, the authors concluded that the ligand-based soil P fractionation method revealed soil depletion patterns over the growing season that correlated to temporal changes in plant P. The authors further argued that such relationships were not detected using Mehlich 3 soil test method; therefore, Mehlich 3 P could not appropriately reflect the availability of immobilized P to pigeon pea and soybean in soils amended with treated manure. Both works by He et al. (He and Honeycutt, 2001; 2004b) and Rao and Dao (2008) indicate that soil P fractionation coupled with enzyme hydrolysis is a useful approach in investigating the internal exchange and biological significance of various manure P pools in soils.

11.8. LONG TERM IMPACTS OF REPEATED MANURE APPLICATION ON SOIL P STATUS

11.8.1. On soils Amended with Poultry Litter

A long-term field experiment in which poultry litter (PL) was applied to pastured soils has been maintained for two decades in the Sand Mountain region of north Alabama, USA (He et al., 2008a; 2009c). Five fields have received PL continuously for 0 (control), 5, 10, 15, and 20 years at rates of 0, 2.27, 2.27, 3.63, and 1.36 Mg ha^{-1} yr^{-1}, respectively. Previous work (He et al., 2009c) observed that the P in these five fields had increased with increasing cumulative PL application, suggesting that residual P from PL had accumulated in these soils. Thus, He et al. (unpublished data) further investigated the P accumulation patterns of this residual PL-P by sequential fractionation coupled by enzymatic hydrolysis (Figure 11.3). The data indicated that water soluble P_i increased rapidly with increasing history of poultry litter application, and reached 30 mg P kg^{-1} soil following 20-y of poultry litter application. Unamended soil contained 5.2 mg P_o per kg soil in the water soluble fraction. This concentration did not change in soils with 5- and 10-y poultry litter application. Water soluble P_o in soils receiving 15 and 20 years of poultry litter was slightly higher than that in unamended soils. The impact of poultry litter application on $NaHCO_3$ extracted P_i and P_o was comparable to the two P forms in the water fraction, however, interchanges between P_{ho} and P_{no} were observed. The change in P_i in the NaOH fraction was consistent with the cumulative amount of poultry litter applied, rather than the years of poultry litter application. Among the four fractions examined, the NaOH fraction contained the highest concentrations of all three P forms, P_i, P_{ho}, and P_{no}. Enzymatically hydrolyzable organic P in the soil NaOH fraction without poultry litter was higher than that with poultry litter, but did not change with increasing application. In contrast, applied poultry litter increased P_{no} in soils, however, concentrations fluctuated between the application years. Although, little P_{no} was detected in the HCl fractions, the concentrations of both P_i and P_{ho} in HCl fractions were much lower than those in the NaOH fractions. For example, P_i ranged from 34 and 50 mg kg^{-1} soil in HCl fractions, compared with 131 to 233 mg kg^{-1} soil in NaOH fractions of soils that received poultry litter for 15- and 20 years. However, the two P forms changed in a pattern similar to that of P_i in the NaOH fractions. Contrary to the P distributed in these soils, analysis of 23 poultry litters indicate more P_i (3082 ± 1135 mg kg^{-1} dry litter) and P_{ho} (3701 ± 1212 mg kg^{-1} dry litter) in the HCl fraction than in the NaOH fraction (231± 103 and 1319 ± 672 mg kg^{-1} dry litter, respectively) (He et al., 2006c). These observations imply that some of the HCl-extractable poultry litter P may have transformed to other P forms once applied to the soil. This observation was consistent with solution ^{31}P NMR spectroscopic observations on the 20-yt soil sample (He et al., 2008b). From these observations, He et al. (unpublished) concluded that long term poultry litter application impacted soil P status in two patterns. The levels of labile P_i extracted by water and NaHCO$_3$ extractants were more related to the number of years receiving poultry litter. The levels of stable P_i extractable by NaOH and HCl were related to the cumulative amount of poultry litter applied. This study indicates that repeated application of PL to soils increases the pools of both labile and stable P, but the stable (hydroxide and acid extractable) P is eventually transformed to labile (water and bicarbonate extractable) P, rendering it available for plant uptake.

11.8.2. On Soils Amended with Dairy Manure

Dao (2004b) conducted an enzymatic hydrolysis and soil P desorption study on water-extractable uncomplexed and complexed P forms (i.e., bioactive P) in Unicorn and Christiana soils (Typic Hapludults) amended with or without 7-yr dairy manure application. Dao (2004b) observed that manure amendments and loading rates influenced soil chemical properties and the desorption of soil bioactive P fractions. In a permanent grass plot, addition of dairy manure to the soil surface over the years increased soil total organic C, storing manure C in the 0- to 10-cm depth of the Christiana soil. At the 0.5 mM concentration, EDTA and CDTA, followed by DTPA, desorbed large quantities of bioactive P in untreated and manure-amended soils, between 22 and 40.8%, compared with 6.6 and 8.6% of soil total P by water alone and 0.5 mM oxalate solution, respectively. The ligand type x manure rate interaction was statistically significant and only the three anthropogenic ligands were able to separate the effects of manure loading rate, showing that the 30 kg ha^{-1} treatment was significantly different from the 0 and 15 kg ha^{-1} treatments. Furthermore, fungal phytase yielded 8 to 15 times more bioactive P with ligands than without added ligands. Across manure rate treatments, the proportion of WEP, EEP, and PHP appeared to shift with an increase in EEP and a slight decrease in the percentage of PHP for the 30 kg P ha^{-1} treatment, in the EDTA and CDTA extracts. This buildup of inorganic ligand-extractable P likely resulted from a direct addition of PO$_4$–P native to the manure. While soil PHP concentrations continued to increase with increasing manure loading, the decline in its proportion of total bioactive P occurred because P_o added with manure was either (i) continually being hydrolyzed and sequestered as complexed mineral EEP, or (ii) increasingly unextractable by water or cannot be mobilized by even these strong ligands to be hydrolyzed by fungal phytase. The latter reason is a more plausible scenario as sequestration and storage of IP6 can come about because of the high insolubility of its complexed forms and resistance to hydrolysis by phytases (Dao, 2004b). Thus, Dao (2004b) concluded that repeated manure applications resulted in soil storage of unextractable complexed PHP and a buildup of inorganic EEP. Accumulation of EEP increases risks of mobilization of bioactive P in the top 10 cm of soils.

11.9. CONCLUSION

Orthophosphate-releasing enzymatic hydrolysis is an alternative means for characterizing P_o in animal manure. The approach is not only simple and fast, but can also provide information difficult to obtain by other methods susceptible to P_o hydrolysis. He et al. developed a stepwise multiple-enzyme approach to classify P in animal manure into five groups as soluble P_i, simple monoester P, polynucleotide-like P, phytate-like P, and nonhydrolyzable P. Alternatively, Dao et al. characterized manure P with a single fungal phytase with or without ligands in terms of water-extractable uncomplexed and complexed P forms (or bioactive P). Both methods have been used to evaluate the distribution patterns of animal manure P and their short- and long-term fates in soil. Data from these studies indicated various labile P forms are present in animal manure and the distribution patterns of these P forms are different in different types of animal manure. Although manure application increases soil P levels, the buildup of P_o is relatively lower than total P in manure-amended soils, indicating the lability of manure P_o. Insight from these enzymatic hydrolysis studies suggests that P nutrient values in animal manure should be given prior consideration during early stages of manure application whereas the impact of manure applications on soil properties and the environment should be gradually weighed in with increasing years of application.

Currently, commercially available phosphatases are mainly used in this type of research. These enzyme preparations are usually crudely purified and have broad substrate specificities: two deficiencies which affect both enzyme efficiency and experimental accuracy. Exploration of other enzymes with more exclusive substrate specificity would enable more accurate and detailed P_o characterization. For instance, all five commercially available enzymes He et al. used (He and Honeycutt, 2001; He et al., 2004a) show relatively high activity on inorganic pyrophosphate, ADP, and ATP, whereas neither purified phytase from *Escherichia coli* (Greiner et al., 1993) nor sugar phosphatase from *Neisseria meningitidis* (Lee et al., 1967) release P from these compounds. Selective inclusion of such enzymes, therefore, would release a more specific group of P_o in an incubation sample It is noticeable that many studies have reported that not more than 50% of P_o is hydrolysable using commercially available enzymes. Apparently, these unhydrolysable P compounds are in relatively complex forms. Hydrolysis by phospholipases (Amini and McKelvie, 2005; Pant and Warman, 2000), teichoic hydrolases (Kusser and Fiedler, 1983), polyphosphate hydrolases (Cardona et al., 2002), and /or abiotical hydrolysis(He et al., 2009a; He et al., 2006f) could shed light on characterizing the previously unidentified portions of P_o. A collaborative effort of biochemists and agricultural/environmental scientists is proposed to fulfill this purpose.

REFERENCES

Amini, N., and I.D. McKelvie. 2005. An enzymatic flow analysis method for the determination of phosphatidylcholine in sediment pore waters and extracts. *Talanta* 66:445-452.

Bohn, L., A.S. Meyer, and S.K. Rasmussen. 2008. Phytate: impact on environment and human nutrition. A challenge for molecular breeding. *Journal of Zhejiang University-Science* B 9:165-191.

Bunemann, E.K. 2008. Enzyme additions as a tool to assess the potential bioavailability of organically bound nutrients. *Soil Biol. Biochem.* 40:2116-2129.

Cardona, S.T., F.P. Chavez, and C.A. Jerez. 2002. The exopolyphosphatase gene from *Sulfolobus solfataricus*: characterization of the first gene found to be involved in polyphosphate metabolism in *Archaea*. *Appl. Environ. Microbiol.* 68:4812-4819.

Cheshire, M.V., and G. Anderson. 1975. Soil polysaccharides and carbohydrate phosphates. *Soil Sci.* 119:356-362.

Corbridge, D.E.C. 2000. *Phosphorus 2000*. Chemistry, Biochemistry and Technology Elsevier Science, Amsterdam, the Netherland.

Cross, A.F., and W.H. Schlesinger. 1995. A literature review and evaluation of the Hedley fractionation: applications to the biogeochemical cycle of soil phosphorus in natural ecosystems. *Geoderma* 64:197-214.

Dail, H.W., Z. He, M.S. Erich, and C.W. Honeycutt. 2007. Effect of drying on phosphorus distribution in poultry manure. *Commun. Soil Sci. Plant Anal.* 38:1879-1895.

Dalal, R.C. 1977. Soil organic phosphorus. *Adv. Agron.* 29:83-117.

Dao, T.H. 2003. Polyvalent cation effects on *myo*-inositol hexa*kis* dihydrogenphosphate enzymatic dephosphorylation in dairy wastewater. *J. Environ. Qu*al. 32:694-701.

Dao, T.H. 2004a. Organic ligand effects on enzymatic dephosphorylation of *myo*-inositol hexa*kis* dihydrogenphosphate in dairy wastewater. *J. Environ. Qual.* 33:349-357.

Dao, T.H. 2004b. Ligands and phytase hydrolysis of organic phosphorus in soils amended with dairy manure. *Agron. J.* 96:1188-1195.

Dao, T.H., and K.Q. Hoang. 2008. Dephosphorylation and quantification of organic phosphorus in poultry litter by purified phytic-acid high affinity *Aspergillus* phosphohydrolases. *Chemosphere* 72:1782-1787.

Dao, T.H., A. Lugo-Ospina, J.B. Reeves III, and H. Zhang. 2006. Wastewater chemistry and fractionation of bioactive phosphorus in dairy manure. *Commun. Soil Sci. Plant Anal.* 37:907-924.

Devorshak, C., and R.M. Roe. 2001. Purification and characterization of a phosphoric triester hydrolase from the tufted apple bud moth, *Platynota idaeusalis* (Walker). *J. Biol. Chem. Mol. Toxicol.* 15:55-65.

Dick, W.A., and M.A. Tabatabai. 1977. Determination of orthophosphate in aqueous solutions containing labile organic and inorganic phosphorus compounds. *J. Environ. Qual.* 6:82-85.

Fox, T.R., and N.B. Comerford. 1992. Rhizospher phosphatase activity and phosphatase hydrolyzable organic phosphorus in two forested spodosol. *Soil Biol. Biochem.* 24:579-583.

George, T.S., R.J. Simpson, P.J. Gregory, and A.E. Richardson. 2007. Differential interaction of *Aspergillus niger* and *Peniophora lycii* phytases with soil particles affects the hydrolysis of inositol phosphates. *Soil Biol. Biochem.* 39:793-803.

Greiner, R., U. Konietzny, and K.-D. Jany. 1993. Purification and characterization of two phytases from *Escherichia coli*. *Arch. Biochem. Biophys.* 303:107-113.

Hayes, J.E., A.E. Richardson, and R.J. Simpson. 2000. Components of organic phosphorus in soil extracts that are hydrolysed by phytase and acid phosphatase. *Biol. Fertil. Soils.* 32:279-286.

He, Z., and C.W. Honeycutt. 2001. Enzymatic characterization of organic phosphorus in animal manure. *J. Environ. Qual.* 30:1685-1692.

He, Z., and C.W. Honeycutt. 2005. A modified molybdate blue method for orthophosphate determination suitable for investigating enzymatic hydrolysis of organic phosphates. Commun. *Soil Sci. Plant Anal.* 36:1373-1383.

He, Z., and Z. Dou. 2010. Phosphorus forms in animal manure and the impact on soil P status, p. 83-114, In C. S. Dellaguardia, ed. Manure: management, use and environmental impacts. Nova Science Publishers, Inc., New York, N.Y.

He, Z., C.W. Honeycutt, and T.S. Griffin. 2003. Enzymatic hydrolysis of organic phosphorus in extracts and resuspensions of swine manure and cattle manure. *Biol. Fertil. Soils.* 38:78-83.

He, Z., T.S. Griffin, and C.W. Honeycutt. 2004a. Enzymatic hydrolysis of organic phosphorus in swine manure and soil. *J. Environ. Qual.* 33:367-372.

He, Z., T.S. Griffin, and C.W. Honeycutt. 2004b. Phosphorus distribution in dairy manures. *J. Environ. Qual.* 33:1528-1534.

He, Z., T.S. Griffin, and C.W. Honeycutt. 2004c. Evaluation of soil phosphorus transformations by sequential fractionation and phosphatase hydrolysis. *Soil Sci.* 169:515-527.

He, Z., T.H. Dao, and C.W. Honeycutt. 2006a. Insoluble Fe-associated inorganic and organic phosphates in animal manure and soil. *Soil Sci.* 171:117-126.

He, Z., T.S. Griffin, and C.W. Honeycutt. 2006b. Soil phosphorus dynamics in response to dairy manure and inorganic fertilizer applications. *Soil Sci.* 171:598-609.

He, Z., Z.N. Senwo, R.N. Mankolo, and C.W. Honeycutt. 2006c. Phosphorus fractions in poultry litter characterized by sequential fractionation coupled with phosphatase hydrolysis. *J. Food Agri. Environ.* 4:304-312.

He, Z., G.S. Toor, C.W. Honeycutt, and J.T. Sims. 2006d. An enzymatic hydrolysis approach for characterizing labile phosphorus forms in dairy manure under mild assay conditions. *Bioresour. Technol.* 97:1660-1668.

He, Z., C.W. Honeycutt, T. Zhang, and P.M. Bertsch. 2006e. Preparation and FT-IR characterization of metal phytate compounds. *J. Environ. Qual.* 35:1319-1328.

He, Z., I.A. Tazisong, Z.N. Senwo, and D. Zhang. 2008a. Soil properties and macro cations status impacted by long-term applied poultry litter. *Commun. Soil Sci. Plant Anal.* 39:858-872.

He, Z., D.C. Olk, C.W. Honeycutt, and A. Fortuna. 2009a. Enzymatically- and ultraviolet-labile phosphorus in humic acid fractions from rice soils. *Soil Sci.* 174:81-87.

He, Z., T. Ohno, B.J. Cade-Menun, M.S. Erich, and C.W. Honeycutt. 2006f. Spectral and chemical characterization of phosphates associated with humic substances. *Soil Sci. Soc. Am. J.* 70:1741-1751.

He, Z., A. Fortuna, Z.N. Senwo, I.A. Tazisong, C.W. Honeycutt, and T.S. Griffin. 2006g. Hydrochloric fractions in Hedley fractionation may contain inorganic and organic phosphates. *Soil Sci. Soc. Am. J.* 70:893-899.

He, Z., C.W. Honeycutt, B.J. Cade-Menun, Z.N. Senwo, and I.A. Tazisong. 2008b. Phosphorus in poultry litter and soil: enzymatic and nuclear magnetic resonance characterization. *Soil Sci. Soc. Am. J.* 72:1425-1433.

He, Z., H.W. Waldrip, M.S. Erich, C.W. Honeycutt, and Z.N. Senwo. 2009b. Enzymatic quantification of phytate in animal manure. *Commun. Soil Sci. Plant Anal.* 40:566-575.

He, Z., I.A. Tazisong, Z.N. Senwo, C.W. Honeycutt, and D. Zhang. 2009c. Nitrogen and phosphorus accumulation in pasture soil from repeated poultry litter application. Commun. *Soil Sci. Plant Anal.* 40:587-599.

He, Z., C.W. Honeycutt, T. Ohno, F. Wu, and R. Zhang. 2009d. Characterization of phosphorus associated with natural organic matter. p. 265-279, *In* F. Wu and B. Xing, eds. Natural Organic Matter and its Significance in the Environment. *Science Press,* Beijing, China.

He, Z., B.J. Cade-Menun, G.S. Toor, A. Fortuna, C.W. Honeycutt, and J.T. Sims. 2007. Comparison of phosphorus forms in wet and dried animal manures by solution phosphorus-31 nuclear magnetic resonance spectroscopy and enzymatic hydrolysis. *J. Environ. Qual.* 36:1086-1095.

He, Z., C.W. Honeycutt, T.S. Griffin, B.J. Cade-Menun, P.J. Pellechia, and Z. Dou. 2009e. Phosphorus forms in conventional and organic dairy manure identified by solution and solid state P-31 NMR spectroscopy. *J. Environ. Qual.* 38:1909-1918.

He, Z., H. Zhang, G.S. Toor, Z. Dou, C.W. Honeycutt, B.E. Haggard, and M.S. Reiter. 2010. Phosphorus distribution in sequentially-extracted fractions of biosolids, poultry litter and granulated products. *Soil Sci.* 175:154-161.

Herbes, S.E., H.E. Allen, and K.H. Mancy. 1975. Enzymatic characterization of soluble organic phosphorus in lake water. *Science.* 187:432-434.

Hill, J.E., D. Kysela, and A.E. Richardson. 2007. Isolation and assessment of phytate-hydrolysing bacteria from the Delmarva Penisula. *Environ. Microbiol.* 9:3100-3107.

Johnson, N.R., and J.E. Hill. 2010. Phosphorus species composition of poultry manure-amended soil using high-throughput enzymatic hydrolysis. *Soil Sci. Soc. Am. J.* 74:1786-191.

Jokanovic, M. 2001. Biotransformation of organophosphorus compounds. *Toxicol.* 166:139-160.

Juma, N.G., and M.A. Tabatabai. 1978. Distribution of phosphomonoesterases in soils. *Soil Sci.* 126:101-108.

Kusser, W., and F. Fiedler. 1983. Teichoicase from *Bacillus subtilis* Marburg. *J. Bacteriol.* 155:302-310.

Lai, K., N.J. Stolowich, and J.R. Wild. 1995. Characterization of P-S bond hydrolysis in organophosphorothioate pesticides by organophosphorus hydrolase. *Arch. Biochem. Biophys.* 318:59-64.

Lee, Y.-P., J. Sowokinos, and M.J. Erwin. 1967. Sugar phosphate phosphohydrolase. 1, substrate specificity, intracellular localization, and purification from *Neisseria meningitidis. J. Biol. Chem.* 242:2264-2271.

Leinweber, P., L. Haumaier, and W. Zech. 1997. Sequential extractions and ^{31}P-NMR spectroscopy of phosphorus forms in animal manures, whole soils and particle-size separates from a densely populated livestock area in northwest Germany. *Biol. Fertil. Soils.* 25:89-94.

Murphy, J., and J.P. Riley. 1962. A modified single solution method for the determination of phosphate in natural waters. *Analytica Chimia Acta* 27:31-36.

Nishino, M., S. Tsujimura, M. Kuba, and A. Kumon. 1994. N omega-phosphoarginine phosphatase from rat renal microsome was alkaline phosphatase. *Arch. Biochem. Biophys.* 312:101-106.

Pant, H.K., and P.R. Warman. 2000. Enzymatic hydrolysis of soil organic phosphorus by immobilized phosphatases. *Biol. Fertil. Soils.* 30:306-311.

Pant, H.K., D. Vaughan, and A.C. Edwards. 1994a. Molecular size distribution and enzymatic degradation of organic phosphorus in root exudates of spring barley. *Biol. Fertil. Soils.* 18:285-290.

Pant, H.K., A.C. Edwards, and D. Vaughan. 1994b. Extraction, molecular fractionation and enzyme degradation of organically associated phosphorus in soil solutions. *Biol. Fertil. Soils.* 17:196-200.

Quinn, J.P., A.N. Kulakova, N.A. Cooley, and J.W. McGrath. 2007. New ways to break an old bond: the bacterial carbon-phosphorus hydrolases and their role in biogeochemical phosphorus cycling. *Environ. Microbiol.* 9:2392-2400.

Rao, S.C., and T.H. Dao. 2008. Relationships between immobilized phosphorus uptake in two grain legumes and soil bioactive phosphorus pools in fertilized and manure-amended soil. *Agron. J.* 100:1535-1540.

Ron Vaz, M.D., A.C. Edwards, C.A. Shand, and M.S. Cresser. 1993. Phosphorus fractions in soil solution: influence of soil acidity and fertiliser additions. *Plant Soil.* 148:175-183.

Ruiz, J.M., A. Delgado, and J. Torrent. 1997. Iron-related phosphorus in overfertilized European soils. *J. Environ. Qual.* 26:1548-1554.

Sannigrahi, P., E.D. Ingall, and R. Benner. 2006. Nature and dynamics of phosphorus-containing components of marine dissolved and particulate organic matter. Geochim. Cosmochim. *Acta* 70:5868-5882.

Scalenghe, R., A.C. Edwards, F.A. Marsan, and E. Barberis. 2002. The effect of reducing conditions on the solubility of phosphorus in a diverse range of European agricultural soils. *Eur. J. Soil Sci.* 53:439-447.

Seeling, B., and A. Jungk. 1996. Utilization of organic phosphorus in calcium chloride extracts of soil by barley plants and hydrolysis by acid and alkaline phosphatases. *Plant Soil.* 178:179-184.

Sethuraman, A., N.N. Rao, and A. Kornberg. 2001. The endopolyphosphatase gene: essential in *Saccharomyces cerevisiae. Proc. Natl. Acad. Sci. USA.* 98:8542-8547.

Shand, C.A., and S. Smith. 1997. Enzymatic release of phosphate from model substrates and P compounds in soil solution from a peaty podzol. *Biol. Fertil. Soils.* 24:183-187.

Tazisong, I.A., Z.N. Senwo, R.W. Taylor, and Z. He. 2008. Hydrolysis of organic phosphates by commercially available phytases: Biocatalytic potentials and effects of ions on their enzymatic activities. *J. Food Agric. Environ.* 6:500-505.

Turner, B.L., I.D. McKelvie, and P.M. Haygarth. 2002. Characterization of water-extractable soil organic phosphorus by phosphatase hydrolysis. *Soil Biol. Biochem.* 34:29-37.

Turner, B.L., E. Frossard, and D.S. Baldwin. 2004. Appendix: Organic phosphorus compounds in the environments. p. 381-389, *In* B. L. Turner, et al., eds. *Organic phosphorus in the environments.* CABI Publishing, Cambridge, MA.

Uusitalo, R., and E. Turtola. 2003. Determination of redox-sensitive phosphorus in field runoff without sediment preconcentration. *J. Environ. Qual.* 32:70-77.

Vadas, P.A., and J.T. Sims. 1998. Redox status, poultry litter, and phosphorus solubility in Atlantic Coastal plain soils. *Soil Sci. Soc. Am. J.* 62:1025-1034.

Vincent, J.B., M.W. Crowder, and B.A. Averill. 1992. Hydrolysis of phosphate monoesters: a biological problem with multiple chemical solutions. *Trends. Biochem. Sci.* 17:105-110.

Waldrip-Dail, H., Z. He, M.S. Erich, and C.W. Honeycutt. 2009. Soil phosphorus dynamics in response to poultry manure amendment. *Soil Sci.* 174:195-201.

Wanner, B.L. 1994. Molecular genetics of carbon-phosphorus bond cleavage in bacteria. *Biodegradation* 5:175-184.

Webb, E.C. 1992. Enzyme nomenclature, p. 318-326. Academic Press, New York, N.Y.

Wodzinski, R.J., and A.H. Ullah. 1996. *Phytase. Adv. Appl. Microbiol.* 42:263-302.

Yadav, R.S., and J.C. Tarafdar. 2003. Phytase and phosphatase producing fungi in arid and semi-arid soils and their efficiency in hydrolyzing different organic P compounds. *Soil Biol. Biochem.* 35:745-751.

In: Environmental Chemistry of Animal Manure
Editor: Zhongqi He

ISBN: 978-1-62808-641-6
© 2013 Nova Science Publishers, Inc.

Chapter 12

CHARACTERIZING PHOSPHORUS IN ANIMAL WASTE WITH SOLUTION ^{31}P NMR SPECTROSCOPY

Barbara J. Cade-Menun[*]

12.1. INTRODUCTION

Animal manure is a rich source of nutrients, particularly nitrogen (N) and P. As such, it has long been applied to soil as a fertilizer. However, in recent years as animal production has intensified, manures have been applied to land for disposal purposes rather than to enhance soil fertility (Leinweber et al., 1997). This has lead to soil P accumulation, and an acceleration of P transfer from land to water, contributing to eutrophication (Sims et al., 2000).

The behavior of manure-applied P in soils depends on the forms of P in the manure, because P forms vary in their bioavailability and in their chemical dynamics in soil (Condron et al., 2005). As such, characterizing manure P forms is essential in order to maximize fertilizer potential and to minimize environmental impacts. One key tool for the speciation of P in manures is solution ^{31}P-NMR. First used to identify P forms in manure by Leinweber et al. (1997), solution ^{31}P-NMR has subsequently been applied to studies of manure from a wide range of animals, and to investigations of a number of manure properties, from P bioavailability to the impact of animal diet on feces composition and the effects of manure storage and handling. The objective of this chapter is to review the use of ^{31}P-NMR to characterize P forms in animal waste, discussing the requirements for successful ^{31}P-NMR experiments using animal manure and feces and reviewing the literature that has focused on method refinement, as well as summarizing the results of experiments investigating feces and

---

[*] Barbara.Cade-Menun@agr.gc.ca Agriculture and Agri-Food Canada, SPARC, Box 1030, Swift Current, SK S9H 3X2 Canada

manure properties, grouping these studies by animal type, and discussing future research needs. Please note that this chapter focuses on P speciation in feces (material collected from the animal at the point of excretion, referred to as dung in some studies), manure (feces deposited on the ground or mixed with bedding material, also known as litter for poultry), partially digested material removed from the animal in some way (e.g. ileal digesta removed from the ileum after slaughter) and feed when it has been analyzed and compared with feces or manure. This chapter does not include ^{31}P-NMR studies of P speciation in soil or manure-amended soils.

12.2. ^{31}P-NMR: Principles, P Forms, Methodology

12.2.1. General Principles of NMR Spectroscopy

It is beyond the scope of this chapter to fully explain NMR spectroscopy. More detailed descriptions are available in recent review articles (Cade-Menun 2005 a, b) or in textbooks (e.g. Wilson, 1987; Canet, 1996).

Briefly, NMR spectroscopy is based on the concept that nuclei in which the sum of the number of neutrons and protons is an odd number have magnetic properties (e.g. ^{1}H, ^{13}C, ^{15}N, ^{27}Al, ^{31}P). Each of these nuclei has a positive charge and a half-integer spin, I (i.e. I = 1/2, 3/2, etc.). This spinning generates a small magnetic field so that the nucleus behaves as a magnetic dipole, analogous to a bar magnet with north and south poles. Each nucleus also has a gyromagnetic ratio, γ, which is a fundamental nuclear constant and is different for every nucleus. When a nucleus with I = 1/2 and positive γ, such as P (I = 1/2, $\gamma = \gamma$ /10.829 10^7rad T^{-1} sec^{-1}), is placed in an external magnetic field of strength B$_0$, it can align itself in one of two ways. The first is parallel to B$_0$, which is the stable low-energy configuration. The other possible alignment is antiparallel to B$_0$, which has a higher energy and is less stable. When a radio-frequency pulse is applied to the sample in the magnetic field, nuclei resonate and absorb energy, 'flipping' from the lower energy state to the higher energy state. After the radio frequency pulse stops, the nuclei relax and emit energy, which can be detected as an emission signal. Resonating nuclei are shielded to varying degrees by the electron cloud from the molecules in which the nuclei are located in a sample. This electron shielding affects the amount of energy required for resonance. Thus, different nuclei absorb and subsequently emit energy in different ways, depending on their chemical bonds.

The data from the emission signals of the P nuclei present in the sample are collected as a free induction decay (FID) plot, which is a time-domain signal because it is recorded as a function of intensity over time. To transform a FID to a spectrum, known as processing, the FID is Fourier-transformed using NMR processing software. This converts the data to a frequency-domain form, with signal intensity (y-axis) recorded as a function of frequency (x-axis). Frequency is expressed as chemical shift, which is the measure of the position of a resonance signal relative to a standard (usually 85% phosphoric acid). Chemical shift values are dimensionless and are expressed in parts per million (ppm) with the external standard set to 0 ppm. Please note that ppm for chemical shift does not refer to concentration, but instead refers to Hz per MHz. Because ^{31}P is the only naturally occurring P isotope (100% natural abundance), all P within a sample is detected by NMR spectroscopy. Thus, the area under each peak is proportional to the number of that particular type of P nucleus, allowing NMR spectroscopy to quantitatively identify different P forms in a sample.

In order for the excited nuclei to relax back to equilibrium after a radio-frequency pulse, they exchange energy with their surroundings (spin-lattice relaxation) or with each other (spin-spin relaxation). These types of relaxation are governed by the exponential time constants T_1 and T_2, respectively. Relaxation is important for quantitative data: if the nuclei do not relax back to equilibrium between radio-frequency pulses, the system becomes saturated, some nuclei can no longer resonate, and the peak intensities will not reflect all the nuclei within each P form in the sample. T_1 is dependent on the γ of the nucleus and the mobility of the lattice. The more mobile the lattice, the greater the interaction it will have with the excited nuclei. Therefore, T_1 is shorter in solutions than in solids. T_1 can also be reduced by the presence of paramagnetic ions such as iron (Fe) and manganese (Mn), which are commonly associated with P in environmental samples such as manures and particularly soils. The unpaired electrons in paramagnetic ions readily transfer energy away from excited nuclei. In spin-spin relaxation (governed by T_2), nuclei exchange energy with neighboring nuclei of the same type, but in a different excitation state. Thus, nuclei in the lower energy level become excited, while nuclei that were excited relax to the lower energy level. This decreases the average time that a nucleus remains in the excited state, and can result in line broadening. In solutions, T_1 and T_2 are equal; T_2 is more important in solids.

12.2.2. Phosphorus Forms

Phosphorus compounds (both organic and inorganic) of interest in studies of feces, feed and manure generally fall between 25 and −25 ppm (Fig. 12.1). These include: phosphonates, with a C-P bond, at 20 ppm; orthophosphate at 5 to 7 ppm; orthophosphate monoesters, with one C moiety per P, at 3 to 6 ppm; orthophosphate diesters (two C moieties per P), including phospholipids and deoxyribonucleic acid (DNA), at 2.5 to −1 ppm; pyrophosphate at −5 ppm; and polyphosphate at −20 ppm. More information about these P forms and compound classes can be found in Condron et al. (2005), and more detailed information on the P forms that can be identified in the orthophosphate monoester and orthophosphate diester regions is discussed later in this chapter. Not all of these P forms will be found in every sample, and will vary with sample type and extractant. Figure 12.1 shows spectra from two different experiments. Sample A is a dairy manure sample extracted with sodium acetate buffer plus sodium dithionite (pH 5) from the study published in He et al. (2009), while sample B is a dairy feces sample extracted in sodium hydroxide plus ethylenediaminetetraacetic acid (NaOH-EDTA; pH 9), from the study published in Toor et al. (2005a). Note that the orthophosphate peak in each spectrum is truncated because the spectra have been enlarged to show the details of other peaks. Sample A shows prominent pyrophosphate and polyphosphate peaks compared with Sample B, while sample B shows more prominent phosphonate and orthophosphate diester peaks. These spectra show the range of P forms that have been detected in solution ^{31}P-NMR experiments on manure samples. The impact of sample preparation demonstrated here will be discussed later in the chapter (section 12.2.3.1).

Table 12.1. A summary of papers investigating some aspect of ^{31}P-NMR methodology for animal feces and manure samples. Papers are listed in chronological order.

Author	Species	Material[†]	Extractant	Parameter Tested
Leinweber et al., 1997	Chicken, swine	Manure (FD)	0.5 M NaOH, 0.1 M NaOH	Extractant concentration
Kemme et al., 1999	Chicken, swine	Feces, digesta, feed (FD)	0.75 M HCl (hot), EDTA	Quantification of inositol phosphates
Crouse et al., 2000	Turkey	Manure (AD)	0.25 M NaOH-0.05 M EDTA, Chelex	Temperature, pH during NMR experiment
Gigliotti et al., 2002	Swine	Lagoon slurry, centrifuged for supernatants	Acidified, resin separation (hydrophobic/hydrophilic)	P forms in dissolved organic matter
Toor et al., 2003, 2005 b	Dairy	Farm dairy effluent, leachate (soil column), FD	0.25 M NaOH-0.05 M Na$_2$EDTA	Concentrate and extract liquid effluent, leachate
Hansen et al., 2004	Dairy	Manure, dry pile and liquid lagoon, FD	0.25 M NaOH-0.05 M Na$_2$EDTA	NaOH-EDTA extractant on soils and manure
Turner, 2004	Swine, beef, chicken	Manure, FD	NaOH and EDTA, varying concentrations	Extractant concentration and extraction length
Turner and Leytem, 2004	Swine, beef, chicken	Manure, FD	Water, NaHCO$_3$, NaOH, HCl	P forms in sequential extractions
McDowell and Stewart, 2005a	Sheep, Deer, Dairy	Feces, fresh	Water	Water soluble P, no pH adjustment for P-NMR
McDowell and Stewart, 2005b	Sheep, deer, dairy	Feces, fresh and AD	0.25 M NaOH-0.05 M Na$_2$EDTA	Effects of drying on extracted P forms
McDowell et al., 2006	Sheep, deer, dairy	Manure, AD	0.25 M NaOH-0.05 M Na$_2$EDTA	Measurement of T$_1$, effects of delay time on quantification of P forms
Bol et al., 2006	Beef	Manure, OD	0.1 M NaOH-0.4 M NaF, dialysis	Extractant
He et al., 2007	Dairy, chicken	Feces (dairy), Manure (chicken), fresh and OD (65°C)	0.25 M NaOH-0.05 M Na$_2$EDTA, 1M HCl (chicken residues)	Wet and dry extraction, re-extraction of residues for unextracted P
McDowell et al., 2008	Dairy	Feed, forage, feces, fresh and OD (55°C)	Water, 0.012 M HCl or 0.25 M NaOH-0.05 M Na$_2$EDTA	Extractants (non-sequential), effects of drying

Author	Species	Material[†]	Extractant	Parameter Tested
Leytem et al., 2008c	Chicken	Ileal digesta, feces, manure, FD	0.5 M NaOH-0.05 M EDTA	^{31}P-NMR versus HPLC to quantify inositol phosphates
He et al., 2008	Chicken	Manure, AD	Water, NaHCO$_3$, NaOH, HCl	Sequential extraction with and without enzyme hydrolysis
He et al., 2009	Dairy	Manure, FD	Water, sodium acetate with dithionite (pH 5), 0.25 M NaOH-0.05 M Na$_2$EDTA	Extractants, solution versus solid-state P-NMR
Shafqat et al., 2009	Beef, swine, turkey	Manure	Water (discarded), 0.4 M NaOH, gel filtration to remove NaOH	Extractant
Ding et al., 2010	Swine	Manure, lagoon, FD	0.25 M NaOH-0.05 M Na$_2$EDTA, plus 8-hydoxyquinoliine to precipitate Fe and Mn after extraction	Post-extraction treatment of extracts

[†] Abbreviations refer to the nature of the extracted material: FD, freeze-dried; AD, air-dried; OD, oven-dried.

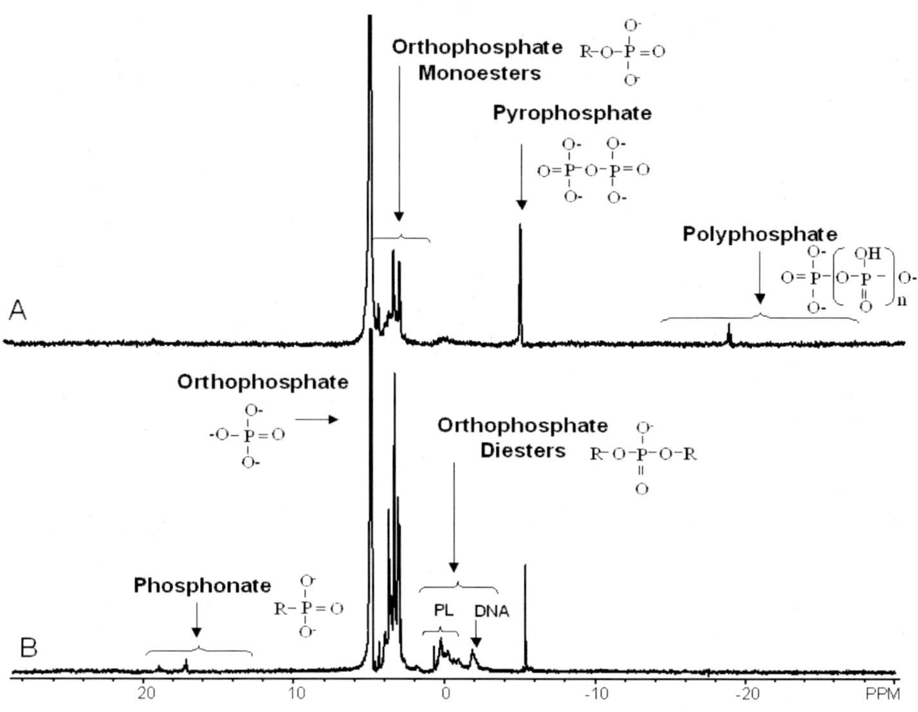

Figure 12.1. Spectra showing the range of P species in typical feces and manure samples. A: dairy manure extracted with sodium acetate plus sodium dithionite (pH 5). This is a previously unpublished spectrum from the study reported in He et al. (2009), 3 Hz line-broadening. B: dairy feces extracted with NaOH-EDTA, 1 Hz line-broadening. This is a previously unpublished spectrum from the study reported in Toor et al. (2005). PL is phospholipids.

Table 12.2. Papers characterizing P forms in poultry feces and manure samples with ^{31}P-NMR spectroscopy.

Author	Species	Material[†]	Extractant	Parameter Tested
Leinweber et al., 1997	Chicken,	Manure (FD)	0.5 M NaOH, 0.1 M NaOH	Extractant concentration
Kemme et al., 1999	Chicken,	Feces, digesta, feed (FD)	0.75 M HCl (hot), EDTA	Quantification of inositol phosphates
Crouse et al., 2000	Turkey	Manure (AD)	0.25 M NaOH-0.05 M EDTA, Chelex	Temperature, pH during NMR experiment
Turner, 2004	Chicken	Manure, FD	NaOH and EDTA, varying concentrations	Extractant concentration and extraction length
Turner and Leytem, 2004	Chicken	Manure, FD	Water, NaHCO$_3$, NaOH, HCl	P forms in sequential extractions
Maguire et al., 2004	Chicken and turkey	Manure, OD (40°C)	0.5 M NaOH-0.05 M EDTA	High and low P diets, feed additives (phytase, 25 OH-D3[‡])
McGrath et al., 2005	Chicken	Manure, OD (60°C)	0.5 M NaOH-0.05 M EDTA	High and low P diets, feed additives (phytase, 25 OH-D3[‡]), storage
Leytem et al., 2006	Chicken	Manure, FD or OD (55°C)	0.5 M NaOH-0.05 M EDTA	Diets with varying P and phytate, effects on P solubility
He et al., 2007	Chicken	Manure fresh and OD (65°C)	0.25 M NaOH-0.05 M Na$_2$EDTA, 1M HCl	Wet and dry extraction, re-extraction of residues for unextracted P
Leytem et al., 2007a	Chicken	Feed, feces	0.5 M NaOH-0.05 M EDTA	Changes in P forms from low-phytate barley
Ajiboye et al., 2007	Chicken	Manure, FD	Water, NaHCO$_3$, NaOH, HCl	Comparison of sequential extractions, P-XANES and P-NMR
Leytem et al., 2007c	Chicken	Ileal digesta, feces, FD	0.5 M NaOH-0.05 M EDTA	Effects of dietary available P, phytase and calcium on manure P
Leytem et al., 2008a	Chicken	Ileal digesta, feces, manure, FD	0.5 M NaOH-0.05 M EDTA	Effects of diets varying in cereal grain, P level and phytase addition
Leytem et al., 2008b	Chicken	Ileal digesta, feces, manure, FD	0.5 M NaOH-0.05 M EDTA	Effects of cereal grains varying in phytate and phytase content
Leytem et al., 2008c	Chicken	Ileal digesta, feces, manure, FD	0.5 M NaOH-0.05 M EDTA	^{31}P-NMR versus HPLC to quantify inositol phosphates
He et al., 2008	Chicken	Manure, AD	Water, NaHCO$_3$, NaOH, HCl	Sequential extraction with and without enzyme hydrolysis
Shafqat et al., 2009	Turkey	Manure	Water (discarded), 0.4 M NaOH, gel filtration to remove NaOH	Extractant

[†] Abbreviations refer to the nature of the extracted material: FD, freeze-dried; AD, air-dried; OD, oven-dried.
[‡] 25 OH-D3:25-hydroxycholecalciferol, a vitamin D$_3$ metabolite.
Papers are listed in chronological order.

12.2.3. Methodological Considerations for ^{31}P-NMR Experiments

This section of the chapter focuses on the requirements for a successful solution ^{31}P-NMR experiment to characterize P forms in animal feces and manure samples. To optimize these experiments, many parameters have been tested; these have been summarized in Table 12.1.

12.2.3.1. Sample Preparation

For solution ^{31}P-NMR experiments, solid samples must first be converted to solutions by extraction. Prior to extraction, it is important to consider the condition of the starting material. As Tables 12.1-12.4 show, samples used in the past have been extracted fresh (McDowell and Stewart, 2005a, b; He et al., 2007; McDowell et al., 2008; Hill and Cade-Menun 2009), air-dried (Crouse et al., 2000; McDowell and Stewart 2005 b; McDowell et al., 2006; He et al., 2008), frozen and subsequently freeze-dried (many papers, including Turner, 2004; Ajiboye et al., 2007; He et al., 2009) and oven-dried at various temperatures from 40-70°C (Maguire et al., 2004; Toor et al., 2005a; McGrath et al., 2005; Leytem et al., 2006; He et al., 2007; McDowell et al., 2008; Leytem and Thacker, 2008; Fuentes et al., 2009).

A few studies have investigated the effects of sample preparation on P forms. McDowell and Stewart (2005b) compared P forms in feces of dairy cattle, deer and sheep extracted fresh and dry, to see how P forms change when fresh feces naturally air-dry, as would occur in the field after deposition. They found that more P (~30%) was extracted from fresh than from dry feces with NaOH-EDTA. McDowell and Stewart (2005b) also observed that drying increased the relative percentage but not the concentration of orthophosphate extracted from feces, but decreased phospholipids and orthophosphate diesters. When He et al. (2007) compared extraction by NaOH-EDTA of fresh versus oven-dried samples, there was no significant difference in total P extracted from dairy feces (P recovery), but drying reduced P recovery from poultry manure (71% of total P was extracted with NaOH-EDTA from wet samples; 65% from dry samples). Drying increased the relative percentage of orthophosphate and phytate (salts of *myo*-inositol hexakisphosphate or *myo*-IP$_6$) in dairy feces, but decreased orthophosphate and did not change phytate in poultry manure extracts (He et al., 2007). McDowell et al. (2008) used water, NaOH-EDTA and dilute HCl to extract fresh and dry dairy feces from four different feeding regimes. In general, McDowell et al. (2008) noted that drying decreased total water-extractable P by about 19% but increased orthophosphate (8 %) and phytate (1.4%) extracted by NaOH-EDTA. However, the results of McDowell et al. (2008) varied among feeding regimes, making it difficult to generalize about the effects of drying on their samples.

No single study has compared the extraction of fresh, oven-dried, air-dried and freeze-dried samples for manure or feces. However, Cade-Menun et al. (2005) compared the extraction of three marine particulate samples that had been extracted with NaOH-EDTA fresh, after freezing (an important step for freeze-drying), after oven drying (40°C) or after storage for six months at 4°C. For all samples, the condition of the sample at extraction produced differences in P forms with ^{31}P-NMR. However, individual samples remained distinct from one another, suggesting that the natural differences in P forms among these samples were greater than the differences produced by sample preparation.

All of these results suggest that sample preparation prior to extraction will affect the P forms observed with ^{31}P-NMR. Thus, all samples in a single study need to be prepared the same way prior to extraction, and care should be taken when comparing results from studies with different preparation methods. For samples that are extracted field-moist, the concentration of the extractant may need to be increased to compensate for dilution by the sample moisture. Careful homogenization of samples prior to subsampling is also important, and is easier to do with air-dried, oven-dried and freeze-dried samples, which can be ground and sieved.

12.2.3.2. Extractants and Extraction Length

As Tables 12.1-12.4 show, a number of extractants have been used for solution ^{31}P-NMR of animal feces and manure samples. An ideal extractant will recover the majority of P from a sample with minimum degradation of P forms. The first studies of animal feces and manure used extractants that were developed and tested for soil extraction. Leinweber et al. (1997) used 0.1 M NaOH and 0.5 M NaOH, based on Newman and Tate (1980), the first paper to report the use of ^{31}P-NMR to characterize soil P forms. Crouse et al. (2000) and Hansen et al. (2004) used 0.25 M NaOH-0.05 M Na$_2$EDTA, based on Cade-Menun and Preston (1996). Other extractants that have been used include water (McDowell and Stewart 2005a; McDowell et al., 2008; He et al., 2009), various concentrations of hydrochloric acid (HCl; Kemme et al., 1999; He et al., 2007; McDowell et al., 2008), 0.1 M NaOH-0.4 M sodium fluoride (NaF) plus dialysis (Bol et al., 2006), sodium acetate buffer plus dithionite, pH 5 (He et al., 2009), sequential extraction with water, bicarbonate (NaHCO$_3$), NaOH and HCl (Turner and Leytem, 2004; Ajiboye et al., 2007; He et al., 2008), and a complicated procedure involving water (discarded), 0.4 M NaOH and gel filtration to remove NaOH (Shafqat et al., 2009). Many of these extractants were chosen because they are meaningful in other non-NMR uses. For example, water and dilute HCl are used to predict P loss in overland flow (McDowell et al., 2008) while sodium acetate buffer plus dithionite is known to have a high affinity for Fe-relevant P (He et al., 2006, 2009). Sequential fractionation, whereby P is extracted from manure and soils with increasingly strong solutions, is used to quantify pools of P with varying degrees of solubility (Turner and Leytem, 2004; Ajiboye et al., 2007). Water and NaHCO$_3$ are thought to remove the readily soluble P fraction, while the NaOH and HCl P fractions are considered to be poorly soluble (Turner and Leytem, 2004).

It is difficult to compare different extractants, because each removes different P concentrations and different P forms. Caution should be taken in interpreting the results from sequential extraction due to possible matrix effects, whereby the preceding extractant alters the sample in some way that affects what can be solubilized by the next extractant (e.g. precipitation), or due to hydrolysis of organic P forms by the harsher reagents (Turner and Leytem, 2004). Additional treatment such as dialysis following extraction may also affect the sample and subsequent NMR results. For example, dialysis following the 0.1 M NaOH-0.4 M NaF extraction used by Bol et al. (2006) removed orthophosphate, biasing the spectrum toward organic P forms. As such, dialysis may be useful to see other P forms in samples dominated by orthophosphate. However, undialyzed samples should also be analyzed to gain a complete picture of the relative proportions of all the P forms in a sample. Ding et al. (2010) added 8-hydroxyquiloline (8-HOQ) to a NaOH-EDTA extract of pig manure. This precipitated Fe and Mn, allowing better resolution of peaks that were overlapping in NaOH-EDTA alone. For the extraction procedure of Shafqat et al. (2009) the sample was extracted first with water, which was discarded. Next, the residue was extracted twice with 0.4 M NaOH, which was pooled. The NaOH extract was then subjected to gel filtration to remove the NaOH, eluting the column with demineralized water. Unfortunately, this produced spectra with broad peaks and very poor resolution, such that almost no useful information could be obtained from the spectra. This may have been due to the sample pH, or it may have been due to loss of P by discarding the water extract and NaOH fraction after gel filtration. This procedure is not recommended for future studies.

The most commonly used extractant for solution ^{31}P NMR studies of animal feces and manure is NaOH (0.5 M or 0.25 M) plus 0.05 M Na_2EDTA (Tables 12.1-12.4). The popularity of this extractant is due to the high recovery of total P from the combination of NaOH and EDTA (Turner, 2004) and chelation of paramagnetic ions (Fe, Mn) by EDTA, which reduces line-broadening and shortens relaxation times (see sections 12.2.1 and 12.2.3.4). This extractant also produces a sample close to the ideal pH for solution ^{31}P-NMR experiments (see 12.2.3.3).

Turner (2004) tested NaOH concentrations from 0.15 M to 0.5 M with various concentrations of Na_2EDTA. A concentration of at least 0.025 M Na_2EDTA was required for maximum recovery of total P, regardless of NaOH concentration, and the inclusion of Na_2EDTA significantly increased the recovery of P from animal manure compared with NaOH alone. Lower concentrations of NaOH reduced the resolution of ^{31}P-NMR spectra for swine and chicken manure, by extracting lower concentrations of P and higher concentrations of Fe and Mn. However, changing the concentration of NaOH had no effect on beef manure, possibly due to the increased viscosity of the beef manure extracts relative to those from swine and chicken (Turner, 2004). Although spectral resolution was improved by using 0.5 M NaOH-0.05 M Na_2EDTA, the relative percentages of phospholipids and polyphosphate were reduced compared with 0.25 M NaOH-0.05 M Na_2EDTA for all three manure types. Leinweber et al. (1997) also reported reduced proportions of orthophosphate diesters, including phospholipids, in 0.5 M NaOH extracts versus 0.1 M NaOH extracts. The choice of extractant NaOH concentration may therefore depend on the objective of the NMR experiment. If the objective is to quantify all P forms including orthophosphate diesters, then 0.25 M NaOH-0.05 M Na_2EDTA should be used. However, if the sole focus of the study is to quantify phytate in feces and manure from animals fed diets with a range of phytate concentrations (e.g. Leytem et al. 2006; Leytem et al. 2007a, b, c; Leytem et al., 2008a, b, c), then 0.5 M NaOH-0.05 M Na_2EDTA may be a better choice due to the improved resolution of peaks in the orthophosphate monoester region, although some of these peaks are more likely to result from degradation of orthophosphate diesters (Turner et al., 2003; Doolette et al., 2009).

Turner (2004) tested the effect of extraction length on P recovery. Because there was no significant difference in total P recovery between 4- and 16-h extraction times, Turner (2004) recommended using 4-h extractions to minimize the potential for degradation of alkali-labile P forms as noted by Turner et al. (2003). No ^{31}P-NMR analysis to determine the differences in P forms from different extraction times was reported by Turner (2004), or elsewhere in the literature. The studies listed in Tables 12.1-12.4 report extraction times from 4- to 16-h.

After extraction, the extracts must be concentrated in some way. The majority of the studies listed in Tables 12.1-12.4 used freezing and freeze-drying to concentrate extracts; the exception was Leinweber et al. (1997), who concentrated samples in a stream of N_2 at 40°C. Extracts of soil samples for ^{31}P-NMR have also been concentrated by rotary evaporation (Cade-Menun, 2005a, b), but this method has not been used for manure extracts. Cade-Menun (2005a, b) recommend using freeze-drying only, due to the risk of hydrolysis of P forms at the higher temperatures required by the other concentration methods (Cade-Menun et al., 2002). However, Cade-Menun et al. (2006) reported degradation of model P compounds, particularly polyphosphates, during freeze-drying at high pH and suggested reducing the pH before freeze-drying. This has not been tested for extraction of animal feces and manures.

12.2.3.3. Preparation of Samples for NMR Experiments

Concentrated samples must be redissolved for solution ^{31}P-NMR analysis. This can be done with a number of reagents, but must include an aliquot of deuterium oxide (D_2O), which is used for a signal lock in the spectrometer to detect and correct for minor fluctuations in B_0. Crouse et al. (2000) demonstrated that peak separation is best, and the position of the orthophosphate peak remains relatively constant, when the pH of the sample in the NMR tube is > 12. Thus, concentrated NaOH is usually added to increase pH. The remaining liquid used to redissolved samples is usually water or the NaOH-EDTA extracting solution (e.g. He et al., 2007, 2009). The total volume of sample required depends on the size of the probe and thus on the size of the NMR tube, and is about 1 mL for a 5-mm tube and about 3 mL for a 10-mm tube. Care should be taken to ensure that the sample is not too viscous after preparation, as this will increase line broadening. However, because one objective of sample extraction and concentration is to place as much P as possible in the NMR tube, care should be taken not to overly dilute the sample, either. Samples should be centrifuged and/or filtered before they are decanted into NMR tubes, because particulates in the sample will also increase line broadening.

For extractants other than NaOH-EDTA, it can be difficult to adjust the sample to pH > 12. As reported in McDowell et al. (2008), HCl extracts can form precipitates as the pH is increased. This can remove P, especially orthophosphate, from solution. Adjusting the pH to >12 is not recommended by all authors: McDowell and Stewart (2005a) recommend that extracts be analyzed at their extraction pH or adjusted to just above pH 7, to avoid hydrolysis or precipitation. However, because chemical shift is pH-dependent, this will cause the location of peaks to change, hampering peak identification (McDowell and Stewart, 2005a; He et al., 2008) and making comparisons among different studies difficult. Adjusting samples to a uniform pH is particularly important when multiple extractants are used in a single study, such as with sequential fractionation (Turner and Leytem, 2004; Ajiboye et al., 2007; He et al., 2008). In studies like these, it is very important to monitor P concentration in extracts before and after sample preparation to determine if P is lost through precipitation.

12.2.3.4. ^{31}P-NMR Experimental Parameters

The field strength (B_0) of the high field super-conduction magnets used for NMR studies is designated in terms of the frequency of the ^{1}H resonance. Those used for studies of animal manures and feces range from 300 MHz (121 MHz for ^{31}P; Shand et al., 2005) to 600 MHz (243 for ^{31}P; He et al., 2008). Because the spectral resolution or signal-to-noise (S/N) ratio will increase by $B_0^{3/2}$ as B_0 increases, an NMR experiment on a 600 MHz magnet should produce better S/N than one on a 300-MHz magnet using the same sample and experimental parameters.

A broadband probe that can be tuned to ^{31}P is also required. The S/N is proportional to the number of P nuclei within the sample tube; thus, a 10-mm tube is preferable to a 5-mm tube for ^{31}P-NMR experiments of environmental samples such as soil and manure where the natural P concentration is low. Because the sample volume is larger in a 10-mm tube, the number of scans required to achieve the same S/N is reduced by a factor of four. Thus, a sample run in a 10-mm probe may need only 2500 scans (about 3 h), while the same sample in a 5-mm probe may require 10,000 scans (about 12 h) to achieve the same S/N. Because many NMR facilities charge for NMR time by the hour, this can be an important cost difference. It will also affect the number of samples and replicates that can be analyzed for a given study. Replication is rare with studies of animal manures and feces using 5-mm probes (Turner, 2004), but is more common for studies using 10-mm probes (e.g. Toor et al., 2005a; He et al., 2007; Hill and Cade-Menun, 2009). Without replication, it is often difficult to tell if the differences observed among treatments are significant and meaningful. Shorter experiment times also reduce the potential for sample degradation that may occur at the high pH used for ^{31}P-NMR experiments on animal feces and manures. Of the studies listed in Table 12.1-12.4, experiments using 10-mm probes collected 256-8000 scans, with the exception of Ajiboye et al. (2007), who collected 10,000-33,000 scans. Experiments using 5-mm probes collected 1,500-33,000 scans. Studies using 5-mm probes rarely report the presence of phospholipids and other orthophosphate diesters, which are known to degrade at high pH more easily than other P forms (Turner et al., 2003). However, phospholipids and orthophosphate diesters are commonly reported for experiments using 10-mm probes, with the exception of most of the manures analyzed by Ajiboye et al. (2007). While it is possible that this is due to differences in extractants rather than degradation from long NMR experiments, it is worth exploring further in future experiments, and suggests that experiment times should be kept as short as possible to minimize potential degradation.

Once the sample is in the magnet, it must be locked, shimmed and tuned to optimize B_0. The temperature should also be regulated to 20-25°C to minimize degradation (Cade-Menun et al., 2002) and for optimal spectral resolution (Crouse et al., 2000). The spectral window or sweep width is set to 50 ppm, centered at 0 ppm using an external standard of 85% phosphoric acid. It should be noted that although the chemical shift of orthophosphate in the phosphoric acid standard is 0 ppm, the orthophosphate peak in samples adjusted to > pH 12 will be located at 5.5-6.0 ppm because chemical shift is pH-dependent and is also affected by sample chemistry such as salt concentration.

The acquisition time, or period over which data are recorded after the radio frequency pulse, is determined by the spectral width and the number of data points. Acquisition times in the experiments listed in Tables 12.1-12.4 range from 0.1 s (Leinweber et al., 1997; Gigliotti et al., 2002; Bol et al., 2006) to 1.99 s (McDowell and Stewart, 2005b). The length of time in which the radio frequency pulse excites the nuclei is the pulse width, and is measured in μs but is usually expressed in terms of pulse angles. Pulse angles used in the papers listed in Tables 12.1-12.4 are almost evenly split between 45° and 90°. Shorter pulse angles are sometimes used to decrease the delay times needed between pulses. However, this brings the risk that not all nuclei will be fully excited.

As previously discussed, delay times are important for quantitation, to ensure that all P nuclei have relaxed to equilibrium between radio frequency pulses. Delay times reported for the papers listed in Tables 12.1-12.4 range from 0.2 s (Leinweber et al., 1997; Gigliotti et al., 2002; Bol et al., 2006) to 8 s (McDowell and Stewart, 2005b; Fuentes et al., 2009). Although T_1 values will vary among samples depending on the nature of the sample and among P compounds within a sample, few studies of animal manures and feces report measuring T_1 or testing delay time length. A test sample from Shand et al. (2005) showed that changing the delay time from 2 s to 0.2 s or 4 s gave less than 5% change in the integrated peak areas.

McDowell and Stewart (2005a) report that unpublished T_1 tests on their samples confirmed that a total delay of 2.99 s (pulse delay of 1 s, acquisition time of 1.99s) was sufficient to meet T_1 requirements of most peaks in their samples except orthophosphate, with errors of less than 10% in peak areas. McDowell et al. (2006) conducted detailed T_1 experiments on soil and manure samples, and suggested that samples with very low concentrations of Fe and Mn, typical for many types of manure, could require delay times of up to 15 s. Ajiboye et al. (2007) also performed T_1 experiments on some of their samples with extremes in Fe content, and showed that the 2-s delay time used for their samples produced no significant differences in peak identification and quantitation when compared to a 20-s delay. Ideally, T_1 experiments should be conducted routinely on samples to determine the required delay times, but this isn't feasible for researchers with limited access to spectrometers. McDowell et al. (2006) suggest that T_1 can be estimated if the concentrations of P, Fe and Mn are known. They showed that T_1 is directly proportional to the ratio of P/(Fe+Mn) for most P forms, and provide a figure that can be used to estimate T_1 if this ratio is known for a sample. Once T_1 is calculated, from the ratio or through T_1 measurement, the delay time is calculated as 3-5 times T_1. Based on the T_1 measurements from McDowell et al. (2006) and Cade-Menun et al. (2002), it is unlikely that the results of experiments with delay times less than 2 s are quantitative (e.g. Leinweber et al., 1997; Gigliotti et al., 2002; Bol et al., 2006).

About half of the ^{31}P-NMR experiments listed in Tables 12.1-12.4 use proton decoupling to suppress nuclear Overhauser enhancements that can distort relative signal area. Although Turner et al. (2003) showed splitting of a phosphonate peak without decoupling and a single peak with decoupling, there has been no detailed investigation into the use of proton decoupling for ^{31}P-NMR studies of animal feces and manure. Some studies (e.g. He et al., 2008; Hill and Cade-Menun, 2009) use a small amount of spinning (15 Hz or less) to decrease broadening and improve S/N during NMR experiments. This is more likely to be useful for larger samples in 10-mm NMR tubes than for smaller volumes of samples in 5-mm tubes.

12.2.3.5. Sample Processing and Peak Identification

Once the NMR experiment has ended, the FID is Fourier-transformed to a spectrum with processing software. A number of different processing software packages are available. These include some produced by manufacturers of NMR spectrometers, such as TopSpin by Bruker and SpinSight by Varian, as well as software from other sources such as NUTS from Acorn NMR and MNova (formerly MestReC) from Mestrelab. All of these programs are capable of the 1-D NMR processing used for solution ^{31}P-NMR of animal wastes.

Processing software will also allow baseline correction to remove baseline artifacts. During processing, line broadening can be applied to the spectrum. This uses an exponential multiplication factor, in Hz, to reduce noise and improve S/N. If line broadening is too low, noise reduction is inefficient; if too high, useful data can be lost. It is often advisable to apply a higher line broadening (e.g. 7-10 Hz) when plotting the entire spectrum, and to process the spectrum again with less line broadening (e.g. 1-3 Hz) when examining smaller regions of the spectrum where there are more peaks closer together (e.g. the orthophosphate monoester region).

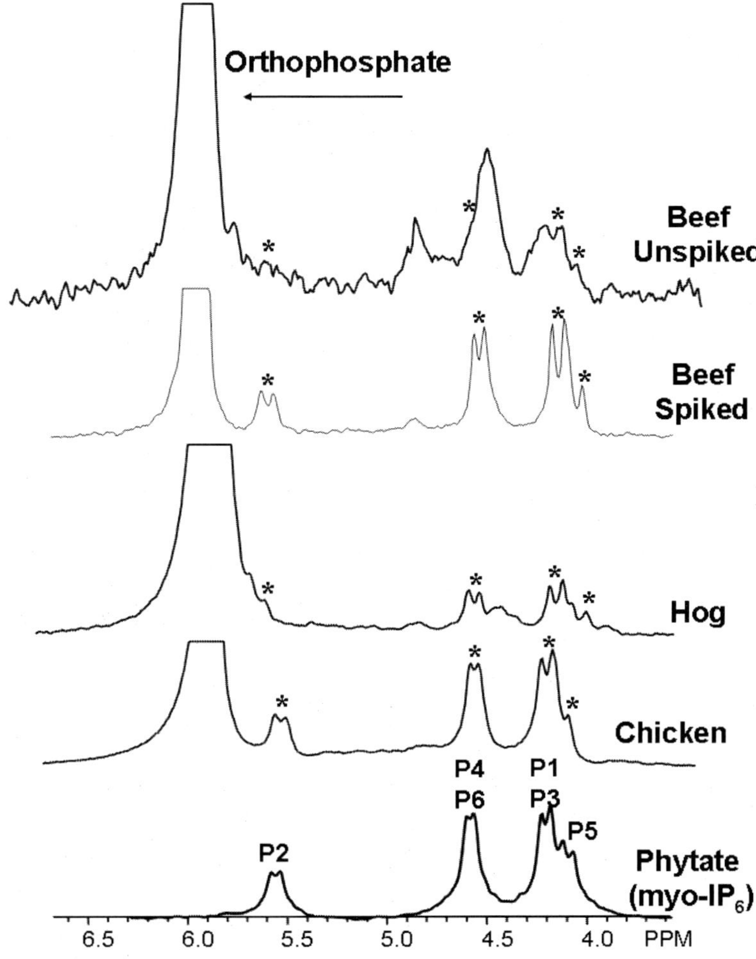

Figure 12.2. Enlargements of the 6.5-3.5 ppm region, which includes orthophosphate and orthophosphate monoesters, in manure from a number of species (unpublished data except the chicken spectrum, which is from the study published in Hill and Cade-Menun, 2009). The "Beef Spiked" sample was spiked with phytate (myo-IP$_6$) to confirm peak identities; more details are available in the text. An * indicates a phytate peak.

Both visual inspection and automated peak picking routines in processing software can be used to distinguish peaks from the background noise. The P forms represented by each peak are then identified in a number of ways. The most common is by reference to the literature, and is best done with papers using a similar matrix or extractant, such as Turner et al. (2003) for NaOH-EDTA extracts. There are no comparable studies examining chemical shifts of P forms in other extractants, such as HCl. Reference to the literature is more than adequate to identify peaks such as phosphonates, pyrophosphate and polyphosphate, which show little variation in chemical shift and are in regions of the spectrum with few other peaks. Peaks in the orthophosphate monoester region are more difficult to identify, because the peaks are very close together with only small differences in chemical shift. However, the identification of peaks in the orthophosphate monoester region has been the focus of the majority of the papers listed in Tables 12.1-12.4, because this is the region when the four peaks for phytate are located. Phytate is of particular concern in studies of animal feces and manure because it is the dominant P compound in most of the cereal grains fed to animals, but cannot be fully digested by most animals, and is only partially digested by monogastric animals, particularly poultry (Turner et al., 2002; Turner, 2004). Due to its high charge density, phytate applied to land from manure application binds tightly to soil, and is not readily available to plants or soil microbes (Condron et al., 2005). It does, however, contribute to eutrophication of nearby water bodies if transported there by runoff or erosion (Sims et al., 2000; Turner et al., 2002).

Figure 12.2 demonstrates the difficulties in peak identification in the orthophosphate monoester region. The four peaks for phytate (with splitting because proton decoupling was not used during ^{31}P-NMR analysis) are clearly visible in the spectrum of the phytate standard at the bottom. It is also fairly easy to identify these peaks in the samples of chicken and hog manure by referring to the peaks in the standard. However, it is much more difficult to distinguish these peaks in the top sample, labeled "Beef Unspiked". In order to identify the phytate peaks in this sample, a small aliquot of sample was removed from the NMR tube after ^{31}P-NMR analysis. This aliquot was replaced by the same volume of phytate dissolved in the NaOH-EDTA extraction solution, as per McDowell et al. (2007). Note that this differs from the spiking procedure of Smernik and Dougherty (2007) and Doolette et al. (2009), who added phytate directly to the sample without first removing anything. This spiked sample was analyzed again with ^{31}P-NMR, producing the spectrum labeled "Beef Spiked". This demonstrates that phytate is present in the sample, but is a much smaller proportion of total P than in the hog and chicken manure samples, and overlaps other peaks. It also demonstrates that other P forms are present in this sample in this region that could easily be mistaken for phytate without spiking. These forms are not present in the hog and chicken manure samples.

Phytate is best identified in samples by identifying and quantifying all four peaks, which should be present in a 1:2:2:1 ratio (e.g. Fig. 12.2). Turner (2004) suggested that phytate in a sample could be quantified by identifying the P2 peak and multiplying the area of that peak by six. This technique has been used by a number of researchers (e.g. Hansen et al., 2004; He et al., 2007; Fuentes et al., 2009). However, it presents difficulties if the P2 peak is not clearly visible, such as in the hog and unspiked beef samples in Figure 12.2.

The identity of other peaks in the orthophosphate monoester region is less certain. Peaks in this region may include sugar phosphates and mononucleotides naturally present in the sample or formed from the breakdown of orthophosphate diesters such as phospholipids and RNA (Turner et al., 2003; Bünemann et al., 2008; Doolette et al., 2009). They may also include lower inositol phosphates that may be naturally formed or produced during sample degradation (Turner and Richardson, 2004). The identity of the other compounds may be determined through spiking experiments analogous to that used to identify phytate (Doolette et al., 2009). The identification of lower inositol phosphates can be confirmed through hypobromite oxidation (Turner and Richardson, 2004), which oxidizes all organic matter except inositol phosphates. However, hypobromite oxidation has only been used on soil extracts to date; there are no published reports of the use of hypobromite oxidation to identify lower inositol phosphates in animal waste samples.

In most of the papers listed in Tables 12.1-12.4, the non-phytate peaks in the orthophosphate monoester region are described as "other monoesters', with peaks assigned to breakdown products of phospholipids (α- and β-glycerophosphate and phosphatidic acid), lower inositol phosphates or mononucleotides (e.g. Turner and Leytem, 2004; Toor et al., 2005a; He et al., 2007; Leytem et al., 2007a; McDowell et al., 2008; He et al., 2008; Hill and

Cade-Menun, 2009; Fuentes et al., 2009; He et al., 2009). In contrast, the non-phytate peaks in the orthophosphate monoester region are all described as lower esters of inositol phosphates in Leytem et al. (2007 b, c) and Leytem et al. (2008 a, b, c). No explanation is given for this shift in peak identification, and for all but Leytem et al. (2008c), no spectra are shown to allow the reader to compare the peaks in these samples to those from other studies. This change in identification may be based on Leytem et al. (2008c), which compared inositol phosphate concentrations determined by high performance liquid chromatography (HPLC) to those determined by ^{31}P-NMR, with the assumption that all orthophosphate monoesters were inositol phosphates. The authors report strong correlations of phytate concentrations determined by both methods, as well as strong correlations of total monoesters by ^{31}P-NMR and total inositol phosphates by HPLC. However, these results cannot be used to conclude that all orthophosphate monoesters in the samples were inositol phosphates. First, the HPLC samples were extracted in HCl, while the NMR samples were extracted in NaOH-EDTA. Other ^{31}P-NMR studies, such as McDowell et al. (2008), have demonstrated that these extractants remove different P pools. Second, only inositol phosphates were used as HPLC standards; no other compounds such as β-glycerophosphate were tested. Finally, the correlations of total monoester to total inositol phosphate shown in Leytem et al. (2008c) include and are strongly influenced by phytate, the identification of which is clear. No correlations without phytate are shown, which may have a weaker relationship. Based on these considerations, it is impossible to conclude from this study that only phytate and lower inositol phosphates are included in these or any other samples. Secondary methods such as spiking or bromination are required to confirm these peak identifications.

Table 12.3. Papers characterizing P forms in bovine feces and manure samples with ^{31}P-NMR. Papers are listed in chronological order.

Author	Species	Material[†]	Extractant	Parameter Tested
Toor et al., 2003, 2005 b	Dairy	Farm dairy effluent, leachate (soil column), FD	0.25 M NaOH-0.05 M Na$_2$EDTA	To develop method to concentrate and extract liquid effluent, leachate, runoff
Hansen et al., 2004	Dairy	Manure, dry pile and liquid lagoon, FD	0.25 M NaOH-0.05 M Na$_2$EDTA	NaOH-EDTA extractant on soils and manure
Turner, 2004	Beef	Manure, FD	NaOH and EDTA, varying concentrations	Extractant concentration and extraction length

Table 12.3. (Continued).

Author	Species	Material[†]	Extractant	Parameter Tested
Turner and Leytem, 2004	Beef	Manure, FD	Water, $NaHCO_3$, NaOH, HCl	P forms in sequential extractions
Toor et al., 2005a	Dairy	Feed, feces, manure, OD (55°C)	0.25 M NaOH-0.05 M Na_2EDTA	Effect of high and low P diets
McDowell and Stewart, 2005a	Dairy	Feces, fresh	Water	Water soluble P, no pH adjustment for P-NMR
McDowell and Stewart, 2005b	Dairy	Feces, fresh, AD	0.25 M NaOH-0.05 M Na_2EDTA	Effects of drying on extracted P forms
Leytem and Westermann, 2005	Dairy, beef	Liquid and solid manure, FD	0.5 M NaOH-0.05 M EDTA	Plant-available P from high and low P diets
McDowell et al., 2006	Dairy	Manure, AD	0.25 M NaOH-0.05 M Na_2EDTA	Measurement of T_1, effects on delay time on quantification of P forms
Bol et al., 2006	Beef	Manure, OD	0.1 M NaOH-0.4 M NaF, dialysis	Temporal changes in P forms with decomposition on soil surface
He et al., 2007	Dairy	Feces fresh and OD (65°C)	0.25 M NaOH-0.05 M Na_2EDTA	Wet and dry extraction
Ajiboye et al., 2007	Dairy	Manure, FD	Water, $NaHCO_3$, NaOH, HCl	Comparison of sequential extractions, P-XANES and P-NMR
McDowell et al., 2008	Dairy	Feed, forage, feces, fresh and OD (55°C)	Water, 0.012 M HCl or 0.25 M NaOH-0.05 M Na_2EDTA	Extractants (non-sequential), effects of drying
Fuentes et al., 2009	Dairy	Feces, raw and aerobically digested, OD (70°C)	0.25 M NaOH-0.05 M Na_2EDTA	Temporal effects of aerobic digestion in a reactor
He et al., 2009	Dairy	Manure, FD	Water, sodium acetate with dithionite (pH 5), 0.25 M NaOH-0.05 M Na_2EDTA	Extractants, solution versus solid-state P-NMR
Shafqat et al., 2009	Beef	Manure	Water (discarded), 0.4 M NaOH, gel filtration to remove NaOH	Extractant

[†] Abbreviations refer to the nature of the extracted material: FD, freeze-dried; AD, air-dried; OD, oven-dried.

The identification of peaks in the orthophosphate diester region between 2.5 and -1 ppm is generally simpler than for the peaks in the orthophosphate monoester region. The peak at approximately -0.7 ppm has been identified as DNA (Turner et al., 2003). Peaks for phospholipids have been identified between 1.8 and 0.8 ppm and for RNA at 0.5 ppm (Turner et al., 2003). However, Turner et al. (2003) demonstrated that phospholipids and RNA degrade rapidly at the alkaline pH range used for ^{31}P-NMR experiments, resulting in peaks in the orthophosphate monoester region from degradation products. Some authors attribute a peak at 1.5 ppm to teichoic acid (e.g. Leinweber et al., 1997; Bol et al., 2006), but there is some uncertainty about this peak assignment (Makarov et al., 2002). Doolette et al. (2009) also note that most phospholipids have low aqueous solubility. As such, phospholipids may be under-represented in studies using water-based extractants, which includes all the studies listed in Tables 12.1-12.4. This may account for some of the unextracted P in samples where P recovery is less than 100%.

Peak areas can be calculated using an integration routine in the processing software. First, the entire spectrum is integrated, and the integral is then divided into regions representing each peak. The height of the integral for each peak is determined as a percent of the total integral, which is the percent of total sample P for each P species. Nuclei that are equivalent magnetically, such as the two P nuclei in pyrophosphate (P_2O_7), will show only a single peak. It can be difficult to determine peak intensities if peaks overlap, but spectral deconvolution routines included with the processing software can separate broad peaks that may include more than one compound, such as those in the orthophosphate monoester region of solution (e.g. Hill and Cade-Menun, 2009).

12.3. Phosphorus Forms in Animal Feces and Manure Characterized with ^{31}P-NMR

The following section summarizes the published studies using ^{31}P-NMR to characterize P forms in animal feces and manures. In this section and in Tables 12.2-12.4, these studies are grouped by species studied. However, many of these studies analyzed more that one type of manure, resulting in some repetition.

12.3.1. Studies Investigating Poultry

The studies characterizing P forms in poultry feces and manure are listed, in chronological order, in Table 12.2. More studies have examined the P forms in poultry waste than for other animal groups. The majority of studies have extracted chicken manure (Leinweber et al., 1997; Turner, 2004; Turner and Leytem, 2004; Maguire et al., 2004; McGrath et al., 2005; Leytem et al., 2006; He et al., 2007; Ajiboye et al., 2007; Leytem et al., 2008b, c; He et al., 2008; Hill and Cade-Menun, 2009). Some studies have looked at P forms in chicken feces (Kemme et al., 1999; Leytem et al., 2007a, c; Leytem et al., 2008a, b, c), while a few also extracted samples of ileal digesta (Kemme et al., 1999; Leytem et al., 2007c; Leytem et al., 2008a, b, c).

Some studies of chicken manure have focused on methodology, either to improve the ^{31}P-NMR experiment or to extract more information about P forms. Leinweber et al. (1997) and Turner (2004) examined the effects of NaOH concentration on P forms, while Turner (2004) also experimented with Na₂EDTA concentration and length of extraction time. He et al. (2007) compared wet versus dry extraction, and the use of HCl to recover unextracted P from residues after NaOH-EDTA extraction. Turner and Leytem (2004), Ajiboye et al. (2007) and He et al. (2008) all used sequential extraction to assess the P forms in pools of different solubilities, with He et al. (2007) adding enzyme hydrolysis and Ajiboye et al. (2007) also using P X-ray Absorption Near-Edge Spectroscopy (P-XANES). Kemme et al. (1999) and Leytem et al. (2008c) both focused on improving the quantification of inositol phosphates.

McGrath et al. (2005) studied the effects of storage on chicken manure P forms. Compared to the initial samples, dry storage decreased total P by 6% but resulted in few changes in P forms, while wet storage increased total P by 10% and orthophosphate by 24% and decreased phytate by 22%. Hill and Cade-Menun (2009) compared fresh chicken manure to that which had been composted for three months, and observed only small changes in P forms. Some studies have compared P forms in waste from different types of animals. Leinweber et al. (1997) and Kemme et al. (1999) compared chicken and swine, Turner (2004) and Turner and Leytem (2004) used chicken, beef and swine manure, He et al. (2007) compared chicken and dairy waste, and Ajiboye et al. (2007) studied manure from chicken, dairy and swine. In general, chicken waste had a higher proportion of P as phytate than waste from other animal species.

The remaining studies of P forms in chicken samples have examined some type of dietary manipulation. These include diets high and low in total P, with the extra P in the high-P diets from non-phytate P sources such as mono-calcium phosphate (Maguire et al., 2004; McGrath et al., 2005), diets with varying phytate concentrations (Leytem et al., 2006; Leytem et al., 2007a; Leytem et al., 2008a, b), the effects of dietary additives such as phytase and 25-hydroxycholecalciferol (25OH-D₃), a vitamin D₃ metabolite (Maguire et al., 2004; McGrath et al., 2005; Leytem et al., 2007c) and the effect of dietary calcium (Ca, Leytem et al., 2007c). Animals fed diets with reduced non-phytate P had lower manure total P and less orthophosphate. Low-phytate grain reduced manure phytate, as did dietary phytase, but 25OH-D₃ had no effect on manure P speciation. An increase in dietary Ca increased manure phytate and reduced soluble orthophosphate.

There are only a handful of studies of P forms in turkey waste, all using manure. Crouse et al. (2000) utilized turkey manure in a study evaluating the effects of temperature and pH during ³¹P-NMR experiments. Maguire et al. (2004) compared forms from diets with high and low total P (with the increase to high P from non-phytate P sources), and the effects of the feed additives phytase and 25OH-D₃ on manure P forms. Low P diets reduced manure orthophosphate and phytase reduced manure phytate, but 25OH-D₃ had little effect on P speciation. Shafqat et al. (2009) compared P forms in turkey manure to those in beef and swine manure. However, the poor quality of their spectra makes it difficult to detect significant differences among samples.

12.3.2. Studies Investigating Beef and Dairy Cattle

Table 12.3 lists the studies characterizing P forms in beef and dairy cattle, in chronological order. Studies of cattle have speciated P in farm dairy effluent (Toor et al., 2003, 2005b), dairy manure (Hansen et al., 2004; Leytem and Westermann, 2005; Toor et al., 2005a; McDowell et al., 2006; Ajiboye et al., 2007; He et al., 2009), dairy feces (Toor et al., 2005a; McDowell et al., 2005a, b; He et al., 2007; McDowell et al., 2008; Fuentes et al., 2009) and beef manure (Turner, 2004; Turner and Leytem, 2004; Leytem and Westermann,

2005; Bol et al., 2006; Shafqat et al., 2009). Some of these studies have focused on methodology: Toor et al. (2003, 2005b) concentrated liquid farm dairy effluent for ^{31}P-NMR analysis; Turner (2004) examined extractants and extraction times; McDowell et al. (2005a) characterized water-soluble P without pH adjustment; Turner and Leytem (2004) and Ajiboye et al. (2007) used sequential extractions; McDowell and Steward (2005a) and He et al. (2007) compared wet and dry extraction; McDowell et al. (2006) measured T_1; and McDowell et al. (2008) and He et al. (2009) tested different extractants.

The effects of high and low P diets on P forms in feed, feces and manure were examined by Toor et al. (2005a) and in liquid and solid manure by Leytem and Westermann (2005), both showing that reducing dietary total P will reduce manure total P and orthophosphate content. In contrast to wastes from other species, beef and dairy manure had higher orthophosphate content and lower phytate content, particularly when compared with poultry manure (Turner 2004; Turner and Leytem, 2004; Ajiboye et al., 2007; He et al., 2007).

12.3.3. Studies Investigating Swine

The papers characterizing P in swine waste are listed in Table 12.4. Using manure, Leinweber et al. (1997) and Turner (2004) tested different concentrations of NaOH, while Turner (2004) also tested different concentrations of EDTA and different extraction lengths. Turner and Leytem (2004) and Ajiboye et al. (2007) used sequential extraction to determine P forms in pools of different solubilities, with Ajiboye also including P-XANES. Kemme et al. (1999) extracted feces, digesta and feed to determine phytate content, Leytem and Thacker (2008) speciated P in feces, and Gigliotti et al. (2002) characterized P forms in hydrophobic and hydrophilic dissolved organic matter. Leytem and Westermann (2005) and Leytem and Thacker (2008) tested the effects of high- and low-phytate diets on P forms in swine waste. Although monogastric animals, the spectra of swine fed high-phytate diets were predominantly orthophosphate, indicating that swine are able to digest dietary phytate, unlike poultry.

12.3.4. Studies Investigating Sheep and Deer

As listed in Table 12.4, solution ^{31}P-NMR spectroscopy has been used to characterize P forms in sheep feces (Shand et al., 2005; McDowell and Stewart, 2005a, b), sheep manure (McDowell et al., 2006), sheep diet, rumen, duodenum and rectum samples (Leytem et al., 2007b), deer feces (McDowell and Stewart, 2005a, b) and deer manure (McDowell et al., 2006). Several of these studies have focused on improving various aspects of methodology for ^{31}P-NMR, such as water extraction without adjusting the sample pH to over 12 (McDowell and Stewart, 2005a), testing the effects of extracting fresh versus dried samples (McDowell and Stewart, 2005b), and measuring the effects of delay times on P quantitation (McDowell et al., 2006). Shand et al. (2005) examined the changes in P forms over time as sheep manure sat on the soil surface in a pasture over 84 days. Orthophosphate was the dominant P form in most spectra, with an increase in pyrophosphate and DNA over time. Leytem et al. (2007b) examined P forms in rumen, duodenum and rectal samples from sheep fed high- and low-phytate barley. Small amounts of phytate were found in feces of sheep fed high-phytate grain, while no phytate was found with low-phytate grain. Other P species in feces included phosphate diesters, pyrophosphate and phosphonates.

12.4. FUTURE RESEARCH NEEDS

A quick glance at Tables 12.1-12.4 will easily reveal that there is no common methodology among the various research groups using solution ^{31}P-NMR to characterize P forms in animal wastes. However, as the results discussed in the methodology section in the first part of the chapter show, the P forms observed in a ^{31}P-NMR experiment will be directly affected by a number of methodological factors such as sample preparation prior to extraction (fresh, freeze-dried, air-dried, over-dried), and by the extractant used. Even for a single factor such as oven drying, there isn't agreement on the temperature to use. These methodological discrepancies hamper comparisons across research groups, and prevent drawing anything but broad conclusions about P forms in animal waste. The effects of other factors require further testing, particularly the effect of long experiments collecting high numbers of scans, to determine if the observed results contain experimental artifacts such as increased phospholipid degradation. More testing to determine T_1, particularly with different extractants, is required to ensure that results are truly quantitative. And the identification of peaks, particularly those in the orthophosphate monoester region, requires more research using secondary techniques such as spiking and bromination.

Table 12.4. Papers characterizing P forms in swine, sheep and deer feces and manure samples with ^{31}P-NMR.

Author	Species	Material†	Extractant	Parameter Tested
Leinweber et al., 1997	Swine	Manure (FD)	0.5 M NaOH, 0.1 M NaOH	Extractant concentration
Kemme et al., 1999	Swine	Feces, digesta, feed (FD)	0.75 M HCl (hot), EDTA	Quantification of inositol phosphates
Gigliotti et al., 2002	Swine	Lagoon slurry, centrifuged	Acidified, resin separation	P forms in hydrophobic and hydrophilic dissolved organic matter
Turner, 2004	Swine	Manure, FD	NaOH and EDTA, varying concentrations	Extractant concentration and extraction length
Turner and Leytem, 2004	Swine	Manure, FD	Water, NaHCO₃, NaOH, HCl	P forms in sequential extractions
Leytem and Westermann, 2005	Swine	Liquid and solid manure, FD	0.5 M NaOH-0.05 M EDTA	Plant-available P from high and low P diets
Ajiboye et al., 2007	Swine	Manure, FD	Water, NaHCO₃, NaOH, HCl	Comparison of sequential extractions, P-XANES and P-NMR
Leytem and Thacker, 2008	Swine	Feces, OD (66°C)	0.5 M NaOH-0.05 M EDTA	Effects of cereals with a range of phytate concentrations
Shafqat et al., 2009	Swine	Manure	Water (discarded), 0.4 M NaOH, gel filtration to remove NaOH	Extractant

Author	Species	Material[†]	Extractant	Parameter Tested
Ding et al., 2010	Swine	Manure, lagoon, FD	0.25 M NaOH-0.05 M Na$_2$EDTA, plus 8-hydoxyquinoliine to precipitate Fe and Mn after extraction	Post-extraction treatment of extracts
Shand et al., 2005	Sheep	Feces, FD	0.25 M NaOH-0.05 M EDTA	Temporal changes in P forms with decomposition on soil surface
McDowell and Stewart, 2005a	Sheep, deer	Feces, fresh	Water	Water soluble P, no pH adjustment for P-NMR
McDowell and Stewart, 2005b	Sheep, deer	Feces, fresh, AD	0.25 M NaOH-0.05 M Na$_2$EDTA	Effects of drying on extracted P forms
McDowell et al., 2006	Sheep, deer	Manure, AD	0.25 M NaOH-0.05 M Na$_2$EDTA	Measurement of T_1, effects on delay time on quantification of P forms
Leytem et al., 2007b	Sheep	Diet, rumen, duodenum and rectum samples, FD	0.5 M NaOH-0.05 M EDTA	High and low phytate barley

[†] Abbreviations refer to the nature of the extracted material: FD, freeze-dried; AD, air-dried; OD, oven-dried.

Papers are listed in chronological order.

Future studies will also need to include replication. Most studies collect replicates from the field, but use only one sample or a composite sample for NMR analysis. This prevents statistical analysis of the forms determined by ^{31}P-NMR, and ultimately limits the usefulness of the data. Although ^{31}P-NMR can be expensive, budgets must be planned such that field replicates can be analyzed.

Many of the main dietary aspects of animal waste, linking dietary P forms and dietary additives to P forms in feces and manure, have been addressed by previous studies. However, the introduction of new feed products will require more research. For example, waste products from biofuel production such as dried distillers grain are now fed to animals, but the P forms that these contain, and their effects on animal waste, is unknown. More research is needed into the effects of manure storage and handling, under a broader range of environmental conditions such as heavy rainfall, and extreme temperatures. The mobility of various P forms under different environmental conditions or when dissolved or in particulate form is also unknown, as are the links to the microbial factors transforming P in animal wastes in the environment.

12.5. CONCLUSION

Solution ^{31}P-NMR spectroscopy has been used to characterize the P forms in feed, ileal digesta, feces and manure from a number of animal species, including beef and dairy cattle, chickens, turkeys, swine, sheep and deer. A variety of P forms have been detected, including orthophosphate, pyrophosphate, polyphosphate, phosphonates, orthophosphate monoesters including phytate, and orthophosphate diesters including DNA. Dietary P, particularly phytate, dietary additives such as phytase, and manure storage and decomposition can all affect P forms. In addition, the forms and concentration of P in a given sample will also be affected by methodology such as sample condition at time of extraction (wet versus dry), extractant, and NMR parameters, making it very difficult to compare the results from different research projects and indicating that further standardization of methods is needed.

REFERENCES

Ajiboye, B., O.O. Akinremi, Y. Hu and D.N. Flaten. 2007. Phosphorus speciation of sequential extracts of organic amendments using nuclear magnetic resonance and x-ray absorption near-edge structure spectroscopies. *J. Environ. Qual.* 36:1563-1576.

Bol, R., W. Amelung and L. Haumaier. 2006. Phosphorus-31-nuclear magnetic-resonance spectroscopy to trace organic dung phosphorus in a temperate grassland soil. *J. Plant Nutr. Soil Sci.* 169:69-75.

Bünemann, E.K., R.J Smernik, A.L. Doolette, P. Marschner, R. Stonor, S.A. Wakelin and A.M. McNeill. 2008. Forms of phosphorus in bacteria and fungi isolated from two Australian soils. *Soil Biol. Biochem.* 40, 1908-1915.

Cade-Menun, B.J. 2005a. Characterizing phosphorus in environmental and agricultural samples by ^{31}P nuclear magnetic resonance spectroscopy. *Talanta* 66, 359-371.

Cade-Menun, B.J. 2005b. Using phosphorus-31 nuclear magnetic resonance spectroscopy to characterize phosphorus in environmental samples. In: B.L. Turner, E. Frossard and D. Baldwin, eds. *Organic Phosphorus in the Environment.* CABI Publishing. pp 21-44.

Cade-Menun, B.J., and C.M. Preston. 1996. A comparison of soil extraction procedures for ^{31}P NMR spectroscopy. *Soil Sci.* 161: 770-785.

Cade-Menun, B.J., J.A. Navaratnam and M.R. Walbridge. 2006. Characterizing dissolved and particulate phosphorus in water with ^{31}P nuclear magnetic resonance spectroscopy. *Environ. Sci. Technol.* 40:7874-7880.

Cade-Menun, B. J., C. W. Liu, R. Nunlist, and J. G. McColl. 2002. Soil and litter phosphorus-31 nuclear magnetic resonance spectroscopy: Extractants, metals, and phosphorus relaxation times. *J. Environ. Qual.* 31:457-465.

Cade-Menun, B.J., C.R. Benitez-Nelson, P. Pellechia and A. Paytan. 2005. Refining ^{31}P nuclear magnetic resonance spectroscopy for marine particulate samples: Storage conditions and extraction recovery. *Mar. Chem.* 97:293-306.

Canet, D. 1996. *Nuclear Magnetic Resonance: Concepts and Methods.* John Wiley & Sons. New York. 270 pp.

Condron, L.M., B.L. Turner and B.J. Cade-Menun. 2005. Chemistry and dynamics of soil organic phosphorus. pp. 87-121. In: J.T. Sims and A.N. Sharpley, eds. *Phosphorus, Agriculture and the Environment. Monograph no 46.* Soil Science Society of America. Madison, WI.

Crouse, D.A., H. Sierzputowska-Gracz, and R. Mikkelsen. 2000. Optimization of sample pH and temperature for phosphorus-31 nuclear magnetic resonance spectroscopy of poultry manure extracts. *Commun. Soil Sci. Plant Anal.* 31:229-240.

Ding, S., D. Xu, B. Li, C. Fan, and C. Zhang. 2010. Improvement of ^{31}P NMR spectral resolution by 8-hydroxyquinoline precipitation of paramagnetic Fe and Mn in environmental samples. *Environ. Sci. Technol.* 44:2555-2561.

Doolette, A.L., Smernik, R.J., Dougherty, W.J. 2009. Spiking improved solution phosphorus-31 nuclear magnetic resonance identification of soil phosphorus compounds. *Soil Sci. Soc. Am. J.* 73, 919-927.

Fuentes, B., M. Jorquera, and M. de la Luz Mora. 2009. Dynamics of phosphorus and phytate-utilizing bacteria during aerobic degradation of dairy cattle dung. *Chemosphere* 74:325-331.

Gigliotti, G., K. Kaiser, G. Guggenberger, and L. Haumaier. 2002. Differences in the chemical composition of dissolved organic matter from waste material of different sources. *Biol. Fertil. Soils* 36:321-329.

Hansen, J.C., B. J. Cade-Menun, and D.G. Strawn. 2004. Phosphorus speciation in manure-amended alkaline soils. *J. Environ. Qual.* 33:1521-1527.

He, Z., T.H. Dao and C.W. Honeycutt. 2006. Insoluble Fe-associated inorganic and organic phosphates in animal manure and soil. *Soil Sci.* 171:117-126.

He, Z., C.W. Honeycutt, B.J. Cade-Menun, Z.N. Senwo and I.A. Tazisong. 2008. Phosphorus in poultry litter and soil: enzymatic and nuclear magnetic resonance characterization. *Soil Sci. Soc. Am. J.* 72:1425-1433.

He, Z., C.W. Honeycutt, T.S. Griffin, B.J. Cade-Menun, P. Pellechia and Z. Dou. 2009. Phsophorus forms in conventional and organic dairy manure identified by solution and solid state P-31 NMR spectroscopy. *J. Environ. Qual.* 38:1909-1918.

He, Z., B.J. Cade-Menun, G.S. Toor, A.M. Fortuna, C.W. Honeycutt and J.T. Sims. 2007. Comparison of phosphorus forms in wet and dried animal manures by solution phosphorus-31 nuclear magnetic resonance spectroscopy and enzymatic hydrolysis. *J. Environ. Qual.* 36:1086-1095.

Hill, J. E. and Cade-Menun, B. J. 2009. Phosphorus-31 nuclear magnetic resonance spectroscopy transect study of poultry operations on the Delmarva Peninsula. *J. Environn Qual.* 38:130-138.

Kemme, P.A., A. Lommen, L.H. De Jonge, J.D. Van der Klis, A.W. Jongbloed, Z. Mroz and A.C. Beynen. 1999. Quantification of inositol phosphates using ^{31}P nuclear magnetic resonance spectroscopy in animal nutrition. *J. Agric. Food Chem.* 47:5116-5121.

Leinweber, P., L. Haumaier and W. Zech. 1997. Sequential extractions and ^{31}P-NMR spectroscopy of phosphorus forms in animal manures, whole soils and particle-size separates from a densely populated livestock area in northwest Germany. *Biol. Fertil. Soils* 25:89-94.

Leytem, A.B., and P.A. Thacker. 2008. Fecal phosphorus excretion and characterization from swine fed diets containing a variety of cereal grains. *J. Anim. Vet. Adv.* 7:113-120.

Leytem, A.B., and D.T. Westermann. 2005. Phosphorus availabiltiy to barley from manures and fertilizers on a calcareous soil. *Soil Sci.* 170:401-412.

Leytem, A.B., P.A. Thacker and B.L. Turner. 2007a. Phosphorus characterization in feces from broiler chicks fed low-phytate barley diets. *J. Sci. Food Agric.* 87:1495-1501.

Leytem, A.B., G.P. Widyaratne, and P.A. Thacker. 2008a. Phosphorus utilization and characterization of ileal digesta and excreta from broiler chickens fed diets varying in cereal grain, phosphorus level, and phytase addition. *Poultry Sci.* 87:2466-2476.

Leytem, A.B., B.P. Willing and P.A. Thacker. 2008b. Phytate utilization and phosphorus excretion by broiler chickens fed diets containing cereal grains varying in phytate and phytase content. *An. Feed Sci. Technol.* 146:160-168.

Leytem, A.B., D.R. Smith, T.J. Applegate and P.A. Thacker. 2006. The influence of manure phytic acid on phosphorus solubility in calcareous soils. *Soil Sci. Soc. Am. J.* 70:1629-1638.

Leytem, A.B., J.B. Taylor, V. Raboy and P.W. Plumstead. 2007b. Dietary low-phytate mutant-M 955 barley grain alters phytate degradation and mineral digestion in sheep fed high-grain diets. *An. Feed Sci. Technol.* 138:13-28.

Leytem, A.B., P. Kwanyuen, P.W. Plumstead, R.O. Maguire and J. Brake. 2008c. Evaluation of phosphorus characterization in broiler ileal digesta, manure and litter samples: ^{31}P-NMR vs. HPLC. *J. Environ. Qual.* 37:494-500.

Leytem, A.B., P.W. Plumstead, R.O. Maguire, P. Kwanyuen and J. Brake. 2007c. What aspect of dietary modification in broilers controls litter water-soluble phosphorus: dietary phosphorus, phytase or calcium? *J. Environ. Qual.* 36:453-463.

Maguire, R.O., J.T. Sims, W.W. Saylor, B.L. Turner, R. Angel and T.J. Applegate. 2004. Influence of phytase addition to poultry diets on phosphorus forms and solubility in litters and amended soils. *J. Environ. Qual.* 33:2306-2316.

Makarov, M.I., L. Haumaier and W. Zech. 2002. Nature of soil organic phosphorus: an assessment of peak assignments in the diester region of ^{31}P NMR spectra. *Soil Biol. Biochem.* 34:1467-1477.

McDowell, R.W., and I. Stewart. 2005a. Phosphorus in fresh and dry dung of grazing dairy cattle, deer, and sheep: sequential fraction and phosphorus-31 nuclear magnetic resonance analyses. *J. Environ. Qual.* 34:598-607.

McDowell, R.W., and I. Stewart. 2005b. Peak assignments for phosphorus-31 nuclear magnetic resonance spectroscopy in pH range 5-13 and their application in environmental samples. *Chem. Ecol.* 21:211-226.

McDowell, R.W., B. Cade-Menun, and I. Stewart. 2007. Organic P speciation and pedogenesis: analysis by ^{31}P nuclear magnetic resonance spectroscopy. *Eur. J. Soil Sci.* 58:1348-1357.

McDowell, R.W., I. Stewart and B.J. Cade-Menun. 2006. An examination of spin-lattice relaxation times for analysis of soil and manure extracts by liquid state phosphorus-31 nuclear magnetic resonance spectroscopy. *J. Environ. Qual.* 35:293-302.

McDowell, R.W., Z. Dou, J.D. Toth, B.J. Cade-Menun, P.J.A. Kleinman, K. Soder and L. Saporito. 2008. A comparison of phosphorus speciation and potential bioavailability in feed and feces of different dairy herds using ^{31}P nuclear magnetic resonance spectroscopy. *J. Environ. Qual.* 37:741-752.

McGrath, J.M., J.T. Sims, R. O. Maguire, W.W. Saylor, C.R. Angel and B.L. Turner. 2005. Broiler diet modification and litter storage: impacts on phosphorus in litters, soils and runoff. *J. Environ. Qual.* 34:1896-1909.

Newman, R.H., and K.R. Tate. 1980. Soil phosphorus characterization by ^{31}P-nuclear magnetic resonance. *Commun. Soil Sci. Plant Anal.* 11:835-842.

Shafqat, M.N., G.M. Pierzynski and K. Xia. 2009. Phosphorus source effects on soil organic phosphorus: A ^{31}P NMR study. *Commun. Soil Sci. Plant Anal.* 40:1722-1746.

Shand, C.A., G. Coutts, S. Hillier, D.G. Lumsdon, A. Chudek and I. Eubeler. 2005. Phosphorus composition of sheep feces and changes in the field determined by ^{31}P NMR spectroscopy and XRPD. *Environ. Sci. Technol.* 39:9205-9210.

Sims, J.T., A.C. Edwards, O.F. Schoumans and R.R. Simard. 2000. Integrating soil phosphorus testing into environmentally based agricultural management practices. *J. Environ. Qual.* 64:525-540.

Smernik, R.J., and W.J. Dougherty. 2007. Identification of phytate in phosphorus-31 nuclear magnetic resonance spectra – the need for spiking. *Soil Sci. Soc. Am. J.* 71:1045-1050.

Toor, G.S., B.J. Cade-Menun and J.T. Sims. 2005a. Establishing a linkage between phosphorus forms in dairy diets, feces and manures. *J. Environ. Qual.* 34:1380-1391.

Toor, G.S., L.M. Condron, B.J. Cade-Menun, H.J. Di and K.C. Cameron. 2005b. Preferential phosphorus leaching from an irrigated grassland soil. *Eur. J. Soil Sci.* 56:155-167.

Toor, G.S., L.M. Condron, H.J. Di, K.C. Cameron and B.J. Cade-Menun. 2003. Characterization of organic phosphorus in leachate from a grassland soil. *Soil Biol. Biochem.* 35:1317-1323.

Turner, B.L. 2004. Optimizing phosphorus characterization in animal manures by solution phosphorus-31 nuclear magnetic resonance spectroscopy. *J. Environ. Qual.* 33:757-766.

Turner, B.L, and A.B. Leytem. 2004. Phosphorus compounds in sequential extracts of animal manures: chemical speciation and a novel fractionation procedure. *Environ. Sci. Technol.* 38:6101-6106.

Turner, B.L., and A.E. Richardson. 2004. Identification of scyllo-inositol phosphates in soil by solution phosphorus-31 nuclear magnetic resonance spectroscopy. *Soil Sci. Soc. Am. J.* 68:802-808.

Turner, B.L., Mahieu, N., and L.M. Condron. 2003. Phosphorus-31 nuclear magnetic resonance spectral assignments of phosphorus compounds in soil NaOH-EDTA extracts. *Soil Sci. Soc. Am. J.* 67:497-510.

Turner, B. L., M.J. Paphazy, P.M. Haygarth, and I.D. McKelvie, I. D. 2002. Inositol phosphates in the environment. *Phil. Trans. R. Soc. London Ser.* B. 357:449-469.

Wilson, M.A. 1987. *NMR Techniques and Applications in Geochemistry and Soil Chemistry.* Pergamon Press. New York. 53 pp.

In: Environmental Chemistry of Animal Manure
Editor: Zhongqi He

ISBN: 978-1-62808-641-6
© 2013 Nova Science Publishers, Inc.

Chapter 13

METAL SPECIATION OF PHOSPHORUS DERIVED FROM SOLID STATE SPECTROSCOPIC ANALYSIS

Olalekan O. Akinremi[1,], Babasola Ajiboye[2] and Zhongqi He[3]*

13.1. INTRODUCTION

Whereas solution-based characterization provides knowledge on manure P solubility and forms (Chapters 10-12), solid-state techniques are more suitable to investigate metal-P interaction and/or metal species of P compounds. These techniques include, but are not limited, to Fourier-transform infrared (FT-IR) spectroscopy (Arai and Sparks, 2001; Bakhmutova-Albert et al., 2004; He et al., 2006), scanning electron microscopy with energy dispersive X-ray (SEM-EDS) spectroscopy (Cooperband and Good, 2002; Seaman et al., 2003; Massey et al., 2010), powder X-ray diffraction (XRD) analysis (Bakhmutova-Albert et al., 2004; Huang and Shenker 2004; Massey et al., 2010), synchrotron radiation based X-ray absorption near edge structure (XANES) spectroscopy (Peak et al., 2002; Ajiboye et al., 2007a; 2007b), and solid state ^{31}P nuclear magnetic resonance (NMR) spectroscopy (Hunger et al., 2004; Jayasundera et al., 2005). The objective of this chapter is to provide the current status of solid state XANES and ^{31}P NMR spectroscopic methods to identify metal species of organic and inorganic P in manure. Such characterization provides an increasing understanding of manure P release mechanism in soil and allows us to better predict the potential of P loss following manure addition to agricultural soils.

---

[*] Corresponding Author: akinremi@cc.umanitoba.ca
[1] Department of Soil Science University of Manitoba, Winnipeg, Manitoba, R3T 2N2, Canada
[2] Soil Science, School of Agriculture, Food and Wine, Waite Campus, the University of Adelaide, Adelaide SA 5005 Australia
[3] USDA-ARS, New England Plant, Soil, and Water Laboratory, Orono, ME 04469, USA

13.2. X-RAY ABSORPTION SPECTROSCOPY

13.2.1. Principles of the Method

X-ray absorption spectroscopy (XAS) is one of the synchrotron radiation-based analytical techniques that are now used in the field of environmental sciences. Synchrotron radiation is an extremely bright electromagnetic radiation that is generated when electron bunches are accelerated at approximately the speed of light around a circular path using electro-magnets and radio frequency waves. When bending magnets, as the electro-magnets are called, alter the course of the accelerated electrons, a natural phenomenon occurs in which a very brilliant and highly focused light is emitted tangentially to the circular orbit. The full spectrum of the emitted light is channelled to the beam lines where the desired wavelengths/energies are selected for different types of experiments (Sham and Rivers 2002). For example, the wavelength for XAS experiments are in the X-ray region ($10^{-2} - 10^{-9}$ m) and are often expressed as energy, E (in electron volts, eV) according to equation [13.1]

$$E = \frac{hc}{\lambda}$$

(13.1)

where h is Planck's constant ($\sim 4.1 \times 10^{-15}$ eV·s) and c, the speed of light ($\sim 3.0 \times 10^{-8}$ m s^{-1}) and λ is the wavelength (m). Based on the energy of the incident X-ray photons, the X-ray region is also divided into soft, tender/intermediate, and hard X-ray. The monochromator, equipped with crystals, is used to diffract the X-ray beam to create a continuous flux radiation over an energy range depending on the crystal cut. Recent advances in X-ray optics, creating very high photon flux, and X-ray detection systems capable of measuring low concentration of the element of interest in a matrix have made XAS suitable for many environmental applications (Kelly et al., 2008; Lombi and Susini, 2009).

In principle, XAS analysis involves the interaction of X-rays with atoms and molecules. This principle was covered extensively in a recent review of the basic principles and methods of XAS (Kelly et al., 2008). From the atomic model theories, electrons surround a nucleus in an atom and move in quantized orbits with discrete energies. When an atom absorbs X-ray photons, a core level electron (from K, L, or M orbitals) is ejected, the atom becomes ionized and is promoted to an excited state. The ejection of the core level electron into the continuum (unoccupied electronic sites that are not localized on the absorbing atom), leaves behind an empty electron level (core hole). The excitation process is accompanied by a relaxation, which involves the filling of the core hole by a higher electronic level electron and subsequently, an emission of X-ray fluorescence or an Auger electron, if the energy released during relaxation ejects another electron into the continuum. The energies of the absorbed photon or the ejected photo-electron, or the intensity of the emitted photon-electron are then measured. The recorded signal over a range of incident photon energy is then used to generate the XAS spectrum.

The resulting XAS spectrum usually consists of an absorption edge, shown as an increased intensity of the absorption coefficient at a photon energy that approximates the binding energy of the orbital (K, L, or M) electrons. The absorption edge gives information about the oxidation states and the local and chemical environment of the element of interest. A typical XAS spectrum consists of two regions: XANES – X-ray absorption near edge structure, which ranges from few eV before the edge to approximately 50 eV post edge, and EXAFS –extended X-ray absorption fine structure, ranging from 50 eV to some 1000 eV post-edge.

13.2.2. Measurement of Phosphorus K-edge XANES Spectra

For P and many other third row element commonly found in the environment, the XANES region is usually used for speciation analysis. The XANES spectra of some environmentally important P compounds, including those found in manure samples, are shown in Figure 13.1. Three signal detection modes are often employed in XAS measurements, fluorescence yield (FY), total electron yield (TEY), and transmission (Stöhr 1992). The FY detection is best suited for less concentrated or thin samples (Stern and Heald, 1979). The intensity of the fluorescence signal (I_f) is proportional to the absorption coefficient as a function of energy, $\mu(E)$ according to Equation 13.2.

$$\mu(E)\alpha\frac{I_f}{I_0} \tag{13.2}$$

where, I_0 is the incident X-ray flux.

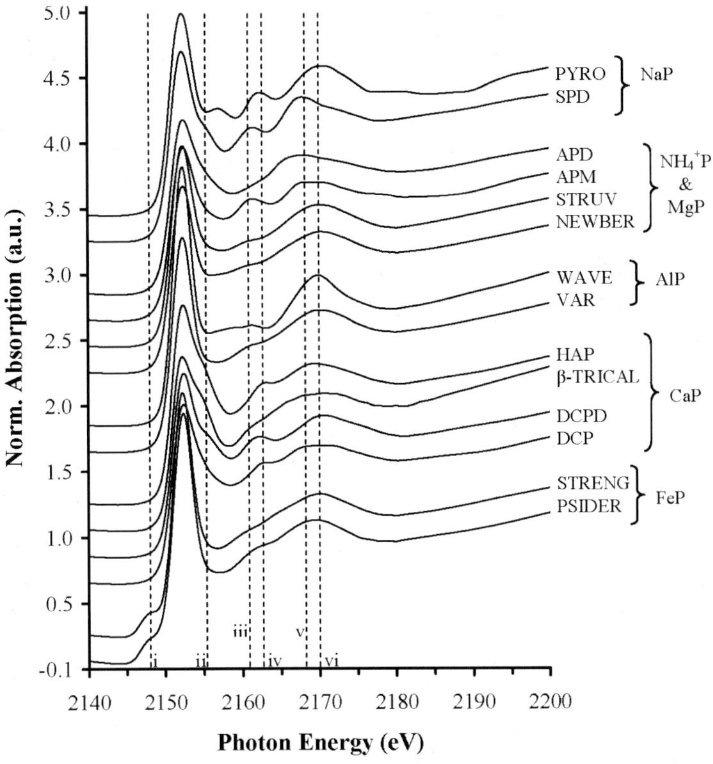

(a)

Figure 13.1. (Continued on next page).

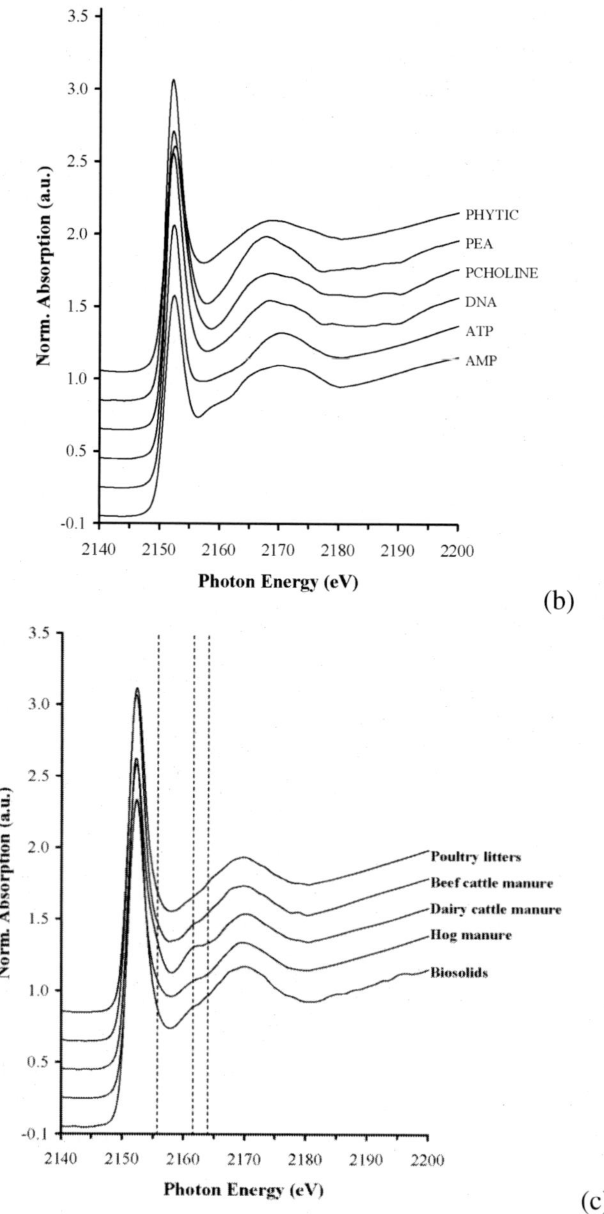

Figure 13.1. Normalized P K-edge X-ray adsorption near-edge structure (XANES) spectra of reference inorganic P (a), organic P compounds (b), and manure samples (c). Inorganic reference compounds include sodium pyrophosphate dibasic (PYRO), sodium phosphate dibasic (SPD), ammonium phosphate dibasic (APD), ammonium phosphate monobasic (APM), struvite (STRUV), newberryite (NEWBER), wavellite (WAVE), variscite (VAR), hydroxyapatite (HAP), β-tricalcium phosphate (β-TRICAL), dicalcium phosphate dihydrate (DCPD), dicalcium phosphate (DCP), strengite (STRENG), and phosphosiderite (PSIDER). Organic compounds include adenosine 5-monophosphate, AMP, adenosine 5-triphosphate, ATP, deoxyribonucleic acid, DNA, phosphatidyl choline, PC, phosphatidyl ethanolamine, PEA, and phytic acid. All spectra were plotted with slight vertical displacement to aid comparison of the features: the pre-edge of Fe phosphate (i), shoulder of Ca phosphate (ii), the 2161-eV peak (iii), the 2163-eV peak (iv), and various O resonances (v and vi). (Reprinted from Ajiboye et al. 2007b; 2008).

The fluorescence signal is often recorded in energy dispersive mode, in which only the intensity of the fluorescence line of interest is recorded. For example, when a P L_{III} subshell electron drops into the K shell hole, a characteristic fluorescence, termed $K\alpha_1$ line is emitted at 2013.7 eV (Kortright and Thompson, 2001). The recording of this signal over the incident energy range is the basis of XANES measurement in fluorescence mode. For soft and tender X-rays, the fluorescence detection is done under a high vacuum or He purged environment due to the attenuation of the low energy fluorescence emission of low (atomic number) Z elements in air.

For experiments in the FY mode, the ground samples can be sprinkled on a tape but must be dilute, homogenous, and finely powdered for a meaningful measurement to be made. Uniform sample thickness and particle size are less important but a homogenous distribution of the element of interest is necessary because only a small area is used for the XAS measurement. At high concentrations of the element of interest, self-absorption may be observed in the fluorescence spectrum. The self-absorption effect can be minimized, in theory, by either decreasing the sample thickness or diluting the sample with an inert material to give a particle size of one absorption length or a total sample thickness of ~ 2.5 absorption lengths (Kelly et al. 2008). Certain XAS data analysis programs such as ATHENA and SixPack contain routines to correct for self-absorption effect in the fluorescence spectra (Ravel and Newville, 2005).

The other main detection method used in soft X-ray applications is TEY, especially for light elements with low fluorescence yield. The total electron collected in this mode includes the initial photoelectron created by the excitation process and any Auger electron created by various decay processes of the core hole excited state (Stöhr 1992). The total electron yield is measured by applying a positive bias voltage to a collection wire located in front of the sample, and then recording the drain current that flow back into the sample. Similar to the FY detection, the intensity of the TEY (I_e) is proportional to $\mu(E)$ of the sample. The TEY is a surface sensitive technique with an escape depth of the electron from the sample of less than 0.1 μm. Because the penetration depth of the incident X-rays is always greater than the escape depth of the electron, self-absorption is not observed in TEY spectra. A common problem with TEY measurements, however, is the charging of samples containing low concentrations of the element of interest. This creates irreproducible background features, which results in a low signal relative to the background.

The third detection mode, commonly used for high atomic number (Z) elements, is transmission. In this mode, the incident X-ray passes though a sample of thickness x and the intensity of the beam is measured before and after it passes through the sample using ionization chambers. The absorbance A of the sample is expressed according to Beer's Law (Equation 13.3)

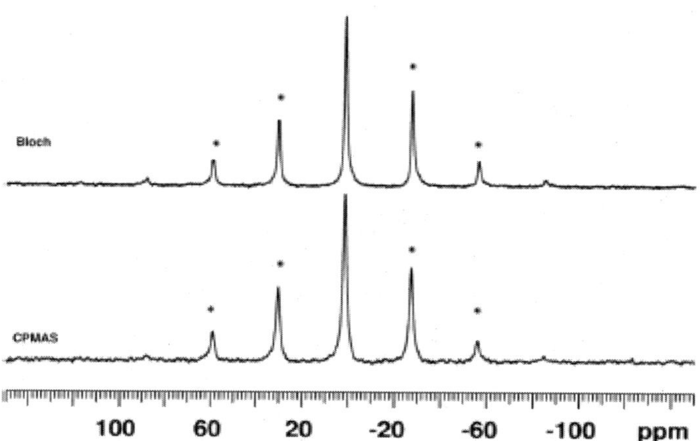

Figure 13.2. Solid state ^{31}P nuclear magnetic resonance (NMR) spectra of calcium saturated cation exchange resin-sand mixture to which monoamonium phosphate was added and incubated for 2 weeks. Central resonance band for the precipitated dicalcium phosphate dihydrate is at 1.3 ppm. Spinning side bands are marked with asterisks. (from Akinremi and Wadu, Unpublished).

$$A = \mu x = \ln(I_0/I_f)\,] \tag{13.3}$$

where μ is the absorption coefficient (cm^{-1}): a product of absorption cross-section (cm^2 g^{-1}) and mass density of the sample ρ (g cm^{-3}), I_0 is the incident X-ray flux, and I_f, the transmitted X-ray flux. The transmission mode is best suited for measurement at high X-ray energies (> 4 keV) that can pass through the sample, and for samples with a reasonably high concentration of between 3 and 5 wt-% of the element of interest (Fendorf and Sparks, 1996). To avoid distortion of the signal for experiments in the transmission mode, it is also important to ensure that the sample is homogenous and its thickness uniform over the entire area irradiated by the X-ray beam. The sample is usually ground alone or mixed with an inert compound of low Z such as boron nitride. The ground sample is then packed into the slot of a sample holder of known thickness and covered front and back with an appropriate adhesive tape.

For all measurement modes in XAS, it is desirable to obtain a high signal-to-noise ratio (S/N) in the spectrum. Given that the random noise in the data decreases with N$^{\frac{1}{2}}$, it is worthwhile to collect multiple scans to get good quality data. In addition, noise in FY is photon related and the counting statistics can be improved by increasing the dwell time (Fendorf and Sparks, 1996).

Analysis of the XAS spectra is crucial in order to identify the chemical species present in the sample. For XANES spectra, qualitative and quantitative analyses are commonly used. Qualitative analysis involves the use of the spectral features of reference compounds (Figure 13.1) as a fingerprint to identify their presence or absence in the spectra of the experimental samples. To obtain an estimate of the proportion of the species present in the sample, quantitative analysis is usually carried out using statistical methods involving a combination of principal components analysis (PCA) and target transformation (TT), and linear least-squares fittings (LLSF). The PCA uses a multivariate statistical procedure to identify the number of independent orthogonal components that constitute the sample spectra while TT identifies the actual chemical species (Ressler et al., 2000; Beauchemin et al., 2002). The success of quantitative analysis of XANES spectra largely depends on the quality and distinctiveness of the spectra of both reference compounds and samples.

13.2.3. Phosphorus Speciation in Manure Derived from P K-edge XANES Analysis

The application of P K-edge XANES spectroscopy to metal speciation of P in manure and manure-amended soils has been reported in last decade (Table 13.1). These studies are reviewed in this section based on the sample types.

Table 13.1. Studies of phosphorus speciation in animal manure and manure-amended soils using phosphorus K-edge X-ray absorption near edge structure spectroscopy.

Reference	Sample	Treatment factors
Peak et al., 2002	Poultry litter	Alum amendments
Sato et al., 2005	Poultry manure, manure-amended soils	Long-term land
Toor et al., 2005	Boiler litter and Turkey manure	application
Maguire et al., 2006	Poultry manure	Diet modifications
Shober et al., 2006	Dairy manure, poultry litter, and biosolids	Liming poultry manure
Ajiboye et al., 2007b	Dairy, beef, hog and poultry manures, and	None
Gungor et al., 2007	biosolids	Different extracts
Ajiboye et al., 2008	Dairy manure	Anaerobic digestion
Seiter et al., 2008	Dairy, and hog manures, and biosolids	Soil incubation
	Poultry litter and alum-amended poultry litter	Different extracts

Table 13.2. Proportion of P species (%) estimated by XANES in calcareous soils; Osborne series (Typic Humicryert) and Lakeland series (Typic Calciudoll) amended with dairy cattle and liquid swine manures.

P species	Unamended soils		Amended soils			
			Osborne		Lakeland	
	Osborne	Lakeland	Dairy	Hog	Dairy	Hog
Adsorbed P	62 (±5)	79 (±5)	65 (±4)	53(±4)	53 (±7)	82 (±3)
Tricalcium phosphate	-	-	31 (±10)	47 (±4)	19 (±9)	-
Hydroxylapatite	38 (±5)	18 (±6)	-	-	-	6 (±1)
Strengite	-	-	9 (±1)	-	28 (±12)	-
Variscite	-	-	-	-	-	-
Phytic acid	-	2 (±1)	-	-	-	13 (±3)

Standard errors of the fit are in parentheses. Adapted from Ajiboye et al. 2008.

In a study aimed at understanding the mechanisms of P removal in poultry litter, Peak et al. (2002) used XANES to identify P species in both unamended and alum ($AlSO_4$)-amended poultry litter. By using the fingerprinting approach, where the XANES spectra of litter samples were compared to those of model compounds, these researchers identified soluble inorganic P in the form of dicalcium phosphate [DCP, $CaHPO_4$] in the unamended samples. In the amended samples, however, P adsorbed onto aluminum hydroxides was identified as the dominant species. Both amended and unamended samples supposedly contained a considerable amount of organic P in the form of phytic acid, which could not be distinguished from the aqueous phosphate expected in the moist samples analyzed in the experiment. Seiter et al. (2008) further advanced the knowledge of P speciation in alum-amended and non-alum-treated poultry litter by XANES analysis of the residues of these samples after each step of a sequential extraction (Chapter 10). Linear least squares fitting (LLSF) analysis of spectra collected from sequentially extracted litters showed that P was present in inorganic (P sorbed on Al oxides, calcium phosphates, Ca-pyrophosphate) and organic (phytate, ADP, orthophosphate diesters and orthophosphate monoesters) forms in alum- and non-alum–amended poultry litter. These results are similar to those of the XANES analysis of the whole litter samples (Peak et al., 2002). However, the data of Seiter et al. (2008) showed that sodium hydroxide (NaOH)-extracted P in alum-amended litters is predominantly organic (80%), whereas in the non-alum–amended samples, >60% of NaOH-extracted P was inorganic P.

The effect of modifying poultry diets, by decreasing mineral P supplementation and phytase in combination with low phytic acid corn, on the chemical composition of manure P was investigated by Toor et al. (2005) using XANES spectroscopy and sequential chemical extraction. Air-dried ground samples were analyzed in a high vacuum chamber in both TEY and FY modes. The LCF of the TEY-XANES spectra were then used to quantify the proportion of different P species. The dominant P species identified in the samples of broilers and turkeys fed with normal diet was DCP (65-77 %), followed by aqueous P (13-18%) and phytate (7-20 %). However, a mixture of DCP (33-45%) and hydroxyapatite [HAP, $Ca_5(PO_4)_3OH$] in almost equal proportion (HAP:35-39%) were identified in the manure from turkey fed with reduced mineral P and phytase. The results from this study suggested that phytase addition, while beneficial to the poultry, also reduced the solubility of P in manure and thus P availability when the manure was applied to most soils. Maguire et al. (2006) used XANES analysis to evaluate the ability of liming materials, i.e., calcium oxide (CaO) and calcium hydroxide [$Ca(OH)_2$] to stabilize P in poultry manure. Quantitative analysis carried out using LCF showed that HAP was the dominant P species in both the lime-stabilized manure and untreated manure with a greater proportion in the lime-stabilized manure (82-86%) than in the untreated manure (76%). Other P species identified in all the samples were adsorbed P on boehmite (11-19%) and phytate P (1-6%). Clearly, addition of Ca, as was done in the study of Maguire et al. (2006) raised the pH, increased the proportion of HAP, and reduced water solubility of P. Sato et al. (2005) applied P K-edge XANES in speciation analysis of P in poultry manure and manure-amended soils to provide an insight on the long-term dynamics of P in soils. A relatively soluble form of Ca-phosphate (CaP), either as DCP or dicalcium phosphate dihydrate, [DCPD, $CaHPO_4 \cdot 2H_2O$], and amorphous CaP were identified to constitute about one-half of total P in the manure, with the remaining one-half identified as aqueous P. However, the subtle differences in the XANES spectra for aqueous P, several organic P and P weakly bound to Al oxides prompted the assignment of the aqueous P spectra to represent free and weakly bound phosphate. In hindsight, the aqueous P might represent any form of organic P, given the organic nature of the poultry manure. Sato et al. (2005) also observed that the application of the poultry manure to an acidic soil in the short term resulted in dissolution of strengite [STRENG, $FePO_4 \cdot 2H_2O$] from the soil, adsorption of P onto the surface of other minerals, and formation of DCP. However, prolonged (>25 yr) application resulted in the disappearance of STRENG and the formation of more stable tricalcium phosphate [TRICAL, $Ca_3(PO_4)_2$].

Shober et al. (2006) combined XANES analysis with sequential extraction to provide a thorough understanding of the complex P speciation of five different slurry pit dairy manures and alum-treated and untreated poultry litters. Using a XANES experimental set-up and quantitative analysis similar to others (Peak et al., 2002; Sato et al., 2005; Maguire et al., 2006), they identified HAP and P sorbed on Al hydroxides and phytic acid in the dairy manures. Overall, P sorbed on Al hydroxides predominated in three out of five dairy manures, and the proportion of HAP was higher in treated poultry litter than in untreated samples. A significant linear relationship was also found between some specific forms (or combinations of species) identified by XANES and those extracted by chemical fractionation, but not always in a 1:1 relationship. Therefore, they cautioned against attributing P in sequential extract to specific chemical species. Gungor et al. (2007) used XANES analysis in combination with XRD to investigate P speciation in raw and anaerobically digested dairy manure and to assess P availability from on-farm digesters. Dried and sieved (25-53 μm) samples were analyzed using XANES with experimental set-up and LCF analysis similar to that of Toor et al. (2005). In the raw dairy manure, XANES analysis indicated that DCP and either struvite [STRUV, $NH_4MgPO_4 \cdot 6H_2O$], aqueous P or newberyite, [NEWBER, $MgHPO_4 \cdot 3H_2O$] were the dominant P species. However, in the anaerobically digested sample, HAP with either STRUV, aqueous P, or NEWBER were identified. By combining XANES with XRD, the uncertainty associated with P species identified by XANES as either STRUV, aqueous P, or NEWBER was resolved, and the species was confirmed by XRD to be struvite. Overall, the undigested manure sample contained soluble (DCP - 57 %) and slow-release P (STRUV - 43%), while the digested samples contained a relatively insoluble HAP (22%) and more STRUV (78%).

In a study aimed at elucidating the forms of P in swine, cattle, and poultry manures, Ajiboye et al. (2007b) used a combination of sequential chemical extraction, solution ^{31}P NMR and XANES to identify different P species in intact and sequential extractions residues of different manures. In the study by Ajiboye et al. (2007b), organic amendments were subjected to sequential extraction. The extracts and residues remaining after extraction were analyzed by solution ^{31}P NMR and XANES spectroscopies, respectively. By using both fingerprinting and LCF analyses of the XANES spectra, readily soluble forms of CaP, DCP and DCPD, were identified in the swine and dairy cattle manures. The DCP and DCPD identified could also be interpreted to be P adsorbed onto $CaCO_3$ given that the spectrum of adsorbed P on $CaCO_3$ also has resonance features similar to, but not as sharp as, DCP and HAP (Peak et al., 2002). Furthermore, STRUV was also identified in beef cattle and poultry manure. The identification of STRUV is of particular agro-environmental importance, considering that it is a source of slow-release P fertilizer that can be used in horticultural crops and pastures grown in high rainfall areas and on coarse-textured soils where the risk of leaching loss of P is high. The recovery of STRUV in livestock manure is one of the technological options currently being pursed to reduce P release from livestock manures into the environment (Zeng and Li, 2006; Huang et al., 2006). In all the manures, HAP was not identified, and this was attributed to the inhibition of P precipitation by organic acids from the manures as indicated by their phytic acid content. It is well understood that the presence of organic acids decreases precipitation of P as HAP due to sorption of organic anions on the HAP seed crystal, thereby blocking the sites for crystal growth (Inskeep and Silvertooth, 1988). Thus, Ajiboye et al. (2008) subsequently evaluated the P species in two calcareous soils (Typic Humicryert and Typic Calciudoll) amended with swine, cattle and poultry manures using XANES spectroscopy with soil incubation experiments. Manures previously characterized by XANES and NMR were applied to two calcareous soils and incubated for a period of 16 weeks. Using LLSF, reasonable estimates of the P species in both amended and unamended (control) soils were obtained, despite the noisiness of the XANES spectra. Adsorbed phosphate (53-82%) and HAP (18-38%) dominated the unamended soils (Table 13.2). However, the addition of manure appeared to have aided the transformation of HAP in the soil to other more soluble P species. For example, in the Osborne soil (Typic Humicryert) the other P species identified was TRICAL with both manure amendment. However, in the Lakeland soil (Typic Calciudoll), a combination of STRENG and tricalcium P was identified following dairy manure amendment, and phytic acid following hog manure amendment.

13.2.4. Future Prospects for Speciation Analysis of Manure P

Although XANES is believed to be a powerful technique for identifying inorganic P species in environmental samples, the lack of distinguishing XANES features in organic P compounds is a major limitation to accurately determining the proportion of organic P species in manures, soils, and other environmental samples. For example, the spectra of various reference compounds, such as aqueous phosphate (Sato et al., 2005), P adsorbed on amorphous $Al(OH)_3$ and gibbsite (Peak, 2002), and organic orthophosphate monoesters (like phytic acid) (Ajiboye et al., 2007b) were all similar, lacking any resonance features. Although Sato et al. (2005) did not detect phytic acid in their poultry manure, but the combined analyses of NMR and LLSF of P XANES of organic amendments in the study by Ajiboye et al. (2007b) showed that phytic acid was present in considerable amounts in all manures. Phytic acid was more in poultry manure than in hog, dairy and beef manures. This is consistent with other P-NMR studies of manures (see Chapter 12). As such, the high amount of aqueous P (especially in dry samples) and phytic acid obtained in other speciation studies of manure employing XANES-LLSF should be interpreted with caution. To improve the organic P identification, reference spectra of more model compounds should be acquired. To do so, Brandes et al. (2007) systematically examined a range of 23 P compounds with different organic and mineral phases using fluorescence XANES spectroscopy. They found that polyphosphates have a broad secondary peak located approximately 2 eV higher in energy than a similar feature in orthophosphate monoesters and diesters and the substitution of aromatic carbon groups in close proximity to P structures in organic compounds generated both pre- and post-peak features as well as a number of secondary peaks. With these reference spectra, Brandes et al. (2007) demonstrated the presence of significant polyphosphate-dominated regions in a marine sediment sample, implying that such phases can play an important role in marine P cycling. Apparently, these reference spectral data could be also used in manure P species identification.

Table 13.3. Phosphorus K-edge XANES white line peaks and solid state ^{31}P NMR spectral features of metal phytate (IP6) compounds.

Compound	XANES		NMR	
	Peak (eV)[a]	Data collection mode[b]	Peak (ppm)	Spinning sideband (ppm)
$Na_{12}IP6$	2.3	TEY and FY	7.5	68.2, -41.7
$K_2H_{10}IP6$	3.0	TEY and FY	-0.7	48.6, -49.8
$KMgH_9IP6$	-[c]	-	-1.9	46.8, -51.4
$K_4Mg_2H_4IP6$	-	-	-0.9	48.1, -49.7
Ca_6IP6	2.8	TEY and FY	0.5	49.8, -48.8
$CaH_{10}IP6$	-	-	-2.0[d]	46.0, -45.0
Mg_6IP6	-	-	-2.5	45.9, -51.8
Ba_6IP6	-	-	1.8	49.9, -47.7
Cd_6IP6	-	-	5.1	53.0, -45.0
$Zn_5H_{12}IP6$	-	-	1.6	50.1, -47.4
Mn_6IP6	3.1	TEY and FY	-	-
Fe_4IP6	3.4	TEY and FY	-	-
Al_4IP6	-	-	-14.8	36.6, -63.0

Adapted from He et al., 2007a; 2007b.

[a] Relative to the nominal P K-edge at 2149.5eV.

[b] Total electron yield mode (TEY) or total fluorescence yield mode (FY).

[c] Not determined.

[d] Both major peak and spinning sidebbands are composited of six distinguishable subpeaks.

Although earlier researchers (Beauchemin et al., 2003) claimed that the P K-edge XANES spectra of phytic acid was mostly featureless, He et al. (2006) pointed out that phytic acid contains a 6-carbon ring with one hydrogen and one phosphate attached to each carbon. Each of the 6 phosphate groups is attached in an ester linkage and retains 2 replaceable hydrogens. These two non-ester hydroxyl groups should impart some inorganic P-like (orthophosphate bond) properties to phytate, thereby leading to interactions of phytate with various metal ions in the environment to form various soluble or insoluble compounds (phytate salts). Thus, He et al. (2006) tested Na, K, Ca, Mn, and Fe phytate compounds, and demonstrated that significant differences can be observed in the intensity, position, and width of the white line at approximately 2153eV among the phosphorus K-edge XANES spectra of five metal phytate compounds (He et al., 2007a). The white line energy positions were in the order of Na<K<Ca<Mn<Fe phytates (Table 13.3). In addition, a little pre-edge feature is present in the spectrum of Fe phytate, which was also observed in the spectra of Fe-relevant inorganic phosphates which could be especially useful for identifying Fe P species (Hesterberg et al., 1999). This work (He et al., 2007a) demonstrated that P K-edge XANES spectroscopy might be used to distinguish metal speciation of phytate compounds in animal manure and other environmental samples although no such report has been published yet so far.

Table 13.4a. Solid-state NMR chemical shift of metal species of model P compounds.

Name	Chemical shift (ppm)	Reference
Tribasic calcium phosphate $Ca_3(PO_4)$	2.9 ± 1	Hunger et al., 2004
Anhydrous dicalcium phosphates / Monetite $CaHPO_4$	-1.5 ± 0.4	Rothwell et al., 1980
Dicalcium phosphate dihydrate / Brushite $CaHPO_4.2H_2O$	1.7 ± 0.3	Rothwell et al., 1980
Octacalcium phosphate $Ca_8H_2(PO_4)_6. 5H_2O$	3.4 ± 0.3	Rothwell et al., 1980
Calcium pyrophosphates $Ca_2P_2O_7$	-6.88	McBeath et al., 2006
Monocalcium phosphate monohydrate $Ca(H_2PO_4)2.H_2O$	-4.6 ± 0.3	Rothwell et al., 1980
Anhydrous mono calcium phosphates $Ca(H_2PO_4)_2$	-0.5 ± 0.4	Rothwell et al., 1980
Amorphous calcium phosphates $Ca_3(PO_4)_2.xH_2O$	2.6	Belton et al. 1988
Hydroxy apatite $Ca_{10}(OH)_2(PO_4)_6$	2.8 ± 2	Rothwell et al.,1980
Newberyite $MgHPO4._3H_2O$	-7.2	Hunger et al. 2004
Bobierrite $Mg_3(PO_4)_2.8H_2O$	4.6	Hunger et al. 2004
$MgHPO_4$	-2.4	Hunger et al. 2004
Magnesium pyrophosphates $Mg_2P_2O_7$	13.1	Mudrakovskii et al. 1986
Struvite $MgNH_4PO_4.6H_2O$	6.35	Bak et al. 2000
$Mg_3(PO_4)_2$	0.6	Mudrakovskii et al. 1986
Variscite $AlPO_4.2H_2O$	-19	Hunger et al. 2004
Wavellite $Al_3(OH)_3(PO_4)_2.5H_2O$	-11	Hunger et al. 2004
Berlinite $AlPO4$	-25	Hunger et al. 2004

Table 13.4b. Solid-state NMR chemical shift of some metal species of P in organic amendments reported in early literature.

Name of compound	Type of manure	Chemical shift (ppm)	Reference
Al-P compounds	Biosolids	-7 to -30	Duffy and van Loon, 1995
P adsorbed to $Al(OH)_3$	Poultry litter	-4 to -10	Hunger at al., 2004
Wavellite	Biosolids	-13	Frossard et al., 1994
Ca octaphosphate and apatite	Biosolids	2 to 3	Frossard et al., 1994
Carbonated, poorly ordered apatite	Biosolids	3	Hinedi et al., 1988
Pyrophosphate [a]	Biosolids	-9	Hinedi et al., 1989
Struvite	Alum amended poultry litter	6.35	Hunger et al., 2004
Organic P [a]	Compost	0 to -4	Frossard et al., 2002

[a] Metal ions not specified.

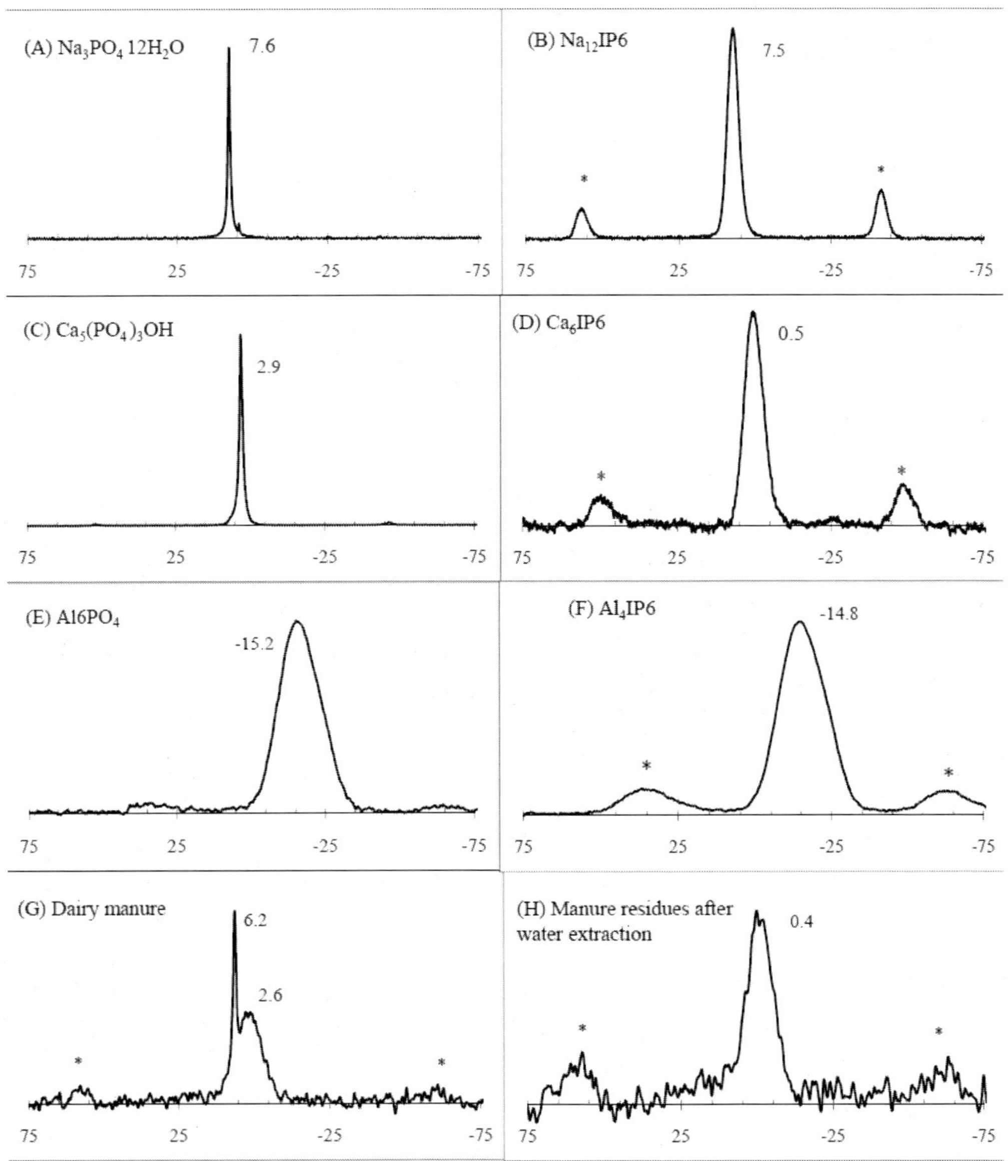

Figure 13.3. Solid-state ^{31}P NMR spectra of inorganic phosphate, phytate compounds, and conventional dairy manure and the residues after extraction with deionized water. The values at the peaks are isotropic chemical shifts in ppm. Spinning side bands are marked with asterisks. Data are from He et al. (2007a, 2007b, 2009a).

Another promising option to improve P identification is the P L$_{2,3}$-edge XANES spectroscopy. The P L$_{2,3}$-edge XANES spectra are richer in spectral features than those of the P K-edge (Ajiboye et al., 2007a; Kruse et al., 2009). Kruse et al. (2009) recently presented a L$_{2,3}$-edge XANES spectra library of mineral P, organic P, and P-bearing minerals for fingerprinting identification. By comparing the L$_{2,3}$-edge spectral features of calcium phytate and sodium phytate hydrate, Kruse et al. (2009) suggested that the spectral differences between phytate compounds are more pronounced than previously reported P K-edge measurements. As such, XANES at the P $L_{2,3}$-edge may be more sensitive to distinguish between metal phytates than P K-edge XANES. However, there are only a couple of studies in which P $L_{2,3}$-edge XANES were applied to P speciation in agro-industrial byproducts and soils (Schefe et al., 2009; Negassa et al., 2010).

13.3. SOLID STATE ^{31}P NUCLEAR MAGNETIC RESONANCE SPECTROSCOPY

13.3.1. Solid State ^{31}P NMR Spectral Features

The principle of solid state ^{31}P NMR spectroscopy is the same as that of solution ^{31}P NMR analysis (Chapter 12). Unlike their solution counterparts, solid state ^{31}P NMR spectra suffer from line broadening as a result of interactions between dipole moments of ^{31}P nuclei (Cade-Menun, 2005). A breakthrough in solid state NMR occurred with the realization that some of the factors causing line broadening could be reduced by rotating the sample rapidly at a given angle to the magnetic field axis, i.e. the magic angle of 54.7^0 (magic angle spinning, MAS) (Cade-Menun, 2005). A side effect of this rapid spinning is the appearance of spinning side bands (SSB) that are located on each side of the central resonance (Figure 13.2). Similar to the isotropic chemical shift, in some cases (such as in the absence of paramagnetic ions), the SSB can provide additional diagnostic information on P species because the patterns of these SSB are characteristic of different P phases (Hinedi et al., 1989). Dougherty et al. (2005) and Shand et al. (1999) observed that inorganic P species rise to sharp resonance with relatively small spinning sidebands, while organic P give rise to broad peaks with prominent spinning sidebands. Dougherty et al. (2005) proposed that this could be used to differentiate inorganic P from organic P. With observations on more phytate model compounds, He et al. (2007a; 2007b) proposed that the strength of symmetric spinning sidebands could be used to distinguish organic and inorganic compounds with similar major isotropic chemical shifts such as sodium phytate and sodium phosphate (Figure 13.3A-F).

13.3.2. Phosphorus Speciation in Manure Derived from Solid State ^{31}P NMR Analysis

The major chemical shift of metal species of model P compounds and some metal P species of organic amendments in early literature are listed in Table 13.4. Alum is added to poultry litter to reduce water soluble P and conserve nitrogen in the manure by reducing ammonia volatilization (Refer to Chapter 15 for more information). Hunger et al. (2004, 2005, 2008) systematically investigated the species of P in the alum-amended poultry litter and the changes induced by various environmental factors (Table 13.5). First, Hunger et al., (2004) used three NMR techniques, solid state MAS, CP-MAS, and ^{31}P $\{^{27}$Al$\}$ TRAPDOR (*TRA*nsfer of *P*opulations in *DO*uble *R*esonance) to compare the P species in alum amended and unamended poultry litter. The TRAPDOR is a solid state NMR technique where the effects of dioplar coupling between a quadrupolar nucleus (^{27}Al) and a dipolar nucleus (^{31}P) can be observed in the spectrum of the dipolar nucleus. The technique can be used to determine whether or not a dipolar nucleus is in close proximity of a quadrupolar nucleus. Hunger et al., (2004) found two major peaks at 6.4 and 2.8 ppm in the single-pulse proton-decoupled MAS ^{31}P NMR spectra of these samples (Figure 13.4). In their initial work, Hunger et al. (2004) assigned the peak at 6.4 ppm to physically-bound, singly-protonated orthophosphate (HPO_4^{2-}). As pointed out by Toor et al. (2005), the chemical shift at 6.4 in the study by Hunger et al. (2004) may be that of struvite which Hunger et al. (2004) also identified in the manure sample using the complementary XRD technique. Struvite has been reported to have a characteristic chemical shift at 6.35 ppm (Bak et al. 2000). The sharp peak at 2.8 was correctly identified as condensed CaP such as HAP, carbonate-apatite and TRICAL. Hunger et al. (2005) further reported their results from the investigation of alum-amended poultry litter by sequential extraction with solid-state ^{31}P NMR spectroscopic analysis of the residues. Aluminum is predominantly found in the fine size separate (<125 μm), indicating that the alum added to the litter in poultry houses hydrolyzed without being completely dispersed in the litter. The NMR spectra confirmed that CaP phases are only dissolved during extraction with dilute acid and phosphate associated with Al is mainly dissolved during extraction with NaOH. However, ^{31}P NMR spectroscopy of the residues indicated that extraction of phosphate associated with Al was incomplete. Later work (Hunger et al., 2008) reported that struvite was identified in five of six poultry litter samples. Furthermore, struvite concentrations were generally lower in dried samples ($\leq 14\%$) than in samples stored moist (23 and 26%). The moist samples also had higher concentrations of phosphate bound to aluminum hydroxides. However, the complex mixtures of organic and inorganic phosphate species were not resolved in these studies apparently in part due to the lack of relevant reference data of model compounds (Table 13.3, 13.4a).

Jayasundera et al. (2005) measured the solid components in 11 dairy manure samples by solid state ^{31}P NMR spectroscopy. These samples represented the normal range and diversity of diets and dairy management techniques used in the USA. These spectra revealed at least two inorganic P (1.5 and 7.7 ppm) and three broad organic P (0.8, 29, and -32 ppm) resonances that are sensitive to cross polarization. Jayasundera et al. (2005) assigned the inorganic P peaks to Mg-P and Ca-P forms, and the organic peaks to phytate without specifying the metal cations. With access to more reference spectra (He et al., 2007a, 2007b), He et al. (2009a) comparatively investigated the P species in dairy manure samples collected from conventional and organic dairy farms by multiple extractions, solution and solid state ^{31}P NMR spectroscopic techniques. Phosphorus in both types of dairy manure was separately extracted with water, Na acetate buffer (100 m*M*, pH 5.0) plus 20 mg ml[-1] Na dithionite (NaOAc-SD), or 0.025 M NaOH with 50 m*M* EDTA (NaOH-EDTA). Whereas the three extractions gave data on solubility (i.e. lability) of manure P, further characterization of the extracts and residues by solution and solid state ^{31}P NMR spectroscopy provided information of P species in the manures. The solid state ^{31}P NMR spectra of the conventional manure and residue after water extractions are presented in Figure 13.3G-H. In the major chemical shift region from -20 to 20 ppm, a sharp peak appeared at 6.2 ppm with a broad peak centered 2.6 ppm in the spectra of conventional dairy manure (Figure 13.3G). With the removal of water soluble P compounds, the chemical shift at 0.4 ppm became the major peak with several shoulders for the conventional dairy manure (Figure 13.3H). In other words, in addition to the disappearance of the peak at 6.2 ppm, the left portion of the peak centered at 2.6 ppm also disappeared after water extraction. These peaks that disappeared following water extraction represent different forms of soluble Ca and Mg phosphate species. The peak in the water-residue spectrum represents water insoluble P species in the conventional dairy manure. Stronger spinning sidebands were observed in the spectrum of water residues than in that of untreated manure, indicating the enrichment of organic P species in the residual pool (Figure 13.3). Combining the data from the multiple extractions and solution ^{31}P NMR spectroscopy, 15 P species were identified in these two manure samples (Table 13.6). These data indicate that conventional dairy manure contained relatively higher contents of soluble inorganic P species and recalcitrant metal P species, and organic dairy manure contained more moderately soluble Ca and Mg P species. However, more data is needed to make a conclusive judgment on the difference in P species between the two types of dairy manure since only one organic dairy manure sample has been studied to date (He and Dou, 2010).

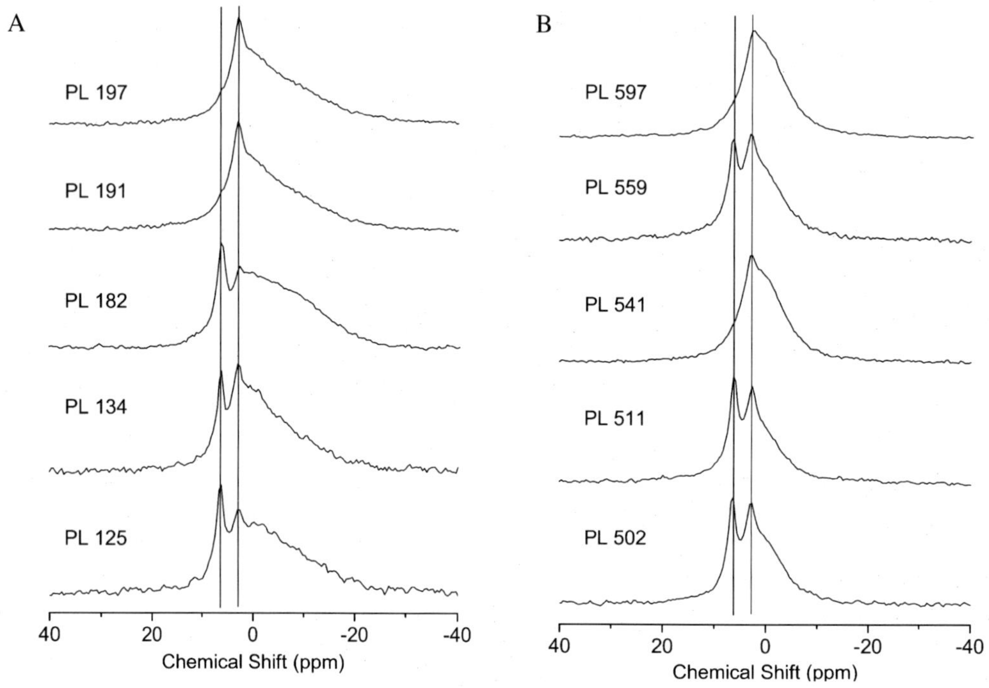

Figure 13.4. Single-pulse proton-decoupled magic angle spinning (MAS) ^{31}P NMR spectra showing two peaks in poultry manure with (a) and without (b) alum amendments. Chemical shifts of the marked peaks are 6.4 and 2.8 ppm. (Reprinted from Hunger at al. 2004).

Table 13.5. Studies of phosphorus speciation in animal manure using solid state ^{31}P NMR spectroscopy.

Reference	Sample	Treatment factors
Hunger et al., 2004	Poultry litter	Alum amendment
Jayasundera et al., 2005	Dairy manure	High- and low-P manures
Shand et al., 2005	Sheep feces	Weathering in the field
Hunger et al., 2005	Poultry litter	Alum amendment, Different extracts
Hunger et al., 2008	Poultry litter	Storage and Drying
He et al., 2009a	Dairy manure	Conventional and organic farming

Table 13.6. P species in the conventional (CD) and organic (OD) dairy manure estimated by the combination of selective extractions (1), solution (2) and solid-state (3) ^{31}P NMR spectroscopies as reported by He et al. (2009a).

P species	Percentage of total P		Identification method
	CD	OD	
Na/K HPO_4^{2-} bonded with H_2O	16.9	24.0	1, 2 and 3
Soluble Ca- and Mg- PO_4^{3-}	24.2	13.5	1, 2 and 3
Fe-related pyrophosphate	5.6	1.7	1, 2 and 3
Moderate hydrogenated Ca-PO_4^{3-}	20.8	-[a]	1, 2 and 3
Moderate tribasic Ca-P minerals	-[a]	35.7	1, 2 and 3
Polyphosphate	2.5	0.9	1 and 2
Soluble DNA	0.8	0.3	1 and 2
Moderate soluble DNA	0.6	0.6	1 and 2
Other diesters (soluble)	2.5	1.7	1 and 2
Moderate Mg and Zn hydrogenated phytate species	6.4	10.0	1, 2 and 3
Other soluble monoesters	5.9	2.8	1 and 2
Moderate soluble other monoesters	3.4	3.8	1 and 2
Al phytate species, some Al phosphate possible	2.0	1.6	1, 2 and 3
Stable divalent metal phytate and phosphate	13.5	6.9	1, 2 and 3
Phosphonates	trace	2.4	3

Reprinted from He and Dou, 2010.

[a] Minor moderate hydrogenated Ca-PO_4^{3-} species or tribasic Ca-P minerals might be present.

In addition to poultry litter, speciation of metal-P compounds in sheep and dairy manure has also been investigated by solid state ^{31}P NMR spectroscopy (Table 13.5). Shand et al. (2005) measured P composition of sheep feces using a combination of ^{31}P MAS and XRD. The ^{31}P MAS NMR spectrum showed resonances and sidebands consistent with DCPD and struvite. When sheep feces were applied to sheep-free pasture in synthetic patches during late summer, Shand et al. (2005) found that the concentration of P in the feces, recovered at intervals up to 84 days, changed little with time but the contribution from DCPD and struvite decreased to < 50% within one week indicating conversion into other forms. Additional XRD analysis confirmed these results and allowed quantification of these minerals, which accounted for 63% of manure P. This observation was in contrast to that of Hunger et al. (2008), who claimed that phosphates associated with Ca and Al were not detected by XRD although they made up a large proportion of P species in their poultry litter samples.

These results point to the uncertainty and pitfalls in using one of two approaches to identify an unknown NMR peak. The first approach is to compare the unknown peak to the peak obtained from a model P compound (Shand et al., 2005). The limitation of this approach is the assumption that the model compound has the same chemical shift as the unknown compound in a complex matrix, such as manure or soil. That is, it is assumed that the complex nature of the manure containing the unknown compound has no effect on its behaviour in the magnetic field. The validity of this assumption has not been unequivocally demonstrated for solid state NMR spectra of P species in manure and soil. The second approach for identifying an unknown compound is to compare the chemical shift produced by the unknown compound to the chemical shifts of known P compounds reported in the literature. The limitation of this approach is that no two manure samples are the same; slight variations in composition can result in significant variations in chemical shift. Indeed, chemical shift is known to be influenced by the crystalinity of the metal species, the degree of condensation and the water content (He et al., 2007b).

13.3.3. Future Prospects of Solid State ^{31}P NMR Spectroscopy in Manure Characterization

Although solid state ^{31}P NMR spectroscopy has been used in characterizing P in environmental samples for decades (Preston et al., 1986; Hinedi et al., 1989), there are relatively few analyses of manure samples using this technique (Table 13.5). However, it is reasonable to expect that solid state ^{31}P NMR spectroscopy will play an integral role in manure P characterization with the increasing sensitivity and analytical capabilities of modern NMR spectrometers. It is clear that more work is needed to understand the complexation of phytate, the major organic P form in many manure samples, with inorganic cations (He et al., 2006). The solid state ^{31}P NMR spectra of model metal phytates reported by He et al. (2007a, 2007b) will facilitate this kind of research. The combination of solution and solid state ^{31}P NMR spectroscopic techniques can provide a complementary identification of P species in manures, as the solution spectra provide information on the anionic phosphate portion and solid state spectra provide information on cation (metal) species of P compounds (He et al., 2009a). Riggle and von Wandruszka (2007) investigated the mobility of inorganic P attached to humic acid and fulvic acid via a metal "anchor" using solid-state ^{31}P NMR spectroscopy. The peak width of the ^{31}P resonance was monitored as an indicator of the degree of attachment of the element to the humic matrix. The same approach could be used to evaluate the degree of association between inorganic P and humic-like materials in animal manure, which has been shown to be rich in humic-like organic matter (He et al., 2009b).

Dougherty et al. (2005) used the technique of spin counting to measure ^{31}P NMR observability (the proportion of P in a sample that is visible to NMR) in pasture soils. They reported that P observability was 4% by CP (cross polarization) and 22% by DP (direct polarization). This poor P observability was attributed to paramagnetic iron in close association with both organic and inorganic P in the soil. Phosphorus observability was increased to >70% after HF pre-treatment. The authors suggested that information contained in non-frequency parameters such as observability, chemical shift anisotropy and relaxation rates can be used to significantly improve information from solid state ^{31}P NMR analysis. Furthermore, the authors suggested that significant improvements in information generated from solid state ^{31}P NMR analysis of soil will come not from improving resolution, but in using information contained in nonfrequency parameters, such as observability, chemical shift anisotropy, and relaxation rate. This prospect could be applied to manure research, too.

13.4. Major Limitations of the Two Spectroscopic Methods

Although NMR and XANES are two powerful techniques for identifying the species of P in manure and other environmental samples, the methods are not without their limitations and drawbacks. There are limitations that are common to the two methods, for example, the lack of measures of errors associated with the anlysis due to the use of a single replicate, and there are limitations that are specific to each of the two spectroscopic techniques.

As a result of cost associated with NMR analysis and limited availability of beam time, most researchers, if not all, have used a single replicate in solid state NMR and XANES analyses. In the work by Ajiboye et al. (2007b), the authors stated that "statistical analysis was not conducted because the high cost of running the NMR spectrometer prevented replication of measurements. However, in a previous study, the standard errors associated with estimating the proportion of P species by peak integration were approximately 5 and 10% for large and small peaks, respectively (Leinweber et al., 1997)". In another study aimed at experimentally validating quantitative XANES analysis for P species, Ajiboye et al. (2007c) reported a relative error in the range of 0.8 to 17% for various binary combinations of HAP and VAR, and 2.9 – 42% for combinations of HAP and PSIDER obtained using linear combination fitting. A measure of standard error or relative error associated with identified P species in manures by these two techniques will go a long way in increasing confidence in their results.

One limitation of these two techniques is that their wide spread use is hampered by availability, especially for XANES. For example, until a few years ago, no synchrotron facility exists in Canada, which means that Canadian scientists had to go to the United States or Europe (Ajiboye et al., 2007b; 2007c; 2008) for synchrotron experiments. With the commissioning of the Canadian Light Source in Saskatchewan, Canada, access may still be limited by geographical location of scientists. While most Chemistry Departments have NMR spectrometer, cost of analysis is high and may limit access to such instruments.

As mentioned previously, the ability of XANES to unequivocally identify organic P species in manures is limited due to the lack of distinguishing XANES features in organic P compounds. This is a major limitation as manures are known to contain considerable proportion of organic P compounds (Refer to Chapter 11 and 12). However, with an increase in access to spectra of model organic P compounds (Table 13.1), the potential exist that P K-edge XANES spectroscopy might be used to distinguish metal speciation of phytate compounds in animal manure and other environmental samples (He et al., 2007a; 2007b).

Another important limitation to the acquisition of ^{31}P NMR spectra of solids such as manures and manure treated soils is the presence of paramagnetic materials (Frossard et al., Hinedi and Chang 1989, Hinedi et al. 1989). These include Fe(III), Cu(II) and Mn(II). These metals are present in manures as they are part of animal feed or are fed as mineral supplements to the animal. Paramagnetic materials in manure and soil drastically reduce the relaxation times (τ_1 and τ_2, see Chapter 12) which can broaden the peaks to the extent that the signal is destroyed or unusable. In solution NMR, these materials are removed by pre-treatment such as extraction using EDTA. Their absence in solution causes τ_1 and τ_2 to be long enough to produce spectra with sharp lines. The implication of this limitation is that, of the four important metals associated with P in manures (Ca, Mg, Al and Fe), only three can be measured using NMR as P associated with Fe cannot be detected due to the paramagnetic nature of Fe.

13.5. SUMMARY

Significant improvements in the last decade have been made to the understanding of the metal species of manure P using XANES and solid-state ^{31}P NMR techniques. Both solid-states technique are particularly sensitive to inorganic forms of P associated with metals in manure samples. In unamended manure samples, the labile P fraction was frequently identified as Ca-P and Mg-P. However, manures amended with P removal chemicals like alum usually consist of Al-P, which has been clearly distinguished from Ca-P and Mg-P by both XANES and NMR spectra. Of the possible metal species of P in manure, iron(Fe)-P cannot be identified by solid state NMR due to the paramagnetic property of Fe(III). In contrast, a unique minor pre-edge feature is present in the P K-edge XANES spectra of Fe-P compounds, which could be especially useful to identify Fe relevant P species in manure. Currently, both solid state spectroscopic techniques are used mainly to identify inorganic P species in manure. Recent literature on the reference spectra of different metal-organic P compounds (mainly phytate compounds) and technical improvement (i. e. P $L_{2,3}$-edge XANES) may facilitate characterizing organic P species in manure. Application of these solid state techniques and/or solution-based approaches in complementary and confirmatory manners could further increase confidence in the experimental results and obtain a full spectrum of manure P characterization.

REFERENCES

Ajiboye, B., O.O. Akinremi, and Y. Hu. 2007a. Phosphorus $L_{2,3}$-edge XANES: A potential soil P speciation technique. p. 75-76. *In* M. Dalzell (ed.) *Canadian Light Source Activity Report 2005-2006*. Canadian Light Source Inc., Saskatoon SK Canada.

Ajiboye, B., O.O. Akinremi, Y. Hu, and D.N. Flaten. 2007b. Phosphorus speciation of sequential extracts of organic amendments using nuclear magnetic resonance and X-ray absorption near-edge structure spectroscopies. *J. Environ. Qual.* 36:1563-1576.

Ajiboye, B., O.O. Akinremi, A. Jurgensen 2007c. Experimental validation of quantitative XANES analysis for phosphorus speciation. *Soil Sci. Soc. Am. J.* 71:1288-1291

Ajiboye, B., O.O. Akinremi, Y. Hu, and A. Jurgensen. 2008. XANES speciation of phosphorus in organically amended and fertilized vertisol and mollisol. *Soil Sci. Soc. Am. J.* 72:1256-1262.

Arai, Y., and D.L. Sparks. 2001. ATR-FTIR spectroscopic investigation on phosphate adsorption mechanisms at the ferrihydrite-water interface. *J. Colloid Interface Sci.* 241:317-326.

Bak, M., J.K. Thomsen, H.J. Jakobsen, S.E. Peterson, T. E. Perterson, and N.C. Nielsen. 2000. Solid state 13C and 31P NMR of urinary stones. *J. Urol.* 164:856-863.

Bakhmutova-Albert, E.V., N. Bestaoui, V.I. Bakhmutov, A. Clearfield, A.V. Rodriguez, and R. Llavona. 2004. A novel cadmium aminophosphonate: X-ray powder diffraction structure, solid-state IR and NMR spectroscopic determination of the fine structure of the organic moieties. *Inorg. Chem.* 43:1264-1272.

Beauchemin, S., D. Hesterberg, and M. Beauchemin. 2002. Principal component analysis approach for modeling sulfur K-XANES spectra of humic acids. *Soil Sci. Soc. Am. J.* 66:83-91.

Beauchemin, S., D. Hesterberg, J. Chou, M. Beauchemin, R.R. Simard, and D.E. Sayers. 2003. Speciation of phosphorus in phosphorus-enriched agricultural soils using X-ray absorption near-edge structure spectroscopy and chemical fractionation. *J Environ. Qual.* 32:1809-1819.

Belton P.S., R.K. Harris, and P.J. Wilkes. 1988. Solid state phosphorus-31 studies of synthetic inorganic calcium phosphates. *J. Phys. Chem. Solids.* 49:21, 21-27.

Brandes, J.A., E.D. Ingall, and D. Paterson. 2007. Characterization of minerals and organic phosphorus species in marine sediments using soft X-ray fluorescence spectromicroscopy. M*ar. Chem.* 103:250-265.

Cade-Menun, B.J. 2005. Characterizing phosphorus in environmental and agricultural samples by ^{31}P nuclear magnetic resonance spectroscopy. *Talanta* 66:359-371.

Cooperband, L.R., and L.W. Good. 2002. Biogenic phosphate minerals in manure: implications for phosphorus loss to surface waters. *Environ. Sci. Technol.* 36:5075-5082.

Dougherty, W. J., R. J. Smernik, and D. J. Chittleborough. 2005. Application of spin counting to the solid-state ^{31}P NMR analysis of pasture soils with varying phosphorus content. *Soil Sci. Soc. Am. J.* 69:2058-2070.

Duffy, S.J., and G.W. van Loon. 1995. Investigations of aluminum hydroxyphosphates and activated sludge by ^{27}Al and ^{31}P MAS NMR. *Can. J. Chem.* 73:1645-1659.

Fendorf, S.E., and D.L. Sparks. 1996. X-ray absorption fine structure spectroscopy. p. 377-416. *In* D.L. Sparks (ed.) Methods of Soil Analysis: Part 3. *Chemical Methods.* SSSA, Madison, WI.

Frossard, E., P. Tekely, and J.Y. Grimal. 1994. Characterization of phosphate species in urban sewage sludges by high-resolution solid-state ^{31}P NMR. Eur. *J. Soil Sci.* 45:403-408.

Frossard, E., P. Skrabal, S. Sinaj, F. Bangerter, and O. Traore. 2002. Forms and exchangeability of inorganic phosphate in composted solid organic wastes. *Nutr. Cycl. Agroecosyst.* 62:103-113.

Gungor, K., A. Jurgensen, and K.G. Karthikeyan. 2007. Determination of phosphorus speciation in dairy manure using XRD and XANES spectroscopy. *J Environ. Qual.* 36:1856-1863.

He, Z., and Z. Dou. 2010. Phosphorus forms in animal manure and the impact on soil P status, p.83-114, In C. S. Dellaguardia, ed. *Manure: Management, Uses and Environmental Impacts.* Nova Science Publishers, Inc., Hauppauge, NY.

He, Z., C.W. Honeycutt, T. Zhang, and P.M. Bertsch. 2006. Preparation and FT-IR characterization of metal phytate compounds. *J. Environ. Qual.* 35:1319-1328.

He, Z., C.W. Honeycutt, T. Zhang, P.J. Pellechia, and W.A. Caliebe. 2007a. Distinction of metal species of phytate by solid state spectroscopic techniques. *Soil Sci. Soc. Am. J.* 71:940-943.

He, Z., C.W. Honeycutt, B. Xing, R.W. McDowell, P.J. Pellechia, and T. Zhang. 2007b. Solid-state Fourier transform infrared and ^{31}P nuclear magnetic resonance spectral features of phosphate compounds. *Soil Sci.* 172:501-515.

He, Z., C.W. Honeycutt, T.S. Griffin, B.J. Cade-Menun, P.J. Pellechia, and Z. Dou. 2009a. Phosphorus forms in conventional and organic dairy manure identified by solution and solid state P-31 NMR spectroscopy. *J. Environ. Qual.* 38:1909-1918.

He, Z., J. Mao, C.W. Honeycutt, T. Ohno, J.F. Hunt, and B.J. Cade-Menun. 2009b. Characterization of plant-derived water extractable organic matter by multiple spectroscopic techniques. *Biol. Fertil. Soils.* 45:609-616.

Hersterberg, D., W. Zhou, K.J. Hutchison, S. Beauchemin, and D.E. Sayers. 1999. XAFS study of adsorbed and mineral forms of phosphate. *J. Synchrotron Radiat.* 6:636-638.

Hinedi, Z.R., A.C. Chang, and J. Yesinowski. 1989. Phosphorus-31 magic angle spinning nuclear magnetic resonance of wastewater sludges and sludge-amended soil. *Soil Sci. Soc. Am. J.* 53:1053-1056.

Hinedi, Z. R., A.C. Chang, and Lee, R.W.K. 1988. Mineralization of phosphorus in sludge-amended soils monitored by phosphorus-31 nuclear magnetic resonance spectroscopy. *Soil Sci. Soc. Am. J.* 52: 1593-1596.

Huang, X.L., and M. Shenker. 2004. Water-soluble and solid-state speciation of phosphorus in stabilized sewage sludge. *J. Environ. Qual.* 33:1895-1903.

Huang, H., D.S. Mavinic, K.V. Lo, and F.A. Koch. 2006. Production and basic morphology of struvite crystals from a pilot-scale crystallization process. *Environ. Technol.* 27:233-245.

Hunger, S., J.T. Sims and D.L. Sparks. 2008. Evidence for struvite in poultry litter: effects of storage and drying. *J. Environ. Qual.* 37:1617-1625.

Hunger, S., J.T. Sims, and D.L. Sparks. 2005. How accurate is the assessment of phosphorus pools in poultry litter by sequential extraction? *J. Environ. Qual.* 34:382-389.

Hunger, S., H. Cho, J.T. Sims, and D.L. Sparks. 2004. Direct speciation of phosphorus in alum-amended poultry litter: solid-state ^{31}P NMR investigation. *Environ. Sci. Technol.* 38:674-681.

Inskeep, W.P. and J.C. Silvertooth. 1988. Inhibition of hydroxyapatite precipitation in the presence of fulvic, humic, and tannic acids. *Soil Sci. Soc. Am. J.* 52:941-946.

Jayasundera, S., W.F. Schmidt, J.B. Reeves III, and T.H. Dao. 2005. Direct ^{31}P NMR spectroscopic measurement of phosphorus forms in dairy manures. *J. Food Agri. Environ.* 3:328-333.

Kelly, S.D., D. Hesterberg, and B. Ravel. 2008. Analysis of soils and minerals using X-ray absorption spectroscopy. p. 387-463. *In* A.L. Ulery, and L.R. Drees (ed.) *Methods of Soil Analysis Part 5 - Mineralogical Methods.* SSSA, Madison, WI.

Kortright, J.B. and A.C. Thompson. 2001. X-ray emission energies. p. 1-8 - 1-27. *In* A.C. Thomspon and D. Vaughan (ed.) *X-ray data booklet* 2nd ed. (LBNL/PUB-490 Rev. 2). Lawrence Berkeley National Laboratory, Berkeley, CA

Kruse, J., P. Leinweber, K. Eckhardt, F. Godlinski, Y. Hu, and L. Zuin. 2009. Phosphorus L2,3-edge XANES: overview of reference compounds. J. Synchrotron Rad. 16:247-259.

Lombi, E. and J. Susini. 2009. Synchrotron-based techniques for plant and soil science: opportunities, challenges and future perspectives. *Plant Soil* 320:1-35.

Maguire, R.O., D. Hesterberg, A. Gernat, K. Anderson, M. Wineland, and J. Grimes. 2006. Liming poultry manures to decrease soluble phosphorus and suppress the bacteria population. *J. Environ. Qual.* 35:849-857.

Massey, M.S., J.A. Ippolito, J.G. Davis, and R.E. Sheffield. 2010. Macroscopic and microscopic variation in recovered magnesium phosphate materials: Implications for phosphorus removal processes and product re-use. *Bioresour. Technol.* 101:877-885.

Mudrakovskii, I.L,V.P. Shmachkova, N. S. Kotsarenko, and V.M. Mastikhin. 1986. ^{31}P NMR study of I-IV group polycrystalline phosphates. *J. Phys. Chem. Solids.* 47:335-339.

McBeath, T.M., Smernik, R.J., Lombi, E., McLaughlin, M.J. 2006. Hydrolysis of Pyrophosphate in a highly calcareous soil: A solid state phosphorus-31 NMR study. *Soil Sci Soc. Am. J.* 70:856-862.

Negassa, W., J. Kruse, D. Michalik, N. Appathurai, L. Zuin, and P. Leinweber. 2010. Phosphorus speciation in agro-industrial byproducts: sequential fractionation, solution P-31 NMR, and P K- and L-2,L-3-edge XANES spectroscopy. *Environ. Sci. Technol.* 44:2092-2097.

Peak, D., J.T. Sims, and D.L. Sparks. 2002. Solid-state speciation of natural and alum-amended poultry litter using XANES spectroscopy. Environ. Sci. Technol. 36:4253-4261.

Preston, C. M., J. A. Ripmeester, S. P. Mathur, and M. Levesque. 1986. Application of solution and solid-state multinuclear NMR to a peat-based composting system for fish and crab scrap. *Can. J. Spectr.* 31:63-69.

Ravel, B., and M. Newville. 2005. Athena, Artemis, Hephaestus: Data analysis for X-ray absorption spectroscopy using IFEFFIT. *J. Synchrotron Rad.* 12:537-541.

Ressler, T., J. Wong, J. Roos, and I.L. Smith. 2000. Quantitative speciation of Mn-bearing particulates emitted from autos burning (methylcyclopentadienyl)manganese tricarbonyl-added gasolines using XANES spectroscopy. *Environ. Sci. Technol.* 34:950-958.

Riggle, J., and R. von Wandruszka. 2007. P-31 NMR peak width in humate-phosphate complexes. *Talanta* 73:953-958.

Rothwell, W.P., J.S. Waugh, and J.P. Yesinowski. 1980. High resolution variable temperature phosphorus-31 NMR of solid calcium phosphates. *J. Am. Chem .Soc.* 102:2637-2643.

Sato, S., D. Solomon, C. Hyland, Q.M. Ketterings, and J. Lehmann. 2005. Phosphorus speciation in manure and manure-amended soils using XANES spectroscopy. *Environ. Sci. Technol.* 39:7485-7491.

Schefe, C.R., P. Kappen, L. Zuin, P.J. Pigram, and C. Christensen. 2009. Addition of carboxylic acids modifies phosphate sorption on soil and boehmite surfaces: A solution chemistry and XANES spectroscopy study. *J. Colloid Interface Sci.* 330:51-59.

Seaman, J.C., J.M. Hutchison, B.P. Jackson, and V.M. Vulava. 2003. In situ treatment of metals in contaminated soils with phytate. *J. Environ. Qual.* 32:153-161.

Seiter, J.M., K.E. Staats-Borda, M. Ginder-Vogel, and D.L. Sparks. 2008. XANES spectroscopic analysis of phosphorus speciation in alum-amended poultry litter. *J. Environ. Qual.* 37:477-485.

Sham, T.K., and M.L. Rivers. 2002. A brief overview of synchrotron radiation. p. 117-147 *In* Fenter, P.A. et al. (ed.) *Reviews in Mineralogy and Geochemistry 49. Mineralogical Soc. Am.*, Washington, DC.

Shand, C.A., M.V. Cheshire, C.N. Bedrock, P.J. Chapman, A.R. Fraser, and J.A. Chudek. 1999. Solid-phase ^{31}P NMR spectra of peat and mineral soils, humic acids and soil solution components: influence of iron and manganese. *Plant Soil.* 214:153-163.

Shand, C.A., G. Coutts, S. Hillier, D.G. Lumsdon, A. Chudek, and J. Eubeler. 2005. Phosphorus composition of sheep feces and changes in the field determined by ^{31}P NMR spectroscopy and XRPD. *Environ. Sci. Technol.* 39:9205-9210.

Shober, A.L., D.L. Hesterberg, J.T. Sims, and S. Gardner. 2006. Characterization of phosphorus species in biosolids and manures using XANES spectroscopy. *J. Environ. Qual.* 35:1983-1993.

Stern, E.A., and S.M. Heald. 1979. X-ray filter assembly for fluorescence measurements of X-ray absorption fine structure. *Rev. Sci. Instru.* 50:1579:1583.

Stöhr, J. 1992. *NEXAFS Spectroscopy.* Springer-Verlag. New York.

Toor, G.S., J.D. Peak, and J.T. Sims. 2005. Phosphorus speciation in broiler litter and turkey manure produced from modified diets. *J. Environ. Qual.* 34:687-697.

Zeng, L., and X. Li. 2006. Nutrient removal from anaerobically digested cattle manure by struvite precipitation. *Environ. Sci. Eng.* 5:285-294.

In: Environmental Chemistry of Animal Manure
Editor: Zhongqi He

ISBN: 978-1-62808-641-6
© 2013 Nova Science Publishers, Inc.

Chapter 14

MODELING PHOSPHORUS TRANSFORMATIONS AND RUNOFF LOSS FOR SURFACE-APPLIED MANURE

Peter A. Vadas[*]

14.1. INTRODUCTION

The USEPA reports that of the nation's 16.3 million acres of lakes, ponds, and reservoirs assessed during the 2004 reporting cycle, 64% were impaired, with nutrients cited as one of the leading causes of impairment and agricultural activities cited as the leading source of impairment (USEPA, 2009). One of the main water quality concerns is accelerated eutrophication of fresh waters by phosphorus (P), which limits water use for drinking, recreation, and industry (Carpenter et al., 1998). The subsequent challenge for the agricultural community, from scientists to producers, is to identify agricultural areas with a high potential for P export, accurately quantify that export, and assess the ability of alternative management practices to minimize P export. Water quality simulation models are seen as one relatively rapid and cost effective way to help achieve these goals (Sharpley et al., 2002).

While significant loss of P from agricultural fields can occur through leaching in sandy soils (Novak et al., 2000), organic soils (Porter and Sanchez, 1992), and soils with artificial drainage (Heckrath et al., 1995; Vadas et al., 2007b), the primary pathway of P loss from the majority of agricultural soils is through surface runoff. The three major sources of P to surface runoff are soil, plant material, and applied fertilizers, manures, or biosolids (Heathwaite and Dils, 2000; Withers et al., 2001). Research has clearly shown that even though soil and plant material can be significant sources of P to runoff, their effect can be overwhelmed by P release from recently applied animal manures, especially if the manures are left unincorporated (Daverede et al., 2004; Kleinman et al., 2002a; Shigaki et al., 2006). Phosphorus loss in runoff from surface-applied manures is often greatest in the first runoff event after application and decreases in subsequent events (Kleinman and Sharpley, 2003).

---

[*] peter.vadas@ars.usda.gov USDA-ARS, Dairy Forage Research Center, Madison, WI 53705, USA

Surface manures can continue to be a source of P loss to runoff for several months after application (Vadas et al., 2007a).

Given the significant degree to which surface-applied manures can contribute to P loss in runoff, it is important for water quality models to be able to reliably simulate surface application of manure P, physical and chemical transformations of manure and its P with time after application, and loss of P from manure to runoff during storm events. However, commonly used water-quality models, such as EPIC (Williams et al., 1983), GLEAMS (Leonard et al., 1987), ANSWERS (Bouraoui and Dillaha, 1996), or SWAT (Arnold et al., 1998), do not simulate these surface manure processes. Instead, the models immediately incorporate any applied manure and P into soil P pools and allow P loss to runoff from the soil. The result is a poor representation and prediction of P loss in runoff (Pierson et al., 2001b; Sharpley et al., 2002). Therefore, such models could be improved by adding routines to simulate surface manure P processes and loss in runoff. This chapter reviews and discusses research on processes controlling surface manure P transformations and loss in runoff and efforts to translate understanding of these processes into simulation models. The chapter also emphasizes gaps in knowledge and needed research.

14.2. CONCEPTUAL MODELING OF SURFACE MANURE AND PHOSPHORUS TRANSFORMATIONS AND LOSS IN RUNOFF

This chapter's discussion of surface manure and P processes is based on the conceptual model depicted in Figure 14.1. In that model, manure is applied to the soil surface, with manure either left unincorporated or only partially incorporated by tillage or injection. Manure can be applied by machine or by grazing animals. If liquid manures are applied, a portion of the applied P will immediate infiltrate into the soil, where its transformations and fate will be controlled by soil processes. When applied, manure P exists in two basic forms, water extractable P (WEP) and non-WEP, which are experimentally defined. Manure WEP is the portion of manure P that is immediately available to be released from manure during a storm, and non-WEP is not immediately available. Both manure WEP and non-WEP exist in inorganic and organic forms, for a total of four manure P pools. Over time, manure mass will decrease as manure organic matter decomposes, and manure non-WEP will transform (through organic P mineralization and other biochemical processes) to WEP. Manure mass and P will also be physically assimilated into soil by macro-invertebrates (bioturbation). During storms, manure P will be released from the WEP pools and will either infiltrate into soil or be lost in surface runoff. The following sections discuss the current research and knowledge gaps related to the processes depicted in Figure 14.1.

14.3. SIMULATING MANURE APPLICATION TO SOILS

The first step in simulating surface manure P processes is the actual application of manure to soils. There are a number of methods available for field application of manure. Traditional methods entail spreading solid, semi-solid, or liquid manures with appropriate machines. These manures could then be tilled in or left unincorporated, as in pasture or no-till production. From a modeling perspective, important information about application would include the percent of a field area covered by manure and the percent of manure incorporated by different tillage implements. These factors would be important when simulating the amount of rain that interacts with manure (e.g., less manure coverage means less rain interacts with manure), the rate of manure assimilation into soil (e.g., less coverage would reduce the rate of assimilation), and the amount of manure left on the soil surface that is available to supply P to runoff. While there are good estimates of the degree to which different tillage implements incorporate manure (Stott et al., 1995), there may be no information on the effect of application method on area covered. A growing number of new methods are available for field application of manure that potentially reduce soil disturbance while incorporating manure. These methods include injection of manure into shallow furrows (McLaughlin et al., 2006), high pressure injection (Morken and Sakshaug, 1998), and enhanced infiltration of liquid manure by pairing manure application with aeration (Bittman et al., 2005; van Vliet et al., 2006). Information on the effect of such application methods on percent area covered and percent incorporation of manure is needed for modeling.

Manure properties are also important when simulating manure application. Animal manures vary widely in their physical and chemical properties due to animal diet and manure management (Kleinman et al., 2005). One important property is dry matter content and how it affects the ability of manure to infiltrate into soils at the time of application. Poultry and beef manures often have dry matter contents great enough that the manures handle as solids rather than flowable liquids. Conversely, dairy and swine manures often have a dry matter content of less than 10-15%, which means they are land-applied essentially as flowable liquids. The liquid portion of such dairy and swine manures, and any fine solids and P associated with the liquid portion, can infiltrate into the soil at the time of application, leaving the coarser manure solids on the soil surface (Vadas, 2006). Manure P that infiltrates at application and becomes part of the soil matrix becomes less available to loss in runoff, while P remaining on the soil surface in manure solids remains highly available to runoff. In developing a model to predict dissolved P loss in runoff from surface manures, Vadas et al. (2004) found it was necessary to account for infiltrating P in liquid manures to achieve reliable predictions. In more controlled experiments, Vadas (2006) observed that 40-65% of P in a liquid manure can infiltrate into soil at application and react with the soil matrix to a depth of about 2 cm. However, these results represented only three manure samples and one soil type. Several studies have investigated P distribution in the liquid and solid fraction of liquid manures, but mostly from the standpoint of separating slurry for anaerobic digestion, transport, re-feeding, or other utilization systems (Hill and Baier, 2000; Holmberg et al., 1983; Moller et al., 2002; Zhang and Westerman, 1997). Beyond the experiments of Vadas (2006), there has apparently been no formal investigation of P distribution in liquid manure with regard to its infiltration into soil at application and its subsequent availability to runoff. Future research into the manure properties or management practices that affect P distribution between liquid and solid phases or soil properties that affect the degree of manure P infiltration could help improve model predictions.

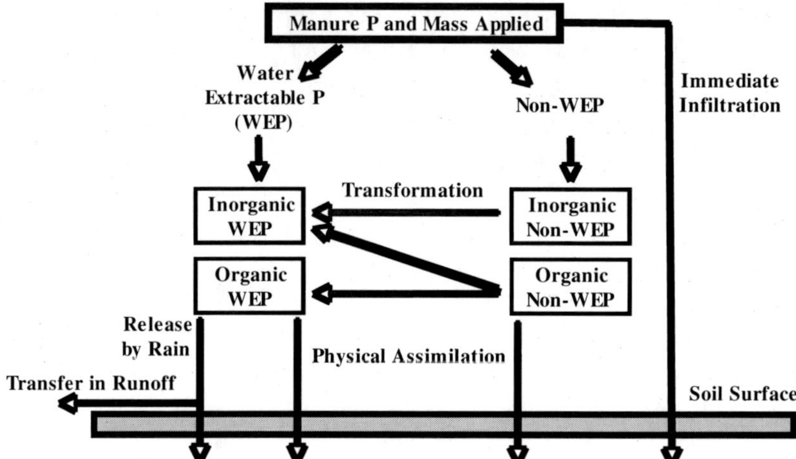

Figures 14.1. Schematic diagram of a conceptual model of P pools and processes for manure applied to the soil surface and left unincorporated.

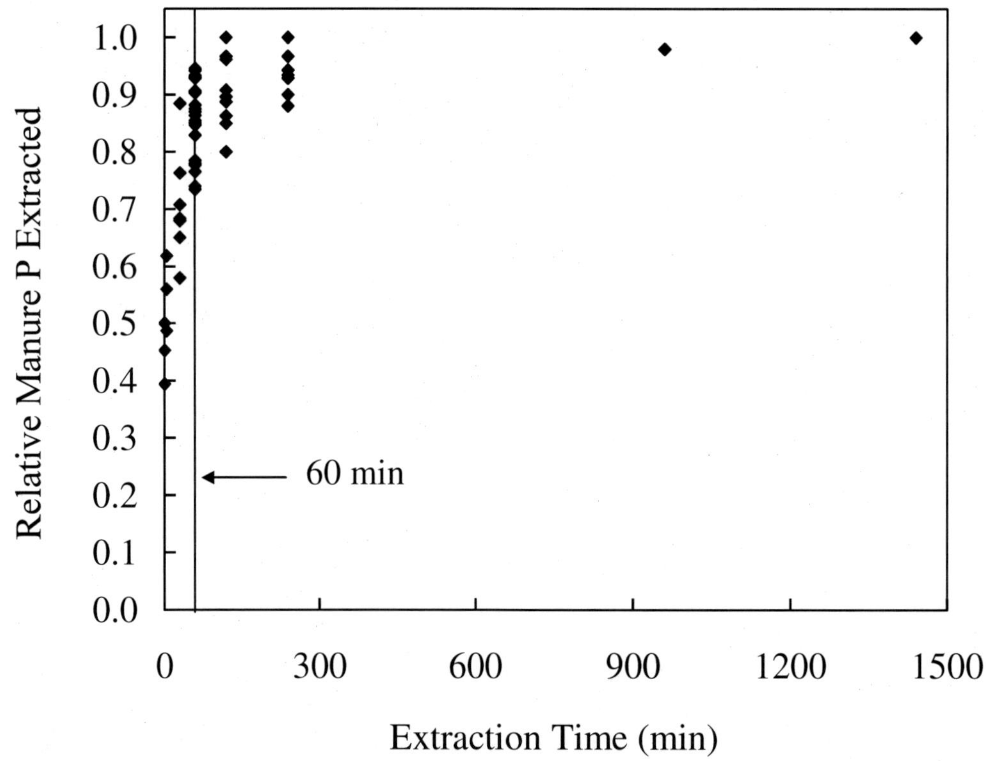

Figure 14.2. Relative dissolved inorganic P release to water from dairy, poultry, and swine manures as a function of extraction time. Data are from Dou et al. (2000), Kleinman et al. (2002b), Mamo et al. (2007), Studnicka (2006), and Tasistro et al. (2004).

14.4. SIMULATING MANURE PHOSPHORUS RELEASE DURING A STORM

The second step in simulating surface manure P processes is prediction of P release from surface manure during a storm. Relatively early studies on the mechanisms of P loss in runoff from surface manures reported good relationships between the solubility of P in manures and dissolved P loss in runoff (Kleinman et al., 2002a; Moore et al., 2000; Withers et al., 2001). These results demonstrated that a pool of readily extractable P in manures (typically referred to as water extractable P (WEP)) was likely the source of manure P to runoff. Early studies used a water extraction to quantify this pool of manure WEP. Kleinman et al. (2002b) evaluated different manure water extraction methods and found that the amount of manure P extracted by water increased as both the time of extraction and the extraction ratio (volume of water to manure dry matter) increased. The amount of manure P extracted increased non-linearly so that increasing extraction times beyond about 60 min and extraction ratios beyond about 250:1 (cm^3 g^{-1}) did not appreciably increase the amount of manure P extracted. Several studies since have observed similar results (Dou et al., 2000; Haggard et al., 2005; Kleinman et al., 2007; Mamo et al., 2007; Tasistro et al., 2004; Toor et al., 2007; Vadas and Kleinman, 2006). These data are shown in Figures 14.2 and 14.3.

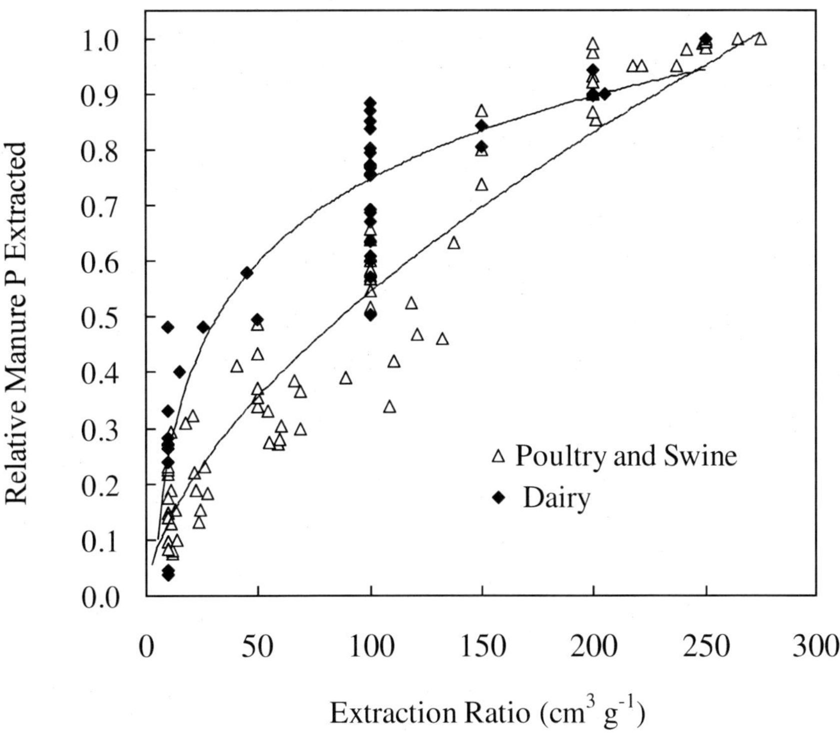

Figure 14.3. Relative dissolved inorganic P release from dairy, poultry, and swine manures as a function of water to manure ratio during extraction. Data are from Haggard et al. (2005), Kleinman et al. (2002b), Kleinman et al. (2007), Mamo et al. (2007), Studnicka (2006), Toor et al. (2007), Vadas and Kleinman (2006).

Vadas et al. (2004; 2005) proposed manure water extraction data could be used to predict the amount of P that is released from surface manure during a storm. Because experimental data consistently showed that increasing extraction times beyond about 60 min did not increase the amount of P that can be extracted from manure by water (Figure 14.2) (Dou et al., 2000; Kleinman et al., 2002b; Tasistro et al., 2004), and because most widely used water quality models operate on a daily time-step and would not simulate rain-manure interactions of less than 24 h, Vadas et al. (2004) proposed that the length of storm was not a critical parameter to include when predicting P release from manure. Instead, Vadas et al. (2004) proposed that variations in the amount of water that interacts with manure during a storm was a more critical parameter to simulate. To do this, Vadas et al. (2004) first proposed that manure WEP (i.e., the maximum amount of P that could be released from manure during a storm) could be reliably quantified with a water extraction conducted for 1 h at an extraction ratio (cm^3 g^{-1}, dry weight equivalent) of 250:1. Both inorganic and organic manure WEP could be quantified with this procedure. This 250:1 ratio was based on the consistent results of several studies (Figure 14.3) showing that increasing the extraction ratio beyond 250:1 did not appreciably increase the amount of P that could be extracted from manure (Haggard et al., 2005; Kleinman et al., 2007; Kleinman et al., 2002b; Mamo et al., 2007; Toor et al., 2007; Vadas and Kleinman, 2006). In the study investigating the widest range of extraction ratios, Studnicka (2006) found that manure P extracted at a ratio of 200:1 was 60 to 100% (average of 80%) of that extracted at a ratio of 1000:1. From a modeling perspective, Vadas et al. (2004) proposed that the extraction ratio (cm^3 g^{-1}) could represent the ratio of rain volume during a storm to the mass of manure solids on the soil surface. This ratio (W) would be calculated as:

$$W = \text{Rain (cm)} / \text{Manure Mass (kg)} \times \text{Area covered (ha)} \times 100000.0 \qquad [14.1]$$

Then, regression equations developed using data from Figure 14.3 could be used to predict the quantity (kg) of total manure WEP that is released from manure solids during a storm. These equations are documented in the SURPHOS model of Vadas et al. (2007c) and are:

$$\text{Dairy: WEP Released} = 1.2 \ [W / (W + 73.1)] \ (\text{Manure WEP}) \qquad [14.2]$$

$$\text{Poultry and Swine: WEP Released} = 2.2 \ [W / (W + 300.1)] \ (\text{Manure WEP}) \qquad [14.3]$$

These SURPHOS equations can be validated using manure P release data reported in the literature (Guo and Song, 2009; Guo et al., 2009a; Guo et al., 2009b; Muck and Ludington, 1979; Robinson and Sharpley, 1995; Sharpley and Moyer, 2000). In these experiments, a known mass of manure was leached in a saturated flow cell, by simulated rainfall, or by natural rainfall, and the amount of dissolved P released was measured. In all cases, the amount of water and manure mass were reported so that Eq. [14.1] - [14.3] could be used to predict the amount of dissolved P released from the manure. A comparison of measured and predicted P release values in Figure 14.4 shows the equations can reliably predict P release from manure during a storm.

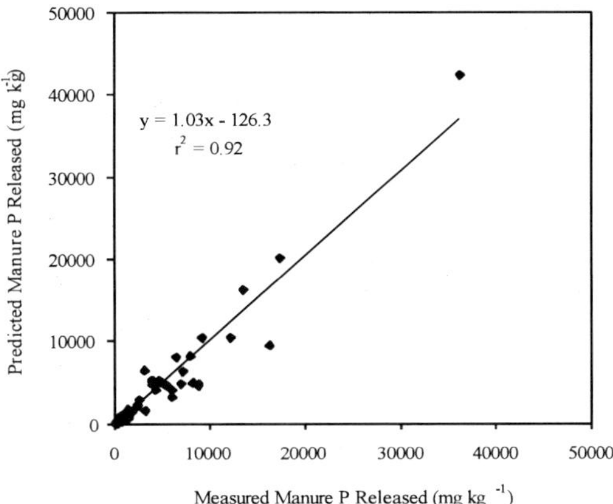

Figure 14.4. Relationship between measured and predicted dissolved inorganic P released from manures (p < 0.001). Data are from six published studies as specified in the text.

The SURPHOS Eq. [14.2] and [14.3] were developed using data from batch manure extraction experiments. Because manure P release during a storm will occur more like flow experiments, such as those conducted by Muck and Ludington (1979), refining these equations based on data from flow experiments might help improve their ability to predict manure P release. Investigating the effect of temperature and manure moisture may also help develop P release equations for all possible climate conditions (Bechmann et al., 2005).

Other approaches have been taken to predict the amount of P that is released from manure during a storm. These approaches, developed by Gerard-Marchant et al. (2005) and documented by Hively et al. (2006), predict manure P release as function of a constant characteristic time or characteristic volume of water. However, this approach has apparently not been fully validated and may not account for the dynamic effect of the ratio of rain volumes and manure solids on manure P release.

Data in Figure 14.3 show that the relationship between extraction ratios and manure P released to water is fairly consistent across a range of different manures analyzed in various studies. There is a clear difference in the relationship between P extracted and the extraction ratio for dairy manures and for poultry and swine manures. This could be due to fundamental animal diet differences, where dairy cattle are fed more forage-based diets and poultry and swine are fed more grain-based diets with greater relative amounts of supplemental inorganic P (Cooperband and Good, 2002; Leytem et al., 2007; McDowell et al., 2008; Toor et al., 2005a). Variability within manure P extraction patterns can also occur due to differences in animal diet (within a species type), bedding, or manure handling and storage (Toor et al., 2005b). For example, Studnicka (2006) performed water extractions on dairy manures relatively low in total P (~5000 mg kg⁻¹) and high in total P (~8500 mg kg⁻¹) and found the relationship between P extracted and the extraction ratio differed between the two groups of manures (Figure 14.5). The high P manures actually released relatively less P to water at the same extraction ratio, with the relationship more resembling that of poultry and swine manures than other dairy manures. This may be because high P dairy manures came from animals fed diets with more supplemental inorganic P, which is more like poultry and swine diets (Leytem et al., 2007). Furthermore, data from Toor et al. (2007), Vadas and Kleinman (2006), and Kleinman et al. (2007) clearly showed that manure management practices that vary significantly from conventional practices can dramatically change the relationship between extraction ratios and manure P released to water. Such practices can include composting, heating, granulation, and addition of chemicals such as alum or urea. A more systematic investigation of how variations in animal diet, manure management, and thus manure chemistry control P solubility in water could help improve empirical equations used to predict manure P release during a storm (Tasistro et al., 2007).

Finally, Vadas et al. (2007c) proposed that a single water extraction conducted for 60 min at an extraction ratio of 250:1 is a reliable way to quantify manure WEP for modeling purposes. Alternative, slightly acidic extractants have been proposed to quantify manure WEP because they may not be as sensitive to differences in manure pH or Ca content, which can affect the amount of manure P released to water (Dou et al., 2007; Tasistro et al., 2007). Further investigation of these extractants in conjunction with controlled, manure P runoff experiments could assess their superiority to water extractions.

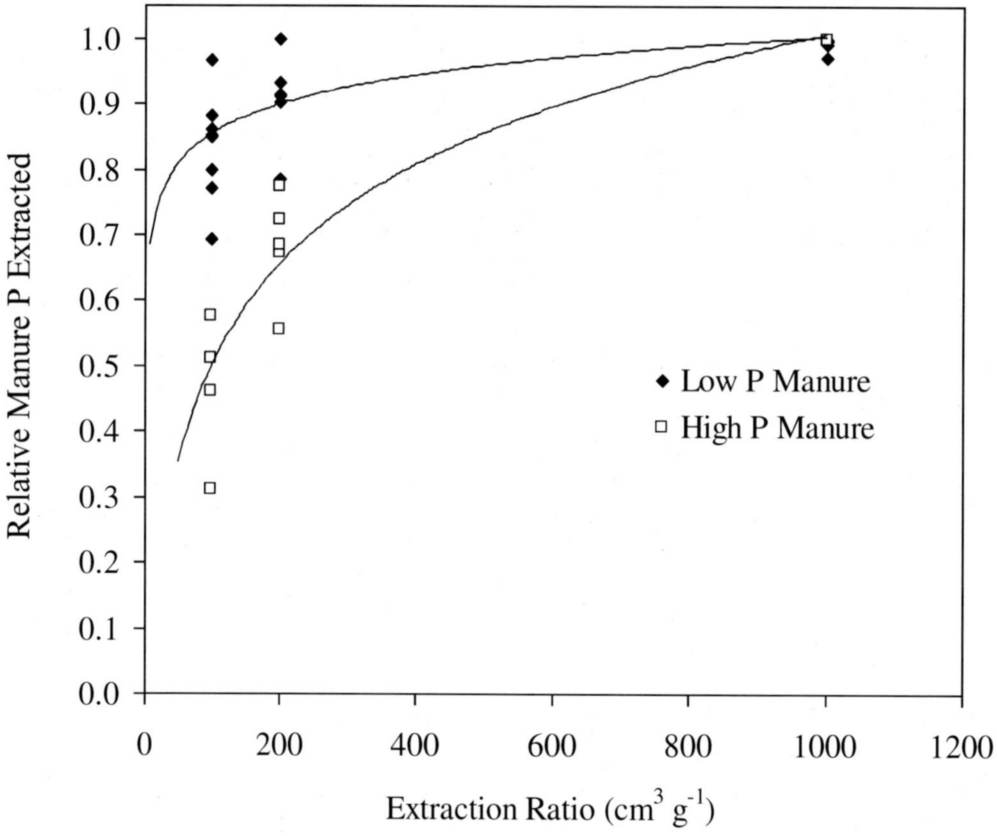

Figure 14.5. Relative dissolved inorganic P release from relatively high and low P dairy manures as a function of water to manure ratio during extraction. Data are from Studnicka (2006).

14.5. SIMULATING MANURE PHOSPHORUS LOSS IN SURFACE RUNOFF

The third step in simulating surface manure P processes is prediction of manure P loss in surface runoff. First, the concentration of P released (mg L^{-1}) from manure can be calculated by dividing the quantity of manure WEP released (calculated using Eq. [14.2] and [14.3]) by the volume of rain. Then, this concentration can be adjusted to calculate the actual concentration of manure P in runoff. Both Sharpley and Moyer (2000) and Muck and Ludington (1979) showed that P release from manure to water decreases with time during a storm. This means that at the beginning of a storm if soils are relatively dry and no runoff occurs, P released from manure, which will be at relatively high concentrations, will infiltrate into the soil. As soils become wetter and runoff begins, P released from manure both infiltrates into soil and moves in runoff. Eventually, when soils become saturated, most of the rainfall may be converted to runoff, and nearly all the P released from manure moves in runoff. At this latter point, however, P concentrations released from manure will be significantly less than at the initial stages of the storm (Tarkalson and Mikkelsen, 2004; Torbert et al., 1999; Vadas et al., 2004; White et al., 2003). Thus, the earlier that runoff occurs during a storm, which in turn typically results in greater runoff to rain ratios for the entire storm (Srinivasan et al., 2007), the greater the concentrations of manure P in runoff will be. Several studies have shown a positive relationship between storm runoff to rain ratios and manure P concentrations in runoff (Hanrahan et al., 2009; Kleinman et al., 2002a; Pierson et al., 2001a; Pote et al., 2001; Sistani et al., 2009; Smith et al., 2007). Vadas et al. (2005) developed an empirical relationship based on storm runoff to rain ratios to account for dynamic P release from manure with time during a storm and the subsequent distribution of manure P between infiltration and runoff. This equation is:

$$\text{Pdfactor} = (\text{Runoff} / \text{Rain})^{0.225} \qquad\qquad [14.4]$$

The Pdfactor, or P Distribution Factor, varies between 0.0 and 1.0, and is multiplied by the concentration of P released from manure to calculate the actual runoff P concentration.

No research has been conducted to determine if specific aspects of manure P chemistry influence the rate of P release from manure during a storm, which could affect the parameters in Eq. [14.4]. Faster initial rates of release may allow more manure P to infiltrate into soil, while slower initial rates of release may allow more manure P to be lost in runoff.

14.6. SIMULATING MANURE AND PHOSPHORUS TRANSFORMATIONS WITH TIME AFTER APPLICATION

Several studies have shown that P loss in runoff from surface-applied manures may be greatest in the first storm after application and will decline in subsequent storms (Kleinman and Sharpley, 2003). However, surface manure can be a significant source of P loss in runoff for months after application (Pierson et al., 2001a; Vadas et al., 2007a). In a 14 to 18-month field study, Vadas et al. (2007a) clearly showed that manure P initially in non-WEP forms transforms to WEP over time and maintains a source for P loss in runoff for months after application. Therefore, it is critical for models to simulate manure and P transformations with time after application. There are several interrelated mechanisms that can affect the availability of manure P to loss in runoff with time after application and also how a model might predict manure P runoff loss. These are:

Decomposition of manure organic matter with time. Decomposition decreases the manure mass and coverage on the soil surface, which will change the rain to manure solids ratio during a storm. This could affect the amount of manure P released from the solids during a storm. However, decomposition of manure solids should not decrease the P content of manure.

Physical assimilation where macro-invertebrates mix manure solids and P into soil (bioturbation). Assimilation decreases the amount of manure P on the soil surface and thus available to loss in runoff. Assimilation also decreases the amount of manure solids and coverage on the soil surface, which may affect manure P loss in runoff similar to the decomposition mechanism above.

Transformation, through mineralization or other biochemical processes, of manure non-WEP into WEP forms that are available for release by rain and loss in runoff.

14.6.1 Decomposition of Manure Organic Matter

Decomposition of manure organic matter and a subsequent decrease in mass is a fundamental biological process that has been described in numerous models (Lewis and McGechan, 2002). If P release from manure during storms is simulated based on the ratio of rain volume to manure mass, as in the SURPHOS model of Vadas et al. (2007c), then manure organic matter decomposition is a critical process to simulate. Floate (1970a; 1970b) found that decomposition of manure organic matter decreased with temperature, but not manure moisture, and measured a daily rate of decomposition of about 0.003, which aggress well with the rate of decomposition measured by Dao and Schwartz (2010) during a 353-day incubation of dairy manure. This 0.003 rate factor is used in the SURPHOS model and is multiplied by the square root of a temperature factor taken from Stroo et al. (1989) as:

$$TFA = [(2)(32^2)(T^2)\text{-}T^4] / 32^4 \qquad\qquad [14.5]$$

where T is average daily air temperature in degrees Celsius. There is evidence that the rate of decomposition is not a function of manure type (Vadas et al., 2007a) or manure management, such as chemical amendment (Gilmour et al., 2004), but a more systematic review of such variables may help develop the most accurate manure decomposition algorithms. It would also seem logical that as manure organic matter decomposes, the area manure covers would also decrease proportionally. This in turn would affect how much rain interacts with manure mass to release P during storms. However, there is apparently no information in the literature verifying that manure cover decreases in proportion with manure mass, although data for grazing cattle suggest the area of fecal pats may decrease more slowly than mass (Sinton et al., 2007).

14.6.2 Physical Assimilation of Manure

Physical assimilation of manure solids into soil by bioturbation should have a similar effect as manure decomposition on altering the ratio of rain volume to manure solids and thus the amount of P released from manure during storms. However, assimilation will also decrease the amount of manure P on the soil surface in proportion to the rate of solids assimilation. Esse et al. (2001) and Gallagher and Wollenhaupt (1997) provide estimates of the rate of manure or plant residue assimilation into soils by macrofauna and earthworms. There have also been numerous studies on the rate of disappearance of fecal dung pats of grazing cattle (Aarons et al., 2004a; Aarons et al., 2004b; Dickinson and Craig, 1990; Dickinson et al., 1981; Dimander et al., 2003; Gittings et al., 1994; Holter, 1974; Holter, 1977; Holter, 1979; Holter and Hendriksen, 1988; Lumaret and Kadiri, 1995; Lumaret et al., 1993), whose data can be used to develop algorithms to simulate manure assimilation into soils. Data from these and other studies demonstrate that manure assimilation proceeds between a maximum rate of about 800 kg ha^{-1} d^{-1} and a minimum rate of about 30 kg ha^{-1} d^{-1}. Differences in daily rates appear to be a positive function of both manure moisture content and temperature. In the SURPHOS model, the assimilation rate is calculated as:

$$\text{Assimilation Rate} = 30.0 * e^{(2.5 * \text{manure moisture})} \qquad [14.6]$$

where manure moisture content is expressed as a decimal between 0.0 and 1.0. This assimilation rate is then multiplied by the temperature factor in Eq. [14.5]. In SURPHOS, changes in manure moisture content are simulated such that manure dries at a daily rate of:

$$\text{Drying Rate} = -0.05 \ (\% \text{ of Manure Mass Applied}) + 0.075 \qquad [14.7]$$

This drying rate is then multiplied by the temperature factor in Eq. [14.5]. The drying rate is empirically based on field-measured manure moisture data from Sinton et al. (2007) and is designed so that freshly applied manure dries slower than more decomposed manure that assumedly has more surface area exposed. When rain occurs in SURPHOS, manure moisture content increases at a rate of:

$$\text{Wetting Rate} = -0.3 \ (\% \text{ manure moisture}) + 0.27 \qquad [14.8]$$

The wetting rate is designed so that dry manure takes on more moisture than wet manures when rain occurs.

14.6.3. Manure P Transformations

Transformation of manure P from non-WEP to WEP forms is a critical process for modeling manure P transformations and loss in runoff. However, data concerning long-term manure and P transformations in the field are limited for surface-applied manures. In a 14 to 18-month field study in both Texas and Pennsylvania, USA, Vadas et al. (2007a) clearly showed that manure P initially in non-WEP forms transforms to WEP over time. Laboratory-scale manure incubations by McGrath et al. (2005) and Warren et al. (2008) showed similar trends of manure P transformation.

The first step in modeling manure P transformation is to determine the distribution of chemical forms of P in manure that is land-applied. If one assumes that manure WEP can be quantified by a 60 min extraction at an extraction ratio of 250:1, then WEP may comprise only 25-50% of manure total P (Kleinman et al., 2005). The remaining manure P is initially in non-WEP forms, both organic and inorganic. Because manure WEP concentrations can depend greatly on animal diet (Hanrahan et al., 2009; Maguire et al., 2005) and manure management practices (Toor et al., 2005a), manure WEP, both organic and inorganic, should be quantified and the data used as specific input to a simulation model. The remainder of manure P is in non-WEP forms and is distributed between inorganic and organic forms of varying extractability. A number of studies have attempted to quantify specific P forms in manure, by a variety of chemical and instrumental procedures (Ajiboye et al., 2004; Dou et al., 2000; He et al., 2003; He et al., 2006a; He and Honeycutt, 2001; McDowell and Stewart, 2005; McGrath et al., 2005; Sharpley and Moyer, 2000; Turner and Leytem, 2004). All of these cited studies used a modified Hedley et al. (1982) fractionation procedure. Other studies have attempted to characterized manure P forms with spectroscopic or HPLC analysis (Gungor et al., 2007; He et al., 2009; Leytem et al., 2008; McDowell et al., 2008; Shand et al., 2005; Toor et al., 2005a). Such data can be used to estimate the distribution of manure non-WEP in organic and inorganic forms. For example, in developing the SURPHOS model, Vadas et al. (2007c) used fractionation data and assumed that NaOH-extractable, HCl-extractable, and residual P represented manure non-WEP. When HCl-extractable P was not fractionated into organic and inorganic P, it was assumed 65% of this P was inorganic (He et al., 2006b). It was also assumed all residual P was organic, although there is a lack of data to confirm this assumption. Based on the data from the manure chemical fractionation studies cited above, the SURPHOS model assumes that 25% of manure non-WEP is inorganic and 75% is organic. This 25/75 manure non-WEP division is an area of active research that requires more investigation. For example, animal species, animal diet, or manure management could alter the distribution of manure P among organic and inorganic forms (He et al., 2009; McDowell et al., 2008; Toor et al., 2005a). If such animal and manure management practices alter manure P in consistent ways, then algorithms could be developed to better simulate the long-term fate of manure P after land application. For example, studies investigating forms of P in manure that has been chemically amended to reduce P solubility suggest that amendments may not change the form of manure non-WEP, but rather change how those P forms are adsorbed to other Ca, Al, or Fe minerals (Peak et al., 2002; Warren et al., 2008). There are a growing number of studies that have characterized P forms in animal manures and the effect of diet and manure management on those P forms, but there has been no systematic review of those studies and translation of their data into usable model algorithms.

The second step in modeling manure P transformation is to determine the rate at which manure non-WEP transforms to WEP. Manure P transformation is essentially the mineralization of organic non-WEP and organic WEP into inorganic WEP, but also includes transformation of inorganic non-WEP into inorganic WEP by desorption or dissolution reactions. Data on the rate of manure P transformation are limited. In a 14 to 18-month field study, Vadas et al. (2007a) observed that manure P initially in non-WEP forms transforms to WEP over time in a manner that enables the manure WEP pool to supply P to runoff for months after manure application. In that study, the rate of manure P transformation was 0.0014 d^{-1} in Texas poultry manure and 0.0013 d^{-1} and 0.0016 d^{-1} in Pennsylvania poultry and dairy manures, respectively. However, the consistency among these transformation rates suggests the nature of manure P transformations is consistent across manure types and regions.

In a 150 d laboratory incubation study, McGrath et al. (2005) presented manure P transformation data representing an average P transformation rate of 0.006 d^{-1} for four poultry manures stored wet under temperatures ranging from 25-30°C. Warren et al. (2008) presented data for a P transformation rate of 0.012 d^{-1} in a poultry manure incubated for 93 d at 25°C. Jalali and Ranjbar (2009) presented data for a transformation rate of 0.008 d^{-1} in poultry and sheep manure incubated for 84 d at 25°C. Dao and Schwartz (2010) measured P transformation rates of about 0.005 for dairy manure incubated for 353 d at 22°C. The rates from these incubation studies are greater than rates from the field study of Vadas et al. (2007a) mostly likely because of more ideal moisture and temperature conditions during the incubation that enabled faster P transformations. Similar to the rates from Vadas et al. (2007a), the strong consistency among these laboratory incubation rates suggest P transformation for a variety of manures can be modeled with the same algorithms.

The rate of manure P transformation can be a function of environmental conditions as well as manure properties. McGrath et al. (2005) clearly showed that the rate of transformation is a function of both temperature and manure moisture. For poultry manures stored dry, the rate of transformation was negligible, and the rate of transformation was much greater when temperatures ranged from 20 to 30°C than from 5 to 15°C. Furthermore, Warren et al. (2008) showed that chemically amending poultry manure with alum did not change the forms of manure P, but it did slow the transformation rate from 0.012 to 0.002 d^{-1}. Thus, manure P transformations will clearly be a function of climate, including temperature and rainfall that will wet manure on the soil surface, and manure management, such as chemical amendment. Data from McGrath et al. (2005) also suggest that the P transformation rate is much greater for organic non-WEP than inorganic non-WEP. Based on data from the studies cited above, the SURPHOS model of Vadas et al. (2007c) uses a transformation rate of 0.01 d^{-1} for organic non-WEP and 0.0025 d^{-1} for inorganic non-WEP. These rates are then tempered by temperature and manure moisture factors that reduce overall transformation rates to values similar to those measured in the cited studies.

Data available for the rates of manure P transformations are apparently limited to the studies of Dao and Schwartz (2010), Jalali and Ranjbar (2009), Warren et al. (2008), McGrath et al. (2005), and Vadas et al. (2007a). There are an ever-increasing number of studies that have investigated the forms of organic and inorganic P in manure. As the analytical ability to determine specific P forms in manures improves, such studies should systematically investigate how differences in animal species, animal diet, or manure management, and especially time affect the nature of manure P transformations after land application (Shand et al., 2005).

14.7. SIMULATING MANURE PHOSPHORUS PROCESSES FOR GRAZING ANIMALS

This chapter has until now focused on manure and P processes for situations where manure is applied to fields by machine. Animal feces can also be spread on fields by grazing animals. While the conceptual model in Figure 14.1 can be used to simulate P process for machine-applied manure and feces deposited by grazing animals, there are some unique aspects to simulating grazing situations. These include physical and chemical properties of feces, fecal deposition quantities and spatial distribution, fecal pat decay, and fecal P transformations.

Machine-applied manure is a mixture of feces, urine, water, and bedding material and can vary widely in its dry matter content and subsequent likelihood to infiltrate into soil at the time of application (Kleinman et al., 2005). Similarly, the consistency of feces from grazing animals can vary due to dietary changes or time of year and subsequent differences in pasture grasses consumed (Hirata et al., 2009). These differences in fecal consistency could affect the ability of manure and its P to infiltrate into soil at the time of fecal deposition. However, there are apparently no studies in the literature that investigate such a possibility and its potential impact on simulating P loss in runoff from grazing manure.

In grazing situations, it is also important to simulate total feces production, distribution, and P content. The amount of feces deposited on pastures from grazing animals will be a function of herd management, animal species, animal diet, and animal size and stage of development (Morse et al., 1994; Wilkerson et al., 1997). This may be especially true for lactating dairy cattle whose dry matter intake and feces production can vary according to stage of lactation and milk production. Distribution of deposited feces may also be an important parameter to simulate as it could affect how much rain interacts with feces and releases nutrients during a storm, especially if feces is concentrated in feeding, watering, shade, or pasture entrance areas (Richards and Wolton, 1976; White et al., 2001). The total P content of feces from grazing manure can also vary significantly due to herd management and animal characteristics and diet (McDowell et al., 2008; Rowarth et al., 1988) .

The decomposition and assimilation of deposited feces (fecal pats) has been investigated in a number of studies (Aarons et al., 2004a; Holter, 1979; Holter and Hendriksen, 1988; Lee and Wall, 2006; Madsen et al., 1990). These studies show that fecal pats may be completely integrated into soil after about 100 days after deposition, although the rate of disappearance can vary widely depending on climate and feces moisture content (Dickinson and Craig, 1990; Dickinson et al., 1981; Floate, 1970a; Floate, 1970b; Rowarth et al., 1985; Wilkerson et al., 1997), and the presence of specific soil fauna. For example, Hirata et al. (2009) observed very slow rates of cattle feces decomposition in grazed forest and proposed that the lack of dung beetles prevented faster decomposition, such as might occur in grazed grasslands. Accurate simulation of fecal pat disappearance is critical, as longer disappearance times allow greater opportunity of P release from fecal pats and possible loss in runoff. However, there appears to be no other existing model of dung pat disappearance other than the routines used in the SURPHOS model. In SURPHOS, fecal organic matter decomposition is calculated using the daily rate of 0.003 and the temperature factor in Eq. [14.5], as described previously. The assimilation rate is calculated:

$$\text{Assimilation rate} \ = 30.0 * e^{[3.5*(\sqrt{\text{manure moisture}})]} \qquad\qquad [14.9]$$

which is similar to Eq. [14.6]. This assimilation rate is then multiplied by the temperature factor in Eq. [14.5], but raised to the power of 0.1. Overall, these equations provide for faster assimilation of fecal dung pats than machine-applied manure. Changes in manure moisture content are calculated using Eq. [14.7] and [14.8].

The release of P from fecal pats during a storm has not been well investigated. Because fecal pats may form a small mound, interaction of rain and dry matter will be different than manure that is distributed across a field by machine. The possibility of feces drying and developing a crust could also play an important role in the ability of rain to infiltrate fecal pats and release P (Dimander et al., 2003; Sinton et al., 2007). There is scant information in the literature concerning the mechanisms of crust formation and its effects on fecal P transformations, although it appears that dry conditions after deposition promote crust formation and wet conditions inhibit it (Dickinson and Craig, 1990; Dickinson et al., 1981).

The transformation of P forms in deposited feces is also a critical process to simulate for predicting P loss in runoff from grazing systems. However, there are limited data on the transformations of feces P with time after deposition in a grazing scenario. Several field studies have shown that the concentration of total P in feces remains fairly stable or declines slightly with time after deposition (Aarons et al., 2004a; Dickinson and Craig, 1990; Rodriguez et al., 2001; Shand et al., 2005). Because the mass of feces is also decreasing during this period, this means that P is being released from feces so that overall P concentrations remain fairly stable (Vadas et al., 2007a). However, these cited studies did not investigate specific forms of fecal P and thus provide data to estimate the rate at which fecal P may transform from non-WEP to WEP forms. Two studies that conducted more controlled laboratory experiments investigated transformations in fecal P with time. Floate (1970c) incubated four sheep feces at 30°C and 100% moisture holding capacity for 12 weeks and measured net transformation of feces organic P ranging from 5 to 30% of original feces total P. Given the variability in feces total P, this translated into an average daily rate of P transformation of 0.014. Given the ideal temperature and moisture incubation conditions, this could be considered an upper potential for P transformation, and it compares well to the 0.01 rate used in the SURPHOS model (Vadas et al., 2007c). In a similar incubation of dairy cattle feces over 105 days, Fuentes et al. (2009) observed organic P transformation at a greater daily rate of 0.023. Given that their incubation temperatures ranged from 40 to 50°C for the first 40 d, this rate could be considered greater than an upper limit that may occur under field conditions. Therefore, it appears from these limited data that the 0.01 P transformation rate used in the SURPHOS model for machine-applied manure could also be used for grazing-deposited feces. Studies to further confirm these fecal P transformation rates would help validate current model parameters.

14.8. CONCLUSION

A broad array of research over the last 10 to 20 years has clearly shown that land-applied animal manures can be a significant source of P loss from agricultural fields to surrounding water bodies. This research has greatly improved understanding of the processes controlling manure P loss in runoff, but this improved understanding has not been adequately translated into computer models that are commonly used to assess the impact of agricultural activities on the environment. In fact, current versions of most models do not simulate surface application of manure and associated P loss in runoff, which research has clearly shown to be a critical process that cannot be overlooked. Advances in modeling manure P transformations and loss in runoff have been made only recently (Vadas et al., 2007c), and more progress is needed. It is clear that research on animal manures and the environmental impact of P needs to shift towards developing and improving models.

Key aspects of manure processes include: i) infiltration of manure P into soil at application, ii) release of manure P during storms and transfer to runoff, and iii) manure P transformations between storms, including decomposition, assimilation into soil, and P transformation. The major research gaps in this process include: i) manure P infiltration into soil at application and availability to runoff, ii) how manure properties control P release, iii) the rate of manure decomposition as a function of manure type or management, iv) assimilation of manure and P into soil as a function of soil fauna, climate, and manure characteristics, v) manure P transformation as a function of climate, animal species, diet, or manure management, and vi) manure P processes for grazing animals.

REFERENCES

Aarons, S.R., C.R. O'Connor, and C.J.P. Gourley. 2004a. Dung decomposition in temperate dairy pastures - I. Changes in soil chemical properties. *Aust. J. Soil. Res.* 42:107-114.

Aarons, S.R., H.M. Hosseini, L. Dorling, and C.J.P. Gourley. 2004b. Dung decomposition in temperate dairy pastures - II. Contribution to plant-available soil phosphorus. *Aust. J. Soil. Res.* 42:115-123.

Ajiboye, B., O.O. Akinremi, and G.J. Racz. 2004. Laboratory characterization of phosphorus in fresh and oven-dried organic amendments. *J. Environ. Qual.* 33:1062-1069.

Arnold, J.G., R. Srinivasan, R.S. Muttiah, and J.R. Williams. 1998. Large area hydrologic modeling and assessment - Part 1: Model development. *J. Am. Water Resour. Assoc.* 34:73-89.

Bechmann, M.E., P.J.A. Kleinman, A.N. Sharpley, and L.S. Saporito. 2005. Freeze-thaw effects on phosphorus loss in runoff froin manured and catch-cropped soils. *J. Environ. Qual.* 34:2301-2309.

Bittman, S., L.J.P. van Vliet, C.G. Kowalenko, S. McGinn, D.E. Hunt, and F. Bounaix. 2005. Surface-banding liquid manure over aeration slots: A new low-disturbance method for reducing ammonia emissions and improving yield of perennial grasses. *Agron. J.* 97:1304-1313.

Bouraoui, F., and T.A. Dillaha. 1996. Answers-2000: Runoff and Sediment Transport Model. *J. Environ. Engin. Asce* 122:493-502.

Carpenter, S.R., N.F. Caraco, D.L. Correll, R.W. Howarth, A.N. Sharpley, and V.H. Smith. 1998. Nonpoint pollution of surface waters with phosphorus and nitrogen. *Ecol. Appl.* 8:559-568.

Cooperband, L.R., and L.W. Good. 2002. Biogenic phosphate minerals in manure: Implications for phosphorus loss to surface waters. *Environ. Sci. Technol.* 36:5075-5082.

Dao, T.H., and R.C. Schwartz. 2010. Mineralizable phosphorus, nitrogen, and carbon relationships in dairy manure at various carbon-to-phosphorus ratios. *Bioresour. Technol.* 101:3567-3574.

Daverede, I.C., A.N. Kravchenko, R.G. Hoeft, E.D. Nafziger, D.G. Bullock, J.J. Warren, and L.C. Gonzini. 2004. Phosphorus runoff from incorporated and surface-applied liquid swine manure and phosphorus fertilizer. *J. Environ. Qual.* 33:1535-1544.

Dickinson, C.H., and G. Craig. 1990. Effects of Water on the Decomposition and Release of Nutrients from Cow Pats. *New. Phytol.* 115:139-147.

Dickinson, C.H., V.S.H. Underhay, and V. Ross. 1981. Effect of Season, Soil Fauna and Water-Content on the Decomposition of Cattle Dung Pats. *New. Phytol.* 88:129-141.

Dimander, S.O., J. Hoglund, and P.J. Waller. 2003. Disintegration of dung pats from cattle treated with the ivermectin anthelmintic bolus, or the biocontrol agent Duddingtonia flagrans. *Acta. Vet. Scand.* 44:171-180.

Dou, Z., J.D. Toth, D.T. Galligan, C.F. Ramberg, and J.D. Ferguson. 2000. Laboratory procedures for characterizing manure phosphorus. *J. Environ. Qual.* 29:508-514.

Dou, Z.X., C.F. Ramberg, L. Chapuis-Lardy, J.D. Toth, Y. Wang, R.J. Munson, Z.G. Wu, L.E. Chase, R.A. Kohn, K.F. Knowlton, and J.D. Ferguson. 2007. A novel test for measuring and managing potential phosphorus loss from dairy cattle feces. *Environ. Sci. Technol.* 41:4361-4366.

Esse, P.C., A. Buerkert, P. Hiernaux, and A. Assa. 2001. Decomposition of and nutrient release from ruminant manure on acid sandy soils in the Sahelian zone of Niger, West Africa. *Agric. Ecosys. Environ.* 83:55-63.

Floate, M.J.S. 1970a. Decomposition of organic materials from hill soils and pastures: IV. The effect of moisture content on the mineralization of carbon, nitrogen and phosphorus from plant materials and sheep faeces. *Soil. Biol. Biochem.* 2:275-283.

Floate, M.J.S. 1970b. Decomposition of organic materials from hill soils and pastures: III. The effect of temperature on the mineralization of carbon, nitrogen and phosphorus from plant materials and sheep faeces. *Soil. Biol. Biochem.* 2:187-196.

Floate, M.J.S. 1970c. Decomposition of organic materials from hill soils and pastures: II. Comparative studies on the mineralization of carbon, nitrogen and phosphorus from plant materials and sheep faeces. *Soil. Biol. Biochem.* 2:173-185.

Fuentes, B., M. Jorquera, and M.D. Mora. 2009. Dynamics of phosphorus and phytate-utilizing bacteria during aerobic degradation of dairy cattle dung. *Chemosphere* 74:325-331.

Gallagher, A.V., and N.C. Wollenhaupt. 1997. Surface alfalfa residue removal by earthworms Lumbricus terrestris L in a no-till agroecosystem. *Soil. Biol. Biochem.* 29:477-479.

Gerard-Marchant, P., M.T. Walter, and T.S. Steenhuis. 2005. Simple models for phosphorus loss from manure during rainfall. *J. Environ. Qual.* 34:872-876.

Gilmour, J.T., M.A. Koehler, M.L. Cabrera, L. Szajdak, and P.A. Moore. 2004. Alum treatment of poultry litter: Decomposition and nitrogen dynamics. *J. Environ. Qual.* 33:402-405.

Gittings, T., P.S. Giller, and G. Stakelum. 1994. Dung Decomposition in Contrasting Temperate Pastures in Relation to Dung Beetle and Earthworm Activity. *Pedobiologia* 38:455-474.

Gungor, K., A. Jurgensen, and K.G. Karthikeyan. 2007. Determination of phosphorus speciation in dairy manure using XRD and XANES Spectroscopy. *J. Environ. Qual.* 36:1856-1863.

Guo, M., and W. Song. 2009. Nutrient value of alum-treated poultry litter for land application. *Poul. Sci.* 88:1782-1792.

Guo, M.X., N. Tongtavee, and M. Labreveux. 2009a. Nutrient dynamics of field-weathered Delmarva poultry litter: implications for land application. *Biol. Fert. Soils* 45:829-838.

Guo, M.X., M. Labreveux, and W.P. Song. 2009b. Nutrient release from bisulfate-amended phytase-diet poultry litter under simulated weathering conditions. *Waste Manage.* 29:2151-2159.

Haggard, B.E., P.A. Vadas, D.R. Smith, P.B. DeLaune, and P.A. Moore. 2005. Effect of poultry litter to water ratios on extractable phosphorus content and its relation to runoff phosphorus concentrations. *Biosys. Engin.* 92:409-417.

Hanrahan, L.R., W.E. Jokela, and J.R. Knapp. 2009. Dairy Diet Phosphorus and Rainfall Timing Effects on Runoff Phosphorus from Land-Applied Manure. *J. Environ. Qual.* 38:212-217.

He, Z., C.W. Honeycutt, and T.S. Griffin. 2003. Comparative investigation of sequentially extracted phosphorus fractions in a sandy loam soil and a swine manure. *Commun. Soil Sci. Plant. Anal.* 34:1729-1742.

He, Z., Z.N. Senwo, R.N. Mankolo, and C.W. Honeycutt. 2006a. Phosphorus fractions in poultry litter characterized by sequential fractionation coupled with phosphatase hydrolysis. *Journal of Food Agriculture and Environment* 4:304-312.

He, Z.Q., and C.W. Honeycutt. 2001. Enzymatic characterization of organic phosphorus in animal manure. *J. Environ. Qual.* 30:1685-1692.

He, Z.Q., A.M. Fortuna, Z.N. Senwo, I.A. Tazisong, C.W. Honeycutt, and T.S. Griffin. 2006b. Hydrochloric fractions in Hedley fractionation may contain inorganic and organic phosphates. *Soil Sci. Soc. Am. J.* 70:893-899.

He, Z.Q., C.W. Honeycutt, T.S. Griffin, B.J. Cade-Menun, P.J. Pellechia, and Z.X. Dou. 2009. Phosphorus Forms in Conventional and Organic Dairy Manure Identified by Solution and Solid State P-31 NMR Spectroscopy. *J. Environ. Qual.* 38:1909-1918.

Heathwaite, A.L., and R.M. Dils. 2000. Characterising phosphorus loss in surface and subsurface hydrological pathways. *Sci. Total. Environ.* 251:523-538.

Heckrath, G., P.C. Brookes, P.R. Poulton, and K.W.T. Goulding. 1995. Phosphorus Leaching from Soils Containing Different Phosphorus Concentrations in the Broadbalk Experiment. *J. Environ. Qual.* 24:904-910.

Hedley, M.J., J.W.B. Stewart, and B.S. Chauhan. 1982. Changes in Inorganic and Organic Soil-Phosphorus Fractions Induced by Cultivation Practices and by Laboratory Incubations. *Soil Sci. Soc. Am. J.* 46:970-976.

Hill, D.T., and J.W. Baier. 2000. Physical and chemical properties of screened-flushed pig slurry waste. *J. Agr. Eng. Res.* 77:441-448.

Hirata, M., N. Hasegawa, M. Nomura, H. Ito, K. Nogami, and T. Sonoda. 2009. Deposition and decomposition of cattle dung in forest grazing in southern Kyushu, *Japan. Ecol. Res.* 24:119-125.

Hively, W.D., P. Gerard-Marchant, and T.S. Steenhuis. 2006. Distributed hydrological modeling of total dissolved phosphorus transport in an agricultural landscape, part II: dissolved phosphorus transport. *Hydrol. Earth Syst. Sci.* 10:263-276.

Holmberg, R.D., D.T. Hill, T.J. Prince, and N.J. Vandyke. 1983. Potential of Solid-Liquid Separation of Swine Wastes for Methane Production. *Trans. ASAE* 26:1803-1807.

Holter, P. 1974. *Food Utilization of Dung-Eating Aphodius Larvae (Scarabaeidae)*. Oikos 25:71-79.

Holter, P. 1977. Experiment on Dung Removal by Aphodius-Larvae (Scarabaeidae) and Earthworms. *Oikos* 28:130-136.

Holter, P. 1979. Effect of Dung-Beetles (Aphodius Spp) and Earthworms on the Disappearance of Cattle Dung. *Oikos* 32:393-402.

Holter, P., and N.B. Hendriksen. 1988. Respiratory Loss and Bulk Export of Organic-Matter from Cattle Dung Pats - a Field-Study. *Holarctic Ecol.* 11:81-86.

Jalali, M., and F. Ranjbar. 2009. Rates of decomposition and phosphorus release from organic residues related to residue composition. *J. Plant. Nutr. Soi.l Sci.* 172:353-359.

Kleinman, P., D. Sullivan, A. Wolf, R. Brandt, Z.X. Dou, H. Elliott, J. Kovar, A. Leytem, R. Maguire, P. Moore, L. Saporito, A. Sharpley, A. Shober, T. Sims, J. Toth, G. Toor, H.L. Zhang, and T.Q. Zhang. 2007. Selection of a water-extractable phosphorus test for manures and biosolids as an indicator of runoff loss potential. *J. Environ. Qual.* 36:1357-1367.

Kleinman, P.J.A., and A.N. Sharpley. 2003. Effect of broadcast manure on runoff phosphorus concentrations over successive rainfall events. *J. Environ. Qual.* 32:1072-1081.

Kleinman, P.J.A., A.N. Sharpley, B.G. Moyer, and G.F. Elwinger. 2002a. Effect of mineral and manure phosphorus sources on runoff phosphorus. *J. Environ. Qual.* 31:2026-2033.

Kleinman, P.J.A., A.N. Sharpley, A.M. Wolf, D.B. Beegle, and P.A. Moore. 2002b. Measuring water-extractable phosphorus in manure as an indicator of phosphorus in runoff. *Soil Sci. Soc. Am. J.* 66:2009-2015.

Kleinman, P.J.A., A.M. Wolf, A.N. Sharpley, D.B. Beegle, and L.S. Saporito. 2005. Survey of water-extractable phosphorus in livestock manures. *Soil Sci. Soc. Am. J.* 69:701-708.

Lee, C.M., and R. Wall. 2006. Cow-dung colonization and decomposition following insect exclusion. *Bull. Entomol. Res.* 96:315-322.

Leonard, R.A., W.G. Knisel, and D.A. Still. 1987. Gleams - Groundwater Loading Effects of Agricultural Management-Systems. *Trans.* ASAE 30:1403-1418.

Lewis, D.R., and M.B. McGechan. 2002. A review of field scale phosphorus dynamics models. *Biosys. Engin.* 82:359-380.

Leytem, A.B., P.W. Plumstead, R.O. Maguire, P. Kwanyuen, and J. Brake. 2007. What aspect of dietary modification in broilers controls litter water-soluble phosphorus: Dietary phosphorus, phytase, or calcium? *J. Environ. Qua*l. 36:453-463.

Leytem, A.B., P. Kwanyuen, P.W. Plumstead, R.O. Maguire, and J. Brake. 2008. Evaluation of phosphorus characterization in broiler ileal digesta, manure, and litter samples: P-31-NMR vs. HPLC. *J. Environ. Qual.* 37:494-500.

Lumaret, J.P., and N. Kadiri. 1995. The influence of the first wave of colonizing insects on cattle dung dispersal. *Pedobiologia* 39:506-517.

Lumaret, J.P., E. Galante, C. Lumbreras, J. Mena, M. Bertrand, J.L. Bernal, J.F. Cooper, N. Kadiri, and D. Crowe. 1993. Field Effects of Ivermectin Residues on Dung Beetles. *J. Appl. Ecol.* 30:428-436.

Madsen, M., B.O. Nielsen, P. Holter, O.C. Pedersen, J.B. Jespersen, K.M.V. Jensen, P. Nansen, and J. Gronvold. 1990. Treating Cattle with Ivermectin - Effects on the Fauna and Decomposition of Dung Pats. J. *Appl. Ecol.* 27:1-15.

Maguire, R.O., J.T. Sims, and T.J. Applegate. 2005. Phytase supplementation and reduced-phosphorus turkey diets reduce phosphorus loss in runoff following litter application. *J. Environ. Qual.* 34:359-369.

Mamo, M., C. Wortmann, and C. Brubaker. 2007. Manure phosphorus fractions: Development of analytical methods and variation with manure types. *Commun. Soil Sci. Plant. Anal.* 38:935-947.

McDowell, R.W., and I. Stewart. 2005. Phosphorus in fresh and dry dung of grazing dairy cattle, deer, and sheep: Sequential fraction and phosphorus-31 nuclear magnetic resonance analyses. *J. Environ. Qual.* 34:598-607.

McDowell, R.W., Z. Dou, J.D. Toth, B.J. Cade-Menun, P.J.A. Kleinman, K. Soder, and L. Saporito. 2008. A comparison of phosphorus speciation and potential bioavailability in feed and feces of different dairy herds using IT nuclear magnetic resonance spectroscopy. *J. Environ. Qual.* 37:741-752.

McGrath, J.M., J.T. Sims, R.O. Maguire, W.W. Saylor, C.R. Angel, and B.L. Turner. 2005. Broiler diet modification and litter storage: Impacts on phosphorus in litters, soils, and runoff. *J. Environ. Qual.* 34:1896-1909.

McLaughlin, N.B., Y.X. Li, S. Bittman, D.R. Lapin, S.D. Burtt, and B.S. Patterson. 2006. Draft requirements for contrasting liquid manure injection equipment. *Can. Biosys. Eng.* 48:2.29-2.37.

Moller, H.B., S.G. Sommer, and B.K. Ahring. 2002. Separation efficiency and particle size distribution in relation to manure type and storage conditions. *Bioresour. Technol.* 85:189-196.

Moore, P.A., T.C. Daniel, and D.R. Edwards. 2000. Reducing phosphorus runoff and inhibiting ammonia loss from poultry manure with aluminum sulfate. *J. Environ. Qual.* 29:37-49.

Morken, J., and S. Sakshaug. 1998. Direct ground injection of livestock waste slurry to avoid ammonia emission. *Nutr. Cycl. Agroeco.* 51:59-63.

Morse, D., R.A. Nordstedt, H.H. Head, and H.H. Vanhorn. 1994. Production and Characteristics of Manure from Lactating Dairy-Cows in Florida. *Trans. ASAE* 37:275-279.

Muck, R.E., and D.C. Ludington. 1979. Simulation of Nitrogen and Phosphorus Leaching from Poultry Manure. *Trans. ASAE* 22:1087-&.

Novak, J.M., D.W. Watts, P.G. Hunt, and K.C. Stone. 2000. Phosphorus movement through a coastal plain soil after a decade of intensive swine manure application. *J. Environ. Qual.* 29:1310-1315.

Peak, D., J.T. Sims, and D.L. Sparks. 2002. Solid-state speciation of natural and alum-amended poultry litter using XANES spectroscopy. *Environ. Sci. Technol.* 36:4253-4261.

Pierson, S.T., M.L. Cabrera, G.K. Evanylo, H.A. Kuykendall, C.S. Hoveland, M.A. McCann, and L.T. West. 2001a. Phosphorus and ammonium concentrations in surface runoff from grasslands fertilized with broiler litter. *J. Environ. Qual.* 30:1784-1789.

Pierson, S.T., M.L. Cabrera, G.K. Evanylo, P.D. Schroeder, D.E. Radcliffe, H.A. Kuykendall, V.W. Benson, J.R. Williams, C.S. Hoveland, and M.A. McCann. 2001b. Phosphorus losses from grasslands fertilized with broiler litter: EPIC simulations. *J. Environ. Qual.* 30:1790-1795.

Porter, P.S., and C.A. Sanchez. 1992. The Effect of Soil Properties on Phosphorus Sorption by Everglades Histosols. *Soil Sci.* 154:387-398.

Pote, D.H., B.A. Reed, T.C. Daniel, D.J. Nichols, P.A. Moore, D.R. Edwards, and S. Formica. 2001. Water-quality effects of infiltration rate and manure application rate for soils receiving swine manure. *J. Soil Water Conserv.* 56:32-37.

Richards, I.R., and K.M. Wolton. 1976. Spatial-Distribution of Excreta under Intensive Cattle Grazing. *J. Brit. Grassland Soc.* 31:89-92.

Robinson, J.S., and A.N. Sharpley. 1995. Release of Nitrogen and Phosphorus from Poultry Litter. *J. Environ. Qual.* 24:62-67.

Rodriguez, I., G. Crespo, V. Torres, and S. Fraga. 2001. Changes in the chemical composition of cattle dung in grasslands. *Cuban J. Agr. Sci.* 35:285-290.

Rowarth, J.S., A.G. Gillingham, R.W. Tillman, and J.K. Syers. 1985. Release of Phosphorus from Sheep Feces on Grazed, Hill Country Pastures. New Zeal. *J. Agric. Res.* 28:497-504.

Rowarth, J.S., A.G. Gillingham, R.W. Tillman, and J.K. Syers. 1988. Effects of Season and Fertilizer Rate on Phosphorus Concentrations in Pasture and Sheep Feces in Hill Country. New Zeal. *J. Agric. Res.* 31:187-193.

Shand, C.A., G. Coutts, S. Hillier, D.G. Lumsdon, and A. Chudek. 2005. Phosphorus composition of sheep feces and changes in the field determined by P-31 NMR spectroscopy and XRPD. *Environ. Sci. Technol.* 39:9205-9210.

Sharpley, A., and B. Moyer. 2000. Phosphorus forms in manure and compost and their release during simulated rainfall. *J. Environ. Qual.* 29:1462-1469.

Sharpley, A.N., P.J.A. Kleinman, R.W. McDowell, M. Gitau, and R.B. Bryant. 2002. Modeling phosphorus transport in agricultural watersheds: Processes and possibilities. *J. Soil Water Conserv.* 57:425-439.

Shigaki, F., A. Sharpley, and L.I. Prochnow. 2006. Source-related transport of phosphorus in surface runoff. *J. Environ. Qual.* 35:2229-2235.

Sinton, L.W., R.R. Braithwaite, C.H. Hall, and M.L. Mackenzie. 2007. Survival of indicator and pathogenic bacteria in bovine feces on pasture. *Appl. Environ. Microb.* 73:7917-7925.

Sistani, K.R., H.A. Torbert, T.R. Way, C.H. Bolster, D.H. Pote, and J.G. Warren. 2009. Broiler Litter Application Method and Runoff Timing Effects on Nutrient and Escherichia coli Losses from Tall Fescue Pasture. *J. Environ. Qual.* 38:1216-1223.

Smith, D.R., P.R. Owens, A.B. Leytem, and E.A. Warnemuende. 2007. Nutrient losses from manure and fertilizer applications as impacted by time to first runoff event. *Environ. Pollut.* 147:131-137.

Srinivasan, M.S., P.I.A. Kleinman, A.N. Sharpley, T. Bud, and W.J. Gburek. 2007. Hydrology of small field plots used to study phosphorus runoff under simulated rainfall. *J. Environ. Qual.* 36:1833-1842.

Stott, D.E., E.E. Alberts and M.A. Weltz. 1995 Chapter 9. Residue decomposition and management. In: D.C. Flanagan and M.A. Nearing (eds.), USDA-Water Erosion Prediction Project hillslope profile and watershed model documentation. NSERL Report No. 10, USDA-ARS National Soil Erosion Research Laboratory, West Lafayette, Indiana.

Stroo, H.F., K.L. Bristow, L.F. Elliott, R.I. Papendick, and G.S. Campbell. 1989. Predicting Rates of Wheat Residue Decomposition. *Soil Sci. Soc. Am. J.* 53:91-99.

Studnicka, J. S. 2006. Factors affecting water extractable phosphorus in manure and its relationship to phosphorus losses in runoff. Master's thesis. Department of Soil Sceince, University of Wisconsin-Madison. Madison, Wisconsin.

Tarkalson, D.D., and R.L. Mikkelsen. 2004. Runoff phosphorus losses as related to phosphorus source, application method, and application rate on a piedmont soil. *J. Environ. Qual.* 33:1424-1430.

Tasistro, A.S., M.L. Cabrera, and D.E. Kissel. 2004. Water soluble phosphorus released by poultry litter: effect of extraction pH and time after application. *Nutr. Cycl. Agroeco.* 68:223-234.

Tasistro, A.S., M.L. Cabrera, Y.B. Zhao, D.E. Kissel, K. Xia, and D.H. Franklin. 2007. Soluble phosphorus released by poultry wastes in acidified aqueous extracts. *Commun. Soil Sci. Plant. Anal.* 38:1395-1410.

Toor, G.S., B.J. Cade-Menun, and J.T. Sims. 2005a. Establishing a linkage between phosphorus forms in dairy diets, feces, and manures. *J. Environ. Qual.* 34:1380-1391.

Toor, G.S., J.D. Peak, and J.T. Sims. 2005b. Phosphorus speciation in broiler litter and turkey manure produced from modified diets. *J. Environ. Qual.* 34:687-697.

Toor, G.S., B.E. Haggard, M.S. Reiter, T.C. Daniel, and A.M. Donoghue. 2007. Phosphorus solubility in poultry litters and granulates: Influence of litter treatments and extraction ratios. *Trans. ASABE* 50:533-542.

Torbert, H.A., K.N. Potter, D.W. Hoffman, T.J. Gerik, and C.W. Richardson. 1999. Surface residue and soil moisture affect fertilizer loss in simulated runoff on a heavy clay soil. *Agron. J.* 91:606-612.

Turner, B.L., and A.B. Leytem. 2004. Phosphorus compounds in sequential extracts of animal manures: chemical speciation and a novel fractionation procedure. *Environ. Sci. Technol.* 38:6101-6108.

Vadas, P.A. 2006. Distribution of phosphorus in manure slurry and its infiltration after application to soils. *J. Environ. Qual.* 35:542-547.

Vadas, P.A., and P.J.A. Kleinman. 2006. Effect of methodology in estimating and interpreting water-extractable phosohorus in animal manures. *J. Environ. Qual.* 35:1151-1159.

Vadas, P.A., P.J.A. Kleinman, and A.N. Sharpley. 2004. A simple method to predict dissolved phosphorus in runoff from surface-applied manures. *J. Environ. Qual.* 33:749-756.

Vadas, P.A., B.E. Haggard, and W.J. Gburek. 2005. Predicting dissolved phosphorus in runoff from manured field plots. *J. Environ. Qual.* 34:1347-1353.

Vadas, P.A., R.D. Harmel, and P.J.A. Kleinman. 2007a. Transformations of soil and manure phosphorus after surface application of manure to field plots. *Nutr. Cycl. Agroeco.* 77:83-99.

Vadas, P.A., M.S. Srinivasan, P.J.A. Kleinman, J.P. Schmidt, and A.L. Allen. 2007b. Hydrology and groundwater nutrient concentrations in a ditch-drained agroecosystem. *J. Soil Water Conserv.* 62:178-188.

Vadas, P.A., W.J. Gburek, A.N. Sharpley, P.J.A. Kleinman, P.A. Moore, M.L. Cabrera, and R.D. Harmel. 2007c. A model for phosphorus transformation and runoff loss for surface-applied manures. *J. Environ. Qual.* 36:324-332.

van Vliet, L.J.P., S. Bittman, G. Derksen, and C.G. Kowalenko. 2006. Aerating grassland before manure application reduces runoff nutrient loads in a high rainfall environment. *J. Environ. Qual.* 35:903-911.

Warren, J.G., C.J. Penn, J.M. McGrath, and K. Sistani. 2008. The impact of alum addition on organic P transformations in poultry litter and litter-amended soil. *J. Environ. Qual.* 37:469-476.

White, S.K., J.E. Brummer, W.C. Leininger, G.W. Frasier, R.M. Waskom, and T.A. Bauder. 2003. Irrigated mountain meadow fertilizer application timing effects on overland flow water quality. *J. Environ. Qual.* 32:1802-1808.

White, S.L., R.E. Sheffield, S.P. Washburn, L.D. King, and J.T. Green. 2001. Spatial and time distribution of dairy cattle excreta in an intensive pasture system. *J. Environ. Qual.* 30:2180-2187.

Wilkerson, V.A., D.R. Mertens, and D.P. Casper. 1997. Prediction of excretion of manure and nitrogen by Holstein dairy cattle. *J. Dairy Sci.* 80:3193-3204.

Williams, J.R., K.G. Renard, and P.T. Dyke. 1983. Epic - a New Method for Assessing Erosions Effect on Soil Productivity. J. *Soil Water Conserv.* 38:381-383.

Withers, P.J.A., S.D. Clay, and V.G. Breeze. 2001. Phosphorus transfer in runoff following application of fertilizer, manure, and sewage sludge. *J. Environ. Qual.* 30:180-188.

Zhang, R.H., and P.W. Westerman. 1997. Solid-liquid separation of animal manure for odor control and nutrient management. *App. Engin. Agric.* 13:8.

Chapter 15

IMPROVING THE SUSTAINABILITY OF ANIMAL AGRICULTURE BY TREATING MANURE WITH ALUM

Philip A. Moore, Jr.[*]

15.1. INTRODUCTION

Two of the biggest environmental problems associated with animal manure management are ammonia (NH_3) emissions and phosphorus (P) runoff. Research conducted during the past two decades has shown that a simple topical application of aluminum sulfate (alum) to manure can greatly reduce the magnitude of both of these problems. The objective of this chapter was to provide a literature review of the research on treating poultry litter and other types of manure with alum.

15.2. AMMONIA EMISSIONS FROM POULTRY LITTER

Aerial emissions of compounds such as NH_3 from animal feeding operations (AFOs) and land receiving manure from AFOs is currently under increased scrutiny and may soon be regulated. In 2002 the National Academy of Science concluded that there was an urgent need to collect data on atmospheric emissions from these facilities (NAS, 2002). Approximately 27% of the total NH_3 emissions in the United States were estimated to originate from poultry manure (Battye et al., 1994), hence it is the atmospheric contaminant of greatest concern from poultry production.

Ammonia emissions from manure, like poultry litter, can cause both production and environmental problems. Ammonia concentrations in poultry barns often reach very high concentrations, causing poor poultry performance (Carlile, 1984). Anderson et al. (1964)

[*] philipm@uark.edu USDA-ARS, the Poultry Production and Product Safety Research Unit, Fayetteville AR 72701, USA

showed that NH_3 concentrations as low as 20 uL L^{-1} compromise the immune system of chickens, which makes them more susceptible to diseases. High NH_3 levels also were shown to damage the respiratory system of birds (Anderson et al., 1964). This may be more important now than ever, due to the threat posed by avian influenza. Kling and Quarles (1974) showed that the incidence of airsaculitis in chickens increased dramatically when the birds were exposed to high NH_3 concentrations. Feed conversion, weight gains and ocular damage in poultry have also been shown to be negatively affected by high NH_3 concentrations in poultry barns (Carlile, 1984; Miles et al., 2004; Miles et al., 2006). Such negative impacts on performance were generally observed at NH_3 levels of approximately 25 uL L^{-1} or above. Hence, Carlile (1984) recommended that NH_3 concentrations be less than 25 uL L^{-1} in poultry barns for this reason. High NH_3 concentrations in animal rearing facilities also pose a major risk to the health of agricultural workers in these facilities and may cause upper respiratory ailments (Donham, 1996, 2000).

Moore et al. (2010) recently conducted a study to determine how much nitrogen (N) is lost as NH_3 in modern poultry facilities during poultry production, during storage and following land application. They found the total NH_3 emission factor for 50 day old broilers was 45.6 g NH_3/bird marketed. The great majority of that loss occurred when the manure was still in the barn (37.4 g NH_3/bird). Nitrogen mass balance data indicated that more N was actually lost via NH_3 emissions when the litter was still in the barn than the amount of N removed in the litter between flocks and at the final cleanout (Moore et al., 2010). Ammonia losses that occurred during storage (0.18 g NH_3/bird) and following land application (7.91 g NH_3/bird) were relatively small when compared to losses that occur in the poultry houses (Moore et al., 2010).

Environmental problems linked to such high levels of NH_3 emissions include soil acidification, formation of fine particulate matter, and N deposition into aquatic systems (Schroder, 1985; Hutchinson and Viets, 1969; van Breemen et al., 1982). Ammonia losses also affect the amount of nitrogen (N) in the manure, as well as the N:P ratio. Poultry litter often has N:P ratios as low 2:1, whereas plant requirements are around 8:1 (Moore et al., 1995a). The N:P ratio can become even lower when manure is composted, due to excessive NH_3 losses, resulting in lower forage yields and increasing P runoff (DeLaune et al., 2004, 2006). Hence, it is no surprise that long-term applications of manure on pastures at levels based on the N needs of crops have resulted in high soil test P levels in some areas (Kingery et al., 1994).

15.3. Phosphorus Runoff from Poultry Litter

One of the most widespread water quality problems effecting US waterways is eutrophication (USEPA, 1992). Research has typically shown that nutrient concentrations in runoff increase with intensity of agricultural use (Carpenter et al., 1998; Pionke et al., 1996). Excessive manure applications have been shown to be one of the greatest potential threats leading to eutrophication (Duda and Finan, 1983).

Accelerated eutrophication can increase the risk of a number of water quality problems, including drinking water taste and odor problems, and the potential production of carcinogenic compounds during water treatment (Kotak et al., 1993; Sharpley et al., 1994). Blue green algae can produce the compound geosmin, causing taste and odor problems in drinking water supplies (Gerber, 1983; Izagguirre et al., 1982). The City of Tulsa, OK, sued eight poultry companies in Arkansas in 2002 over non-point source pollution resulting in taste and odor problems because of increases in geosmin in the Eucha/Spavinaw Watershed (Tulsa's main source of drinking water). Currently the state of Oklahoma is suing the same poultry companies for non-point source P pollution in the Illinois River Watershed.

Phosphorus concentrations in runoff water from fields can be very high after manure applications (Edwards and Daniel, 1992a,b). Phosphorus is generally the nutrient that limits eutrophication in freshwater, whereas in brackish and saltwater, N is more likely to limit the growth of algae (Schindler, 1977). Phosphorus in runoff can be in the particulate or in the dissolved (soluble) form, with the latter being more available for algal uptake (Sonzogni et al., 1982). The majority of P in runoff from pastures and land in minimal tillage is in the soluble form (Edwards and Daniel, 1993; Sharpley et al., 1992; Shreve et al., 1995).

15.4. USING ALUM TO REDUCE PHOSPHORUS SOLUBILITY AND RUNOFF FROM MANURE

Prior to the early 1990s, no attempts had been made to precipitate P in animal manure using metal compounds that react with P, such as alum. Moore and Miller (1994) hypothesized that soluble P levels in manure caused high levels of soluble P in runoff from pastures. They speculated that the addition of compounds such as alum ($Al_2(SO_4)_3 \cdot 14H_2O$) or slaked lime ($Ca(OH)_2$) would reduce soluble P in manure through either chemical precipitation or adsorption onto metal oxides or hydroxides, thus reducing the amount of P available for runoff or leaching. Moore and Miller (1994) evaluated 100 chemical treatments in a laboratory study, which included ten calcium (Ca) treatments, four aluminum (Al) treatments) and six iron (Fe) treatments; all of which were evaluated at five different rates. They found that all of the Al, Ca and Fe amendments would reduce soluble P in poultry litter, depending on the rate of application. Water soluble P was reduced from 2,000 mg P/kg litter in untreated litter to less than 1 mg P/kg litter with many of the amendments studied, including alum (Figure 15.1).

Moore and Miller (1994) also reported that acid-forming compounds, like alum, reduced the amount of soluble organic C (SOC) in litter and suggested these compounds may reduce SOC levels in runoff water from land fertilized with manure, thus reducing the biological oxygen demand (BOD) of water. They also suggested lower SOC may lead to reductions in heavy metals in runoff, such as copper (Cu) and zinc (Zn). They stated that chemical addition to litter to reduce P solubility may be a best management practice (BMP) to reduce P runoff. Moore and Miller (1994) concluded that more research was needed to determine: (1) if these results could be verified in field (runoff) studies, (2) if the P minerals formed by this process are stable in various geochemical environments, (3) if this is an economically feasible practice, and (4) if there are any other beneficial or detrimental side effects from this practice.

Shreve et al. (1996) conducted another laboratory study to determine the effects of soil pH on the solubility of P in soils at various pH levels which were incubated for 296 days with either untreated poultry litter or litter treated with alum, slaked lime or ferrous sulfate. They found that alum and ferrous sulfate amendments resulted in the lowest soluble P levels. The study was conducted because some had speculated that the P associated with Al in alum-treated litter would suddenly become available if significant changes in soil pH were to occur which either made the soil more acid or more basic. On the contrary, their study found the lowest P solubility in soils incubated with alum-treated litter occurred under the most acid (pH 4) and basic (pH 8) conditions, indicating P associated with Al in alum-treated litter is geochemically stable over a wide range of conditions. The highest P solubility in their study was found with Ca amendments under acid soil conditions (Shreve et al., 1996). This was expected, since Ca phosphate minerals would be expected to dissolve under acidic conditions. Shreve et al. (1996) did not evaluate the effect of flooding or low redox potentials on P solubility, however, it is well know that under reducing conditions P will be released from Fe phosphate minerals due to redox reactions. Aluminum phosphates are not affected by redox reactions.

Shreve et al. (1995) conducted a rainfall simulation study to test the theory proposed by Moore and Miller (1994) that the addition of metal salts would reduce P runoff from poultry litter. They found that both alum and ferrous sulfate ($FeSO_4 \cdot 7H_2O$) additions to litter reduced P runoff, with alum reducing P concentrations in runoff water by 87% compared to normal litter (Figure 15.2). They also found that SOC levels were significantly lower in runoff from alum-treated litter. Shreve et al. (1995) also found that forage production was significantly greater with alum-treated litter than normal litter and speculated the increase in yield was likely due to a reduction in NH_3 volatilization, which would increase N availability in alum-treated litter.

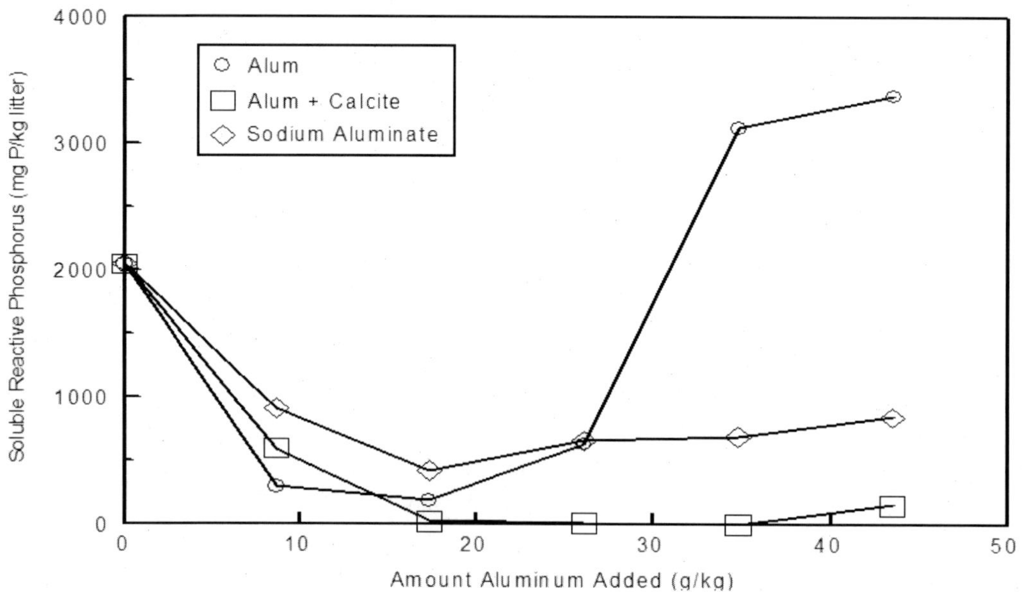

Figure 15.1. Effect of Al amendments to poultry litter on water soluble reactive P (taken from Moore and Miller, 1994).

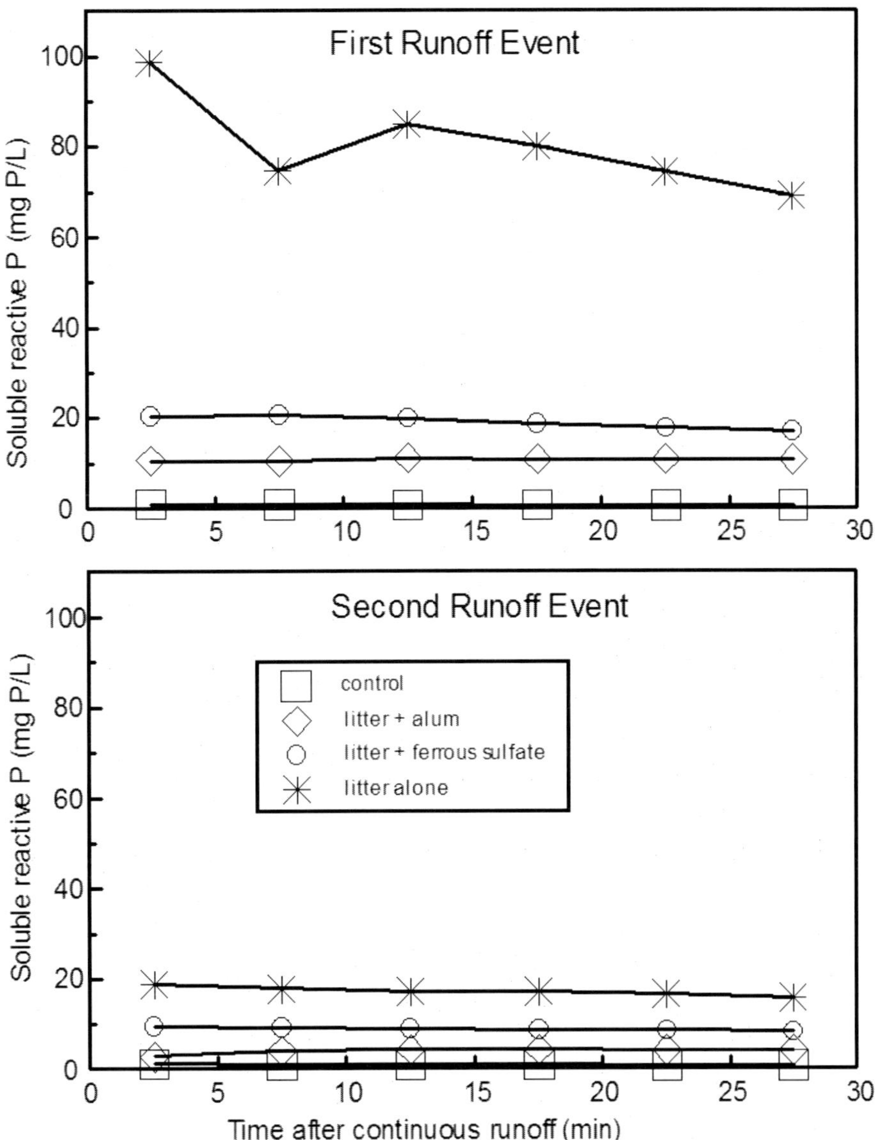

Figure 15.2. Effect of alum and ferrous sulfate amendments on P runoff from small plots fertilized with poultry litter (taken from Shreve et al., 1995).

In 1994 a small watershed study was initiated as part of an EPA 319 grant on a commercial broiler/beef farm in northwest Arkansas to evaluate the effects of alum on P runoff from the edge of field (Moore et al. 1999, 2000a; Moore and Edwards, 2007). Paired watersheds (1 acre each) were equipped with flumes and automatic watersamplers and fertilized with either normal or alum-treated litter obtained from poultry houses located on the same farm. Rates of alum application were equivalent to 0.09 kg/bird or approximately 10% by weight of the litter (moist litter basis). Chemical characteristics of poultry litter from the first year of this study are shown in Table 15.1.

Table 15.1. Chemical characteristics of poultry litter used for runoff study.

Parameter	Alum-treated Litter		Untreated Litter	
Parameter	*Average*	*Std. Dev.*	*Average*	*Std. Dev.*
pH	7.59	0.77	8.04	0.18
EC (uS/cm)	10833	471	6611	311
Total metals				
g/kg				
N	38.5	1.1	34.5	2.7
S	33.9	9.8	6.8	0.4
Ca	29.4	3.6	34.1	4.2
K	27.4	2.7	26.4	1.6
P	18.9	1.8	22.4	1.7
Al	18.7	6.0	1.18	0.2
Na	7.54	0.6	7.84	0.6
Mg	5.79	0.7	6.57	0.4
mg/kg				
Fe	1717	310	1095	155
Mn	893	216	956	134
Cu	679	93	748	102
Zn	598	51	718	69
B	46	4	51	4
Ti	31	11	44	19
As	20	8	43	4
Ni	21	5	15	2
Pb	8	2	11	2
Co	6	2	6	1
Mo	5	0.5	6	0.5
Cd	3	0.4	3	0.2

Taken from Moore et al. (1998).

After three years, Moore et al. (1999, 2000a) found P runoff from alum-treated litter was 75% lower than normal litter. Likewise, after ten years Moore and Edwards (2007) found the cumulative P load from normal litter was 15.0 kg P ha^{-1}, while that from alum-treated litter was 4.45 kg P ha^{-1} (Figure 15.3). The differences in P loads in runoff were mainly due to differences in P concentrations in runoff. Moore and Edwards (2007) showed that cumulative P runoff was highly correlated to the amount of soluble P applied (Figure 15.4). This relationship was not surprising, since several researchers had found that the amount of water soluble P applied in manure governs P runoff from pastures (DeLaune et al., 2004a,b; Sharpley et al., 2001a; Shreve et al., 1995; Moore et al., 1999, 2000a; Kleinman et al., 2002a,b). Other researchers have also observed significant reductions in P runoff with alum-treated litter compared to normal litter (Smith et al., 2001, 2004a, b; Warren et al., 2006a; Sistani et al., 2006). As a direct result of these environmental benefits of treating poultry litter with alum, the USDA/NRCS has made the use of alum a conservation practice standard that is eligible for funding under the environmental quality incentives program (EQIP) in many states (USDA NRCS, 2009). Fact sheets on treating poultry litter with alum are available have been produced by various organizations (Moore, 2006; Moore et al., 2004; Penn and Zhang, 2009).

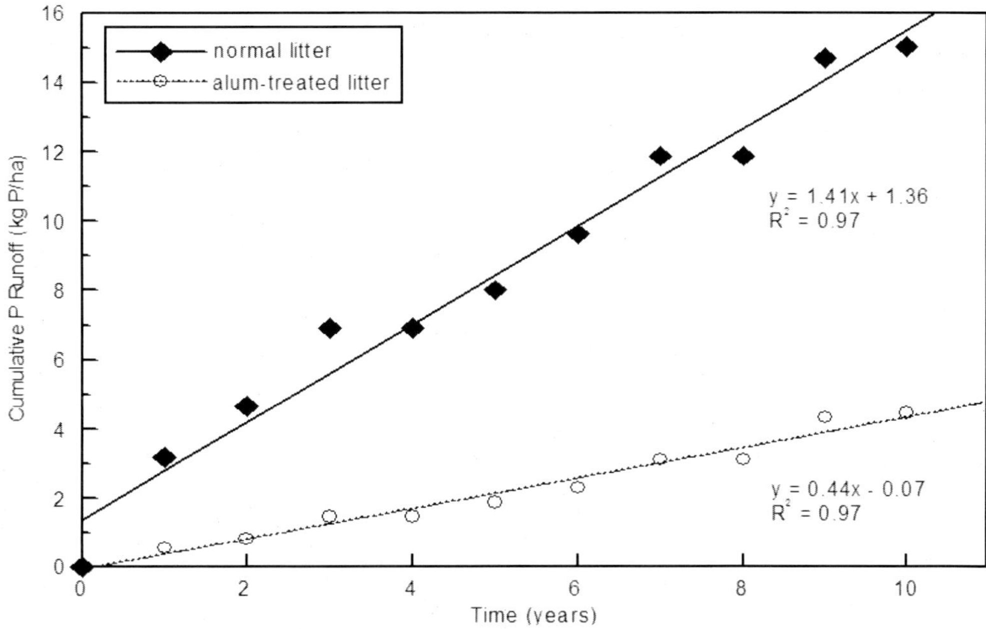

Figure 15.3. Phosphorus loads from small watersheds fertilized with alum-treated or normal litter as a function of time (taken from Moore and Edwards, 2007).

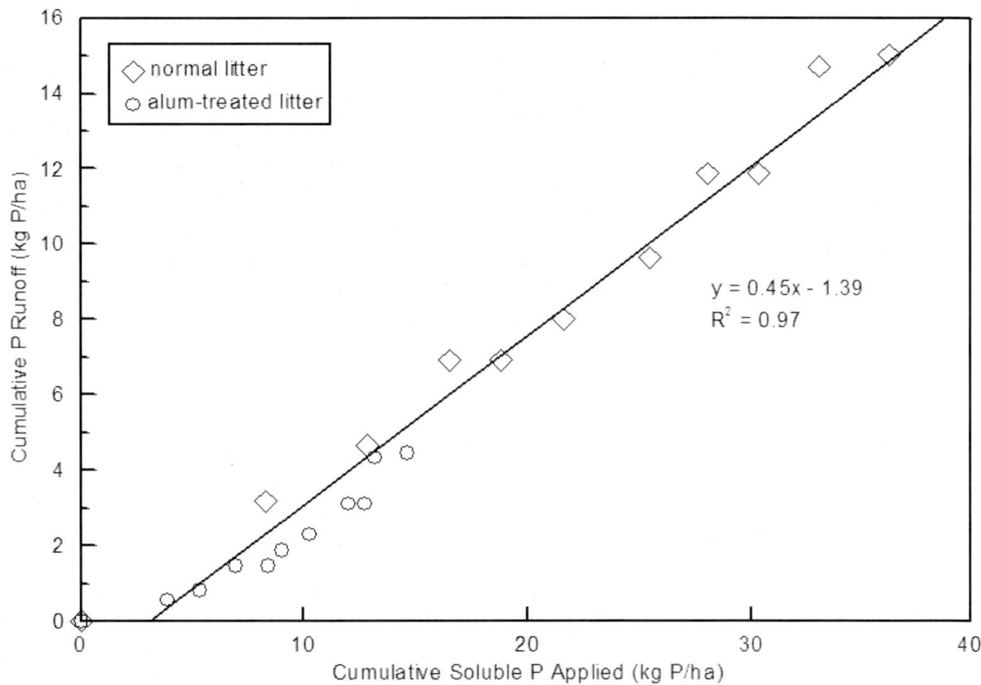

Figure 15.4. Cumulative P runoff as a function of cumulative soluble P applied (taken from Moore and Edwards, 2007).

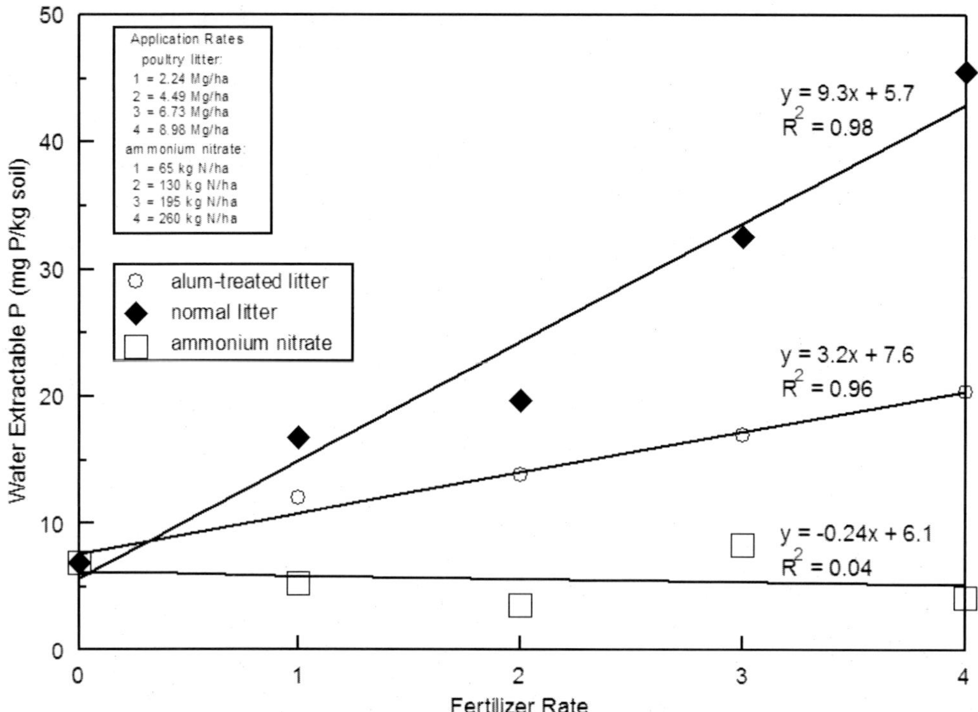

Figure 15.5. Effect of various rates of poultry litter, alum-treated litter and ammonium nitrate on water soluble P in soil after seven years of applications (take from Moore and Edwards, 2007).

In 1995 a long-term small plot study was initiated by Moore and co-workers to evaluate the effects of alum-treated litter on soil chemistry, forage production and runoff water quality (Moore et al., 1999, 2000a; Moore and Edwards 2005, 2007). This study was conducted using 52 small plots on a Captina silt loam soil cropped to tall fescue. There were 13 treatments: unfertilized control plots, four rates of alum-treated litter, four rates of normal litter and four rates of ammonium nitrate applied at the same N rates as alum—treated litter. The litter rates were 1, 2, 3 and 4 tons/acre-yr. After only 7 years there were large differences in water soluble P in the soil at 0-5 cm depth due to the rate of litter application and the type of litter, with alum-treated litter resulting in much lower soluble P concentrations in soil than normal litter (Figure 15.5).

While soluble P levels were much higher in the top 5 cm of soils fertilized with alum-treated litter, Mehlich III extractable P was found to be slightly higher in plots fertilized with alum-treated litter (Figure 15.6) (Moore and Edwards, 2007). They hypothesized that this was due to P leaching in plots fertilized with normal litter and took soil cores to a depth of 50 cm and analyzed for water soluble and Mehlich III P. They found that while Mehlich III P was slightly higher in the surface soil samples when alum was used, it was higher in the subsurface samples with normal litter, indicating that much more P leaching had occurred with normal litter than alum-treated litter (Figure 15.6). Moore and Edwards (2007) found the results for water soluble P were even more pronounced, clearly showing alum additions to manure decreased P leaching (Figure 15.7). This is important because in some areas of the world, particularly where sandy soils predominate, P transport occurs mainly through processes controlled by leaching.

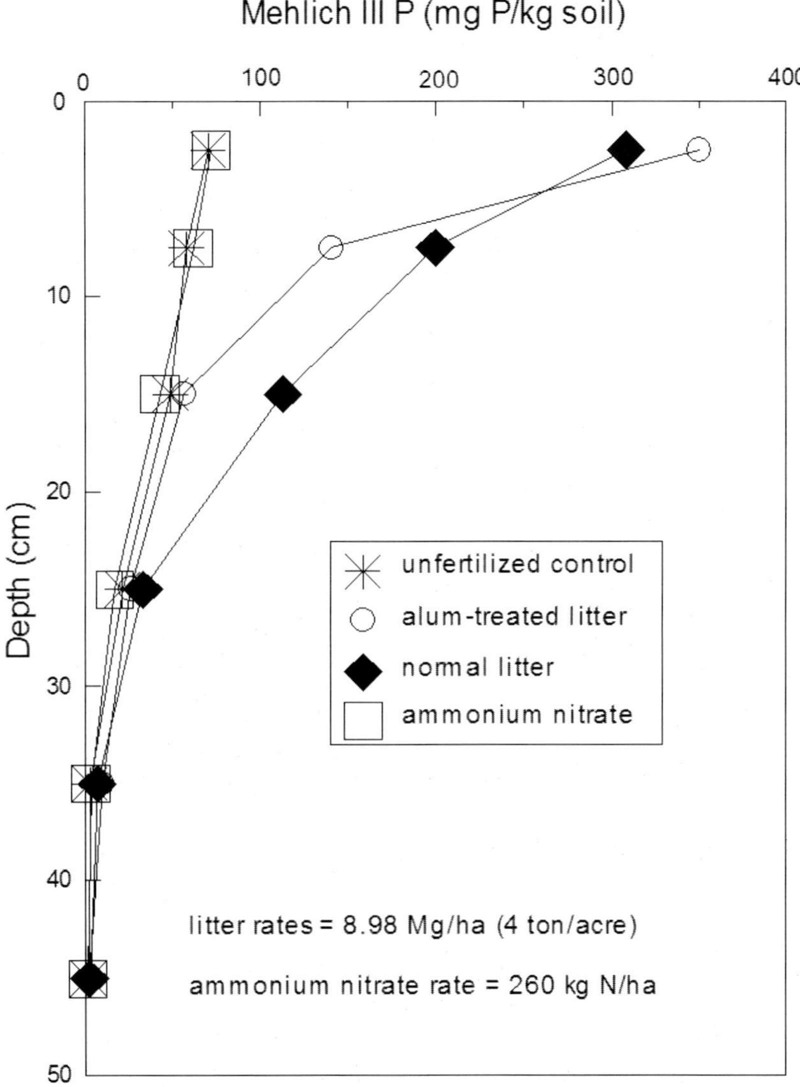

Figure 15.6. Mehlich 3 extractable P in soil as a function of depth in soil fertilized with normal litter, alum-treated litter and ammonium nitrate (taken from Moore and Edwards, 2007).

A large-scale evaluation of alum additions to poultry litter was conducted in 1999-2000 in Delaware, Maryland and Virginia as part of EPA Consent Decree (Sims and Luka-McCafferty, 2002). In this study, alum was added to the litter in 100 poultry houses, while another 100 houses were untreated. Alum was added to the houses at a rate of approximately 0.09 kg/bird or 1 kg alum/m^2. Although this would have resulted in an application rate of 10% alum had the litter been cleaned out in each of the houses at the beginning of the study, that did not happen. In fact, the average number of flocks since the last cleanout had occurred equaled twelve, so the alum concentration was quite dilute in this study (Sims and Luka-McCafferty, 2002). While Moore et al. (2000a) reported an Al to P mole ratio of approximately one in alum-treated litter, the average in this trial was only 0.57, and ranged from 0.14 to 1.13 (Sims and Luka-McCafferty, 2002). However, alum additions still resulted in soluble P levels in litter that were, on average, 67% lower than normal litter. The main conclusions from this study were that the results obtained were very similar to earlier research by Moore and Miller (1994), Moore et al. (1995a, b), Moore et al. (1999), Moore et al. (2000a), and Shreve et al. (1995) indicating that this BMP is transferable and can be implemented under a wide range of conditions and management practices and on a very large scale.

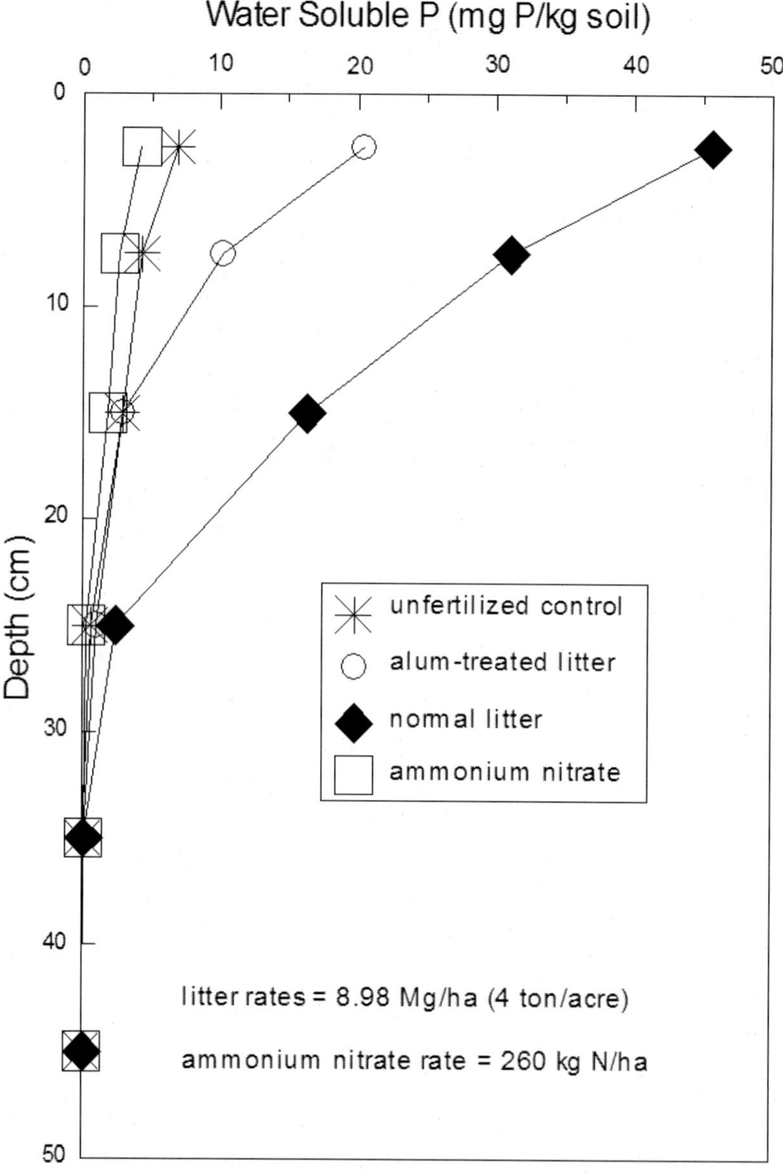

Figure 15.7. Water extractable P in soil as a function of depth in soil fertilized with normal litter, alum-treated litter and ammonium nitrate (taken from Moore and Edwards, 2007).

Smith et al. (2004a) evaluated the effects of alum additions to broiler litter from broilers fed either a normal diet or with diets that contained reduced P levels via the use of high available P (HAP) corn, phytase enzyme additions or the combination of HAP corn and phytase. Alum additions reduced the amount of soluble P in litter by 47%, whereas when alum was applied to litter from birds fed HAP corn and phytase diets it reduced soluble P levels by 74%. Phosphorus runoff from small plots was 52% lower with alum alone and 69% lower when alum was used in conjunction with both HAP corn and phytase. The increased efficiency of alum with modified diets was attributed to a higher Al:P ratio, which further lowered soluble P in manure. Smith et al. (2004a) also found that both soluble and total P in runoff were highly correlated to soluble P in the poultry litter, but not to the total P content of the litter. In a similar study conducted by Miles et al. (2003), phytase diets actually increased soluble P in manure, although phytase in combination with alum treatment of the litter reduced soluble P.

Smith et al. (2005) conducted a laboratory study in which the litter from the four diets used by Smith et al. (2004a) treated with or without alum were incubated in soils which had either low soil test P (STP) levels or high STP levels. In that study soluble P levels in the soils were increased by virtually all of the fertilizer treatments except where diet modification was used in conjunction with alum treatment, in which case the soluble P levels in the soils were similar to unfertilized soils. They also found that the P sorption ratio (PSR), which is the molar ratio of P/(Fe +Al) in Mehlich III extracts, for soils fertilized with alum-treated litter was similar to unfertilized soils. The PSR has been used as a good indicator of P runoff and leaching (Maguire and Sims, 2002).

15.5. MECHANISM OF ACTION OF ALUM ON SOLUBLE P IN MANURE

Several investigators have attempted to determine what P mineral is formed when alum is added to manure (Jaynes et al., 1999; Moore et al., 2000a; Hunger et al., 2004; Peak et al., 2002; Seiter et al., 2008; Shoberg et al., 2006; Staats et al., 2004; Warren et al., 2008). Most of this research has focused on transformations of inorganic P in manure. Moore et al. (1999, 2000a) stated the mechanism of action with respect to lowering P solubility was unknown. They hypothesized that Al from alum is transformed to $Al(OH)_3$, which adsorbs P and with time an amorphous aluminum phosphate mineral would form. They also suggested that an amorphous aluminum phosphate may form immediately after alum additions in litter. This was based on evidence from Jaynes et al. (1999) who were unable to detect crystalline aluminum phosphate compounds in alum-treated litter using x-ray diffraction and thermal analyses, even though ion activity product calculations indicated supersaturation with respect to variscite had occurred in alum-treated litter. Evidence of P adsorption onto $Al(OH)_3$ in alum-treated litter was provided by Peak et al. (2002) who used X-ray absorption near edge structure (XANES) spectroscopy. However, they concluded that more research was needed to determine if the adsorbed P will slowly convert to aluminum phosphate solid phases after incorporation in soils.

Subsequent research by Hunger et al. (2004) using both solid state MAS and CP-MAS ^{31}P nuclear magnetic resonance (NMR) spectroscopy showed that P associated with Al in alum-treated poultry litter was probably in the form of poorly ordered wavellite or surface complexes with $Al(OH)_3$. They also found that there was a complex mixture of organic and inorganic phases in litter and suggested that calcium phosphate precipitants existed as a surface precipitant on calcium carbonate in both alum treated and normal litter. They concluded that about 40% of the total P in alum-treated litter in the samples they analyzed was associated with Al.

15.6. INCORPORATED ALUM-TREATMENT OF MANURE INTO THE PHOSPHORUS INDEX

In 1999 the USDA/NRCS recommended that states develop a P driven approach for developing nutrient management plans for manure spreading (USDA and US EPA, 1999). Since then, 47 states opted to use a P index for writing nutrient management plans (Sharpley et al., 2003). The P index is a field-scale risk assessment tool for predicting the risk of P runoff. It accounts for the risk of P runoff from both source and transport factors (Lemunyon and Gilbert, 1993).

Recently, the Arkansas P index was revised to incorporate changes in soil test P methods and methods for water extractable P (WEP) in manure. The new index also takes into account P release from manure over time, using the following equation:

$$\text{P Sources} = \text{WEP}_{coef} *(\text{WEP}_{litter} + \text{MNRL}_{coef}*(\text{TP}_{litter} - \text{WEP}_{litter})) + \text{STP}_{coef}*\text{STP} \quad \text{Eq. 1}$$

where WEP is water extractable P, WEP_{coef} is the weighting factor for WEP, MNRL_{coef} is the mineralization factor for manure, TP is the total P in litter and STP is the Mehlich III extractable P in the soil. The mineralization coefficients takes into account mineralization of organic P in manure that is not measured as WEP prior to application, can contribute an additional P in runoff (Vadas et al., 2009). For normal manures a mineralization factor of 0.05 (5% of non-WEP total litter P) was utilized, while for alum-treated manure the MNRL_{coef} was 0.005 (0.5% of non-WEP total litter P) was added to the WEP value. The lower mineralization factor for alum treated litter reflects the fact that addition of alum to litter binds P with Al in a mineral rather than organic form. Thus, there is a lower potential for organic P mineralization in alum-treated litter. Liquid manures treated with aluminum chloride to reduce WEP would also use the 0.005 mineralization factor. Inclusion of a mineralization potential addresses the total P content of applied litter and its ability to produce WEP following application.

Six states (Florida, Georgia, Maryland, Pennsylvania, Tennessee and Virginia) address the differences in P solubility from organic sources by using P source coefficients (PSCs) in their P indices (Shober and Sims, 2007). Since alum-treated manure has lower soluble P, the PSCs for these materials are lower (Leytem et al., 2004). For example the PSC for liquid swine manure, poultry manure, semisolid dairy manure and alum-treated poultry manure are 1, 0.8, 0.8 and 0.5, respectively (Coale et al., 2005).

15.7. EFFECTS OF ALUM ON AMMONIA EMISSIONS FROM MANURE

Moore and Miller (1994) suggested that additions of alum to poultry litter may reduce NH_3 emissions from poultry litter because of reductions in litter pH, which would reduce the ratio of NH_3 to NH_4^+ (ammonium) in the litter. They speculated this may result in reductions in NH_3 concentrations in poultry houses, which may result in heavier and healthier birds. They also suggested that reducing NH_3 volatilization would increase the N/P ratio in litter.

Moore et al. (1995b) conducted a laboratory study in which NH_3 volatilization was measured from untreated poultry litter and litter amended with ten different treatments, including slaked lime at two rates, alum at two rates either with or without calcium carbonate additions, ferrous sulfate at two rates and Multi-purpose litter treatment (MLT), a commercial product consisting of ethylene glycol. They found that litter amendments which did not decrease litter pH, such as MLT and slaked lime, had no effect on NH_3 emissions. Ferrous sulfate reduced emissions by 11 and 58% when applied at 10 and 20% by weight, respectively. Alum reduced NH_3 losses better than the other amendments with the 10 and 20% alum treatments resulting in 36 and 99% lower NH_3 emissions, respectively, over the 42 day incubation period (Figure 15.8). Alum additions also resulted in significant increases in total kjeldahl N (41.4 g N/kg versus 26.1 g N/kg for the control) and NH_4-N in the litter (17.6 g N/kg versus 3.72 g N/kg for the control).

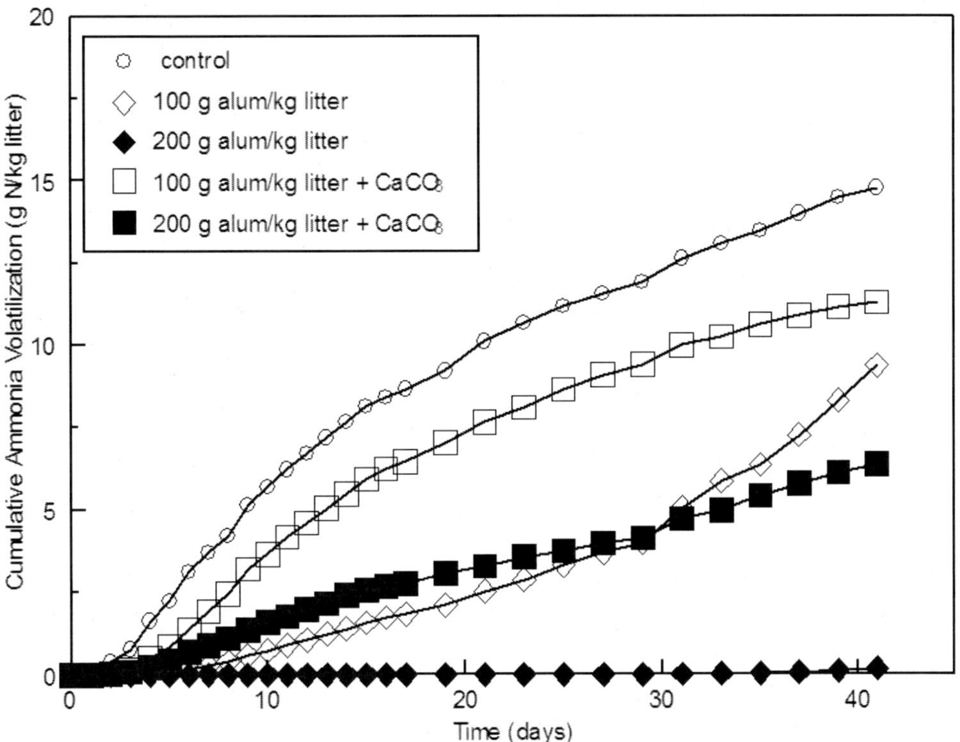

Figure 15.8. Effects of alum additions to poultry litter on ammonia volatilization in a laboratory study (taken from Moore et al., 1995).

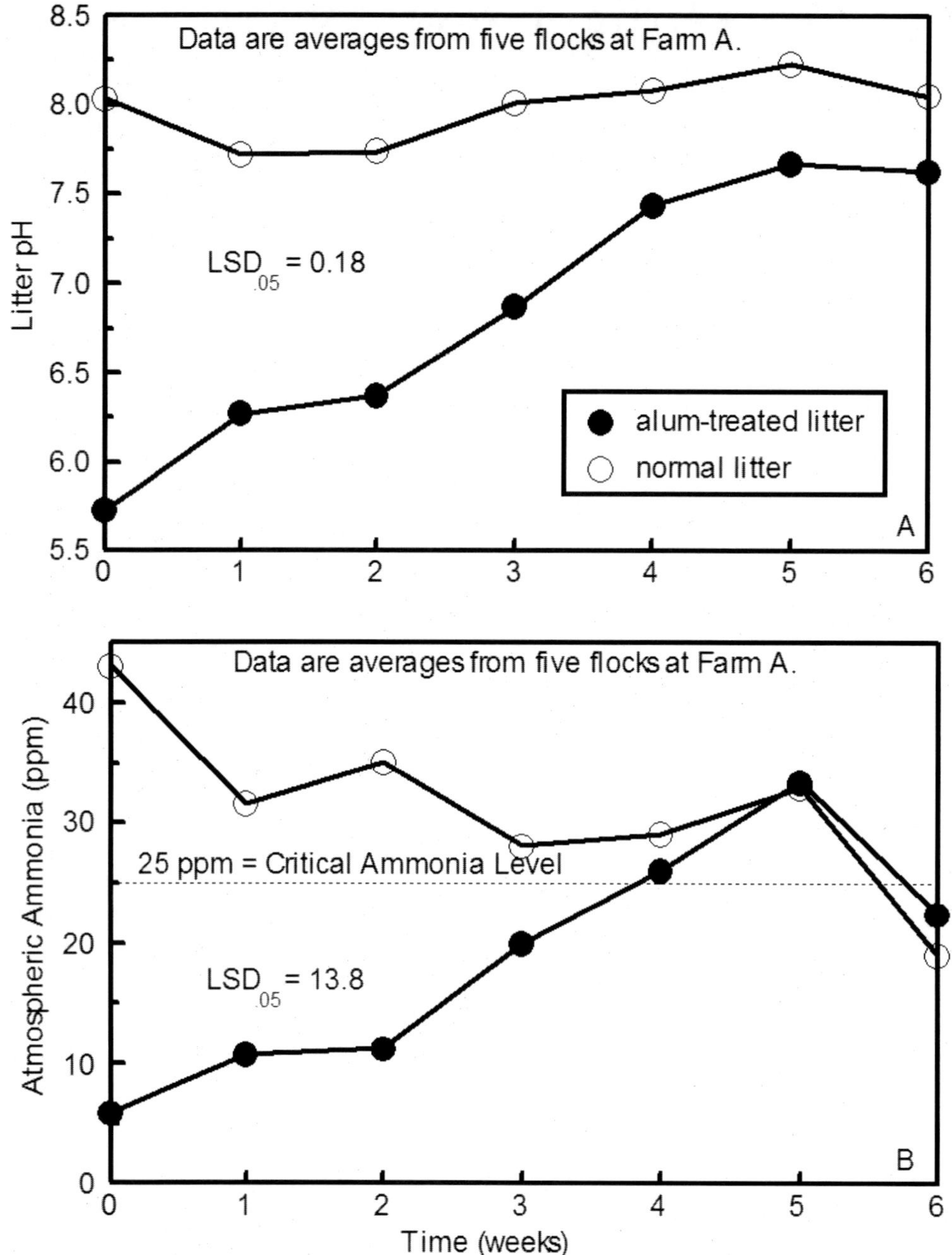

Figure 15.9. Effect of alum additions to poultry litter on: (A) soil pH, and (B) atmospheric ammonia in broiler houses, plotted as a function of time. Taken from Moore et al.(1999).

Moore et al. (1995b) showed there was an inverse relationship between final N content of the litter and the cumulative NH_3 loss, as would be expected, with alum-treated litter having the highest total N and the lowest NH_3 loss. The final N content of the litter treated with alum was higher than the original litter, which would make it a better fertilizer source, since N is the most common limiting nutrient for crop production (Moore et al., 1995b). They also found alum additions resulted in significantly higher N/P ratios in the litter. Soluble P, SOC and soluble Cu were also reduced in the litter treated with alum, as reported by Moore and Miller (1994).

Moore et al. (1996) conducted another lab study evaluating the effects of ten treatments on NH_3 volatilization from litter, including alum, ferrous sulfate, ferric chloride, phosphoric acid, a commercial product composed of calcium-iron silicate with a phosphoric acid coating and sodium bisulfate. The results of that study were comparable to that of Moore et al. (1995b) in that the compounds that reduced litter pH (alum, ferrous sulfate, ferric chloride and phosphoric acid) resulted in significant reductions in NH_3 volatilization and higher concentrations of NH_4-N in the litter. However, they concluded that ferrous sulfate would make a poor litter amendment since Wallner-Pendleton et al. (1986) and Pescatore and Harter-Dennis (1989) had found ferrous sulfate can cause catastrophic mortality for chickens (iron toxicity) when used as a litter amendment. It should be noted that Huff et al. (1996) laced chicken feed with alum in an attempt to cause Al toxicity. While alum did reduce growth at very high levels in the feed, Huff et al. (1996) concluded that chicks would never consume enough alum in alum-treated litter for toxicity to occur.

The first field trials of alum additions to poultry litter in commercial broiler houses were performed by Moore et al. (1999, 2000a). Alum was applied at a rate of 0.091 kg/bird to houses which contained 20,000 birds. Two farms were used and half of the houses at each farm were treated with alum while the other half served as controls. Alum additions significantly decreased litter pH for the first 3 to 4 weeks after the beginning of each growout, but after 5 weeks or so the pH leveled off at around 7.5, while the litter pH of the control manure remained relatively constant at pH 8 (Figure 15.9). Moore et al. (2000a) indicated that the reduction in litter pH is due to the following reaction:

$$Al_2(SO_4)_3.14H_2O + 6H_2O \rightarrow 2Al(OH)_3 + 3SO_4^{2-} + 6H^+ + 14H_2O \qquad \text{Eq. 2}$$

Lower litter pH shifts the NH_3/NH_4^+ equilibria towards NH_4^+, which is not volatile, as shown in Eq. 3:

$$NH_3 + H^+ \rightarrow NH_4^+ \qquad \text{Eq. 3}$$

Alum additions significantly reduced atmospheric NH_3 concentrations in the houses for the first three weeks (Figure 15.9). The average NH_3 concentrations in the houses were above 25 ppm for the first five weeks of the growout in the control barns, whereas NH_3 concentrations were low in the alum-treated barns. The differences in NH_3 emissions from these houses could not be determined by simply measuring NH_3 concentrations because each of the houses had different ventilation rates and the growers significantly increased ventilation in control houses due to high NH_3 concentrations. Hence, Moore et al. (2000a) measured NH_3 fluxes from the litter, which were found to be 99% lower from houses treated with alum during the first four weeks of the growout (basically zero until litter pH increased above 7). During the entire six week growout, fluxes were 75% lower for alum-treated litter than normal litter (2.14 vs. 8.27 mg NH_3/m^2-hr).

Moore et al. (2008, 2009) evaluated the effects of alum added to manure when very large broilers were being grown. They found NH_3 emissions were reduced by 75% for the first two weeks of the flock, 50% for the third week and 20-30% thereafter. They indicated that the main reason emissions from alum-treated litter were higher than that reported by Moore et al. (2000a) was the size of birds (2.72 kg at market age compared to 1.82 kg), since larger birds produce much more manure, which in turn requires more acid to decrease litter pH. Moore et al. (2009) also reported dry alum was more effective at reducing NH_3 losses than liquid alum.

McCubbin et al. (2002) estimated the impact of alum additions to manure on NH_3 emissions and subsequent fine particulate matter formation in the atmosphere. They concluded that alum was a very cost effective BMP for reducing NH_3 emissions from broiler litter.

15.8. EFFECTS OF ALUM ON POULTRY PRODUCTION

Moore et al. (1999, 2000) found that alum treatment of litter in commercial broiler houses decreased propane use compared to control houses (3,020 vs. 3,357 L/growout) due to reduced ventilation requirements during the cool periods of the year. Likewise electricity usage was lower in alum-treated houses compared to controls (7,320 vs. 8,330 kW/growout). Broilers grown on alum-treated litter were also heavier than those grown on normal litter (1.73 vs. 1.66 kg/bird). This is not surprising since high concentrations of NH_3 have been shown to negatively affect weight gains in broilers (Carlile, 1984). Feed conversion was also improved by the addition of alum to the litter (1.98 vs. 2.04 kg feed/kg bird), which was also probably due to lower NH_3 levels. Mortality was slightly lower in alum-treated houses than controls (3.90 vs. 4.2%), as was the percent of total liveweight that was rejected at the processing plant (1.5 vs. 2.0%). While most of these poultry production benefits were assumed to be due to lower NH_3 concentrations, reductions in bacterial pathogens (which will be discussed later) may also have contributed to better growth. Subsequent work by McWard and Taylor (2000) also demonstrated that alum additions improved feed conversion and weight gains, while improving the foot pad score of broilers. Another benefit of alum treatment reported by Moore et al. (1999, 2000) was higher N content of the litter (3.85 vs. 3.45%).

Savings to the integrator (company) due to alum use calculated for improved feed conversion, lower mortality and higher total weight accepted at the plant were $480.00, $12.48, and $128.00 per flock, respectively (total of $632.48). Savings to the grower due to alum use calculated for lower propane use, lower electricity use, heavier birds, lower mortality and higher litter N content were $106.80, $6.60, $150.00, $21.00, and $24.00 per flock, respectively (total of $307.86). Hence, the economic benefits of treating with alum were approximately $940.34 per growout. Alum cost, including shipping and application were $480.00 per growout. Hence, the benefit/cost ratio calculated by Moore et al. (1999) was 1.96, indicating this BMP is very cost effective. As a result of these production benefits, approximately one billion broilers are currently grown with alum in the United States.

Subsequent research conducted in laboratory or small chamber studies has shown that alum additions to poultry litter reduced NH_3 emissions (Do et al., 2005; Kim and Choi, 2009;

Choi and Moore, 2008a, b). Do et al. (2005), who treated small rooms in commercial broiler houses, found alum reduced NH_3 concentrations by 86% during the flock. Kim and Choi (2009) showed alum additions increased the NH_4-N content of litter, while decreasing soluble P concentrations. Choi and Moore (2008a) found in a lab study that dry alum, liquid alum, high acid alum, aluminum chloride and sodium bisulfate all decreased litter pH, resulted in lower NH_3 emissions and higher total N contents in litter. Alum additions to litter reduced emissions by 77 to 96%, while $AlCl_3$ only reduced NH_3 emissions by 48 to 92%. In a small pen trial, Choi et al. (2008b) found $AlCl_3$ additions to litter reduced NH_3 fluxes by up to 76% during a 6 week growout. Total volatile fatty acid (odor precursors) contents in litter were reduced by 20 to 51% with $AlCl_3$ additions (Choi and Moore, 2008b).

Worley et al. (1999) found that alum additions to litter in commercial broiler houses reduced propane use by 863L per house. Birds grown on alum-treated litter were heavier than controls (1.831 vs. 1.772 kg/bird) and had better feed conversion than those grown on normal litter (1.84 vs. 1.89 kg feed/kg bird). They also found that alum additions were as affective as a commercial product costing $100/house-flock for suppressing darkling beetle populations. McWard and Taylor (2000) also found darkling beetle counts were significantly reduced with alum. Worley et al. (2000) indicated that much of the economic benefits of alum can be achieved using half of the recommended rate (5%, rather than 10% by weight). However, they indicated that soluble P reductions and reductions in NH_3 emissions would not be as great with lower rates of alum. They also found the higher rate of alum did a better job reducing darkling beetle numbers than the half rate.

Moore et al. (2000b) designed and developed a liquid alum delivery system for high rise hen houses. In this system, liquid alum (25% w/w) was sprayed from PVC pipes suspended above each row of manure. Ammonia concentrations on the second floor where the hens were located were reduced from 75-80 ppm NH_3 to less than 25 ppm. Testing over an 8 month period showed egg production was improved when the system was on compared to when it was off. Fogiel and Powers (2009) also found alum additions to laying hen manure reduced NH_3 emissions.

15.9. Effects of Alum Additions on Pathogens and Microbially-mediated Soil Processes

Moore and coworkers were quick to realize that reductions in manure pH caused by alum were likely to cause dramatic changes in the microbiology of litter. Scantling et al. (1995) showed that *E. coli* and total coliform counts were significantly lower when alum was added to litter, while fungal counts were higher. Recent work by Cook et al. (2008) has shown that bacterial populations are at least 50% lower in alum-treated litter, while fungal populations are greater. Cook et al. (2008) also indicated that the shift in microbial communities was believed to be due to lower pH and higher N contents in alum-treated litter, which favor fungal growth. They speculated that reduction in NH_3 emissions from alum-treated litter may be due to biological (inhibition of ureolytic bacteria) and chemical effects (shifting the chemical equilibria from NH_3 to NH_4). Rothrock et al. (2008) also found that additions of alum to poultry litter also caused shifts in the microbiology. Using PCR analysis, they showed that the *Actinomycetes* community was relatively unaffected by alum. However, they found that pathogens such as *Campylobacter jejuni* and *Escherichia coli* were by reduced 3 and 2 logs, respectively. They also found *Clostridium/Eubacterium* were undetectable in alum-treated litter after 8 weeks.

Gangrenous dermatitis (GD) in poultry (also known as wing rot or red leg) is caused by *Clostridium septicum* and *Clostridium perfringens* (Clark et al., 2004). Birds with GD have small pimples on the skin, which become raw and dark due to blood-tinged fluid under the skin. During an acute outbreak of GD, large populations of birds will die. Clark et al. (2004) indicated that there were a number of field reports showing alum use reduced the incidence of GD. In fact, a big increase in alum sales occurred in 1999 after a poultry grower in Alabama observed catastrophic mortality in two broiler houses which were not treated with alum and few dead birds in two houses which were treated with alum. Data from Rothrock et al. (2008) on *Clostridium* persistence in alum-treated litter supports the reports indicating alum helps control GD outbreaks.

Although this shift in microbial communities may result in improvements in poultry production, it may have much larger impact by reducing the incidence of food-borne illnesses in humans due to bacterial pathogens. While there are no known foodborne fungal pathogens associated with poultry production that cause human health problems, there are a number of bacterial pathogens, such as *Campylobacter* and *Salmonella* on poultry carcasses which cause an untold numbers of illnesses and death each year. Line (2002) found that both *Campylobacter* and *Salmonella* numbers were greatly reduced both in litter and on bird carcasses when alum was added to the litter. In fact, Line (2002) found that *Campylobacter* was totally eliminated from bird carcasses when the litter had been treated with alum.

Tomlinson et al. (2008) measured alkaline phosphatase activities in the 52 plots that Moore and coworkers were using for the long-term alum study described above. They found alkaline phosphatase activities were higher in soils fertilized with high rates of untreated litter compared to soils fertilized with the same rate of alum-treated litter, even though the soil receiving alum-treated litter had lower soluble P concentrations. They concluded that some factor beyond increases in soil C or microbial biomass contributed to elevated alkaline phosphatase activities in soils fertilized with normal litter, despite high concentrations of soluble P. These results imply that alum-treated litter may be more environmentally friendly, since organic P release due to phosphatase enzymes may be much slower.

Gilmour et al. (2004) compared litter decomposition in soils treated with either alum-treated litter or normal litter. They found that C release from the two litters was virtually identical, indicating that decomposition was not affected by the addition of alum. The only significant effect noted for alum-treated litter by Gilmour et al. (2004) was an increase in net N mineralization, which would benefit crop production.

15.10. EFFECTS OF ALUM ON BACTERIA, HEAVY METAL AND ESTROGEN RUNOFF FROM MANURE

As mentioned earlier, alum additions to litter reduce bacterial pathogen numbers, while increasing fungal growth. Reductions in *E. Coli* and total coliform counts in runoff water were measured by Moore (Unpublished data from 1998) with low and high rates of alum reducing E. *coli* numbers by 68 and 94%, respectively, in runoff water while total coliforms numbers were reduced by 62 and 94%. Gandhapudi et al. (2006) found fecal coliform and fecal streptococci numbers were lower in soil slurries amended with alum-treated litter than normal litter and concluded that alum additions could reduce the risk for environmental pollution by reducing fecal bacteria loss, in addition to reducing P runoff. Obviously more research is needed to determine if that is the case.

Moore et al. (1998) measured metal concentrations and loads in runoff from small plots fertilized with various rates of alum-treated and normal poultry litter. They found that the use of alum reduced concentrations and loads of As, Cu, Fe and Zn in runoff by roughly 50% compared to normal litter, whereas alum increased Ca and Mg concentrations in runoff. These findings confirm the hypothesis of Moore and Miller (1994), Moore et al. (1995) and Sims and Luka-McCafferty (2002) that reductions in soluble trace metal concentrations in litter treated with alum would reduce trace metal runoff. Concentrations of Al, K and Na in runoff water were unaffected by litter type. Moore et al. (2000a) showed there was no significant difference in Al concentrations in runoff from small watersheds fertilized with either alum-treated or normal litter. Moore et al. (1998) concluded that reductions in heavy metal runoff appeared to be related to reductions in SOC runoff, particularly in the case for Cu. Most (>95%) of the heavy metals in runoff were present in the soluble form. Moore et al. (1998) concluded that metal runoff from alum-treated litter is less likely to cause environmental problems than normal litter, since threats to the aquatic environment posed by Ca and Mg are much lower than those posed by As, Cu, and Zn.

Aquatic loading of hormones such as estrogen from animal manures may cause disruptions in the health or reproduction of animals. Nichols et al. (1997) conducted a rainfall simulation study to determine the magnitude of beta-17 estradiol in runoff water from small plots fertilized with poultry litter. They found estrogen concentrations in runoff increased linearly with litter application rate, with the highest concentrations observed in runoff water from normal litter. Additions of alum reduced mean beta-17 estradiol concentrations by 42% and loads by 46%.

15.11. EFFECTS OF ALUM-TREATED MANURE ON ALUMINUM AVAILABILITY IN SOILS

One of the most frequently asked question regarding the treatment of manure with alum is will there be an effect on Al availability in soils, Al runoff and/or Al uptake by plants. As mentioned earlier, Moore et al. (1998) found no effect of alum on Al runoff from small plots and Moore et al. (2000a) found Al runoff from small watersheds was unaffected by alum use. Likewise, no effect on Al runoff was noted by Smith et al. (2004a), who conducted a rainfall simulations study using swine manure treated with aluminum chloride.

In order to determine the long term effects of alum use on Al availability, Moore and coworkers began the 20 year study in 1995 described earlier (Moore and Edwards, 2005, 2007). Perhaps the most critical finding of the study by Moore and Edwards (2005) regarding alum treatment of manure was that it does not cause soil acidification like inorganic N fertilizers that contain ammonium (Figure 15.10). Although normal poultry litter is a better liming amendment, alum-treated litter still significantly increased soil pH compared to unfertilized control soils after seven years of application (Moore and Edwards, 2005). Applications of ammonium nitrate resulted in severe soil acidification, with pH values at the highest rates of application less than 4.0. It has long been known that ammoniacal fertilizers, like ammonium nitrate, result in soil acidification as a result of nitrification of ammonium (Pierre, 1928). Although both alum-treated and normal litter contain large amounts of ammoniacal N, they also contain large amounts of bases, such as calcium carbonate, which originate from the feed (Moore and Edwards, 2005). The bases in the manure neutralize the acidity formed during nitrification.

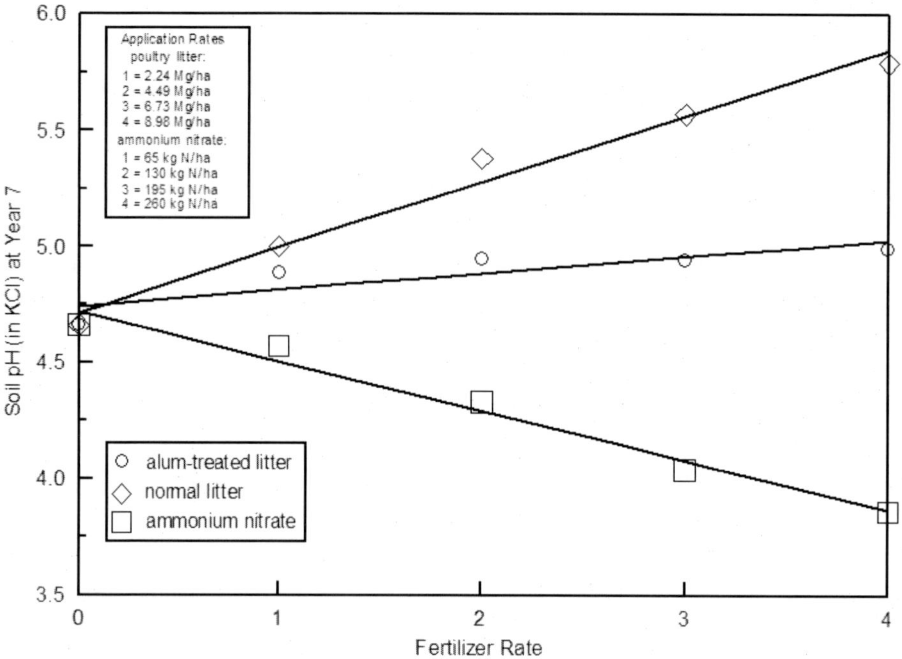

Figure 15.10. Soil pH as a function of fertilizer treatment and rate after 7 years (taken from Moore and Edwards, 2005).

Moore and Edwards (2005) also showed that long term use of alum-treated litter did not cause Al availability in soil to increase, however, after 4 or 5 years of fertilizer applications, plots fertilized with ammonium nitrate had very high exchangeable Al (Figure 15.11). After seven years of fertilizer application plots fertilized with the high rate of ammonium nitrate had exchangeable Al concentrations that were in excess of 100 mg Al kg soil^{-1}. In contrast, exchangeable Al concentrations decreased as the rate of alum-treated or normal litter increased, with concentrations lower than the unfertilized control plots (Figure 15.12). The explanation for these data is simple. Aluminum availability in soils is virtually independent of the total amount of Al in soils. Instead it is controlled by the geochemical conditions in the soil with soil pH being the main factor controlling Al availability (Moore and Edwards, 2005).

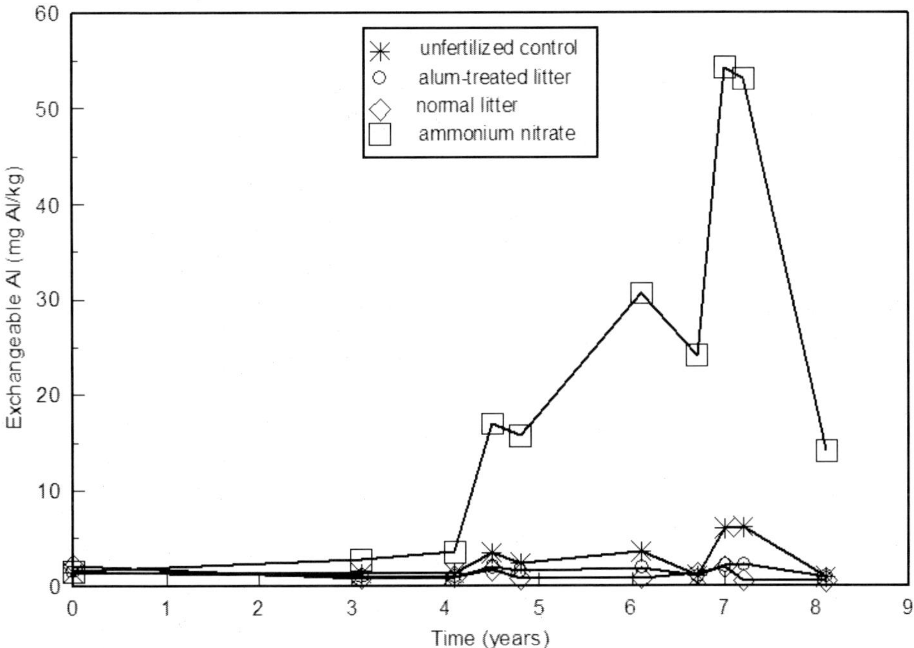

Figure 15.11. Temporal variability in exchangeable Al as a function of fertilizer type (taken from Moore and Edwards, 2005).

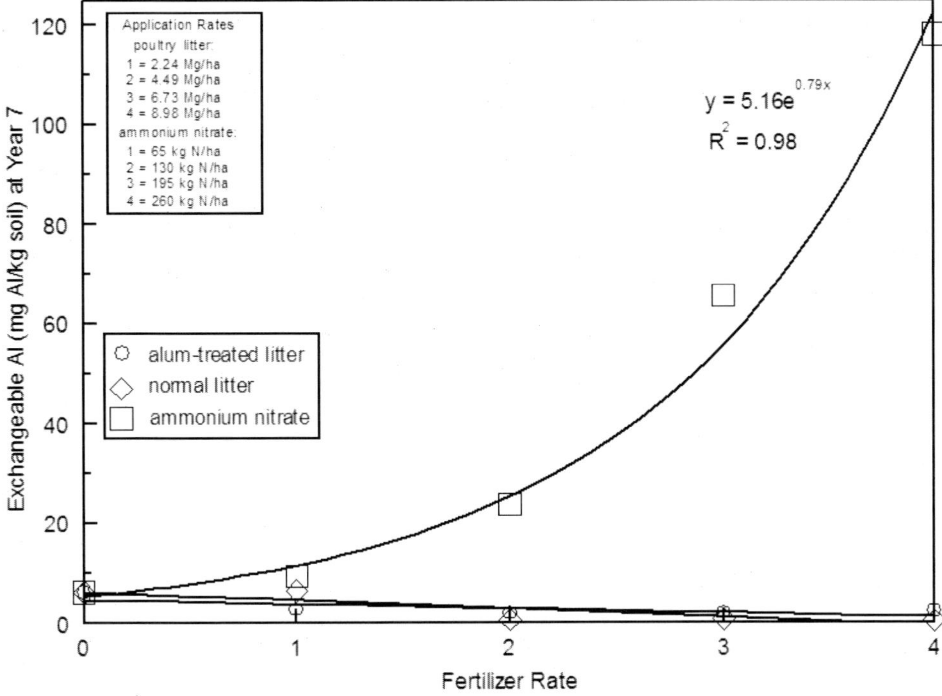

Figure 15.12. Exchangeable Al as a function of fertilizer treatment and rate after 7 years of applications (taken from Moore and Edwards, 2005).

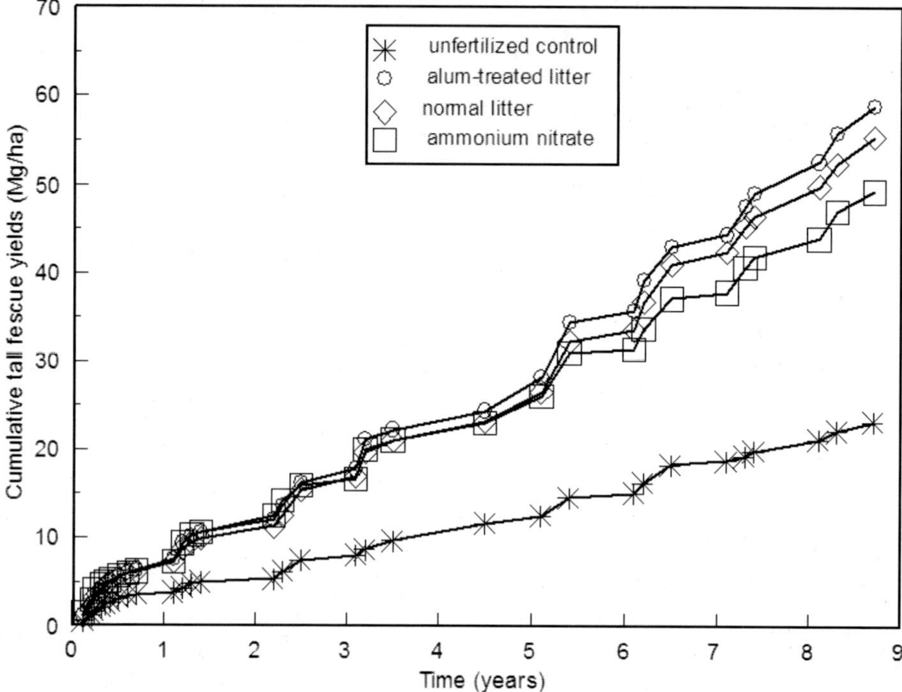

Figure 15.13. Cumulative tall fescue yields as a function of time for the various fertilizer types (taken from Moore and Edwards, 2005).

Warren et al. (2006a) found significantly higher soil pH in soils near Painter, VA that were fertilized with alum-treated and normal poultry litter compared to commercial fertilizers, which resulted in lower concentrations of exchangeable Al. Warren et al. (2006a) also found that normal poultry litter was a better liming amendment than alum-treated litter. They found that alum-treated litter had no effect on the slope of the line between pH and exchangeable Al, indicating that it does not affect available Al at any given pH. Warren et al. (2006a) also found that alum-treated litter had no effect on ear-leaf Al concentrations in corn grown at Painter and Orange, VA.

Soils are comprised of alumino-silicate minerals, hence is not surprising that Al is the most abundant metal in most soils. Total Al concentrations vary in soils from 1 to 30%, with an average of around 7% (Lindsay, 1979). Therefore, in a typical soil containing 7% Al, there is roughly 156,800 kg Al/ha (140,000 lbs Al/acre) in the top 15 cm alone. Alum-treated litter typically contains about 1% Al or less (Sims and Luka-McCafferty, 2002; Moore and Edwards, 2005). Assuming alum-treated litter is applied at a rate of 5.6 Mg/ha (2.5 tons/acre) it would result in the application of only 56 kg Al/ha (50 lbs Al/acre). Thus to increase the soil Al content from 7% to 8% would take 22,400 kg Al/ha or roughly 400 years of annual applications of alum-treated litter. However, those applications would only increase total Al in soil. Data shown in Figure 15.12 indicate that the available Al would probably be decreased, due to the net liming potential of alum-treated litter. Moore and Edwards (2005) could not detect differences in total Al in soils after seven years of applications of alum-treated litter. Likewise, they were unable to detect differences in total or soluble Al in runoff from plots fertilized with normal or alum-treated litter. Aluminum uptake by tall fescue was unaffected by alum treatment. Moore and Edwards (2005) concluded that poultry litter, particularly alum-treated litter, may be a more sustainable fertilizer than ammonium nitrate.

Guo and Song (2009) also concluded that Al toxicity would not be expected from alum-treated litter, based on the results of a laboratory leaching study. They showed that leachate concentrations of Al were highest in the initial leachate (4.6 mg Al/L), were only 1.4 mg Al/L in the second event, and undetectable after the 12th event. Subsequent leaching studies conducted by Guo and Song (2010) under field conditions showed that Al concentrations in leachate were always low (<4.6 mg/L), even though they had applied alum-treated litter at an extremely high rate (110 ton/ha). They indicated that Al phytotoxicity would not be expected, since calculated concentrations of Al^{3+} in solution would be less than 0.02 uM.

15.12. EFFECTS OF ALUM-TREATED MANURE ON SOIL QUALITY AND CROP PRODUCTION

One of the best measures of a fertilizer's effect on sustainability is crop yield, particularly when measurements are taken over long time periods. Moore and Edwards (2005) showed that tall fescue yields were 6% greater with alum-treated litter than with normal litter and 16% greater than with ammonium nitrate (Figure 15.13). Higher yields with alum-treated litter compared to normal litter were attributed to the greater N availability, due to less NH_3 loss (Moore and Edwards, 2005), while low yields with ammonium nitrate were attributed to soil acidification. Higher yields with alum-treated litter were also reported by Shreve et al. (1995). Moore et al. (2000a) showed nitrate leaching was unaffected by alum additions. These data indicate that alum-treated litter is the most sustainable of the three fertilizers evaluated.

Another question often asked regarding the treatment of manure with alum is will it cause P deficiencies in crops. Moore and Edwards (2007) showed there was no difference in P concentrations or uptake by tall fescue fertilized with alum-treated and normal litter. When poultry litter is applied to land there is an excess of P applied relative to the amount of N (N/P ratio of 2:1). Crops typically have a N:P ratio of 8:1. Moore and Edwards (2007) argued that addition of alum reduces the "effective" N:P to roughly 8:1, which fits the chemical composition that is more closely tailored to the needs of the plants.

All minerals, including aluminum phosphate minerals, will form or dissolve in soils depending on the geochemical environment and whether or not the soil solution is undersaturated, supersaturated or in equilibrium with respect to that mineral phase (assuming no kinetic limitations). As stated earlier, aluminum phosphates are not affected by redox reactions or wide changes in soil pH. Hence, they should be more geologically stable than calcium or iron phosphates. However, that does not mean that the P associated with Al in alum-treated litter will never become available. As long as the concentration of P in soil solution remains high (i.e. − soil test P concentrations are high or annual applications of manure are occurring), then the Al-P mineral would remain stable. However, if the soil solution concentration of P drops below the equilibrium concentration for that particular mineral phase, then it would slowly dissolve until it reached equilibrium conditions. Hence, under P deficient soil conditions it is highly likely that the P in alum-treated litter would slowly be released into soil solution where it would be available for plant uptake.

Warren (2005) and Warren et al. (2006b) evaluated the effects of normal and alum-treated poultry litter on tall fescue growth, P uptake and changes in soil P fractions in a Davidson soil which had low to moderate concentrations of soil P. They found that alum use did not negatively impact P uptake or growth of tall fescue. They also found that while Mehlich-1 extractable concentrations were unaffected by alum additions to litter, the NaOH extractable P fraction was increased, while water soluble P in soil was decreased. Warren et al. (2006b) concluded that while alum use does reduce the P solubility in the litter (and P runoff), it does not negative affect P availability to crops.

Warren et al. (2006a) found that alum-treated litter did not affect P uptake or yields of corn compared to normal litter at two locations in Virginia, although alum use did decrease water soluble P concentrations in soil. They also found that P runoff from small plots was 61-71% lower than that from normal litter. They concluded that alum-treated poultry litter can reduce the environmental impact of litter applications primarily through minimizing the P status of soils receiving long-term applications of litter and by reducing P losses in runoff.

Sistani et al. (2006) evaluated the effects of alum-treated and untreated poultry litter on tall fescue yields and nutrient uptake at a site in the Sand Mountain region of Alabama. They found that alum treatment of litter did not significantly affect yields or nutrient uptake. However, during 2002 they observed that P concentrations in tall fescue fertilized with alum-treated litter using a high rate of alum were actually higher than that in plants fertilized with commercial fertilizer. Sistani et al. (2006) found significant reductions in total and soluble P runoff due to alum-treated litter and concluded that this BMP should receive serious consideration as a method of reducing the adverse environmental impact of broiler chicken production when litter is land applied.

15.13. EFFECTS OF TREATING SWINE, DAIRY, AND BEEF MANURE WITH ALUM

The effects of alum additions on P runoff from swine manure were evaluated by Smith et al. (2001). They hypothesized that the sulfate in alum may be reduced to hydrogen sulfide under anaerobic conditions found in liquid manures and suggested that aluminum chloride be used for liquid manures rather than alum. They conducted a rainfall simulation study in which they tested a total of six treatments: (1) unfertililized control, (2) swine manure, (3) swine manure with alum at a rate of 215 mg Al/L, (4) swine manure with alum at a rate of 430 mg Al/L, (5) swine manure with aluminum chloride at a rate of 215 mg Al/L, and (6) swine manure with aluminum chloride at a rate of 430 mg Al/L. The high rates of both alum and aluminum chloride reduced P concentrations in runoff by 84% to 0.87 mg P/L, which was not significantly different from the unfertilized control plots. Although they had hypothesized that the Al compounds would increase yield since they reduced NH_3 emissions and increased the N content of the manure, yields and N uptake by tall fescue were not affected by Al additions (Smith et al., 2001).

Subsequent research by Smith et al. (2004b, c) evaluated the effects of using aluminum chloride manure treatments in combination with phytase additions to swine diets on P runoff and NH_3 emissions. Using nursery swine, they evaluated the effects of phytase or normal diets with liquid aluminum chloride manure treatments of 0, 0.25, 0.50, and 0.75% v/v.

Soluble P in manure was reduced by as much as 73% with aluminum chloride additions (Smith et al., 2004b). Ammonia emissions were reduced by up to 52% with $AlCl_3$ alone and by 60% when it was added to manure from pigs receiving phytase in their diets (Smith et al., 2004c). Reductions in NH_3 emissions were mirrored by increases in total N in the manure. Phosphorus runoff was highly correlated to soluble P in manure. Phytase additions alone actually increased P runoff. However, when phytase was used in conjunction with $AlCl_3$ there was a synergistic effect and reductions in P runoff were greater than with $AlCl_3$ alone. As noted earlier, they found there were no differences in Al runoff due to any of the treatments.

Smith and Moore (2005) found that when swine manure had been treated with $AlCl_3$, the increase in water soluble P in the soil was not significantly different from unfertilized soils with either low or high soil test P concentrations. Aluminum chloride additions to swine manure also significantly decreased Mehlich 3 and Morgan extractable P concentrations in soils (Smith and Moore, 2005).

Alum additions to liquid dairy manure have also been shown to decrease P solubility as well as improve mechanical separation of the solid and liquid fractions. Dao and Daniel (2002) found that alum additions to dairy manure alone reduced soluble P concentrations from 1165 mg P/kg to 397 mg P/kg. They found that solid separation increased from around 10% in untreated manure to over 20% when alum was added. When polymers were added with alum there was a synergistic effect, resulting in solid separation of approximately 60% (Dao and Daniel, 2002). These results were similar to Worley and Das (2000), who showed alum additions to swine manure improved solid separation efficiency by 70%. Adding alum also increased the P removal efficiency and N/P ratio in swine manure increased from 8 to 16.7 with alum addition. After the solid separation step, Worley and Das (2000) composted the swine manure and found alum additions had no effect on the composting process. DeLaune et al. (2004) found that alum additions to composting poultry litter significantly reduced NH_3 emissions which decreased C/N ratios compared to normal composted litter. DeLaune et al. (2006) showed that P runoff was lower from composted alum-treated litter compared to normal litter.

Timby et al. (2004) evaluated the effectiveness of eight different polymers in conjunction with various concentrations of $AlCl_3$ on solid separation and P removal from liquid dairy manure. They found that polymers alone had no effect on P removal. However there was a synergistic effect of adding polymers with $AlCl_3$ on solid separation. They concluded that low rates of $AlCl_3$ (>200 mg Al/L) in combination with cationic polymers with medium to high charge densities was the most effective combination for solid separation.

Lefcourt and Meisinger (2001) used wind tunnels to evaluate the effects of various rates (0.4, 1, 2.5, and 6.25%) of alum to dairy slurry on NH_3 emissions following land application. They found alum additions decreased NH_3 volatilization by up to 60% and reduced soluble P in manure by up to 75%. It should be noted that at the very highest rates of alum (2.5 and 6.25%) the pH of the manure was lowered below 5.0, which resulted in small amounts of soluble Al present in the manure (0.2 to 15.5 mg Al/kg manure). This can be avoided by simply applying the correct rate. Although alum can be over-applied to any manure source, it is the experience of the author that dairy manure is much less buffered than poultry or swine manure. As a result, relatively small additions of acid can cause fairly large reductions in manure pH. Therefore, care should exercised in determining rates of alum or aluminum chloride addition to dairy manure. To avoid potential hydrogen formation, dairy and swine manure should be treated with aluminum chloride, rather than alum, as mentioned earlier.

Alum has also been shown to be effective in reducing soluble P concentrations in cattle manure. Dao (1999) evaluated the effects of caliche, alum and fly ash amendments on composted and stockpiled beef manure. He found the three amendments reduced soluble P concentrations by 21, 60, and 85% in stockpiled manure and by 50, 83, and 93%, respectively, in composted manure. Dao (1999) also evaluated the effects of these amendments on the ratio of N to water soluble P (WSP) and found N/WSP ratios of 214, 1500, 270 and 525 for untreated, alum, caliche and fly ash in stockpiled manure and 93, 1375, 192, and 570 in composted manure, indicating that alum is the most environmentally friendly of the treatments.

Dou et al. (2003) evaluated the effects of alum and coal combustion by-products on various forms of P in dairy, swine and broiler manure. They found alum was the most effective compound for reducing soluble P in manure (reductions of up to 99%). They found that the NaOH-extractable and $NaHCO_3$-extractable P forms were increased with alum additions and indicated that those P mineral phases would be more stable. They indicated that treating manure with alum or coal combustion products before land application of manure could be useful management tactic for reducing negative environmental impacts.

Shi et al. (2001) conducted a laboratory study to determine the effectiveness of alum, Ammonia Hold, calcium chloride, brown humate, black humate, and NBPT on NH_3 emissions from beef cattle manure and reported the lowest emissions from alum-treated manure, which were 1.7 to 8.5% of the untreated control.

15.14. CONCLUSION

Additions of alum to animal manure, such as poultry litter, have been shown to improve air quality by significantly reducing NH_3 emissions into the atmosphere. Improvements in poultry production in houses treated with alum (heavier chickens which have better feed conversion and lower mortality) make this practice cost-effective. Alum additions to litter also reduce bacterial pathogens in litter responsible for both poultry diseases (dermititus) and foodborne illness in humans (*Campylobacter* and *Salmonella*). Energy use (propane and electricity) is also reduced when alum is used due to decreased ventilation requirements in cooler months. Phosphorus runoff and leaching from manure are also significantly reduced when alum additions are made. Furthermore, water quality is improved by alum because it reduces soluble organic C, estrogen, and heavy metals (As, Cu, and Zn) in runoff water. Crop yields have been shown to be greater with alum-treated manure when compared to normal manure because of the additional N present as a result of lower NH_3 emissions. As a result of these benefits, approximately one billion broilers are being grown in the U.S. each year with alum. In conclusion, alum treatment of manure appears to be a very sustainable practice, since it is a cost-effective BMP which improves air and water quality, while improving both animal and crop production.

REFERENCES

Anderson, D.P., C.W. Beard, and R.P. Hanson. 1964. The adverse effects of ammonia on chickens including resistance to infection with Newcastle Disease virus. *Avian Dis.* 8:369-379.

Battye, R., W. Battye, C. Overcash, and S. Fudge. 1994. *Development and selection of ammonia emission factors: Final Report.* EC/R Inc. Durham, N.C. EPA Contract Report #68-D3-0034. U.S. EPA, Research Triangle Park, N.C. Pp. 111.

Carlile, F.S. 1984. Ammonia in poultry houses: A literature review. World's Poul. *Sci. J.* 40:99-113.

Carpenter, S.R., N.F. Caraco, D.L. Correll, R.W. Howarth, A.N. Sharpley, and V.H. Smith. 1998. Nonpoint pollution of surface waters with phosphorus and nitrogen. *Ecol. Appl.* 8:559-568.

Clark, D., S. Watkins, F. Jones and B. Norton. 2004. Understanding and control of Gangrenous Dermatitis in poultry houses. United Poultry Concerns. June 25, 2004. http://www.upc-online.org/poultry_diseases/62504gangrenous.htm. Confirmed on June 2, 2010.

Coale, F.J., T. Basden, D.B. Beegle, R.C. Brandt, H.A. Elliott, D.J. Hansen, P.J.A. Kleinman, G. Mullins, and J.T. Sims. 2005. Development of regionally-consistent phosphorus source coefficients for use in phosphorus index evaluation in the Mid-Atlantic Rigion. Available at http://www.mawaterquality.org/publications/pubs/PSIWhitePaper03-29-05.pdf (verified May 26, 2010). *USDA-CREES Mid-Atlantic Regional Water Program.* College Park, MD.

Choi, I.H. and P.A. Moore, Jr. 2008a. Effects of various litter amendments on ammonia volatilization and nitrogen content of poultry litter. *J. App. Poultry Res.* 17:454-462.

Choi, I.H. and P.A. Moore, Jr. 2008b. Effects of liquid aluminum chloride additions to poultry litter on broiler performance, ammonia emissions, soluble phosphorus, total volatile fatty acids and nitrogen contents of litter. *Poultry Sci.* 87:1955-1963.

Cook, K.L., M.J. Rothrock, Jr., J.G. Warren, K. Sistania and P.A. Moore, Jr. 2008. Effect of alum treatment on the concentration of total and ureolytic microorganisms in poultry litter. J. *Environ. Qual.* 37:2360-2367.

Dao, T.H. 1999. Coamendments to modify phosphorus extractability and nitrogen/phosphorus ratio in feedlot manure and composted manure. *J. Environ. Qual.* 28:1114-1121.

Dao, T.H. and T.C. Daniel. 2002. Particulate and dissolved chemical separation and phosphorus release from treated dairy manure. *J. Environ. Qual.* 31:1388-1398.

DeLaune, P.B., P.A. Moore, Jr., T.C. Daniel, and J.L. Lemunyon. 2004a. Effect of chemical and microbial amendments on ammonia volatilization from composting poultry litter. *J. Environ. Qual.* 33:728-734.

DeLaune, P.B., P.A. Moore, Jr., D.K. Carman, A.N. Sharpley, B.E. Haggard, and T.C. Daniel. 2004b. Evaluation of the phosphorus source component in the phosphorus index for pastures. *J. Environ. Qual.* 33:2192-2200.

DeLaune, P.B., P.A. Moore, Jr., and J.L. Lemunyon. 2006. Effect of chemical and microbial amendments on phosphorus runoff from composted poultry litter. *J. Environ. Qual.* 35:1291-1296.

Do, J.C., I.H. Choi, and K.H. Nahm. 2005. Effects of chemically amended litter on broiler performaces, atmospheric ammonia concentration, and phosphorus solubility in litter. *Poultry Sci.* 84:679-686.

Donham, K.J. 1996. Air quality relationships to occupational health in the poultry industry. Pp 24-28 In (J.P. Blake and P.H. Patterson, eds) P*roc. 1996 National Waste Management Symposium.* Auburn University Press, Auburn, AL.

Donham, K.J. 2000. Occupational health hazards and recommended exposure limits for workers in poultry buildings. Pp. 92-109 In (J.P. Blake and P.H. Patterson, eds) *Proc. 2000 National Waste Management Symposium.* Auburn University Press, Auburn, AL.

Dou, Z. G.Y. Zhang, W.L. Stout, J.D. Toth, and J.D. Ferguson. 2003. Efficacy of alum and coal combustion by-products in stabilizing manure phosphorus. *J. Environ. Qual.* 32:1490-1497.

Duda, A.M., and D.S. Finan. 1983. Influence of livestock on non-point source nutrient levels of streams. *Trans. ASAE.* 26:1710-1716.

Edwards, D.R., and T.C. Daniel. 1992a. Potential runoff quality effects of poultry litter slurry applied to fescue plots. *Am. Soc. Agri. Eng.* 35:1827-1832.

Edwards, D.R., and T.C. Daniel. 1992b. Environmental impacts of on-farm poultry waste disposal - A review. *Bioresource Tech.* 41:9-33.

Edwards, D.R., and T.C. Daniel. 1993. Effects of poultry litter application rate and rainfall intensity on quality of runoff from fescuegrass plots. *J.Environ. Qual.* 22:361-365.

Fogiel, A.C. and W.J. Powers. 2009. *Impact of amendments on air emissions from stored animal manures.* Paper # 096123, 2009 ASABE International Meeting, Reno, NV. ASABE, St. Joseph, MI.

Gandhapudi, S.K., M.S. Coyne, E.M. D'Angelo, and C. Matocha. 2006. Potential nitrification in alum-treated soil slurries amended with poultry manure. *Bioresource Technology* 97:664-670.

Gilmour, J.T., M.A. Koehler, M.L. Cabrera, L. Szajdak, and P.A. Moore, Jr. 2004. Alum treatment of poultry litter: Decomposition and nitrogen dynamics. *J. Environ. Qual.* 33:402–405.

Guo, M. and W. Song. 2009. Nutrient value of alum-treated poultry litter for land application. *Poultry Sci.* 88:1782-1792.

Guo, M. and W. Song. 2010. *Nutrient and aluminum availability of alum-amended poultry litter: a long-term weathering study.* Plant Soil

Huff, W.E., P.A. Moore, Jr., J.M. Balog, G.R. Bayyari, and N.C. Rath. 1996. Evaluation of the toxicity of alum (aluminum sulfate) in young broiler chicks. *Poultry Sci.* 75:1359-1365.

Hunger, S., H. Cho, J. T. Sims, and D. L. Sparks. 2004. Direct speciation of phosphorus in alum-amended poultry litter: Solid-state 31P NMR investigation. *Environ. Sci. Technol.* 38:674–681.

Hutchinson, G.L. and F.G. Viets, Jr. 1969. Nitrogen enrichment of surface water by absorption of ammonia volatilized from cattle feedlots. *Science* 166:514-515.

Izagguirre, G., C.J. Hwang, S.W. Krasner, and M.J. McGuire. 1982. Geosmin and 2-Methylisoborneol from cyanobacteria in three water supply systems. *Applied and Environmental Microbiology* 43:708-714.

Jaynes, W.F., P.A. Moore, Jr., and D.M. Miller. 1999. Effect of aluminum sulfate applications on forms of phosphorus in poultry litter. In K.F. Steele (ed.) *Proc. Of Ark.*

Water Res. Conf. On Quality of Surface and Ground Water and Best Management Practices. Fayetteville, AR, April 8-9, 1998.

Kim, Y.J. and I.H. Choi. 2009. Effect of alum and liquid alum on pH, EC, moisture, ammonium and soluble phosphorus contents in poultry litter during short term: A laboratory experiment. *J. Poult. Sci.* 46:63-67.

Kingery, W.L., C.W. Wood, D.P. Delaney, J.C. Williams, and G.I. Mullins. 1994. Impact of long-term land application of broiler litter on environmentally related soil properties. *J. Environ. Qual.* 23:139-147.

Kleinman, P.J.A., A.N. Sharpley, B.G. Moyer, and G.F. Elwinger. 2002a. Effect of mineral and manure phosphorus sources on runoff phosphorus. *J. Environ. Qual.* 31:2026–2033.

Kleinman, P.J.A., A.N. Sharpley, A.M. Wolf, D.B. Beegle, and P.A. Moore, Jr. 2002b. Measuring water-extractable phosphorus in manure as an indicator of phosphorus in runoff. *Soil Sci. Soc. Am. J.* 66:2009-2015.

Kling, H.F. and C.L. Quarles. 1974. Effect of atmospheric ammonia and the stress of infectious bronchitis vaccination on Leghorn males. *Poultry Science* 53:1161-1167.

Kotak, B.G., S.L. Kenefick, D.L. Fritz, C.G. Rousseaux, E.E. Prepas, and S.E. Hrudey. 1993. Occurrence and toxicological evaluation of cyanobacterial toxins in Alberta lakes and farm dugouts. *Water Res.* 27:495-506.

Lefcourt, A.M. and J.J. Meisinger. 2001. Effect of adding alum or zeolite to dairy slurry on ammonia volatilization and chemical composition. *J. Dairy Sci.* 84:1814-1821.

Lemunyon, J.L., and R.G. Gilbert. 1993. Concept and need for a phosphorus assessment tool. *J. Prod. Agric.* 6:483-486.

Leytem, A.B., J.T. Sims, and F.J. Coale. 2004. Determination of phosphorus source coefficients for organic phosphorus sources: laboratory studies. *J. Environ. Qual.* 33:380–388.

Lindsay, W.L. 1979. *Chemical equilibria in soils.* John Wiley & Sons, New York.

Line, J.E. 2002. Campylobacter and Salmonella populations associated with chickens raised on acidified litter. *Poultry Sci.* 81:1473-1477.

Maguire, R.O. and J.T. Sims. 2002. Measuring agronomic and environmental soil phosphorus saturation and predicting phosphorus leaching with Mehlich 3. *Soil Sci. Soc. Am. J.* 66:2033-2039.

McCubbin, D.R., B.J. Apelberg, S. Roe, and F.Divita, Jr. 2002. Livestock ammonia management and particulate-related health benefits. *Environ. Sci. Tech.* 36:1141-1146.

McWard, G.W. and D.R. Taylor. 2000. Acidified clay litter amendment. *J. Appl. Poultry Res.* 9:518-529.

Miles, D. M., P. A. Moore Jr., D. R. Smith, D. W. Rice, H. L. Stilborn, D. R. Rowe, B. D. Lott, S. L. Branton, and J. D. Simmons. 2003. Total and water-soluble phosphorus in broiler litter over three flocks with alum litter treatment and dietary inclusion of high available phosphorus corn and phytase supplementation. *Poult. Sci.* 82:1544–1549.

Miles, D.M., S.L. Brannon, and B.D. Lott. 2004. Atmospheric ammonia is detrimental to the performance of modern commercial broilers. *Poultry Sci.* 83:1650-1654.

Miles, D.M., W.W. Miller, S.L. Branton, W.R. Maslin, and B.D. Lott. 2006a. Ocular responses to ammonia in broiler chickens. *Avian Diseases* 50:45-49.

Moore, P.A., Jr. 2006. *Treating Poultry Litter with Aluminum Sulfate (Alum).* SERA-17 Factsheet. http://www.sera17.ext.vt.edu/SERA_17_Publications.htm. Website confirmed June 3, 2010.

Moore, P.A, Jr., T.C. Daniel, J.T. Gilmour, B.R. Shreve, D.R. Edwards, and B.H. Wood, 1998. Decreasing metal runoff from poultry litter with aluminum sulfate. *J. Envrion. Qual.* 27: 92-99.

Moore, P.A., Jr. and D.R. Edwards. 2005. Long-term effects of poultry litter, alum-treated litter and ammonium nitrate on aluminum availability in soils. *J. Environ. Qual.* 34:2104-2111.

Moore, P.A., Jr. and D.R. Edwards. 2007. Long-term effects of poultry litter, alum-treated litter and ammonium nitrate on phosphorus availability in soils. *J. Environ. Qual.* 36:163-174.

Moore, P.A., Jr., and D.M. Miller. 1994. Reducing phosphorus solubility in poultry litter with aluminum, calcium, and iron amendments. *J. Environ. Qual.* 23:325-330.

Moore, P.A, Jr., T.C. Daniel, A.N. Sharpley, and C.W. Wood. 1995a. Poultry manure management: Environmentally sound options. *J. Soil and Water Cons.* 50:321-327.

Moore, P.A., Jr., T.C. Daniel, D.R. Edwards, and D.M. Miller. 1995b. Effect of chemical amendments on ammonia volatilization from poultry litter. *J. Environ. Qual.* 24:293-300.

Moore, P.A., Jr., T.C. Daniel, D.R. Edwards, and D.M. Miller. 1996. Evaluation of chemical amendments to inhibit ammonia volatilization from poultry litter. *Poultry Sci.* 75:315-320.

Moore, P.A., T.C. Daniel, J.T. Gilmour, B.R. Shreve, D.R. Edwards, and B.H. Wood. 1998. Decreasing metal runoff from poultry litter with aluminum sulfate. *J. Environ. Qual.* 27:92-99.

Moore, P.A., T.C. Daniel, and D.R. Edwards. 1999. Reducing phosphorus runoff and improving poultry production with alum. *Poultry Science* 78:692-698.

Moore, P.A., Jr., T.C. Daniel, and D.R. Edwards. 2000a. Reducing phosphorus runoff and inhibiting ammonia loss from poultry manure with aluminum sulfate. *J. Environ.* Qual. 29:37-49.

Moore, P.A., Jr., M.G. Wilson, T.C. Daniel and D.R. Edwards. 2000b. *Effects of liquid alum applications to hen manure on ammonia volatilization and phosphorus runoff.* Pp. 52-63 In 2000 Proceedings of the Arkansas Poultry Symposium held in Springdale, AR on April 11-12, 2000. Published by the Poultry Federation. Little Rock, AR.

Moore, P.A., Jr., S.E. Watkins, D.K. Carmen, and P.B. DeLaune. 2004. Treating poultry litter with alum. *Fact Sheet.* University of Arkansas, Cooperative Extension Service, FSA8003-PD-1-04N.

Moore, P.A., Jr., D.M. Miles, R.T. Burns, D.H. Pote, and W.K. Berg. Evaluation of ammonia emissions from broiler litter. 2008. Pp. 33-40 In *Proc. Of the Eight Int. Symposium Livestock Environment.* Iguassu Falls, Brazil. ASAE, St. Joseph, MI.

Moore, P.A., Jr., A.N. Sharpley, W. Delp, B. Haggard, T.C. Daniel, K. VanDevender, A. Baber, and M.D. Daniels. 2009. *The revised Arkansas Phosphorus Index.* Pp. 22. Arkansas Natural Resources Commission.

Moore, P.A., Jr., Dana Miles, Robert Burns, Dan Pote, Kess Berg, and In Hag Choi. 2010. Ammonia emission factors from broiler litter in barns, storage, and after land application. Accepted for publication in *J. Environmental Quality May* 10, 2010.

National Academy of Science. 2002. The scientific basis for estimating emissions from animal feeding operations: Interim report. http://www.bob.nap.edu/books/03090846X/html/.

Nichols, D.J., T.C. Daniel, P.A. Moore, Jr., D.R. Edwards, and D.H. Pote. 1997. Runoff of estrogen hormone 17 beta-estradiol from poultry litter applied to pastures. *J. Environ. Qual.* 26:1002-1006.

Peak, D., J.T. Sims, and D.L. Sparks. 2002. Solid-state speciation of natural and alum-amended poultry litter using XANES spectroscopy. *Environ. Sci. Technol.* 36:4253-4261.

Penn, C. and H. Zhang. 2009. *Alum-treated poultry litter as a fertilizer source.* Oklahoma Cooperative Extension Facts sheet PSS-2254.

Pescatore, A.J. and J.M. Harter-Dennis. 1989. Effects of ferrous sulfate consumption on the performance of broiler chicks. *Poultry Sci.* 68:1063-1067.

Pierre, W.H. 1928. Nitrogenous fertilizers and soil acidity. I. Effect of various nitrogenous fertilizers on soil reaction. *J. Am. Soc. Agron.* 20:2-16.

Pionke, H. B., W.J. Gburek, A.N. Sharpley, and R.R. Schnabel. 1996. Flow and nutrient export patterns for an agricultural hill-land watershed. *Water Resour. Res.* 32:1795-1804.

Rothrock, M.J., Jr., K.L. Cook, J.G. Warren, and K. Sistani. The effect of alum addition on microbial communities in poultry litter. *Poultry Sci.* 87:1493-1503.

Scantling, M., A. Waldroup, J. March and P.A. Moore, Jr. 1995. Microbiological effects of treating poultry litter with aluminum sulfate. *Poultry Sci.* 74(Suppl. 1):216.

Schindler, D.W. 1977. The evolution of phosphorus limitation in lakes. Science195:260-262.

Schroder, H. 1985. Nitrogen losses from Danish agriculture-trends and consequences. *Agri. Ecosyst. Environ.* 14:279-289.

Seiter, J.M., K.E. Staats-Borda, M. Ginder-Vogel, and D.L. Sparks. 2008. XANES spectroscopic analysis of phosphorus speciation in alum-amended poultry litter. J. *Environ. Qual.* 37:477-485.

Sharpley, A.N., S.J. Smith, O.R. Jones, W.A. Berg, and G.A. Coleman. 1992. The transport of bioavailable phosphorus in agricultural runoff. *J. Environ. Qual.* 21:30-35.

Sharpley, A.N., S. C. Chapra, R. Wedepohl, J.T. Sims, and T. C. Daniel. 1994. Managing agricultural phosphorus for protection of surface waters: issues and options. *J. Environ. Qual.* 23:437-451.

Sharpley, A.N., J.L. Weld, D.B. Beegle, P.J.A. Kleinman, W.J. Gburek, P.A. Moore, Jr., and G. Mullins. 2003. *Development of phosphorus indices for nutrient management.*

Shi, Y., D.B. Parker, N.A. Cole, B.W. Auvermann, and J.E. Mehlhorn. 2001. *Surface amendments to minimize ammonia emissions from beef cattle feedlots.* Trans. ASAE 44:677-682.

Shober, A.L. and J.T. Sims. 2007. Integrating phosphorus source and soil properties into risk assessments for phosphorus loss. *Soil Sci. Soc. Am. J.* 71:551-560.

Shober, A.L., D.L. Hesterberg, J.T. Sims, and S. Gardner. 2006. Characterization of phosphorus species in biosolids and manures using XANES Spectroscopy. *J. Environ. Qual.* 35:1983-1993.

Shreve, B.R., P.A. Moore, Jr., T.C. Daniel, D.R. Edwards, and D.M. Miller. 1995. Reduction of phosphorus in runoff from field-applied poultry litter using chemical amendments. *J. Environ. Qual.* 24:106-111.

Shreve, B. R., P. A. Moore Jr., D. M. Miller, T. C. Daniel, and D. R. Edwards. 1996. Long-term phosphorus solubility in soils receiving poultry litter treated with aluminum, calcium, and iron amendments. Commun. *Soil Sci. Plant Anal.* 27:2493–2510.

Sims, J.T. and N.J. Luka-McCafferty. 2002. On-farm evaluation of aluminum sulfate (alum) as a poultry litter amendment: effects on litter properties. *J. Environ. Qual.* 31:2066-2073.

Sistani, K.R., D.A. Mays, and R.A. Dawkins. 2006. Tall fescue fertilized with alum-treated and untreated broiler litter: runoff, soil, and plant nutrient content. *J. Sustainable Agri.* 28:109-119.

Sonzogni, W.C., S.C. Chapra, D.E. Armstrong and T.J. Logan. 1982. Bioavailability of phosphorus inputs to lakes. J. *Environ. Qual.* 11:555-563.

Smith, D.R. and P.A. Moore, Jr. 2005. Soil extractable phosphorus changes with time after application of fertilizer: II. Manure from swine-fed diets. *Soil Sci.* 170:640-651.

Smith, D.R., P.A. Moore, Jr., C.L. Griffis, T.C. Daniel, D.R. Edwards, and D.L. Booth. 2001. Effects of alum and aluminum chloride on phosphorus runoff from swine manure. *J. Environ. Qual.* 30:992-998.

Smith, D.R., P.A. Moore, Jr., D.M. Miles, B.E. Haggard, and T.C. Daniel. 2004a. Decreasing phosphorus runoff losses from land-applied poultry litter with dietary modifications and alum addition. *J. Environ. Qual.* 33:2210-2216.

Smith, D.R., P.A. Moore, Jr., C.V. Maxwell, B.E. Haggard, and T.C. Daniel. 2004b. Reducing phosphorus runoff from swine manure with dietary phyase and aluminum chloride. *J. Environ. Qual.* 33:1048-1054.

Smith, D.R., P.A. Moore, Jr., B.E. Haggard, C.V. Maxwell, T.C. Daniel, K. VanDevander, and M.E. Davis. 2004c. Effect of aluminum chloride and dietary phytase on ammonia loss from swine manure. *J. Anim. Sci.* 82:605-611.

Smith, D.R., P.A. Moore, Jr., and D.M. Miles. 2005. Soil extractable phosphorus changes with time after application of fertilizer: I. Litter from poultry-fed modified diets. *Soil Sci.* 170:530-542.

Staats, K.E., Y. Arai, and D.L. Sparks. 2004. Alum amendment effects on phosphorus release and distribution in poultry litter-amended sandy soils. *J. Environ. Qual.* 33:1904-1911.

Timby, C.G., T.C. Daniel, R.W. New, and P.A. Moore, Jr. 2004. Polymer type and aluminum chloride affect screened solids and phosphorus removal from liquid dairy manure. *ASAE* 20:57-64.

Tomlinson, P.J., M.C. Savin, and P.A. Moore, Jr. 2008. Phosphatase activities in soil after repeated untreated and alum-treated poultry litter applications. *Biol. Fertil. Soils.* 44:613-622.

USDA NRCS. 2009. National Conservation Practice Standards. http://www.nrcs.usda.gov/technical/Standards/nhcp.html.

USDA and USEPA. 1999. Unified national strategy for animal feeding operations. Available at www.epa.gov/npdes/pubs/pubs/finafost.pdf. USDA and USEPA, Washington, D.C.

USEPA. 1992. National Water Quality Inventory: 1992 Report to Congress. U.S. EPA Office of Water, EPA841-R-001. US EPA. Office of Water (4503F), U.S. Government Printing Office. Washington, DC.

Vadas, P.A., L.W. Good, P.A. Moore, Jr., and N. Widman. 2009. Estimating phosphorus loss in runoff from manure and fertilizer for a phosphorus loss quantification tool. *J. Environ. Qual.* 38:1645-1653.

van Breemen, N., P.A. Burrough, E.J. Velthorst, H.F. van Dobben, T. de Wit, T.B. Ridder, and H.F.R. Reijinders. 1982. Soil acidification from atmospheric ammonium sulphate in forest canopy throughfall. *Nature* 299:548-550.

Wallner-Pendleton, E, D.P. Froman and O. Hedstrom. 1986. Identification of ferrous sulfate toxicity in a commercial broiler flock. A*vian Dis.* 30:430-432.

Warren, G.G., S.B. Phillips, G.L. Mullins, D. Keahey, and C.J. Penn. 2006a. Environmental and production consequences of using alum-amended poultry litter as a nutrient source for corn. *J. Environ. Qual.* 35:172-183.

Warren, G.G., S.B. Phillips, G.L. Mullins, and L.W. Zelazny. 2006b. Impact of alum-treated poultry litter applications on fescue production and soil phosphorus fractions. *Soil Sci. Soc. Am. J.* 70:1957-1966.

Warren, J.G. 2005. Management of alum-treated poultry litter. *Ph.D. dissertation.* Virginia Polytechnic Institute and State University, Blacksburg, VA. Pp. 91.

Warren, J.G., C.L. Penn, J.M. McGrath, and K. Sistani. 2008. The impact of alum addition on organic P transformations in poultry litter and litter amended soil. *J. Environ. Qual.* 37:469-476.

Worley, J.W. and K.C. Das. 2000. Swine manure separation and composting using alum. *Appl. Eng. Agric.* 16:555-561.

Worley, J. W., L. M. Risse, M. L. Cabrera, and M. P. Nolan, Jr. 1999. Bedding for broiler chickens: two alternative systems. *Appl. Eng. Agric.* 15:687-693.

Worley, J.W., M.L. Cabrera, and L.M. Risse. 2000. Reduced levels of alum to amend broiler litter. *Applied Eng. in Agric.* 16:441-444.

PART IV. HEAVY ELEMENTS AND ENVIRONMENTAL CONCERNS

In: Environmental Chemistry of Animal Manure
Editor: Zhongqi He

ISBN: 978-1-62808-641-6
© 2013 Nova Science Publishers, Inc.

Chapter 16

SOURCES AND CONTENTS OF HEAVY METALS AND OTHER TRACE ELEMENTS IN ANIMAL MANURES

Jackie L. Schroder[1,], Hailin Zhang[1],
Jaben R. Richards[1] and Zhongqi He[2]*

16.1. INTRODUCTION

Animal manures are available in many parts of the world and serve as abundant sources of macro and micronutrients for crop and grass production. Besides providing valuable nutrients to the soil, manure supplies organic matter to improve physical, chemical and biological properties of soils, thus improving water infiltration, enhancing retention of nutrients, reducing wind and water erosion, and promoting growth of beneficial organisms. Confined animal feeding operations (CAFO) are the major source of animal manures in most countries. Virtually all animal manures are land applied with approximately 2.2×10^9 wet tons of manure being produced annually in the United States and approximately 80×10^6 wet tons produced annually in the United Kingdom (Wright, 1998; Bolan et al., 2004).

Until recently, the majority of concerns associated with land application of manure have focused on the contamination of groundwater or surface waters with N and P (Sims and Wolf, 1994; Moore et al., 1995). However, animal manures also contain substantial amounts of potentially toxic trace elements such as As, Cu, and Zn (Bolan et al., 2004). Nicholson et al. (1999) estimated that approximately 25 to 40% of the total annual inputs of Cu, Ni, and Zn to soil came from animal manures. Farm gate balance experiments in Sweden, where all inputs and outputs were examined, found the most important source of trace elements in manure came from purchased feedstuffs (Bengtsson et al., 2003; Öborn et al., 2005). Substantial amounts of As are introduced to the environment of the Delaware-Maryland-Virginia Peninsula by the use of As containing compounds such as roxarsone in poultry feed (Christen,

---

[*] Corresponding Author: jackie.schroder@okstate.edu
[1]Oklahoma State University, Department of Plant and Soil Sciences, Stillwater, OK 74078, USA
[2]USDA-ARS, New England Plant, Soil, and Water Laboratory, Orono, ME 04469, USA

2001). While several researchers have shown that the application of manures increases trace element concentrations in plants (Kornegay, 1976; Bomke and Lowe, 1991; Bibak, 1994; Dufera, 1999; Bolan et al., 2003), only a few reports have indicated phytotoxicity due to land application of manure (Bolan et al., 2004). For example, McGrath et al. (1980, 1982) found that growth of perennial ryegrass (*Lolium perenne*) seedlings was retarded with the addition of approximately 200 ppm of Cu to soil from pig manure slurries. In another study, Cresswell et al. (1990) observed the yield of mushrooms was decreased by excessive amounts of B and Cu in poultry manure compost. Several researchers have reported metal toxicity to ruminants grazing on pastures which had received manure applications (Bremner, 1981; Lamand, 1981; Poole, 1981; Eck and Stewart, 1995). However, in most cases, the toxicity was due to a direct intake of trace element rich manure directly from the soil or through contaminated herbage (Batey et al., 1972; Bolan et al., 2004). Conversely, several researchers have evaluated applying copper rich pig manure slurries to sheep pastures and have reported no adverse effects on sheep grazing the manure amended fields (Gracey, 1976; Kneale and Smith, 1977; Bremner, 1981).

Elevated concentrations of As, Cu, and Zn have been observed in soils that have received long-term application of manures (Kingery et al., 1994; van der Watt et al., 1994; He et al. 2009). Additionally, researchers have reported high concentrations of metals in runoff from soils that had received manure applications (Edwards et al., 1997; Moore et al., 1998). Thus, a potential exists for manure-treated soils to serve as a non-point source of metal pollution through leaching, runoff or erosion. Edwards and Somershwar (2000) indicated that rather than focusing on only one component (i.e. N and/or P concentration), land application guidelines should consider the total composition of animal manures.

The Part 503 rule that limits land application of chemicals in biosolids is based on a risk assessment framework that originally evaluated 14 exposure pathways for humans, animals, plants, and soil organisms (U.S. EPA, 1995a, National Research Council, 2002). The inorganic chemicals evaluated in the exposure pathways included As, Cd, Cr, Cu, Hg, Pb, Ni, Mo, Se, and Zn. According to the U.S. EPA Part 503 risk assessment, other trace elements in biosolids do not present potential risk to human health or the environment when applied at typical rates (U.S. EPA, 1995a). While much literature exists dealing with metal inputs from biosolids and inorganic fertilizers, few comprehensive studies have been conducted on the sources and distributions of metals and other trace elements in animal manure. The purpose of this chapter is to examine the sources and distributions of metals and other trace elements in animal manures.

16.2. SOURCES OF TRACE ELEMENTS IN ANIMAL MANURES

The assimilation of trace metals by livestock is important for many physiological functions including enzyme formation, vitamin formation, metabolism, and electron transport (Table 16.1). Both growth and health of poultry and livestock may be adversely affected if the supply of trace metals is inadequate in the diet (Hostetler et al., 2003). Although livestock get certain amounts of trace elements via their diets, many times the levels of metals in plants is too small to meet dietary requirements (Sistani and Novak, 2006). Additionally, the bioavailabilty of these plant-derived (or bound) trace metals to livestock may be low. The improvement of feed efficiency and health of livestock is important in concentrated animal feeding operations (Bolan et al., 2004). Therefore, livestock diets are typically supplemented with the trace elements As, Co, Cu, Fe, Mn, Mo, Se, and Zn to prevent diseases, improve weight gains and feed conversion, and increase egg production for poultry (Miller et al., 1991; Tufft and Nockles, 1991; Sims and Wolf, 1994; Moore et al., 1995) (Table 16.2). Copper compounds such as copper sulfate and copper hydroxide are often utilized as growth promoters in swine and poultry and as foot baths to prevent hoof warts and treat lameness in dairy cattle (Poulsen, 1998; Bolan et al., 2003; Jokela et al., 2010). Zinc oxide is added to premixes, supplementary feeds, and mineral feeds (Sager, 2006). Broiler chickens and swine are frequently fed organoarsenical compounds (e.g., roxarsone and *p*-arsanilic acid) for the control of coccidiosis and growth promotion (Chapman and Johnson, 2002; Makris et al., 2008a). Additionally, roxarsone and *p*-arsanilic acid are used with antibiotics in the swine industry to control dysentery, bacterial respiratory diseases, and/or promote growth (Carlson and Fangman, 2000; Makris et al., 2008b).

Table 16.1. Physiological functions of Trace Elements in livestock diets.

Trace Elements	Physiological Function	Reference
As	Growth promoter, control of coccidiosis	Bolan et al. (2004)
B	Increases bone strength and feed efficiency in swine	Sistani and Novak (2006)
Co	Vitamin formation	Sistani and Novak (2006)
Cr	Metabolism of glucose, lipids, and proteins	Bolan et al. (2004)
Cu	Enzyme formation, growth promoter in swine and poultry, reproductive processes	Poulsen (1998) Bolan et al. (2004) Sistani and Novak (2006)
F	major constituent of bone and teeth	Bolan et al. (2004)
Fe	cytochrome functions and electron transfer	Bolan et al. (2004)
I	major constituent of thyroid hormone	Bolan et al. (2004)
Mn	Enzyme formation, reproductive processes	Bolan et al. (2004) Sistani and Novak (2006)
Mo	Enzyme formation	Bolan et al. (2004)
Ni	Increases bone strength in poultry	Bolan et al. (2004)
Se	Enzyme formation, growth promoter	Bolan et al. (2004)
Zn	Enzyme formation, fetus development	Bolan et al. (2004) Sistani and Novak (2006)

Table 16.2. Mean or median (depends on the particular publication cited) trace element concentrations (mg kg^{-1}) in cattle, poultry, and swine feeds.

Sample	As	Cd	Co	Cr	Cu	Hg	Mn	Mo	Ni	Pb	Se	Zn	Reference
Cattle Feed													
1	<0.40	<0.035	0.46	<0.19	14.5	-[†]	60.6	0.59	<0.80	<0.30	-	130	Sager (2006)
2	<0.40	0.13	4.35	2.58	56.2	-	162	0.95	4.46	0.64	-	354	Sager (2006)
3	0.99	0.30	28.9	9.70	281	-	1071	0.72	6.70	2.00	-	2030	Sager (2006)

Table 16.2. (Continued).

Sample	As	Cd	Co	Cr	Cu	Hg	Mn	Mo	Ni	Pb	Se	Zn	Reference
Cattle Feed													
4	0.10	0.05	0.10	0.75	3.00	<0.01	16.8	<2.50	<QL[‡]	0.36	0.19	20.0	Caper et al. (1978)
5	0.88	0.28	1.70	20.0	24.0	0.05	117	29.9	<QL	2.10	0.35	115	Caper et al. (1978)
6	2.20	0.24	2.20	31.0	21.0	0.03	111	42.9	<QL	3.28	0.32	86.0	Caper et al. (1978)
7	3.03	1.79	-	42.0	1484	<BDL[§]	-	-	9.0	5.50	-	2900	Nicholson et al. (1999)
8	0.49	0.27	-	1.66	34.6	<BDL	-	-	3.10	<1.00	-	189	Nicholson et al. (1999)
9	0.02	0.22	0.36	13.7	19.1	0.001	85.4	1.01	2.30	5.76	-	109	Cang et al. (2004)
Swine Feed													
10	<0.40	0.065	0.49	1.30	25.0	-	73.5	0.96	1.78	0.30	-	119	Sager (2006)
11	<0.40	0.17	2.21	4.75	87.0	-	215	2.58	5.94	0.68	-	398	Sager (2006)
12	1.67	1.12	16.2	28.8	747	-	1674	0.82	12.9	2.80	-	3170	Sager (2006)
13	0.38	0.16	-	1.31	28.5	BDL	-	-	2.70	<1.00	-	177	Nicholson et al. (1999)
14	0.43	0.13	-	0.75	161	BDL	-	-	2.30	<1.00	-	834	Nicholson et al. (1999)
15	0.39	<0.10	-	0.54	159	BDL	-	-	3.10	<1.00	-	356	Nicholson et al. (1999)
16	0.28	<0.10	-	0.80	128	BDL	-	-	2.80	<1.00	-	308	Nicholson et al. (1999)
17	3.20	-	-	-	-	-	-	-	-	-	-	-	Li and Chen (2005)
18	0.09	0.57	0.59	25.7	105	0.006	134	0.94	7.85	10.7	-	144	Cang et al. (2004)
Poultry Feed													
19	<0.40	0.099	0.69	2.53	18.2	-	118	2.15	2.33	<0.30	-	116	Sager (2006)
20	-	-	-	-	-	-	-	-	-	-	-	-	Sager (2006)
21	<0.40	2.04	13.7	39.1	291	-	3489	1.11	12.6	2.10	-	3660	Sager (2006)
-	-	0.39	-	0.76	23.0	BDL	-	-	2.60	<1.00	-	153	Nicholson et al. (1999)
-	-	0.12	-	0.22	32.6	BDL	-	-	2.10	<1.00	-	135	Nicholson et al. (1999)
24	1.27	1.66	-	1.81	887	BDL	-	-	4.40	10.5	-	6980	Nicholson et al. (1999)
25	0.13	0.64	0.52	30.0	22.6	0.005	190	1.34	12.6	7.21	-	154	Cang et al. (2004)
26	15	-	-	-	9	-	-	-	-	-	-	94	Dao and Zhang (2007)

[†] Not measured.

[‡] Less than quantitation limit.

[§] Below detection limit.

1= complete feed, 2 = supplemental feed, 3 = mineral feeds, 4 = feedlot diet, 5 = low-fiber manure diet, 6 = high-fiber manure diet, 7 = dairy minerals, 8 = beef cattle cake, 9 = milch cow feed, 10 = complete feed, 11 = supplemental feed, 12 = mineral feeds, 13 = dry sow feed, 14 = rearer-weaner compound feed, 15 = rearer-grower complete feed, 16 = rearer-finisher complete feed, 17 = average of 6 types of pig feed, 18 = pig feed, 19 = chicken complete feed, 20 = chicken supplemental feeds, 21 = chicken mineral feeds, 22 = layer chicken feed, 23 = broiler-finisher chicken feed, 24 = broiler-breeder chicken supplement and minerals, 25 = chicken feed, 26 = poultry feed.

The U.S. National Research Council (1989) estimates that a 300 kg heifer ingests approximately 6 kg of feed per day and recommends feeds contain 40 mg Zn kg^{-1} to achieve a daily intake of 240 mg per day for dairy cattle (U.S. NRC, 1989). Similarly, the U.S. NRC recommends that a dairy cattle diet contains 10 mg Cu kg^{-1} (U.S. NRC, 1989) to obtain a daily intake of 60 mg per day and recommends between 3.5 and 6.0 mg Cu kg^{-1} in the diet of swine (Sistani and Novak, 2006). Unfortunately, many times trace metals added to livestock diets by producers or feed companies exceed recommended intake amounts (Jondreville, 2003). For example, in China, Zn was used as feed additive and ranged from 28 to 378 mg kg^{-1} in milk cow feed (Chang et al., 2004). In another study, Sager (2006) reported median concentrations of 2,030 mg Zn kg^{-1} and 281 mg Cu kg^{-1} in calf mineral feeds (Table 16.2). Daily dietary concentrations of 150 to 250 mg CuSO$_4$ kg^{-1} and 2,500 to 3,000 mg ZnSO$_4$ kg^{-1} have been used to stimulate swine growth (Brumm, 1998; Poulsen, 1998). However, 5-6 mg Cu kg^{-1} feed and 80-100 mg Zn kg^{-1} has been reported to be adequate for growing swine (National Research Council, 1998). Similarly, the recommended amounts of Cu and Zn in poultry feed are 4 and 50 mg kg^{-1} (Toor et al., 2007), respectively, but Cu in poultry diets has exceeded 30 mg kg^{-1} while Zn has exceeded 150 mg kg^{-1} (Nicholson, 1999). Roxarsone is added to poultry feed at a rate of approximately 50 mg kg^{-1} while *p*-arsanilic acid is added at a rate of approximately 100 mg kg^{-1} (Calvert, 1975).

The accumulation of trace elements in manures may also be due to ingestion of contaminated soil by animals or by manure being mixed with soil on barn floors. Significant relationships between Al or Fe and Pb and Ba concentrations in dairy cattle manure showed that Pb and Ba were partially derived from ingestion of soil or from mixing of soil with manure (McBride and Spiers, 2001). Similarly, ingestion of soil has been shown to be an important source of Cd for grazing sheep and cattle in New Zealand and Australia (Lee et al., 1996; Loganathan et al., 1999). Stocking rates, grazing management, and pasture status affect surface soil ingestion. Often manure from feedlots contains as much soil as manure (Bolan et al., 2004). Broiler litter (a mixture of chicken manure and bedding material) is composed of approximately 30% bedding materials (Nicholson et al., 1999). However, because these materials contain metal concentrations equivalent to background levels of plant material, they have little effect on the overall metal concentration of the litter. Recent work by several researchers has shown the disposal of waste copper sulfate foot-bath solution into liquid manure pits significantly increased Cu contents in dairy manure (McBride and Spears, 2001; Thomas, 2001; Stehouwer and Roth, 2004; Jokela et al., 2010). For example, Jokela et al. (2010) examined more than 2,300 dairy manure samples collected in Vermont from 1992 to 2006 and found a four-fold increase of Cu in dairy manure mostly after 1998. Their study attributed the increase of Cu in dairy manure to the increased use of copper sulfate foot baths in dairy operations.

Metal concentrations of manure from literature vary greatly between different animal feeds (e.g., cattle feed versus swine feed versus poultry feed) as well as greatly between feeds fed at different growth stages to the same animal (Table 16.2). Both Nicholson et al. (1999) and Sager (2006) reported the highest trace element concentrations were Cu and Zn in mineral supplements provided to cattle. Nicholson et al. (1999) reported typical concentrations of Cu and Zn in mineral supplements for cattle were 1,484 and 2,900 mg kg^{-1}, respectively (Table 16.2). Sager (2006) reported somewhat lower concentrations of 281 mg Cu kg^{-1} and 2,030 mg Zn kg^{-1} in mineral supplements (Table 16.2). Additionally, both studies found that mineral supplements provided to cattle contained greater concentrations of As, Cd,

Cr, Pb, and Ni as compared to other types of feed. Sager (2006) also reported elevated levels of Mn in cattle, swine, and poultry feed. Similarly, some of the highest concentrations were Cu and Zn found in mineral feeds provided to swine. High concentrations of Cu, Mn, and Zn were reported in chicken mineral feeds. Reported mean/median values for Zn ranged from 86 to 2,900 mg kg^{-1} in cattle feed, from 119 to 3,170 mg kg^{-1} in swine feed, and from 94 to 6,980 mg kg^{-1} in poultry feed. Concentrations of Cu ranged from 3.0 to 1,484 mg kg^{-1} in cattle feed, from 25.0 to 747 mg kg^{-1} in swine feed, and from 9.0 to 887 mg kg^{-1} in poultry feed (Table 16.2). Similarly, huge variations in Mn were observed in different feeds. The largest concentration of Mn reported (i.e. 3,489 mg kg^{-1}) was in a chicken mineral feed. Concentrations of As were highly variable in all of the feeds but the highest concentrations occurred in poultry feed. This is consistent with the reported additions of roxarsone and p-arsanilic acid to poultry feed for the control of coccidiosis and as a growth promoter (Chapman and Johnson, 2002; Makris et al., 2008a). The largest concentrations of As in poultry feed were observed in the study conducted by Dao and Zhang (2007) (Table 16.2). The samples analyzed in their study were collected from the eastern shore of Maryland in the United States. The concentrations of As that were reported in poultry feed samples from other countries (i.e. samples 19-25) were much lower than the ones reported in the study by Dao and Zhang (2007). This is probably due to the fact that organoarsenical compounds are not added to poultry feeds in other countries. Mercury was not documented in many studies, but was very low (i.e. ppb levels) when it was reported. Similarly, Caper et al. (1978) only found ppb to low ppm levels of Se in cattle feed (Table 16.2).

Because most of the trace elements ingested by livestock are excreted via the feces and urine, the concentrations of trace elements in manures are dependent on the concentrations of these metals in the animal's diet (Krishnamachari, 1987; Miller et al., 1991). Researchers have reported that swine excrete approximately 85-90% of the total daily Cu and Zn in dietary supplements (Unwin, 1977; Parkinson and Yells, 1985; Brumm, 1998). Similarly, Kunkle et al. (1981) observed that Cu concentrations in broiler litter were linearly related to Cu in the bird's diet and were typically concentrated 3.25 times. Several studies have shown that increased trace element content in livestock feed results in increases in trace elements in animal manures (Sims and Wolfe, 1994; Mikkelsen, 2000; Nahm, 2002). Thus, elevated concentrations of trace minerals are found in manured soils (Li et al., 1997). The primary concern associated with manure-borne metals is that they do not degrade, thus metals will build up in soil with repeated manure application and over time would affect the soil environment (Bolan et al., 2004).

16.3. LEVELS OF TOTAL TRACE ELEMENTS IN ANIMAL MANURES

Animal manures contain many trace elements beneficial to plants. However, they also contain many nonessential plant trace elenzts such as As that are often added to diets as health supplements. Trace element concentrations of animal manures vary greatly due to the large number of feed and manure management systems (Table 16.3-5). Concentrations of trace elements in manure vary depending on livestock type as well as other factors such as animal age, feed source, housing and bedding difference, trace element supplements, and waste management practices (Bolan et al., 2004; Sistani and Novak, 2006). This large variation in trace element contents makes it difficult to substitute micronutrients in animal manures for chemical fertilizers (Eck and Stewart, 1995; Longhurst et al., 2000).

Table 16.3. Mean or median trace element concentrations (mg kg^{-1}) in cattle manure.

Sample	As	B	Cd	Co	Cr	Cu	Hg	Mn	Mo	Ni	Pb	Se	Zn	Reference
1	-[†]	-	-	-	15	29	-	372	-	9.0	8.6	-	67	de Abreu and Berton (1996)
2	1.3	8.1	0.2	2.5	4.6	139	0.02	-	2.5	0.8	2.2	3	191	McBride and Spiers (2001)
3	-	-	-	-	-	200	-	700	-	-	-	-	800	Eneji et al. (2001)
4	6.8	-	0.7	2.23	-	17.5	<0.4	172	-	9.6	7.5	-	-	Raven and Loeppert (1997)
5	3.0	-	0.5	3.55	-	-	<0.4	186	-	6.2	2.6	-	-	Raven and Loeppert (1997)
6	5.2	-	0.4	3.57	14.4	-	<0.4	357	-	8.7	5.4	0.48	164	Raven and Loeppert (1997)
7	-	-	-	-	-	16.5	-	149	-	-	-	-	6480	Wallingford et al. (1975)
8	-	-	-	-	-	0.51	-	0.29	-	-	-	-	1.8	Wallingford et al. (1975)
9	1.63	-	0.38	-	5.32	37.5	-	-	-	3.7	3.61	-	153	Nicholson et al. (1999)
10	1.44	-	0.33	-	5.64	62.3	-	-	-	5.4	5.87	-	209	Nicholson et al. (1999)
11	0.79	-	0.13	-	1.41	16.4	-	-	-	2.0	1.65	-	81	Nicholson et al. (1999)
12	2.6	-	0.26	-	4.69	33.2	-	-	-	6.4	7.07	-	133	Nicholson et al. (1999)
13	0.013	-	0.70	1.67	46.9	46.0	0.039	472	2.91	8.91	9.74	-	186	Cang et al. (2004)
14	1.15	-	0.42	-	2.58	31.4	-	-	-	2.8	2.24	-	145	Chambers et al. (1998)
15	0.71	-	0.14	-	1.50	15.6	-	-	-	2.1	1.40	-	63	Chambers et al. (1998)
16	0.3	-	-	-	1.8	27	-	-	-	-	-	0.6	90	Combs et al. (1998)
17	0.5	-	-	-	4.4	191	-	-	-	-	-	1.4	186	Combs et al. (1998)
18	-	-	0.18	-	-	37.1	-	-	-	-	3.77	-	162	Menzi and Kessler (1998)
19	-	-	0.16	-	-	19.1	-	-	-	-	2.92	-	123	Menzi and Kessler (1998)
20	-	-	0.17	-	-	23.9	-	-	-	-	3.77	-	118	Menzi and Kessler (1998)
21	-	-	0.17	-	-	52.5	-	-	-	-	2.98	-	245	Menzi and Kessler (1998)

Table 16.3. (Continued).

Sample	As	B	Cd	Co	Cr	Cu	Hg	Mn	Mo	Ni	Pb	Se	Zn	Reference
22	-	-	0.15	-	-	22.0	-	-	-	-	2.81	-	91.1	Menzi and Kessler (1998)
23	BDL[‡]	-	0.06	0.88	4.72	13.8	0.07	193	-	4.54	1.94	-	136	He (unpublished)
24	BDL	-	BDL	0.82	1.60	19.0	0.07	224	-	5.60	2.03	-	235	He (unpublished)
25	BDL	-	0.10	0.69	2.10	11.4	0.06	136	-	2.70	1.98	-	119	He (unpublished)
26	-	-	0.4	-	6.1	48	-	-	-	7.7	8.9	-	305	Schultheib et al. (2004)
27	-	-	0.3	-	6.1	25	-	-	-	4.1	5.2	-	122	Schultheib et al. (2004)
28	-	-	-	-	-	23	-	273	-	-	<0.05	-	262	Walker et al. (2004)
29	0.33	-	0.27	2.10	6.6	51	-	180	3.5	6.3	4.1	0.59	164	Sager (2007)

[†]Not measured.

[‡]Below detection limit.

1 = dairy manure, 2 = dairy liquid and solid manure, 3 = cow dung, 4 = cow manure, 5 = composted cattle manure, 6 = composted cattle manure, 7 = feedlot manure, 8 = feedlot lagoon, 9 = dairy cattle feedyard manure, 10 = dairy cattle slurry, 11 = beef cattle feedyard manure, 12 = beef cattle slurry, 13 = milch cow manure, 14 = dairy cattle feedyard manure, 15 = beef cattle feedyard manure, 16 = dairy solid manure, 17 = dairy liquid manure, 18 = dairy cattle liquid manure, 19 = dairy cattle slurry manure, 20 = dairy cattle solid manure, 21 = beef cattle liquid manure, 22 = beef cattle solid manure, 23 = dairy cattle solid manure, 24 = dairy cattle liquid manure, 25 = dairy cattle slurry manure, 26 = cattle manure slurry, 27 = cattle farmyard manure, 28 = fresh cow manure, 29 = cattle manure

Similar to livestock feeds, the greatest concentrations of trace elements found in cattle manure were Cu, Mn, and Zn (Table 16.3). Examination of literature values indicates mean/median values in cattle manure for Zn ranged from 1.8 to 6,480 mg kg^{-1} while Cu ranged from 0.51 to 200 mg kg^{-1} (Table 16.3). Mean/median values of Mn in cattle manure ranged from 0.29 to 700 mg kg^{-1}. Observed mean/median ranges for other trace elements in cattle manure were: As (0.013 to 6.8 mg kg^{-1}), Cd (0.06 to 0.70 mg kg^{-1}), Co (0.69 to 3.57 mg kg^{-1}), Cr (1.41 to 46.9 mg kg^{-1}), Hg (0.02 to <0.4 mg kg^{-1}), Mo (2.5 to 3.5 mg kg^{-1}), Ni (0.8 to 9.6 mg kg^{-1}), Pb (< 0.05 to 9.74 mg kg^{-1}), and Se (0.48 to 3 mg kg^{-1}). Only one study reported B (i.e. 8.1 mg kg^{-1}) in cattle manure (McBride and Spiers, 2001), although B toxicity to plants has been attributed to manure land application. Some of the large variation in ranges for trace elements reported in cattle manure is due to physical condition of the manure (i.e. composted versus non-composted, dry versus fresh, liquid versus solid). For example, Sistani et al. (2001) reported that composting of poultry litter decreased concentrations of Mn and Zn in the litter. Conversely, Hsu and Lo (2001) observed that composting of swine manure increased Cu, Mn, and Zn by approximately 2.7-fold.

Table 16.4. Mean or median trace element concentrations (mg kg^{-1}) in swine manure.

Sample	As	B	Cd	Co	Cr	Cu	Hg	Mn	Mo	Ni	Pb	Se	Zn	Reference
1	-[†]	-	0.25	-	33	1338	-	869	12.4	-	14.0	-	1440	de Abreu and Berton (1996)
2	-	-	-	-	-	1000	-	2100	-	-	-	-	2900	Eneji et al. (2001)
3	-	17.8	-	-	-	1279	-	197	-	-	-	-	231	Mullins et al. (1982)
4	0.86	-	0.37	-	1.98	374	-	-	-	7.50	2.94	-	431	Nicholson et al. (1999)
5	1.68	-	0.30	-	2.82	351	-	-	-	10.4	2.48	-	575	Nicholson et al. (1999)
6	0.012	-	0.80	2.11	46.2	399	0.033	452	1.60	9.51	12.8	-	506	Cang et al. (2004)
7	2.14	-	-	-	-	-	-	-	-	-	-	-	-	Makris et al. (2008)
8	6.73	-	-	-	-	-	-	-	-	-	-	-	-	Makris et al. (2008)
9	2.04	-	-	-	-	-	-	-	-	-	-	-	-	Makris et al. (2008)
10	0.73	-	0.68	-	1.87	346	-	-	-	5.0	2.83	-	387	Chambers et al. (1998)
11	1.33	-	0.30	-	2.44	364	-	-	-	7.8	<1.00	-	403	Chambers et al. (1998)
12	-	-	0.17	-	-	115	-	-	-	-	2.53	-	517	Menzi and Kessler (1998)
13	-	-	0.23	-	-	71.1	-	-	-	-	2.54	-	554	Menzi and Kessler (1998)
14	19.2	-	-	-	-	-	-	-	-	-	-	-	-	Li and Chen (2005)
15	-	-	0.4	-	10	531	-	-	-	12	5.7	-	1508	Schultheib et al. (2004
16	-	-	0.4	-	7.1	1165	-	-	-	16	3.4	-	1884	Schultheib et al. (2004
17	-	-	0.4	-	14	206	-	-	-	4.9	1.9	-	465	Schultheib et al. (2004
18	-	-	-	-	-	343	-	-	-	-	-	-	577	Hsu and Lo (2001)
19	0.88	-	0.46	4.0	6.9	282	-	358	5.3	12.5	1.9	3.37	1156	Sager (2007)
20	0.51	-	0.33	2.3	7.8	84	-	317	2.1	8.9	2.6	1.34	399	Sager (2007)

[†] Not measured.

1 = swine manure, 2 = swine dung, 3= Cu-enriched swine manure, 4 = pig feedyard manure, 5 = pig slurry manure, 6 = pig manure, 7 = swine lagoon, 8 = swine sludge (bottom of lagoon), 9 = swine manure, 10 = pig feedlot manure, 11 = pig slurry, 12 = pig manure (fattening pigs), 13 = sow with piglets, 14 = average of 6 types of pigs ranging from piglets to lactating sows, 15 = pig slurry (mixed), 16 = weaners/growers slurry, 17 pig farmyard manure, 18 = separated swine manure, 19 = pig manure. 20 = pig dung

Table 16.5. Mean or median trace element concentrations (mg kg^{-1}) in poultry manure.

Sample	As	B	Cd	Co	Cr	Cu	Hg	Mn	Mo	Ni	Pb	Se	Zn	Reference
1	-[†]	-	-	-	-	400	-	1800	-	-	-	-	2300	Eneji et al. (2001)
2	-	-	-	-	-	313	-	246	-	-	-	-	327	Wood et al. (1996)
3	-	-	-	313	-	-	-	246	-	-	-	-	327	Wood et al. (1996)
4	34.6	-	4.93	-	9.9	6.1	-	501	-	2.46	0	1.23	743	Jackson and Miller (2000)
5	0.57	-	-	2.0	6.0	30.7	0.06	166	5.0	<QL[‡]	-	0.38	158	Caper et al. (1978)
6	9.01	-	0.42	-	17.2	96.8	-	-	-	5.4	3.62	-	378	Nicholson et al. (1999)
7	0.46	-	1.06	-	4.57	64.8	-	-	-	7.1	8.37	-	459	Nicholson et al. (1999)
8	43	51	3	6	-	748	-	956	6	15	11	-	718	Moore et al. (1998)
9	-	19	2	8	6	19	-	271	-	14	13	-	252	Bomke and Lowe (1991)
10	-	-	0.48	-	7.3	54.3	-	465	7.69	7.0	2.3	-	550	Ihnat and Fernandes (1996)
11	-	390	-	-	-	-	-	-	-	-	-	-	-	Wilkinson (1997)
12	18.8	-	<0.2	<0.2	3.2	410	-	356	<0.2	1.0	BDL[§]	<2.0	371	Kpomblekou-A et al. (2002)
13	0.75	-	0.38	-	7.53	92.4	-	-	-	4.9	2.94	-	403	Chambers et al. (1998)
14	0.45	-	1.03	-	4.79	65.6	-	-	-	6.1	9.77	-	423	Chambers et al. (1998)
15	-	-	0.31	-	-	35.2	-	-	-	-	2.22	-	425	Menzi and Kessler (1998)
16	-	-	0.20	-	-	43.8	-	-	-	-	2.25	-	512	Menzi and Kessler (1998)
17	32	-	-	-	-	708	-	-	-	-	-	-	549	Dao and Zhang (2007)
18	35.8	-	-	-	-	571	-	-	-	-	-	-	541	Dao and Zhang (2007)
19	43	69	-	-	-	599	-	678	-	-	-	-	615	Toor et al. (2007)
20	15.7	-	0.25	-	-	479	-	449	-	11.1	2.06	1.18	373	Jackson et al. (2003)
21	39	-	-	-	-	-	-	-	-	-	-	-	-	Makris et al. (2008)
22	0.047	-	1.84	2.29	81.2	89.1	0.024	624	3.80	17.5	11.1	-	417	Cang et al. (2004)
23	47.8	-	-	-	-	345	-	390	-	-	-	-	386	Arai et al. (2003)
24	28.7	-	0.22	0.80	9.2	76.9	-	310	-	9.7	<0.5	2.0	320	Gabarino et al. (2003)

Sample	As	B	Cd	Co	Cr	Cu	Hg	Mn	Mo	Ni	Pb	Se	Zn	Reference
25	0.12	-	0.43	1.7	10.7	66	-	339	3.3	8.5	5.4	1.4	314	Sager (2007)

[†] Not measured.

[‡] Less than quantitation limit.

[§] Below detection limit of 2.0 mg kg^{-1}.

1 = poultry droppings, 2 = broiler litter, 3 = broiler litter, 4 = broiler litter, 5 = poultry waste with litter, 6 = broiler/turkey litter, 7 = layer manure, 8 = poultry litter, 9 = deep-pit poultry litter, 10 = composted poultry manure, 11 = poultry manure, 12 = broiler litter, 13 = broiler/turkey litter, 14 = layer manure 15 = layer hens, 16 = = broiler/turkey litter, 17 = broiler litter, 18 = layer litter, 19 = poultry litter, 20 = poultry litter, 21 = poultry litter, 22 = chicken manure, 23 = poultry litter, 24 = poultry litter, 25 = poultry dung.

Likewise to cattle manure, elevated concentrations of Cu, Mn, and Zn were reported in swine manure (Table 16.4). Mean/median values in swine manure for Zn ranged from 231 to 2,900 mg kg^{-1} while Cu ranged from 71.1 to 1,338 mg kg^{-1} (Table 16.4). Mean/median values of Mn in swine manure ranged from 197 to 2,100 mg kg^{-1}. Reported mean/median ranges for other trace elements in swine manure were: As (0.012 to 19.2 mg kg^{-1}), Cd (0.23 to 0.80 mg kg^{-1}), Co (2.11 to 4.0 mg kg^{-1}), Cr (1.98 to 46.2 mg kg^{-1}), Mo (1.60 to 12.4 mg kg^{-1}), Ni (5.00 to 16.0 mg kg^{-1}), Pb (< 1.0 to 14.0 mg kg^{-1}), and Se (1.34 to 3.37 mg kg^{-1}). Single studies determined Hg and B in swine manure at 0.033 and 17.8 mg kg^{-1}, respectively (Table 16.4). The observed ranges for mean/median values of trace elements in swine manure are quite broad and mostly dependent on the age of pigs and quantities of trace metals added are greater than those found in cattle manure. This agrees well with the results of Steineck et al. (1999) who evaluated trace elements in livestock wastes from Sweden and reported that concentrations in pig manure were higher because they were fed more easily digestible food.

Poultry manure contained elevated concentrations of Cu, Mn, and Zn (Table 16.5). Copper in poultry manure ranged from 6.1 to 748 mg kg^{-1} and Zn ranged from 158 to 2,300 mg kg^{-1}. Reported mean/median ranges for other trace elements in poultry manure were: As (0.05 to 47.8 mg kg^{-1}), B (19 to 390 mg kg^{-1}), Cd (< 0.2 to 4.93 mg kg^{-1}), Co (< 0.2 to 313 mg kg^{-1}), Cr (3.2 to 81.2 mg kg^{-1}), Mo (< 0.2 to 7.69 mg kg^{-1}), Ni (1.0 to 17.5 mg kg^{-1}), Pb (0 to 13 mg kg^{-1}), and Se (0.38 to 2.0 mg kg^{-1}). Arsenic concentrations were much greater in poultry manure as compared to other types of manures. Although the swine industry utilizes the organoarsenical compounds of roxarsone and p-arsanilic, their use is sporadic and not as consistent as in poultry (Carlson and Fangman, 2000; Makris et al., 2008b).

The ceiling pollutant concentrations for trace elements in biosolids that may be land applied in the United States are As (75 mg kg^{-1}), Cd (85 mg kg^{-1}), Cr (3,000 mg kg^{-1}), Cu (4,300 mg kg^{-1}), Hg (57 mg kg^{-1}), Mo (75 mg kg^{-1}), Ni (420 mg kg^{-1}), Pb (840 mg kg^{-1}), Se (100 mg kg^{-1}), and Zn (7,500 mg kg^{-1}) (U.S. EPA, 1993). Thus, the concentrations of trace elements in manure (i.e. the mean/median values) are typically less than their respective ceiling pollutant concentrations that may be land applied in biosolids. However, there are some exceptions in which unusually high concentrations occur. For example, Dao and Zhang (2007) reported concentrations of As as high as 101 mg kg^{-1} in some poultry litter samples.

Jackie L. Schroder , Hailin Zhang, Jaben R. Richards et al.

Table 16.6. Total elemental content (mg kg^{-1}) and the corresponding water-soluble percentage (in parentheses) of cattle, swine, and poultry manure.

Sample	As	B	Cd	Co	Cr	Cu	Hg	Mn	Mo	Ni	Pb	Se	Zn	Reference
1	-†	-	0.87	-	11.2	356	-	345	-	8.67	7.53	-	765	Bolan et
	-	-	(13.8)	(0.68)	(8.75)	(31.7)	-	(5.10)	-	(14.2)	(4.52)	-	(16.1)	al.
														(2003)
2	16.8	-	0.10	0.31	2.59	656	-	275	-	7.97	0.74	0.95	247	Jackson
	(92.3)		(20.0)	(-)	(16.2)	(48.0)	-	(2.35)	-	(69.4)	(2.70)	(40.0)	(7.37)	and
														Bertsch
														(2001)
3	35.1	-	-	-	-	-	-	-	-	-	-	5.20	-	Jackson
	(72.4)	-	-	-	-	-	-	-	-	-	-	(BDL)	-	and Miller
														(1999)
4	15.5	-	-	-	-	213	-	-	-	-	-	-	124	Brown et
	(36.8)	-	-	-	-	(20.9)	-	-	-	-	-	-	(18.5)	al. (2005)
5	29	-	-	-	-	400	-	-	-	-	-	-	430	Rutherford
	(75.9)	-	-	-	-	(50.0)	-	-	-	-	-	-	(20.0)	et al.
														(2003)
6	26.9	-	-	-	-	-	-	-	-	-	-	-	-	Han et al.
	(55.8)	-	-	-	-	-	-	-	-	-	-	-	-	(2004)
7	-	-	0.48	-	7.3	54.3	-	465	7.69	7.0	2.3	-	550	Ihnat and
	-	-	(83.3)	-	-	(26.2)	-	(6.19)	(15.3)	(14.3)	(21.7)	-	(1.27)	Fernandes
														(1996)
8	15.7		0.25	-	-	479	-	449	-	11.1	2.06	1.18	373	Jackson et
	(70.7)		(16.0)	-	-	(40.9)	-	(2.00)	-	(49.0)	(1.00)	(39.8)	(6.01)	al. (2003)
9	43	69	-	-	-	599	-	678	-	-	-	-	615	Toor et al.
	(70.0)	(72.5)	-	-	-	(40.7)	-	(5.16)	-	-	-	-	(8.78)	(2007)
10	24	122	-	-	-	339	-	453	-	-	-	-	417	Toor et al.
	(54.2)	(82.0)	-	-	-	(37.5)	-	(2.87)	-	-	-	-	(5.52)	(2007)
11	-	-	0.25	-	13.2	419	-	865	-	12.3	13.4	-	1210	Bolan et
	-	-	(28.0)	-	(9.32)	(31.0)	-	(1.69)	-	(26.3)	(9.40)	-	(1.95)	al.
														(2003)
12	3.35	-	-	-	-	-	-	-	-	-	-	-	-	Makris et
	(25.1)	-	-	-	-	-	-	-	-	-	-	-	-	al. (2008b)
13	0.91	-	-	-	-	-	-	-	-	-	-	-	-	Makris et
	(68.1)	-	-	-	-	-	-	-	-	-	-	-	-	al. (2008b)
14	1.87	-	-	-	-	-	-	-	-	-	-	-	-	Makris et
	(18.7)	-	-	-	-	-	-	-	-	-	-	-	-	al. (2008b)
15	-	-	-	-	-	976		331	-	-	-	-	1,540	Hsu and
	-	-	-	-	-	(3.00)		(2.00)	-	-	-	-	(2.00)	Lo (2001)

† Not measured.

‡ Below detection limit.

1= dairy cattle manure, 2 = poultry manure, 3 = broiler poultry litter, 4 = poultry litter extracted with 0.1 M NaCl, 5 = poultry litter, 6 = poultry waste, 7 = poultry manure, 8 = poultry litter, 9 = raw poultry litter, 10 = ground poultry litter, 11 = swine manure, 12 = swine manure, 13 = swine manure, 14 = swine manure, 15 = composted swine manure.

16.4. EXTRACTABLE LEVELS OF TRACE ELEMENTS IN ANIMAL MANURES

Selective sequential chemical extractions have been used to examine extractability of trace elements in animal manures and their distribution in soils following land application (Miller et al., 1986). The primary metal forms in animal manures include soluble, exchangeable, adsorbed, organic-bound, oxide-bound, and precipitated (Mullins et al., 1982; Miller et al., 1986; Payne et al., 1988; Canet et al., 1997). The distribution of trace elements in animal manures varies widely and is dependent on the chemical properties of the individual metal, the characteristics of the manure, and the efficiency of feed conversion by the animals (Nicholson et al., 1999; Bolan et al., 2004). The characteristics of the manure are determined by the animal feed and the manure treatment process. It is important to remember that selective sequential extraction procedures identify operationally defined manure fractions because they extract more or less the fraction they are supposed to extract and redistribution of trace elements may occur as the extraction proceeds (Shuman, 1991; Sauve, 2002).

Most studies indicate manure Cd, Cu, and Zn exist primarily in the organically complexed form (Bolan et al., 2004). For example, Miller et al. (1986) used sequential and nonsequential extractions to evaluate manure from swine that had been fed high levels of copper. Their study found that most of the Cu in the manure samples was extracted with NaOH or $K_4P_2O_7$ indicating the majority of Cu was associated with the organic fraction. In another study, L'Herroux et al. (1997) used the sequential extraction procedure of Tessier et al. (1979) to evaluate trace element distribution in swine manure and reported the following fractions: 66.5% Cu organic-bound, 76.7% Co oxide-bound, 67.2% Zn oxide-bound, and 69.4% Mn carbonate-bound. Hsu and Lo (2001) used sequential extraction to evaluate the effect of composting on the leaching of Cu, Mn, and Zn from swine manure. In their study, the majority of Cu was in the organically bound fraction, the majority of Mn was in the solid particulate fraction, and most of the Zn was in the organically complexed fraction.

The bioavailability, toxicity, and mobility of trace elements in the soil-water-plant ecosystems are largely determined by their speciation and distribution. Several studies have evaluated water-soluble trace elements in animal manures (Table 16.6). Only one study was found that evaluated water soluble trace elements in cattle manure. Water soluble percentages of the total elemental concentrations in cattle manure were Cd (13.8%), Co (0.68%), Cr (8.75%), Cu (31.7%), Mn (5.10%), Ni (14.2%), Pb (4.52%), and Zn (16.1%) (Table 16.6). In swine manure, the water soluble percentage ranges for the different trace elements were As (18.7 to 68.1%), Cd (28.0%), Cr (9.32%), Cu (3.00 to 31.0%), Mn (1.69 to 2.00%), Ni (26.3%), Pb (9.40%), and Zn (1.95 to 2.00%). The water soluble percentage ranges for the different trace elements in poultry manure were As (36.8 to 92.3%), B (72.5 to 82.0%), Cd (16.0 to 83.3%), Cr (9.32 to 16.2%), Cu (20.9 to 50.0), Mn (2.00 to 6.19%), Mo (15.3), Ni (14.3 to 69.4%), Pb (1.00 to 21.7%), Se (39.8 to 40.0), and Zn (1.27 to 20.0%). Overall, among trace elements in animal manures, As, B, Cu, and Ni exhibit the greatest water solubility while Zn clearly displays some of the lowest water solubility. The study by Jackson and Bertsch (2001) found that >90% of the As in poultry litter was water soluble. Their study reported the As species in poultry litter was the original roxarsone and *p*-arsanilic acid as well as the metabolites aresenate, arsenite, dimethyl arsenic acid, and monomethyl arsenic acid (Jackson and Bertsch, 2001). The major species in 50% of the poultry litter samples was roxarsone but As (V) was the major species in the other 50% of the poultry litter samples. Several other studies have found that As in poultry litter was highly water soluble (Jackson and Miller, 1999; Jackson et al., 2003; Rutherford et al., 2003; Han et al., 2004). The study by Jackson and Bertsch (2001) indicated the combination of the high solubility of As, the mineralization potential of roxarsone, and the large amount of poultry litter that is land applied suggested a potential long-term detrimental effect on soil and water quality.

16.5. TRACE ELEMENT ACCUMULATION IN SOIL DUE TO ANIMAL MANURE APPLICATION

Repeated land applications of animal manure often occur because of high transportation costs for manures and limited land availability for manure application (Sistani and Novak, 2006). Another reason for accumulation of trace elements in manure-treated soils is the low trace element requirement of crops. For example, animal manures are often applied to bermudagrass (*Cynodon dactylon* L.), a popular pasture grass in southern and southeastern United States. However, bermudagrass only removes 0.009 kg Cu and 0.218 kg Zn per 7.3 Mg ha^{-1} yield (Zublena, 1991). Novak et al. (2004) evaluated the application of swine effluent to bermudagrass in North Carolina. Their study calculated a mass balance and estimated approximately 10.5 kg Cu ha^{-1} and 15.6 kg Zn ha^{-1} was added via swine effluent. The mass balance estimated that 10.49 kg Cu ha^{-1} and 15.38 kg Zn ha^{-1} would remain in the field after grass removal, thus showing the annual removal rate by bermudagrass was much less than the rate of addition.

Several studies have shown the application of animal manures to soil increases trace elements in soil. Nicholson et al. (1999) estimated annual inputs of trace elements to land due to manure application in Wales and England totaled 1,821 Mg Cu, 225 Mg Ni, and 5,247 Mg Zn, which represented 25 to 40% of the total inputs. A study performed in Sydney, Australia found that elevated levels of Cd in soils and vegetables were due to repeated applications of poultry manure (Jindasa et al., 1997). In another study, the long-term (i.e. 15 to 28 yrs) application of poultry litter to Ultisols in Alabama increased soil Cu from control concentration of 0.75 mg kg^{-1} to 2.5 mg kg^{-1} and increased soil Zn from a control concentration of 2.2 mg kg^{-1} to 10 mg kg^{-1} (Kingery et al., 1994). Brock et al. (2006) evaluated the long-term application of liquid dairy manure and solid poultry manure to 109 fields in southern New York. Application rates of liquid dairy manure ranging from 149,600 to 168,300 L ha^{-1} significantly increased soil Cu by as much as 627% and soil Zn by as much as 72% as compared to unamended fields that had not received any manure in the past 10 yr. Likewise, in the same study, solid poultry manure application rates ranging from 26.9 to 31.4 Mg ha^{-1} increased soil Cu by as much as 318% and soil Zn by as much as 267% as compared to the unamended fields. Novak et al., (2004) found elevated levels of Zn in soils that had received annual applications of swine effluent for 10 yrs. Kirchmann et al. (2009) estimated the mean annual fluxes of trace element in soils of Swedish long-term fertility experiments that received 1 Mg ha^{-1} yr^{-1} dry weight solid cattle manure and reported the manure application enriched soils with Pb, Cr, Zn, Cu, and Mn.

Keller and Schulin (2003) modeled P, Cd, and Zn fluxes in rural agricultural soils of Switzerland for different farm types and crop types. Their study evaluated arable farms, dairy and mixed farms, and animal husbandry farms. Their study showed the largest fluxes of trace elements were on the animal husbandry farms where the largest production of manure occurred. Sheppard et al. (2009) modeled trace element additions to Canadian agricultural soils via the atmosphere, fertilizers, manures, and biosolids and reported that century soil concentrations (i.e. concentrations after 100 yrs of application) were increased by as much as threefold as compared to background soil concentrations.

In a long-term research study conducted at the Experiment Station in Guymon, Oklahoma, Richards et al. (2010) found that application of beef feedlot manure increased total soil Cu by approximately 23%, total soil Mo by approximately 19%, and total soil Zn by approximately 33% (Table 16.7). Similarly, long-term application of swine effluent increased total soil Cu by 17% and total soil Zn by 27% (Table 16.7). Due to the fact that the behavior and effects of trace elements added to soils in animal manures are generally very similar to that added as biosolids, Bolan et al. (2004) suggested it would be prudent to adopt similar guidelines for manure application as have been utilized for biosolids. A significant increase in the concentration of a pollutant does not necessarily result in increased risk to plants, animals, or humans. Risk assessments were used to determine cumulative pollutant loading rates (CPLRs) as part of the Part 503 regulations for biosolids. The CPLRs are the maximum amount of a pollutant that can be added to a soil without causing adverse effects on plants, animals, and humans (USEPA, 1993). A comparison of measured cumulative loading rates of pollutants to their representative CPLRs may be used to evaluate the potential risk to human health and the environment. The assumption that the soil mass is 2×10^6 kg ha^{-1}, which is equal to a bulk density of 1.33 g cm^{-3}, is used in the measurement of cumulative loading rates (USEPA, 1995a). The recommended N agronomic rate for corn assuming a yield goal of 120 bushel acre^{-1}, which is equivalent to 7,560 kg ha^{-1}, is approximately 145 kg ha^{-1} (Zhang and Raun, 2006). The N application rates at the Guymon site ranged from 40 to 350% of the recommended agronomic rate. The calculated CPLRs for trace elements for highest N rate at the Guymon study ranged from 1.40% for Mo to 6.08% for Zn indicating manure application did not significantly increase risk to plants, animals, or humans (Table 16.8). Similarly, the study by Brock et al. (2006) found that it would take approximately 163 yr of dairy manure application and 1,070 yr of poultry manure application to reach the CPLR for Cu. The study also reported the CPLR for Zn would be reached after 518 and 317 yr for dairy and poultry manure application, respectively.

Table 16.7. Total elemental content after 14 yrs application of feedlot beef manure and swine effluent to the Richfield soil.

Site	Manure	N rate (kg ha^{-1})	Total B (mg kg^{-1})	Total Cu (mg kg^{-1})	Total Fe (g kg^{-1})	Total Mn (mg kg^{-1})	Total Mo (mg kg^{-1})	Total Zn (mg kg^{-1})
Guymon	Beef	0	56.3a	12.0a	14.8a	396a	0.26a	63.9a
		56	54.9a	12.2a	14.5a	398a	0.25a	4.9a
		168	55.7a	12.9a	14.2ab	385a	0.27a	73.1a
		504	54.9a	14.7b	13.4b	380a	0.31b	85.1b
Guymon	Swine	0	55.7a	11.9a	15.5a	398a	0.26a	65.9a
		56	54.6a	12.4ab	15.1a	399a	0.24a	70.8ab
		168	55.5a	13.3bc	15.1a	404a	0.28a	77.1bc
		504	56.1a	13.9c	14.9a	400a	0.28a	83.8c

Adapted from Richards (2010).

Table 16.8. Measured mean pollutant loading rates per hectare for a 15-cm depth of feedlot beef manure or swine effluent incorporation in the greatest rate of treatment (504 kg N ha^{-1}) and the percentage of the cumulative pollutant loading rates.

Site	Manure	Trace Element	CPLR[†] (kg ha^{-1})	Measured Pollutant Loading Rate (kg ha^{-1})	Percentage of CPLR
Guymon	Beef	Cu	1500	29.4	1.96
		Mo	40[‡]	0.62	1.55
		Zn	2800	170.2	6.08
Guymon	Swine	Cu	1500	27.8	1.85
		Mo	40	0.56	1.40
		Zn	2800	167.6	5.99

Adapted from Richards (2010).

[†] Cumulative Pollutant Loading Rates (USEPA, 1995a).

[‡] Value obtained from O'Connor et al. (2001).

16.6. Impact of Manure Land Application on Micronutrient Availability

Micronutrients are elements essential for plant growth which are absorbed in small (micro) quantities (Sposito, 1989) and are a subset of the trace elements present in animal manures. Identified essential micronutrients include B, Cu, Cl, Fe, Mn, Mo, and Zn. Micronutrient availability is critical for a balanced plant nutrition to achieve optimum crop growth. Micronutrients play key roles in many plant processes. Zinc promotes growth hormones, starch formation, seed maturation and production and is involved in a number of metallo-enzymes (Adriano, 2001). Iron is involved in chlorophyll formation while copper is vital in photosynthesis as well as protein and carbohydrate metabolism and is a constituent of a number of plant enzymes (Adriano, 2001). Manganese plays a role in photosynthesis, nitrogen metabolism and assimilation (Brady and Weil, 1999). Manganese also functions in chlorophyll formation (Adriano, 2001). Although the necessity of B in plant nutrition has been established, its exact role remains somewhat vague (Adriano, 2001). However, it is suspected that B functions at the membrane level and is essential for cell division and development. Thus, adequate levels of available micronutrients in soil are essential for proper plant nutrition and maximization of yield.

The adequacy levels of micronutrients in soil for most plant growth are well established. For example, adequate concentrations of B for most plant growth are between 0.05 and 0.10 mg kg^{-1} (Adriano, 2001). Chelating agents [e.g., ethylenediamine-tetraacetic acid (EDTA) and diethylenetriamine-pentaacetic acid (DTPA)] are more effective in removing soluble metal–organic complexes that are potentially bioavailable and have often been found to be more reliable in predicting plant availability (Sims and Johnson, 1991). DTPA-extractable Zn levels of 0.5 to 1.0 mg kg^{-1} are considered adequate for most crops, while concentrations below 0.5 mg kg^{-1} Zn are considered deficient (Adriano, 2001). Lindsay and Norvell (1978) found that the critical level of DTPA-extractable soil Cu is 0.20 mg kg^{-1} and soil concentrations below this value are considered deficient. DTPA-extractable Mn levels of greater than 1.0 mg kg^{-1} are sufficient and levels less than this are considered deficient (Zhang et al., 2009). Although the adequate concentrations are small, many soils in the world are deficient in one or more micronutrients due to unfavorable soil chemical conditions or naturally low contents of micronutrients in the parent material. Thus, animal manures may serve as sources of micronutrients for crop and grass production in addition to other benefits. Wallingford et al. (1975) reported that beef-feedlot manure increased DTPA-extractable Cu, Fe, Mn, and Zn in soils and enhanced uptake of Mn and Zn in corn. Benke et al. (2008) reported the application of cattle manure for 25 consecutive yrs increased EDTA-extractable Cu and Zn in Canadian soils. Tewolde et al. (2005) utilized broiler litter in a greenhouse study to effectively demonstrate that a one time application of broiler litter was able to supply adequate amounts of Fe, Cu, and Mn to cotton. Adeli et al.(2008) examined the application of broiler litter to no-till cotton over a 3 yr period and reported that litter application significantly increased soil Cu and Zn. Similarly, Franzluebbers et al. (2004) found the application of broiler litter over 5 yrs to bermuda grass increased Cu, Mn, and Zn in soil. In another study, Berenguer et al. (2008) applied liquid swine manure for 6 consecutive yrs to maize and found that extractable Cu and Zn soil concentrations increased more than 60%. In this study, they also evaluated Cu and Zn concentrations in whole maize plants and grain and found concentrations were lower than threshold values for animal and human ingestion. Several other studies have found that application of swine manures increased DTPA-extractable metals (Payne et al., 1988; Anderson et al., 1991; Zhu et al., 1991; Narwahl and Singh, 1998; Arnesen and Singh, 1998). Rutherford et al. (2003) evaluated soil samples from a pasture in Oklahoma that had received long-term (i.e. > 30 yrs) poultry litter application and reported elevated water-soluble concentrations of As, C, P, Cu, and Zn as compared to an unamended field.

The results of long-term research conducted at the Experiment Station in Guymon, Oklahoma (Richards, 2010) agree well with these studies. In this long-term research, plots had received annual applications of beef feedlot manure or swine effluent to corn at N rates of 0, 56, 168, and 504 kg ha^{-1} for 14 yrs. Long-term application of beef feedlot manure significantly increased concentrations of DTPA-extractable Cu, Mo, and Zn (Figure 16. 1). Similarly, long-term application of swine effluent increased DTPA-extractable Cu and Zn (Figure 16.2). Long-term application of beef feedlot manure increased total micronutrients (Figure 16.3) while the application of swine effluent increased total B, Cu, Mo, and Zn (Figure 16.4).

16.7. METHODS TO DECREASE ENVIRONMENTAL PROBLEMS OF MANURE APPLICATION

Different approaches have been utilized to decrease the environmental problems associated with the land application of animal manures (Bolan et al., 2004). Chemical amendments have shown to be useful in reducing the solubility of trace elements. Moore et al. (1998) reported that treatment of poultry litter with alum [Al$_2$(SO$_4$)$_3$] reduced As, Cu, Fe, and Zn in runoff as compared to untreated litter. Sims and Luka-McCafferty (2002) conducted a study in the Delmarva (Delaware–Maryland–Virginia) peninsula evaluating the effect of alum on poultry litter properties and elemental composition and on the solubility of several elements in litter that are of particular concern for water quality (Al, As, Cu, P, and Zn). Their study reported alum decreased water soluble As by approximately 63%, water soluble Cu by approximately 37%, and water soluble Zn by approximately 48%. Additionally, Westerman and Bicudo (2000) found reductions of Cu and Zn by approximately 87% in swine waste effluent treated with lime slurry, ferric chloride, or polymer. Wang et al. (2010) found the addition of bauxite residue (i.e. red or brown mud) significantly reduced water-soluble As, Cu, and Zn in both cattle manure and chicken litter. At a 50% mixing rate, Wang et al. (2010) reported reductions of 60 to 70% fo water-soluble Zn, 64 to 65% for water-soluble Cu, and 55 to 60% for water-soluble As in chicken litter. Furthermore, they reported reductions of approximately 84 to 89% Zn in cattle manure. Thus, treatment of animal manures with chemical amendments has the potential to immobilize trace elements in waste streams.

Other approaches include the reduction of trace element concentrations in livestock and poultry diets and increasing the availability of trace elements in feed, thus reducing concentrations in manures (Ashmead et al., 2008). Jondreville et al. (2003) estimated that reducing copper from 100 mg kg^{-1} to 20 mg kg^{-1} in swine feed would reduce the annual accumulation of Cu in top soils by 35%. Their study also indicated reducing Zn from 250 mg kg^{-1} to 100 mg kg^{-1} in swine feed would decrease Zn by 35% in top soils. In another study, Spears et al. (1999) reduced Cu from 15 to 5 mg kg^{-1} and Zn from 100 mg kg^{-1} to 25 mg kg^{-1} in swine feed and effectively decreased Cu and Zn in swine manure by 40% without adversely affecting the swine. Similarly, van Heugten et al. (2002) reported reducing trace mineral levels in diets for grower-finisher pigs reduced fecal mineral excretion of Zn, Cu, Fe, and Mn by 68, 36, 34, and 45% respectively, without negatively affecting carcass characteristics.

Organic forms of trace elements such as chelates and complexes have been suggested as a means to increase availability of trace elements in livestock and poultry feed (Chowdhury et al., 2004). Bowland et al. (1961) showed that the total concentration of Cu in feed was not as important for swine growth as the availability of Cu from the feed. Baker and Ammerman (1995) reported the relative bioavailability of organic Cu sources as compared to cupric sulfate ranged from 88% to 147% in poultry, swine, sheep, and cattle. Creech et al. (2004) reported replacing 50% of the supplemental Zn, Cu, Fe, and Mn with chelated forms in pigs from weaning effectively decreased fecal trace element excretion without negatively affecting pig performance. Similarly, a study that fed amino acid chelated (ACC) Cu and Zn instead of inorganic salts as a source of Cu and Zn to mature sows reduced concentrations in manure (Ashmead et al., 2008). The study by Ashmead et al. (2008) concluded that feeds could be potentially formulated with lower levels of Cu and Zn because of the increased bioavailability of ACC.

Phytate, the salt of phytic acid and the main form of P storage in plants, is responsible for complexing divalent and trivalent cations and reducing their availability to mono-gastric animals such as swine and poultry (Pallauf and Rimbach, 1997). Increases in Cu and Zn bioavailability may be accomplished by hydrolyzing phytate or by providing dietary sources that are less susceptible to phytate interactions (Jondreville et al., 2003). Phytase is an enzyme that can break down phytate, thus releasing digestible P. With this decompsition, Ca and other metals associated with phyate become free bioavailable forms (Sebastian et al., 1998). Lei et al. (1993) showed that supplementing corn-soybean meal diets of weanling pigs with microbial phytase significantly improved bioavailability of Zn. The addition of microbial phytase to cereal poultry diets has been shown to increase digestibility and availability of phytate-P and bound Ca, Zn and Cu (Sebastian et al., 1998). Thus, the utilization of microbial phytase in swine and poultry diets has the potential to decrease Cu and Zn in manures by increasing their bioavailability to the animals.

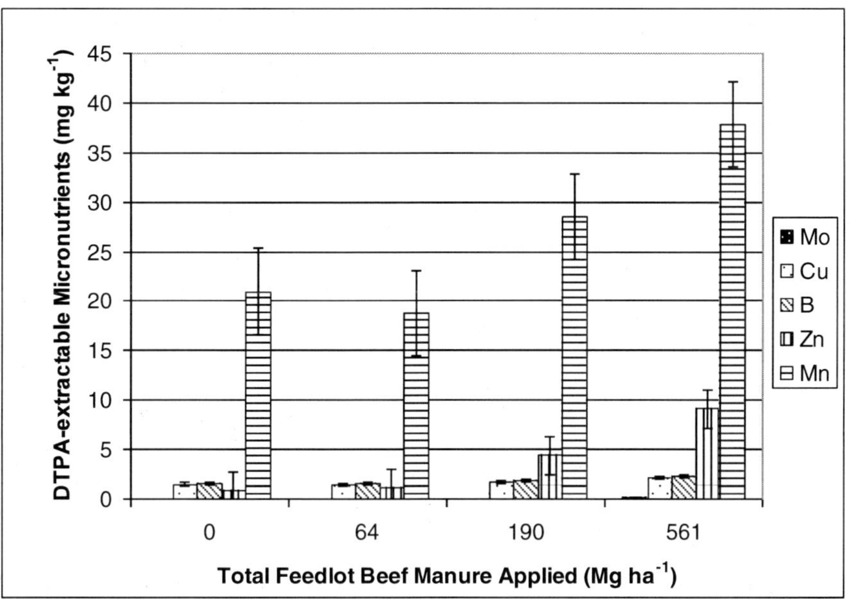

Figure 16.1. The relationship between DTPA-extractable micronutrients and total feedlot beef manure applied annually for 14 years to the Richfield soil [adapted from Richards (2010)].

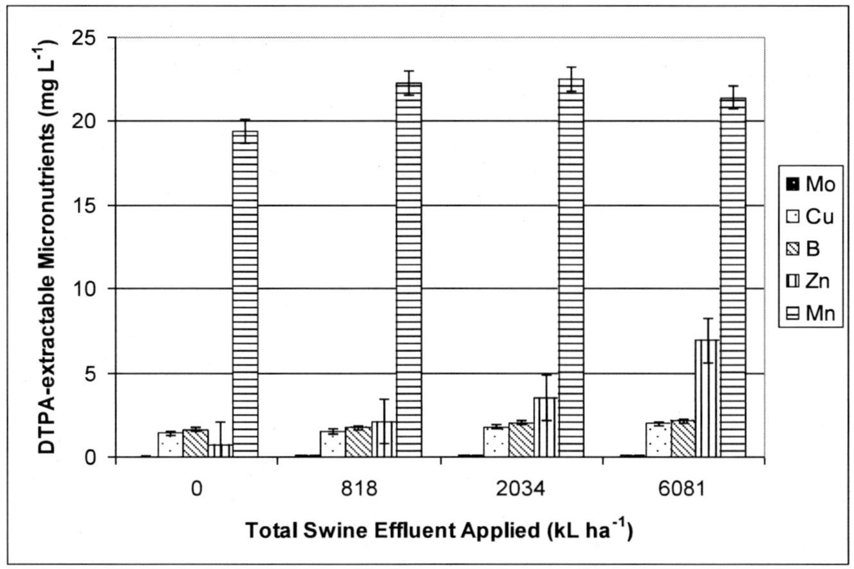

Figure 16.2. The relationship between DTPA-extractable micronutrients and total swine effluent applied annually for 14 years to the Richfield soil [adapted from Richards (2010)].

 Jackie L. Schroder , Hailin Zhang, Jaben R. Richards et al.

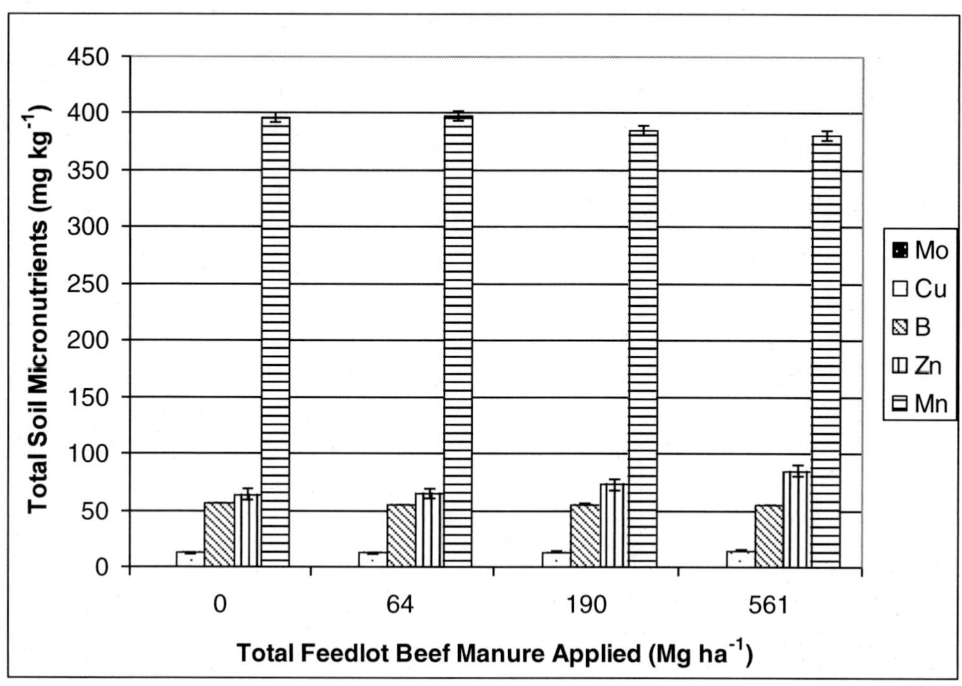

Figure 16.3. The relationship between total micronutrients and total feedlot beef manure applied annually for 14 years to the Richfield soil [adapted from Richards (2010)].

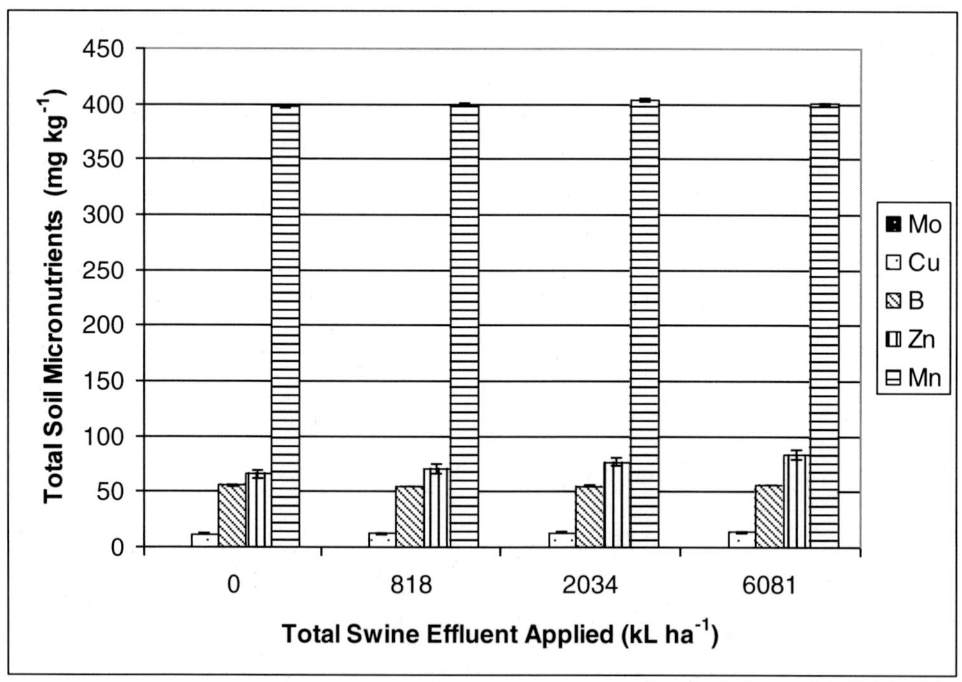

Figure 16.4. The relationship between total micronutrients and total swine effluent applied annually for 14 years to the Richfield soil [adapted from Richards (2010)].

Another method for reducing trace elements in animal manures is the use of best management practices (BMPs) on the farm. For example, several researchers have suggested BMPs for the reduction of Cu in dairy manure that is elevated as a result of dumping used hoof baths containing copper sulfate (Thomas, 2001; Moore and Ippolito, 2009; Rankin, 2010; Stehouwer and Roth, 2010). The suggested BMPs include: (1) reducing the amount of copper sulfate used in the foot baths, (2) reducing the overall frequency of foot baths, (3) improving hoof trimming and stall surfaces, and (4) disposing of the Cu waste in another location instead of the lagoon.

16.8. CONCLUSION

Trace elements are natural and added components of livestock and poultry feeds. Appropriate amounts of these trace elements in the diet of livestock and poultry ensures both health and reproduction. Unfortunately, many times trace metals that are added to livestock diets by producers or feed companies exceed dietary requirements. Because most of the trace elements ingested by livestock and poultry are excreted via the feces and urine, the concentration of metals in manures is dependent on the concentrations of these metals in the animal's diet. Large quantities of animal manures are generated worldwide and virtually all animal manures are land applied. Crops do not remove large amounts of trace elements from soil. Recently the application of animal manures has been identified as a major input of trace elements to soils and repeated applications has increased concentrations of trace elements in soil.

Approaches to decrease the environmental problems associated with the land application of animal manures include the reduction of the water soluble fraction of the trace elements, the reduction of trace elements concentrations in animal feed, and increasing the availability of trace elements in feed. Chemical amendments such as alum, lime slurries, ferric chloride, and polymers have been successfully used to reduce trace elements in waste streams. Organic forms of trace elements such as chelates and complexes have been utilized to increase availability of trace elements in livestock and poultry feed. Additionally, the addition of microbial phytase to animal diets has been shown to increase the availability of trace elements. However, perhaps the most significant way to reduce trace elements in animal manures, is the reduction of trace element supplementation in livestock and poultry feed. Several livestock experts believe additional research is needed to safely accomplish the reduction of trace elements in livestock and poultry diets.

REFERENCES

Adeli, A., M. Shankle, H. Tewolde, K.R. Sistani, and D.E. Rowe. 2008. Nutrient dynamics from broiler litter applied to no-till cotton in an upland soil. *Agron. J.* 100:564-570.

Ashmeade, H.D., R.A. Samford, and S.D. Ashmeade. 2008. Feeding amino acid chelated copper and zinc to reduce mineral pollution from swine manure. Intern. *J. Appl. Res. Vet. Med.* 6:31-37.

Adriano, D.C. 2001. *Trace elements in terrestrial environments: biogeochemistry, bioavailability, and risks of metals.* 2nd edition. Springer-Verlag, New York, NY, USA.

Anderson, M.A., J.R. Mckenna, D.C. Martens, S.J. Donohue, E.T. Kornegay, and M.D. Lindemann. 1991. Long-term effects of copper rich swine manure application on continuous corn production. *Commun. Soil Sci. Plant Anal.* 22:993-1002.

Arai. Y., A. Lanzirotti, S. Suton, J.A. Davis, and D.L. Sparks. 2003. Arsenic speciation and reactivity in poultry litter. *Environ. Sci. Technol.* 37:4083-4090.

Arnesen, A.K.M. and B.R. Singh. 1998. Plant uptake and DTPA-extractability of Cd, Cu, Ni and Zn in a Norwegian alum shale soil as affected by previous addition of dairy and pig manures and peat. *Can. J. Soil Sci.* 78:531-539.

Baker, D.H. and C.B. Ammerman. 1995. Copper bioavailability. p. 127-156. In C.B. Ammermann. D.H. Baker, and A.J. Lewis (eds.), *Bioavailability of Nutrients for Animals. Amino Acids, Minerals, and Vitamins,* Academic Press, San Diego, CA.

Batey, T., C. Berryman and C. Line. 1972. The disposal of copper-enriched pig-manure slurry on grassland. *J. Br. Grassland Soc.* 27:139-143.

Bengtsson, H., I. Öborn, S. Jonsson, I. Nilsson and A. Andersson. 2003. Field balances of some mineral nutrients and trace elements in organic and conventional dairy farming--a case study at Öjebyn, *Sweden. Eur. J. Agron.* 20:101-116.

Benke, M., S.P. Indraratne, X. Hao, C. Chang, and T.B. Goh. 2008. Trace element changes in soil after long-term cattle manure applications. *J. Environ. Qual.* 37:798-807.

Berenguer, P., S. Cela, F. Santiveri, J. Boixadera, and J. Lloveras. 2008. Copper and zinc soil accumulation and plant concentration in irrigated maize fertilized with liquid swine manure. *Agron. J.* 100:1056-1061.

Bibak, A. 1994. Uptake of cobalt and manganese by winter wheat from a sandy loam soil with and without added farmyard manure and fertilizer nitrogen. *Commun. Soil. Sci. Plant Anal.* 25:2675–2684.

Bolan, N.S., M.A. Khan, D.C. Donaldson, D.C. Adriano, and C. Matthew. 2003. Distribution and bioavailability of copper in farm effluent. *Sci. Total Environ.* 309:225–236.

Bolan, B.S., D.C. Adriano, and S. Mahimairaja. 2004. Distribution and bioavailability of trace elements in livestock and poultry manure by-products. *Critical Reviews in Environmental Science and Technology.* 34:291-338.

Bomke, A.A., and L.E. Lowe. 1991. Trace-element uptake by two British Columbia forages as affected by poultry manure application. *Can. J. Soil Sci.* 71:305-312.

Bowland, J.P., R. Braode, A.G. Chamberlain, R.F. Glascock, and K.G. Mitchell. 1961. The absorption, distribution, and excretion of labeled copper in young pigs given different quantities, as sulphate or sulfide, orally or intravenously. *Br. J. Nutr.* 15:59-72

Brady, N.C. and R.R. Weil. 1999. Micronutrient elements. p. 585-611. *In The Nature and Properties of Soils.* 12th ed. Prentice-Hall, Inc. Upper Saddle River, NJ.

Bremner, I. 1981. Effects of the disposal of copper-rich slurry on the health of grazing animals. In P.L. Hermite, and J. Dehandtschutter, D. and Reidel (eds.) *Copper in Animal Wastes and Sewage Sludge,* pp. 245-255. Dordrecht, the Netherlands.

Brock, E.H., Q.M. Ketterings, and M. McBride. 2006. Copper and zinc accumulation in poultry and dairy manure-amended fields. *Soil Sci.* 171:388-399.

Brumm. M.C. 1998. Sources of manure: swine. p. 49. In J.L. Hatfield and B.A. Stewart (eds.), *Animal Waste Utilization, effective use of manure as a soil resource,* Ann Arbor Press, Ann Arbor, MI.

Brown, B.L., A.D. Slaughter, and M.E. Screiber. 2005. Controls on roxarsone transport in agricultural watersheds. *Applied Geochemistry* 20:123-133.

Calvert, C.C. 1975. In E.A. Woolson (ed.), *Arsenical Pesticides. American Chemical Society,* Washington, DC.

Canet, R., F. Pomares, and F. Tarazona. 1997. Chemical extractability and availability of heavy metals after seven years application of organic wastes to a citrus soil. *Soil Use Manage.* 13:17–121.

Cang, L., Y.J. Yang, D.M. Zhou, and Y.H. Dong. 2004. Heavy metals pollution in poultry and livestock feeds and manures under intensive farming in Jiangsu Province, China. *J. Environ. Sci.* 16:371-374.

Caper, S.G., J.T. Tanner, M.H. Friedman, and K.W. Boyer. 1978. Multiple element analysis of animal feed, animal wastes, and sewage sludge. *Environ. Sci. Technol.* 12:785-790.

Carlson, M.S. and T.J. Fangman. 2000. Swine antibiotics and feed additives: Food safety considerations. E*xt. Publ.* No. 2353. Univ. of Missouri-Columbia, Columbia, MO.

Chapman, H.D., and Z.B. Johnson. 2002. Use of antibiotics and roxarsone in broiler chickens in the USA: Analysis for the years 1995 to 2000. *Poult. Sci.* 81:356–364.

Chambers, B.J., F.A. Nicholson, D.R. Solomon, and R.J. Unwin. 1998. Heavy metal loadings from animal manures to agricultural land in England and Wales. In Martinez J. an d Maudet M.N. (eds): *Proc. 8th International Conference on the FAO ESCORENA Network on Recycling of Agricultural,* Municipal and Industrial Residues in Agriculture. available online http://www.ramiran.net/doc98/FIN-ORAL/CHAMBERS.pdf (accessed December 8, 2009: verified December 8, 2009).

Christen, K. 2001. Chickens, manure, and arsenic. *Environ. Sci. Technol.* 35:184A-185A.

Chowdhury, S.D., I.K. Paik, H. Namkung, and H.S. Lim. 2004. Responses of broiler chickens to organic copper fed in the form of copper-methionine chelate. *Anim. Feed Sci. Technol.* 115:281-293.

Combs, S.M., J.B. Peters, and L.S. Zhang. 1998. Manure: Trace Element Content and Impact on Soil Management. 10p. New Horizons Soil Sci. Number 7. Univ. Wisconsin, Madison, WI. Also available at http://www.uwex.edu/ces/forage/wfc/proceedings2001/micronutrient_status_of_manure.htm)

Creech, B.L., J.W. Spears, W.L. Flowers, G.M. Hill, K.E. Lloyd, T.A. Armstrong, and T.E. Engle. 2004. Effect of dietary trace mineral concentration and source (inorganic vs. chelated) on performance, mineral status, and fecal mineral excretion in pigs from weaning through finishing. *J. Anim. Sci.* 82:2140–2147.

Cresswell, G.C., N.G. Nair, and J.C. Evans. 1990. Effect of boron and copper contaminants in poultry manure on the growth of the common mushroom, *Agaricus Bisporus.* Aust. *J. Exp. Agric.* 30, 707–712, 1990.

de Abreu, M.F. and R.S. Berton. 1996. Comparisons of methods to evaluate heavy metals in organic wastes. *Commun. Soil Sci. Plant Anal.* 27:1125-1135.

Dao, T.H. and H. Zhang. 2007. Rapid composition and source screening of heterogeneous poultry litter by x-ray fluorescence spectrometry. *Annals of Environ. Sci.* 1:69-79.

Duffera, M. W., P. Robarge and R.L. Mikkelsen. 1999. Estimating the availability of nutrients from processed swine lagoon solids through incubation studies. Bioresour. *Technol.* 70:261-268

Eck, H.V., and B.A. Stewart. 1995. Manures. In J.E. Richigl (ed.) *Soil amendments and environmental quality.* Lewis Publishers, Boca Raton, FL.

Edwards, D.R., P.A. Moore Jr., T.C. Daniel, P. Srivastava, and D.J. Nichols. 1997. *Vegetative strip removal of metals in runoff from poultry litter-amended fescue grass plots. Trans.* ASAE 40:121-127.

Edwards, J.H. and A.V. Someshwar. 2000. Chemical, physical, and biological characteristics of agricultural and forest byproducts for land application. p. 1-62. In W.A. Dick (ed.) Land application of agricultural, industrial, and municipal by-products. *Soil Sci. Soc. of Am.*, Madison, WI.

Eneji. E.A., S. Yamamoto, T. Honna, and A. Ishiguoro. 2001. Physico-chemical changes in livestock feces during composting. *Commun. Soil Sci. Plant Anal.* 24:477-489.

Franzluebbers, A.J., S.R. Wilkinson, and J.A. Stuedemann. 2004. Bermudagrass management in the southern piedmont, USA:IX trace elements in soil with broiler litter application. *J. Environ. Qual.* 33:778-784.

Garbarino, J.R., A.J. Bednar, D.W. Rutherford, R.S. Beyer, and R.L. Wershaw. 2003. Environmental fate of roxarsone in poultry litter. I. Degradation of roxarsone during composting. *Environ. Sci. Technol.* 37:1509-1514.

Gracey, H.I., T.A. Stewart, J.D. Woodside and R.H. Thompson. 1976. The effect of disposing high rates of copper-rich pig slurry on grassland on the health of grazing sheep. *J. Agric. Res.* 87:617-623.

Han, F.X., W.L. Kingery, H.L. Selim, P.D. Gerard, M.S. Cox, and J.L. Oldham. 2004. Arsenic solubility and distribution in poultry waste and long-term amended soil. *Sci. Total Environ.* 320:51-61.

He, Z., D.M. Endale, H.H. Schomberg, and M.B. Jenkins. 2009. Total phosphorus, zinc, copper, and manganese concentrations in Cecil soil through ten years of poultry litter application. *Soil Sci.* 174:687-695.

Hostetler, C.E. R.L. Kincaid, and M.A. Mirando. 2003. The role of essential trace elements in embryonic and fetal development in livestock. *Vet. J.* 166:125-139.

Hsu, J.H. and S.L. Lo. 2001. Effect of composting on characterization and leaching of copper, manganese, and zinc from swine manure. *Environ. Pollut.* 114:119-127.

Ihnat, M. and L. Frenandes. 1996. Trace elemental characterization of composted poultry manure. *Biores. Technol.* 57:143-156.

Jackson, B.P. and W.P. Miller. 1999. Soluble arsenic and selenium species in fly ash/organic waste-amended soils using ion chromatography-inductively coupled mass spectrometry. *Environ. Sci. Technol. 33*:270-275.

Jackson, B.P., and W.P. Miller. 2000. Soil solution chemistry of a fly ash-, poultry litter-, and sewage sludge-amended soil. *J. Environ. Qual.* 29, 430-436.

Jackson, B.P. and P.M. Bertsch. 2001. Determination of arsenic speciation in poultry wastes by IC-ICP-MS. *Environ. Sci. Technol.* 35:4868–4873.

Jackson, B.P., P.M. Bertsch, M.L. Cabrera. J.J. Camberato, J.C. Seaman, and C.W. Wood. 2003. Trace element speciation in poultry litter. *J. Environ. Qual.* 32:535-540.

Jindasa, K.B., P.J. Milham, C.A. Hawkins, P.S. Cornish, P.A. Williams, C.J. Kaldor and J.P. Conroy. 1997. Survey of cadmium levels in vegetables and soils of greater Sydney, Australia. *J. Environ. Qual.* 26:924-933.

Jokela, W.E., J.P. Tilley, and D.S. Ross. 2010. Manure nutrient content on Vermont dairy farms: long-term trends and relationships. *Commun. Soil Sci. Plant Anal.* 41:623-637.

Jondreville, C., P.S. Revy, and Y.J. Dourmad. 2003. Dietary means to better control the environmental impact of copper and zinc by pigs from weaning to slaughter. *Livestock Prod. Sci.* 84:147-156.

Keller, A. and R. Schulin. 2003. Modelling heavy metal and phosphorus balances for farming systems. *Nutr. Cycl. Agroecosyst.* 66:271-284.

Kingery, W.L., C.W. Wood, D.P. Delaney, J.C. Williams, and G.L. Mullins. 1994. Impact of long-term land application of broiler litter on environmentally related soil properties. *J. Environ. Qual.* 23:139-147.

Kirchmann, H., L. Mattsson J. Eriksson. 2009. Trace element concentration in wheat grain: results from the Swedish long-term soil fertility experiments and national monitoring program. *Environ. Geochem. Health* 31:561-571.

Kornegay, E.T., J.D. Hedges, D.C. Martens, and C.Y. Kramer. 1976. Effect of soil and plant mineral levels following application of manures of different copper levels. *Plant Soil* 45:151–162.

Kneale, W.A. and P. Smith. 1977. The effects of applying pig slurry containing hig levels of copper to sheep pastures. *Expl. Husb.* 32:1-7.

Krishnamachari, K.A.U.R. 1987. Fluorine. In W. Mertz (ed.) *Trace elements in human and animal health,* vol. 1. 5th Ed., pp. 265-416. Academic Press, San Diego, CA.

Kunkle, W.E., L.E. Carr, T.A. Carter, and E.H. Bossard. 1981. Effect of flock and floor type on levels of nutrients and heavy metals in broiler litter. *Poult. Sci.* 60: 1160-1164.

Kpomblekou. A, K., R,O. Ankumah, and H.A. Ajwa. 2002. Trace and nontrace element contents of broiler litter. Commun. *Soil Sci. Plant Anal.* 33:1799-1811.

Lamand, M. 1981. Copper toxicity in sheep. p. 261–268. In P.L. Hermite et al. (eds.) *Copper in animal wastes and sewage sludge.* Dordrecht, the Netherlands.

Lee, J., J.R. Rounce, A.D. Mackay, and N.D. Grace. 1996. Accumulation of cadmium with time in Romney sheep grazing ryegrass white clover pasture: Effect of cadmium from pasture and soil intake. *Aust. J. Agric. Res.* 47:877-894.

Lei, X., P.K. Ku, E.R. Miller, D.E. Ullrey, and M.T. Yokoyama. 1993. Supplemental microbial phytase improves bioavailability of dietary zinc to weanling pigs. *J. Nutrit.* 123:1117-1123.

L'Herroux, L., S.L. Roux, P. Apprioua, and J. Martinez. 1997. Behavior of metals following intensive pig slurry applications to a natural field treatment process in Brittany (France) *Environ. Pollut.* 97:119-130.

Li, M., N.V. Hue, and S.K.G. Hussain. 1997. Changes of metal forms by organic amendments to Hawaii soils. *Comm. Soil Sci. Plant Anal.* 28:381-394.

Li, Y.X. and T.B. Chen. 2005. Concentrations of additive arsenic in Beijing pig feeds and the residues in oig manure. *Resources, Conservation, and Recycling* 45:356-367.

Lindsay, W.L. and W.A. Norvell. 1978. Development of a DTPA soil test for zinc, iron, manganese, and copper. *Soil Sci. Soc. Am. J.* 42:421-428.

Longathan, P., K. Louie, J. Lee, M.J. Hedley, A.H.C. Roberts, and R.D. Longhurst. 1999. A model to predict kidney and liver cadmium concentrations in grazing animals. *N. Z. J. Agric. Res.* 42: 423-432.

Longhurst, R.D., A.H.C. Roberts, and M.B. O'Connor. 2000. Farm dairy effluent: A review of published data on chemical and physical characteristics in New Zealand. *N. Z. J. Agric. Res.* 43:7-14.

Makris, K.C., J. Salazar, S. Quazi, S.S. Andra, D. Sarkar, S.B.H. Bach, and R. Datta. 2008a. Controlling the fate of roxarsone and inorganic arsenic in poultry litter. *J. Environ. Qual.* 2008 37: 963-971.

Makris, K.C., S. Quazi, P. Punamiya, D. Sarkar, Rupali Datta. 2008b. Fate of arsenic in swine waste from concentrated animal feeding operations. *J. Environ. Qual.* 2008 37: 1626-1633.

McBride, M.B. and G. Spiers. 2001. Trace element content of selected fertilizers and dairy manures as determined by ICP-MS. *Comm. Soil Sci. Plant Anal.* 32:139-156.

McGrath, D., D.B.R. Poole, and G.A. Fleming. 1980. Hazards arising from the application to grassland of copper rich pig faecal slurry. pp. 420-431. In J.K.R. Gasser (ed.) *Effluents from Livestock. Applied* Science Publ., London.

McGrath, D., R.F. McCormack, G.A. Fleming and D.B.R. Poole. 1982. Effects of applying copper-rich pig slurry to grassland. 1. Pot experiments. *Ir. J. Agric. Res.* 21:37-48.

Menzi, H., and J. Kessler. 1998. Heavy metal content of manures in Switzerland. In Martinez J. and Maudet M.N. (eds): Proc. 8th International Conference on the FAO ESCORENA Network on Recycling of Agricultural, Municipal and Industrial Residues in Agriculture.

Available online http://www.ramiran.net/doc98/FIN-ORAL/MENZI.pdf (accessed December 8, 2009: verified December 8, 2009).

Miller, W.P., D.C. Martens, L.W. Zelazny, and K.T. Kornegay. 1986. Forms of solid-phase copper in copper-enriched swine manure. *J. Environ. Qual.* 15:69-72.

Miller, R.E., X. Lei, and D.E. Ullrey. 1991. Trace elements in animal nutrition. p. 593– 662. *In* J. J. Mortvedt (ed.) Micronutrients in agriculture. 2nd ed. *Soil Sci. Soc. of Am.,* Madison, WI.

Mikkelsen, R.l. 2000. Beneficial uses of swine products: Opportunities for the future. p. 425-480. In W.A. Dick (ed.). Land application of agricultural, industrial, and municipal by-products. *Soil Sci. Soc. of Am.,* Madison, WI.

Moore, P.A., T.C. Daniel, and A.N. Sharpley, and C.W. Wood. 1995. Poultry manure management – Environmentally sound options. *J. Soil Water Conserv.* 50: 321-327.

Moore, Jr., P.A., T.C. Daniel, J.T. Gilmore, B.R. Shreve, D.R. Edwards, and B.H. Wood. 1998. Decreasing metal runoff from poultry litter with aluminum sulfate. *J. Environ. Qual.* 27:92-99.

Moore, A. and J. Ippolito. 2009. Dairy Manure Field Applications – How Much Is Too Much? University of Idaho Extension CIS 1156, University of Idaho, Boise, ID. Available at http://www.cals.uidaho.edu/edComm/pdf/CIS/CIS1156.pdf

Mullins, G.L., D.C. Martens, W.P. Miller, E.T. Kornegay, and D.L. Hallock. 1982. Copper availability, form, and mobility in soils from three annual copper-enriched hog manure applications. *J. Environ. Qual.* 11:316-320.

Narwal, R.P. and B.R. Singh. 1998. Effect of organic materials on partitioning extractability and plant uptake of metals in an alum shale soil. *Water Air Soil Pollut.* 103:405-410.

National Research Council. 1989. *Nutrient Requirements of Dairy Cattle.* 6th ed., Natl. Acad. Sci., Washington, D.C.

National Research Council. 1998. *Nutrient requirements of Swine.* 10th ed., National Academy Press, Washington, D.C.

National Research Council. 2002. *Biosolids applied to land: Advancing standards and practices.* National Academy Press, Washington, DC.

Nahm, K.H. 2002. Efficient feed nutrient utilization to reduce pollutant in poultry and swine manure. Crit. Rev. *Environ. Sci. Technol.* 32:1-16.

Nicholson, F.A., B.J. Chambers, J.R. Williams, and R.J. Unwin. 1999. Heavy metals contents of livestock feeds and manures in England and Wales. *Bioresour. Technol.* 23:23-31.

Novak, J.M., D.W. Watts, and K.C. Stone. 2004. Copper and zinc accumulation, profile distribution, and crop removal in Coastal Plain soils receiving long-term, intensive applications of swine manure. *Transactions ASAE* 47:1513-1522.

Öborn, I., A.K. Modin-Edman, H. Bengtsson, G.M. Gustafson, E. Salomon, S.I. Nilsson, J. Holmqvist, S. Jonsson, and H. Sverdrup. *A systems approach to assess farm-scale nutrient and trace element dynamics: a case study at the Öjebyn dairy farm.* Ambio:64:301-310.

O'Connor, G.A., R.B. Bropst, R.L. Chaney, R.L. Kincaid, L.R. McDowell, G.M. Pierzynski, A. Rubin, and G.G. Van Riper. 2001. A modified risk assessment to establish molybdenum standards for land application of biosolids. *J. Environ. Qual.* 30:1490-1507.

Pallauf, J. and G. Rimbach. 1997. Nutritional significance of phytic acid and phytase. *Arch. Anim. Nutr.* 50:301-319.

Parkinson, R.J. and R. Yells. 1985. Copper content in soil and herbage following pig slurry application to grassland. *J. Agric. Sci.* 105:183-185.

Payne, G.G., D.C Martens, E.T. Kornegay, and M.D. Lindemann. 1988. Availability and form of copper in three soils following eight annual applications of copper-enriched swine *manure. J. Environ. Qual.* 17:740–746.

Poole, D.B.R. 1981. Implications of applying copper rich pig slurry to grassland-effects on the health of grazing sheep. pp. 273-282. In P.L. Hermite et al. (eds.) *Copper in animal wastes and sewage sludge.* Dordrecht, the Netherlands.

Poulsen, H.D. 1998. Zinc and copper as feed additives, growth factors or unwanted environmental factors. *J. Anim. Feed Sci.* 7:135-142.

Rankin, M. 2010. *Agronomic and Environmental Issues with Foot Bath Solution Land Spreading. University of Wisconsin Extension,* Madison, WI. Available at http://128.104.248.62/ces/crops/4StateCopper.pdf

Raven, K.P. and R.H. Loeppert. 1997. Trace element composition of fertilizers and soil amendments. *J. Environ. Qual.* 26:551-557.

Richards, J.R. 2010. Trace elements in Oklahoma soils: factors affecting total and bioavailable concentrations. Masters thesis. Oklahoma State University, Stillwater, OK.

Rutherford, D.W., A.J. Bednar, J.R. Gabarino, R. Needham, K.W. Staver, and R.L. Wershaw. 2003. Environmental fate of roxarsone in poultry litter: Part II. Mobility of arsenic in soils amended with poultry litter. *Environ. Sci. Technol.* 37:1515-1520.

Sager, M. 2006. Micro- and macro-element composition of animal feedstuffs sold in Austria. *Ernährung/Nutrition* 30: 455-473.

Sager, M. 2007. Trace and nutrient elements in manure, dung, and compost samples in Austria. *Soil Biol. Biochem.* 39:1383-1390.

Sauve, S. 2002. Speciation of metals in soils. p. 7-37. In H. Allen (ed.) Bioavailability of metals in terrestrial ecosystems: *Importance of partitioning for bioavailability to invertebrates, microbes, and plants.* Setac Press, Pensacola, FL.

Sebastian, S., S.P. Touchburn, and E.R Chavez. 1998. Implications of phytic acid and supplemental microbial phytase in poultry nutrition: a review. *World's Poultry Science Journal* 54:27-47.

Schultheib, U., H. Döhler, U. Roth, and H. Eckel. 2004. *Heavy metal fluxes in livestock farming and input reduction strategies.* 2004 - ramiran.net available online http://www.ramiran.net/doc04/Proceedings%2004/Schultheiss.pdf (accessed December 8, 2009: verified December 8, 2009).

Sheppard, S.C., C.A. Grant, M.I. Sheppard, R. de Jong and J. Long. 2009. Risk indicator for agricultural inputs of trace elements to Canadian soils. *J. Environ. Qual.* 38:919-932.

Shuman, L. Chemical forms of micronutrients in soils. p. 113-144. In J.J. Mortvedt (ed.) *Micronutrients in agriculture,* 2nd ed., Soil Science Society of America, Madison, WI.

Sims, J.T., and G.V. Johnson. Micronutrient soil test. 1991. p. 427–476. In J.J. Mortvedt (ed.) *Micronutrients in agriculture,* 2nd ed. Soil Sci. Soc. Am. Madison, WI.

Sims. J.T. and D.C. Wolf. 1994. Poultry waste management: Agricultural and environmental issues. *Adv. Agron.* 52: 1-83.

Sims, J.T., and N. J. Luka-McCafferty. 2002. On-farm evaluation of aluminum sulfate (alum) as a poultry litter amendment: effects on litter properties. *J. Environ. Qual.* 31: 2066-2073.

Sistani. K.R., D.E. Rowe, D.M. Miles, and J.D. May. 2001. Effects of drying method and rearing temperature on broiler manure content. *Commun. Soil Sci. Plant Anal.* 32:2307-2316.

Sistani. K.R. and J.M. Novak. 2006. Trace element accumulation, movement, and remediation in soils receiving animal manure. p. 689-706. In M.N.V. Prasad, K.S. Sajwan, and R. Naidu (eds.) *Trace elements in the environment: biogeochemistry, biotechnology, and bioremediation,* CRC/Taylor and Francis, Boca Raton, FL.

Spears, J. W., B. A. Creech, and W. L. Flowers. 1999. Reducing copper and zinc in swine waste through dietary manipulation. *Proc. 1999 Animal Waste Management Symposium* (Ed. G. B. Havenstein): p. 179-185.

Sposito, G. 1989. *The chemistry of soils.* Oxford University Press, Inc. New York.

Stehouwer, R. and G. Roth. Copper sulfate hoof baths and copper toxicity in soils. Penn State Field Crop News 4:1. Available at http://fcn.agronomy.psu.edu/2004/FCN0401.pdf (accessed 29 March 2010: Verified 29 March 2010).

Steineck, S, G. Gustafson, A. Andersson, and J. Tersmeden Moch Bergström. 1999. *Plant nutrients and trace elements in livestock wastes in Sweden–Swedish Environmental Protection Agency, report* 5111.

Tessier, A., P.G.C. Campbell, and M. Bisson, M. 1979. Sequential extraction procedure for the speciation of particulate trace metals. *Anal. Chem.* 51:844-851.

Tewolde, H., K. R. Sistani, and D. E. Rowe. 2005. Broiler litter as a micronutrient source for cotton: concentrations in plant parts. *J. Environ. Qual.* 34:1697–1706.

Thomas, E.D. 2001. Foot bath solutions may cause crop problems. *Hoard's Dairyman* 146:458-459.

Toor, G.S., B.E. Haggard, and A.M. Donogue. 2007. Water extractable trace elements in poultry litters and granulated products. *J. Appl. Poul. Res.* 16:351-360.

Tufft, L.S., and C.F. Nockels. 1991. The effects of stress, Escherichia coli, dietary ethylene diamine tetraacetic acid, and their interaction on tissue trace elements in chicks. *Poult. Sci.* 70:2439-2449.

Unwin. R.J. 1977. Copper in pig slurry: some effect and consequences of spreading on grassland. p. 306. In Inorganic pollution in agriculture. *MAFF Reference Book* 326, H.M. Stationery Office, London.

U.S. EPA. 1993. Standards for the use or disposal of sewage sludge, Final Rules. 40 CFR Part 503. *Fed. Regist.* 58:9248–9415.

U.S. EPA. 1995a. *A guide to the biosolids risk assessments for the EPA Part 503 Rule.* EPA 832-B-93-005. Office of Wastewater Management, Washington, DC.

U.S. EPA. 1995b. *Process design manual: Land application of sewage sludge and domestic septage.* EPA/625/R-95/001. Office of Research and Development, Cincinnati, OH. Available at http://www.epa.gov/nrmrl/pubs/625r95001/landapp.pdf (accessed 11 Jan. 2010: Verified 11 Jan. 2010).

van der Watt, H.V.H., M.E. Sumner, and M.L. Cabrera. 1994. Bioavailability of copper, manganese and zinc in poultry manure. *J. Environ. Qual.* 23:43-49.

van Heugten, E., P.R. O'Quinn, D.W. Funderburke, W.L. Flowers, and J.W. Spears. 2002. *Effects of supplemental trace mineral levels on growth performance, carcass characteristics, and fecal mineral excretion in growing-finishing swine.* North Carolina State University, A&T State University Cooperative Extension Annual Swine Report 2002.

Walker, D.J., R. Clemente, and M.P. Bernal. 2004. Contrasting effects of manure and compost on soil pH, heavy metal availability and growth of *Chenopodium album* L. in a soil contaminated by pyritic mine waste. *Chemosphere* 57: 215-224.

Wallingford, G.W., L.S. Murphy, W.l. Powers, and H.L. Manges. 1975. Effects of beef-feedlot manure and lagoon water on iron, zinc, manganese, and copper content in corn and in DTPA soil extracts. *Soil Sci. Soc. Am. J.* 39:482-487.

Wang, J.J., H. Zhang, J.L. Schroder, T.K. Udeigwe, Z. Zhang, D. Dodla, and M.H. Stietiya. Reducing potential leaching of phosphorus, heavy metals, and fecal coliform using bauxite residues. *Water Air Soil Pollut.* DOI 10.1007/s11270-010-0420-2.

Wilkinson, S.R.. 1997. Response of Tifway 2 Bermuda grass to fresh or composted broiler litter containing boric acid-treated paper bedding. *Commun. Soil Sci. Plant Anal.* 28:259-279.

Wright, R.J., W.D. Kemper, P.D. Millner, J.F. Power, and R.F. Korcak. 1998. Agricultural uses of municipal, animal, and industrial byproducts. *Conserv. Res. Rep.* 44. U.S. Dept. of Agriculture, ARS, Beltsville. MD.

Westerman, P.W. and J.R. Bicudo. 2000. Tangential flow separation and chemical enhancement to recover swine manure solids, nutrients and metals. *Bioresource Technology* 73:1-11.

Wood, B.H., C.W Wood, K.H Yoo, K.S. Yoon, and D.P. 1996. Nutrient accumulation and nitrate leaching under broiler litter amended corn fields. Commun. *Soil Sci. Plant Anal.* 27:2875-2894.

Zhang, H., and W.R. Raun. 2006. *Oklahoma Soil fertility handbook. 6th ed. Oklahoma Coop.* Ext. Service, Oklahoma State University, Stillwater, OK.

Zhang, H., W.R. Raun, and B. Arnall. 2009. OSU Soil Test Interpretations. Oklahoma.

Cooperative Extension Service. Factsheet PSS-2225. Oklahoma State Univ. Stillwater, OK. Available at http://osufacts.okstate.edu/docushare/dsweb/Get/Document-1490/PSS-2225web.pdf

Zhu, Y.M., D.F. Berry, and D.C. Martens. 1991. Copper availability in two soils amended with 11 annual applications of copper-enriched hog manure. *Commun. Soil Sci. Plant Anal.* 22:769–783.

Zublena, J.P. 1991. Soil facts: nutrient removal by crops in North Carolina. *North Carolina Cooperative Extension Service Bulletin AG-439-16.* North Carolina State University, Raleigh, NC.

In: Environmental Chemistry of Animal Manure   ISBN: 978-1-62808-641-6
Editor: Zhongqi He   © 2013 Nova Science Publishers, Inc.

Chapter 17

FATE AND TRANSPORT OF ARSENIC FROM ORGANOARSENICALS FED TO POULTRY

Clinton D. Church[1,], Jane E. Hill[2] and Arthur L. Allen[3]*

17.1. INTRODUCTION

Recently, there has been growing concern over the use of roxarsone (3-nitro-4-hydroxyphenylarsonic acid) by the poultry industry, since roxarsone contains arsenic (As) that may ultimately accumulate in the environment (Christen, 2001). Roxarsone is an organic acid feed additive that has historically been fed to broiler chickens to improve weight gain, feed consumption and manage coccidial parasites (Chapman and Johnson, 2001). Since it is not metabolized by poultry and therefore does not accumulate in broiler meat, the bulk of roxarsone added to broiler feed is excreted in feces, which, when mixed with bedding, forms poultry litter (Anderson and Chamblee, 2001; O'Connor et al., 2005; Arai et al., 2003). As poultry manure is rich in nutrients, more than 90% of poultry litter is land-applied as fertilizer (Jackson and Bertsch, 2001; Morrison, 1969; Moore et. al., 1998). Rutheford et al. (2003) showed that soils receiving long-term poultry litter application had high levels of water extractable As and that the As appeared to be sorbed to iron oxyhydroxides in the soil types they investigated. However, the ultimate fate of As derived from roxarsone as well as the controls on As movement in the environment are largely unknown (Christen, 2001).

The use of organoarsenicals, primarily Roxarsone, has increased as the poultry industry has grown (Anderson, 1983). Concentrations of roxarsone normally added to feed are relatively low, between 45 and 90 g ton^{-1} (Howie, 2003; NRC, 1999; USFDA, 2000). However, Garbarino et al. (2003) estimated that in 2000 alone, when 8.3×10^6 broiler

---

[*] Corresponding Author: USDA-ARS, Curtin Road, Building 3702, University Park, PA, 16802. Telephone: 814-863-8760. Email: Clinton.Church@ars.usda.gov
[1] USDA-ARS, Pasture Systems and Watershed Management Research Unit, , University Park, Pennsylvania, 16802, USA
[2] School of Engineering, University of Vermont, Burlington, Vermont, 05405, USA
[3] Dept. Agriculture, University Maryland Eastern Shore, Princess Anne, MD 21853, USA

chickens were grown in the U.S., approximately 2.5×10^5 kg of As were likely applied in litter to U.S. soils, equivalent to an annual load of 60 to 250 g As ha^{-1} (Garbarino et al., 2003). In areas of intensive poultry production, such as the Delmarva (Delaware, Maryland, Virginia) Peninsula, the application of roxarsone-derived As to agricultural soils can be quite acute. In 2000, approximately 620,000,000 broilers were raised on the Peninsula resulting in an estimated annual load of As to Delmarva soils of 2.6×10^4 kg from land application of poultry litter (Seiter and Sparks, 2005; 2006). Some sources have estimated that as much as 5×10^4 kg of arsenic is spread on Delmarva soils every year (Christen, 2001; Hileman, 2007). Given the proximity of the Delmarva Peninsula to the Chesapeake Bay, the nation's largest estuary, the fate of As from poultry litter amended soils is a potential water quality concern.

The American Cancer Society recently recognized that As causes cancer of the liver, lung, and skin (Heath and Fontham, 2001). In 2001, the U.S. Environmental Protection Agency lowered the maximum contaminant level (MCL) of elemental As in drinking water to 10 µg/L (U.S. Environmental Protection Agency, 2000), which also is the current provisional guide used by the World Health Organization (World Health Organization, 1999). Arsenic also is toxic to fish and many aquatic organisms (Canadian Council of Ministers of the Environment, 2002) and, if present in soil water, it may be taken up by plants—rendering them unsuitable for human or animal consumption (Marin et al., 1993; Krishnamurti and Naidu, 2002).

17.2. DEGRADATION OF ROXARSONE

The excretion of organoarsenicals by livestock has generated an interest in their fate and transport following land application of manure. Primary treatment of manures, such as composting for poultry litter and lagoon systems for swine waste, followed by application to agricultural soils, can result in the degradation of the organoarsenicals into products such as arsenate. While more research is needed to understand the controls on these transformations, it is clear that microbial action mediates organoarsenical degradation. This section reviews the role of microorganisms in the fate of organoarsenicals, primarily roxarsone, in manures.

17.2.1. Biodegradation of Roxarsone in Poultry Litter Compost and Swine Lagoon Systems

Fresh poultry litter contains 15-48 mg/kg of roxarsone (Jackson and Bertsch, 2001; Pavkov and Goessler, 2001; Jackson et al., 2003), however, composted litter predominantly contains inorganic As (Arai et al., 2003; Garbarino et al., 2003). Garbarino et al. (2003) examined poultry litter from a variety of sources in which arsenicals were initially dominated by roxarsone (91%). They observed that roxarsone was transformed to inorganic As species after composting, with strong evidence of biotic transformation mechanisms. Elsewhere, a study by Cortinas and colleagues (2006) of spiked anaerobic sludge demonstrated the rapid transformation of roxarsone via an initial reductive step to 4-hydroxy-3-aminophenylarsonic acid (HAPA) (see Figure 17.1a), followed by a much slower degradation to arsenite under anaerobic conditions (see Figure 17.1b). Degradation of roxarsone in swine lagoon systems has also been shown to occur under anaerobic conditions and appears to be a function of suspended solid concentration (Makris et al., 2008). Degradation of roxarsone was also documented under aerobic conditions, but only when there was a high solids concentration. Roxarsone degradation under low suspended solid concentrations only occurred under anaerobic conditions. Some evidence by Makris and colleagues suggests an abiotic degradation mechanism for roxarsone under special circumstances, but this mechanism not been thoroughly elucidated.

17.2.2. Microbial Isolates Capable of Degrading Roxarsone

Poultry excreta and litter contain an array of aerobic and anaerobic microbial species (Amit-Romach et al., 2004). *Clostridium* species are commonly identified in poultry excreta (Alexander, 1968; Jean, 1995: Rothrock, 2008; Ngodigha, 2009) Anoxic enrichments of poultry litter using lactate as a carbon source and containing roxarsone, were capable of degrading roxarsone and producing 3-nitro-4-hydroxybenzene arsonic acid and As(V) (Figure 17.2, Stolz et al., 2007). Biphasic growth on the lactate-roxarsone combination was characteristic of enrichment cultures but not of the arsenate-respiring strain (*Clostridium.* sp. strain OhILAs, isolated from the Ohio River; Fisher et al., 2008) enrichment. Degradation of roxarsone to 3A4HBAA and As(V) by the river isolate was faster than the enrichment cultures, with total depletion of roxarsone occurring at approximately 84 hours (compared to approximately 204 hours in the enrichment culture). However, it is unclear whether roxarsone is acting as an electron acceptor or is acting as sink for fermentation-derived reducing equivalents. Importantly, if an oxidative pathway were possible, the nitrophenol product would be expected, but that compound has not been observed or measured in any published study on roxarsone degradation thus far (Stolz et al., 2007; Fisher et al., 2008; Arai et al., 2003; Garbarino et al., 2003).

Alkalipilus oremlandii sp. nov. strain OhILAs, a strict anaerobe, was isolated using an enrichment culture containing arsenate and lactate (Fisher et al., 2008). In addition to being able to reduce arsenate to arsenite, the isolate demonstrated the ability to degrade roxarsone, a process coupled to the oxidation of lactate. Growth on roxarsone and lactate was rapid after an initial lag period and the degradation products 4-hydroxy-3-aminophenylarsonic acid (HAPA) and As(V) were detected (although not explicitly measured). The reaction proceeded according to the following formula:

$$\text{roxarsone} + \text{lactate} \longrightarrow CO_2 + \text{HAPA} + \text{acetate} + H_2O$$

A variety of enzymes have been implicated in the degradation of roxarsone, including nitroreductase, respiratory arsenate reductase, and arsenite-activated ATPase. In addition, a study of the *A. oremlandii* proteome expressed when grown with different electron acceptors revealed a potential role for aldehyde ferredoxin oxidoreductase (for roxarsone-grown cultures) (Chovanec et al., 2010).

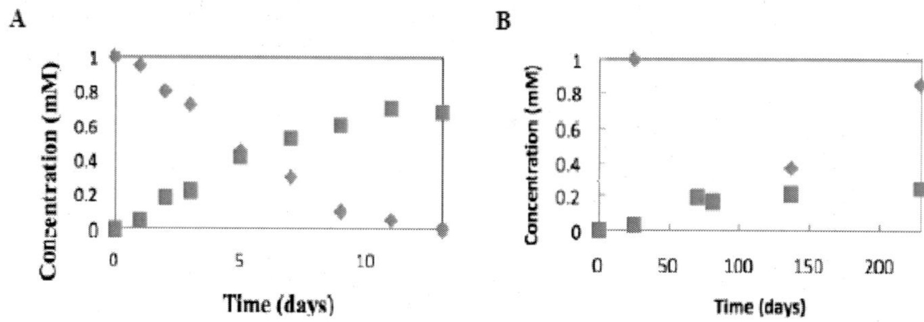

Figure 17.1. Degradation of roxarsone via an initial reductive step to HAPA (a), followed by a much slower degradation of HAPA to arsenite under anaerobic conditions (b). adapted from Cortinas *et al.*, 2006.

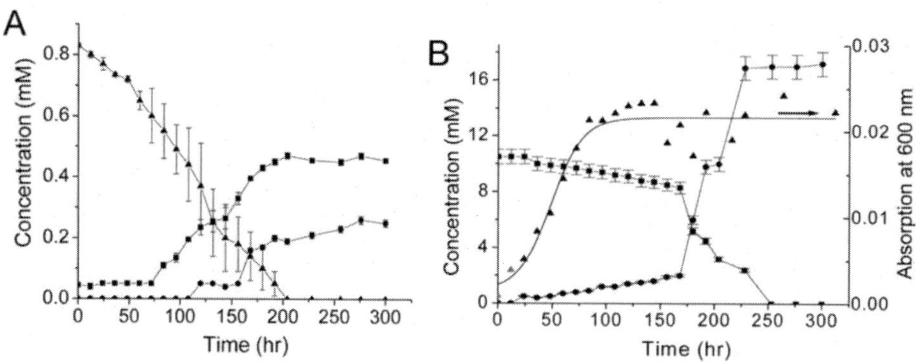

Figure 17.2. Roxarsone transformation in chicken litter enrichments (with permission from Stolz et al., 2007). A. Consumption of roxarsone (▲) and production of 3-amino-4-hydroxybenzene arsonic acid (■) and As(V) (●). (B) First transformation step in the degradation of roxarsone to HABA under anaerobic conditions: consumption of lactate (■), production of acetate (●), and cell growth as optical density at 600 nm (▲).

17.3. TRANSPORT OF ARSENIC

Arsenic occurs in multiple species in the environment, with each form having a different toxicity (Rosal et al., 2005; Villaescusa and Bollinger, 2008), and the chemistry of As is complicated and often difficult to predict. Garbarino et al. (2003) found that all of the As in poultry litter is transformed from roxarsone into As(V), an oxyanion, although, as seen above, the mechanisms of that transformation are not completely known. Major environmental processes that control the chemistry of As are ion exchange, precipitation, adsorption/desorption, and biological activity (Jain and Ali, 2000; Robertson, 1989; Han, et al., 2004). Factors such as pH, Eh and competition for adsorption sites also add to As's complexity (Robertson, 1989).

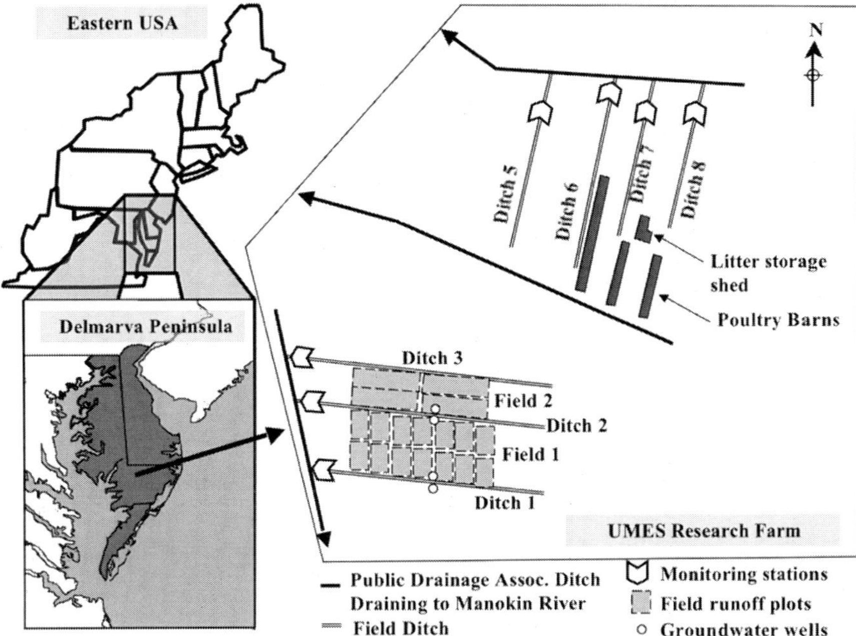

Figure 17.3. Map of research site, showing location of research farm on DelMarVa Peninsula, layout of buildings, fields, and ditches; and distribution of monitoring stations, field plots, and groundwater monitoring sites.

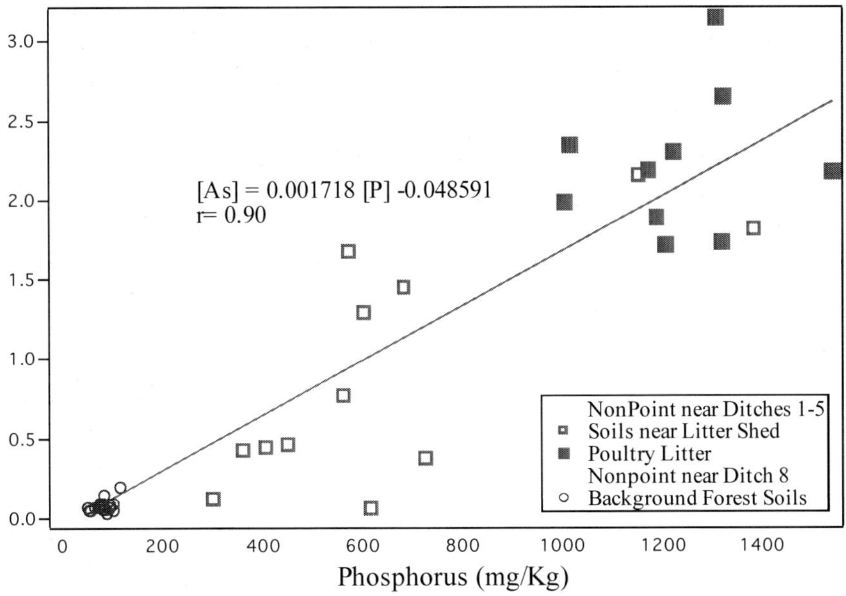

Figure 17.4. Correlation between As and P in litter, farm soils, and background forest soils.

Table 17.1. Occurrence of arsenic and phosphorus in field soils and ditch flow on University of Maryland Eastern Shore Research Farm.

Probable Ditch As, P Sources	Watershed Area	Mehlich-3 P Mean a	Total As Mean a	Ditch flow 2005-2006	Ditch flow 2006-2007	Flow Weighted Average Concentration 2005-2006 As	DP	2006-2007 As	DP	Annual Loss 2005-2006 As	DP	2006-2007 As	DP
		—— mg kg^{-1} ——		—— m^3 ——		———— mg L^{-1} ————				———— kg ha^{-1} ————			
1 Fields	2.74	421	0.29	6164	7880	0.004	0.19	0.005	0.363	0.004	0.33	0.033	1.62
2 Fields	1.77	447	0.37	11556	15341	0.002	0.31	0.005	0.448	0.011	2.16	0.059	3.46
3 Fields	2.00	449	0.31	12827	14138	0.002	0.36	0.004	0.209	0.007	1.85	0.040	1.19
5 Fields	1.10	467	0.26	3174	6857	0.002	0.71	0.003	1.005	0.006	1.52	0.027	5.48
6 Fields, barn	1.04	457; 661 b	0.34; 0.65 b	1414	2149	0.002	2.02	0.009	1.880	0.004	2.34	0.024	3.23
7 Fields, 2 barns	1.20	382	0.27	3450	6215	0.008	1.95	0.008	1.914	0.021	4.84	0.044	8.16
8 Fields, 2 barns, litter shed	0.82	466; 3113 c	0.44; 0.91 c	3345	4425	0.012	3.37	0.026	3.067	0.055	14.99	0.071	18.56
Poultry Litter	0.11	19765	2.208										
Precipitation (mm)				922	1196	922		1196		922		1196	

DP = Dissolved Phosphorus.

a Number of samples varied between 10 and 40.

b Second value applies to soil in immediate vicinity of poultry barn.

c Second value applies to soil in immediate vicinity of litter storage shed.

It is likely that As interacts with P, a dominant anion in poultry litter that is of water quality concern due to its importance to eutrophication, the biological enrichment of surface waters. Arsenic and P can compete for adsorption sites in soils, as they belong to the same periodic family and have similar properties (Tassi et al., 2004). Competition for sorption sites between the As and P has been shown to enhance the mobilization of As in soil (Peryea and Kammereck, 1997; Han, et al., 2004). This competition is one reason why As can be toxic; As prevents the phosphorylation of adenosine diphosphate (ADP) into adenosine triphosphate (ATP), the energy source for most biological systems (Crafts, 1977).

Little is known about patterns of As and P transport at the field scale, though one study (Han et al., 2004) documented the accumulation and solubility of arsenic in poultry wastes and waste-amended soils. That study found that after 25 years of annual applications of As-containing poultry wastes, As in the amended soils had a mean concentration of 8.4 mg kg^{-1} as compared to 2.68 mg kg^{-1} for a non-amended soil, and was found to be strongly correlated with iron oxides, clay and hydroxy interlayered vermiculite concentrations and negatively correlated with Mehlich III-P, mica and quartz content of soils. The study further demonstrated by sequential extraction of 10 representative poultry waste samples that As was primarily found in the water soluble fraction (36-75% of total As) of poultry wastes and that water-soluble As concentrations were correlated with total As ($r^2 = 0.78$). In contrast, sequential extractions of amended soils found As to be present primarily in the residual or, the least bioavailable fraction, while the water-soluble fraction was negligible. The authors concluded that quick leaching of water-soluble As into surface water probably occurs shortly after wastes are applied to fields. A single study examining the rapid transport of As from the field to ditches is examined in detail below.

17.3.1. Case Study: The University of Maryland Eastern Shore (UMES) Research Farm

The poultry-intensive region of the Delmarva Peninsula rears roughly 600 million broiler poultry annually and relies upon surface ditches to drain fields for crop production. A monitoring study was conducted on the UMES research farm in Princess Anne, Maryland (38° 12' 22" N and 75° 40' 35" W), part of the 24,000 ha Manokin River Watershed (Figure 17.3). Prior to its purchase by UMES in 1997, the farm had been a commercial broiler operation for roughly 25 years. The farm includes three broiler houses and a storage shed where litter from the broiler houses is stored prior to land application. These buildings represent key point sources on the farm. Seven drainage ditches on the UMES research farm were monitored for flow and water quality from July 1, 2005 to June 30, 2007 (Figure 17.3). Details on the farm and the study can be found in Kleinman et al., 2007 and Church et al., 2010).

Arsenic transport from the drainage ditches varied widely (Table 17.1). Annual discharges were greatest from ditches 1-3, and ditches 1-3 also had greater contributions from shallow ground water. Annual As concentrations in effluent from these ditches and ditch 5 were lowest, in general, averaging less than 0.003 mg L^{-1}. In contrast, effluents in ditches draining point sources were at least twice as high (0.007 mg L^{-1}) with flow from ditch 8 having As concentrations 4 to 8 fold greater (0.012 mg L^{-1} and 0.026 mg L^{-1}). Average annual concentrations of As in flow from ditches draining point sources exceeded the U.S. EPA

MCL for As in drinking water of 10 µg L^{-1} (U.S. EPA, 2000) for both years of the study. In the case of ditch 8, compaction of soils around the litter shed provided a direct hydrologic connection between the spilled litter outside the shed and ditch 8. In the case of ditches 6 and 7, no spilled litter source was observed and therefore there was less potential for enrichment of runoff with litter-derived As.

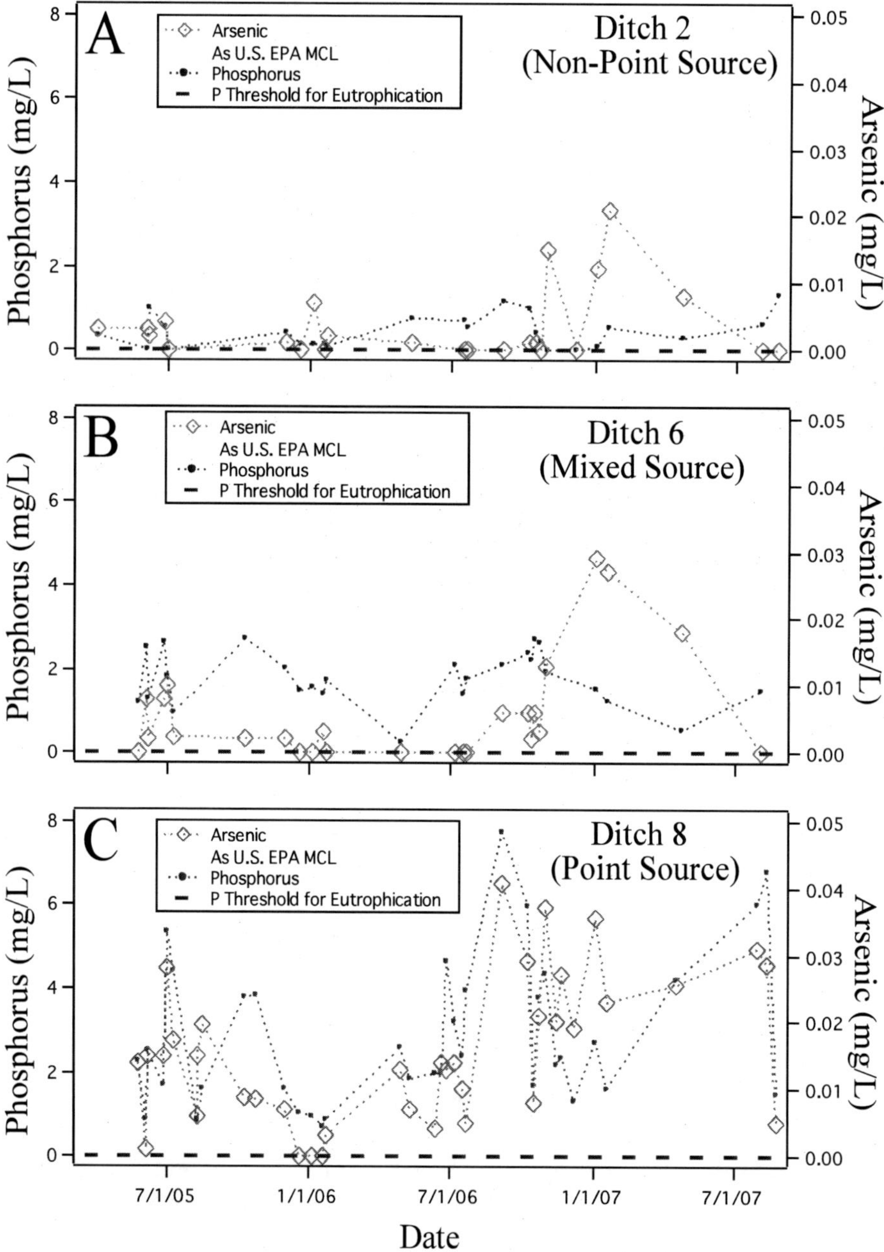

Figure 17.5. Average monthly arsenic and phosphorus concentrations in runoff from selected fields on the research farm over the period of two water years (July 2005 – June 2007).

Patterns of average annual P concentrations in effluent from the UMES research farm ditches were similar to those observed for As, suggesting a general similarity in the behavior of the two elements in ditch water. Average annual concentrations of P in flow from non-point source ditches were lower than from point and mixed point/non-point source areas. End Member analysis between poultry litter, farm soils, and background forest soils indicated that virtually all of the soils on the farm could be described as diluted poultry litter with respect to As and P with a strong correlation (r = 0.90) (Figure 17. 4). This finding highlights the likely role of historical litter applications as the primary origin of these two elements in soils. The highest total As and P concentrations in soils around the drainage ditches were found in association with the litter storage shed at the upper end of ditch 8, reflecting the extensive presence of spilled litter outside of this point source (Table 17.1). Total As concentrations in soils around the storage shed were more than double the concentrations in field soils.

Monthly average concentrations of As and P in ditch flows temper the generalization that the two elements can be managed identically. In the case of ditch 8, where spilled litter from the storage shed was essentially extracted by runoff feeding the ditch, a strong correlation (r = 0.99) between monthly average As and P concentrations exists (Figure 17. 5c). This highlights the near-term role of litter as a source of the two elements. In contrast, no significant correlation was found in monthly average concentrations in ditches draining non-point source areas where soils played a key role in (Figure 17. 5a and b). In this cases, yearly applications of litter, combined with the 30 year legacy of litter application, were the ultimate sources of both As and P, and competitive sorption/desorption between As and P were likely the controlling factors of concentrations seen in the ditches. Because of the competitive sorbtion between As and P in these areas, it is likely that managing As fate and transport would require different strategies than those used for P.

17.4. Conclusion

Given the available literature, roxarsone biotransformation is highly likely to occur in most confined animal feeding operation waste treatment systems, and furthermore, anaerobic degradation seems to be the most important mechanism. With the exception of a few isolates having been studied however, little is known about the variety of organisms that transform roxarsone or the mechanisms they employ to degrade roxarsone. As indicated, the production of the more toxic inorganic species of arsenic (e.g., arsenate) is of serious environmental concern and should be the greater foci of future studies.In the case study, where flow was monitored from ditches draining both point and non-point sources of As and P, the results showed significant concentrations (greater than the U. S. EPA MCL) and loads of both elements in ditch flow, and there were clear associations between the nature of potential sources of As and P and the magnitude of the concentrations seen. The study also showed considerable temporal variability within ditches, with no clear seasonal trends or associations with current management. The study demonstrated the importance of controlling point sources and suggested that certain management practices (such as proper, dry storage of poultry litter and avoidance of spillage outside the storage facility) could be highly effective at reducing transfers.

REFERENCES

Alexander, D. C., J. A. J. Carriere, and K. A. Mckay. 1968. Bacteriological studies of poultry litter fed to livestock. *Canadian Veterinary Journal* 9: 127-131.

Amit-Romach, E., D. Sklan, et al. 2004. Microflora ecology of the chicken intestine using 16S ribosomal DNA primers. *Poultry Science* 83: 1093-1098.

Anderson, C. E.1983. Arsenicals as feed additives for poultry and swine. 89-98. *In* W. H. Lederer and R. J. Fensterheim (ed.) *Arsenic – industrial, biomedical, environmental perspectives.* Van Nostrand Reinhold Co., New York.

Anderson, B. and T. Chamblee. 2001. The effect of dietary 3-nitro-4-hydroxyphenylarsonic acid (Roxarsone) on the total arsenic level in broiler excreta and broiler litter. *Journal of Applied Poultry Research* 10: 323-328.

Arai, Y., A. Lanzirotti, S. Sutton, J. A. Davis, and D. L. Sparks. 2003. Arsenic speciation and reactivity in poultry litter. *Environmental Science and Technology* 37: 4083-4090.

Canadian Council of Ministers of the Environment. 2002. Canadian environmental quality guidelines: Winnipeg, summary tables, 2002 [variously paged], available on the Web at *http://www.ccme.ca/publications/can_guidelines.html*

Chapman, H., and Z. Johnson. 2001. Use of antibiotics and roxarsone in broiler chickens and the USA: Analysis for the years 1995 to 2000. *Poultry Science* 81: 356–364.

Chovanec, P., J. F. Stolz, et al. 2010. A proteome investigation of roxarsone degradation by Alkaliphilus oremlandii strain OhILAs. *Metallomics* 2: 133-139.

Cortinas, I., J. A. Field, et al. 2006. Anaerobic biotransformation of roxarsone and related N-substituted phenylarsonic acids. *Environmental Science and Technology* 40: 2951-2957.

Crafts, A. 1977. Biologic Effects of Arsenic on Plants and Animals, In O. Levander, ed. Arsenic. National Academy of Sciences, Washington, D.C.

Christen, K. 2001. Chickens, manure, and arsenic. *Environmental Science and Technology* 35: 184A–185A.

Church, C. D., P. J. A. Kleinman, R. B. Bryant, L. S. Saporito and A. L. Allen. 2010. Occurrence of arsenic and phosphorus in ditch flow from litter-amended soils and barn areas. *Journal of Environmental Quality* 39:2080-2088.

Fisher, E., A. M. Dawson, et al. 2008. Transformation of inorganic and organic arsenic by Alkaliphilus oremlandii sp nov strain OhILAs. In J. Wiegel, R. Maier and M. Adams Incredible Anaerobes: from Physiology to Genomics to Fuels. *Annals of the New York Academy of Sciences.* 1125: 230-241.

Garbarino, J., A. Bednar, D. Rutherford, R. Beyer, and R. Wershaw. 2003. Environmental fate of roxarsone in poultry litter. I. Degradation of roxarsone during composting. *Environmental Science and Technology* 37: 1509–1514.

Han, F. X., W. L. Kingery, H. M. Selim, P. D. Gerard, M. S. Cox, and J. L. Oldham. 2004. Arsenic solubility and distribution in poultry waste and long-term amended soil. *The Science of the Total Environment* 320: 51-61.

Heath, C. W., Jr., and E. T. H. Fontham. 2001. Cancer etiology: Chap. 3. Lenhard, R. E., Jr., Osteen, R. T., and T. Ganslet, eds., *Clinical oncology:* Atlanta, Ga., American Cancer Society, 37–54.

Hileman, B. 2007. Arsenic in chicken production. *Chemical and Engineering News* 85: 34–35.

Howie, M., (ed.) 2003. *Feed Additive Compendium.* Miller Publishing Company, Minnetonka, MN.

Jackson, B. P. and P. M. Bertsch. 2001. Determination of arsenic speciation in poultry wastes by IC-ICP-MS. *Environmental Science and Technology* 35: 4868–4873.

Jackson, B. P., P. M. Bertsch, et al. 2003. Trace element speciation in poultry litter. *Journal of Environmental Quality* 32: 535-540.

Jain, C., and I. Ali. 2000. Arsenic: occurence, toxicity and speciation techniques. *Water Research* 34: 4304–4312.

Jean, D., F. Gilles, D. Scott, R. Higgins, and S. Quessy. 1995. Clostridium botulinum type C in feedlot steers being fed ensiled poultry litter. *Canadian Veterinary Journal* 36: 626-628.

Kleinman, P., A. Allen, B. Needelman, A. Sharpley, P. Vadas, L. Saporito, G. Folmar, and R. Bryant. 2007. Dynamics of phosphorus transfers from heavily manured Coastal Plain soils to drainage ditches. *Journal of Soil and Water Conservation* 64: 225–235.

Krishnamurti, G. S. R., and R. Naidu. 2002. Solid-solution speciation and phytoavailability of copper and zinc in soils: *Environmental Science and Technology* 36: 2645–2651.

Makris, K. C., J. Salazar, et al. 2008. Controlling the fate of roxarsone and inorganic arsenic in poultry litter. *Journal of Environmental Quality* 37: 963-971.

Marin, A. R., S. R. Pezeshki, P. H. Masscheleyn, and H. S. Choi. 1993. Effect of dimethylarsenic acid (DMAA) on growth, tissue arsenic, and photosynthesis of rice plants: *Journal of Plant Nutrition* 16: 865–880.

Moore, P., T. Daniel, B. Gilmour, B. Shreve, D. Edwards, and B. Wood. 1998. Decreasing metal runoff from poultry litter with aluminum sulfate. *Journal of Environmental Quality* 27: 92–99.

Morrison, J. 1969. Distribution of arsenic from poultry litter in broiler chickens, soil and crops. *Journal of Agriculture and Food Chemistry* 17: 1606–1614.

Ngodigha, E. M., and O. J. Owen. 2009. Evaluation of the bacteriological characteristics of poultry litter as feedstuffs for cattle. *Scientific Research and Essay* 4: 188-190.

NRC. 1999. *The use of drugs in food animals.* Washington, D.C.: National Academy Press. 253.

O'Connor, R., M. O'Connor, K. Irgolic, J. Sabrsula, H. Gurleyuk, R. Brunette, C. Howard, J. Garcia, J. Brien, J. Brien, and J. Brien. 2005. Transformations, air transport, and human impact of arsenic from poultry litter. *Environmental Forensics* 6: 83–89.

Pavkov, M. and W. Goessler. 2001. Determination of organoarsenic compounds in finishing chicken feed and chicken litter by HPLC-ICP-MS. In: *Arsenic Exposure and Health Effects IV,* Chappell, W. R., Abernathy, C. O., Calderon R. W. Eds.

Peryea, F., and R. Kammereck. 1997. Phosphate-enhanced movement of arsenic out of lead arsenate-contaminated topsoil and through uncontaminated subsoil. *Water, Air, and Soil Pollution* 93: 243–254.

Robertson, F. 1989. Arsenic in ground water under oxidizing conditions, south-west United States. *Environmental Geochemistry and Health* 11: 171–185.

Rosal, C., G. Momplaisir, and E. Heithmar. 2005. Roxarsone and transformation products in chicken manure: Determination by capillary electrophoresis-inductively coupled plasma-mass spectrometry. *Electrophoresis* 26: 1606–1614.

Rothrock, M. J. Jr., K. L. Cook, J. G. Warren, and K. Sistani. 2008. The effect of alum addition on microbial communities in poultry litter. *Poultry Science* 87:1493-1503.

Rutheford, D. W., A. J. Bednar, J. R. Garbarino, R. Needham, K. W. Staver, and R. L. Wershaw. 2003. Environmental Fate of roxarsone in poultry litter. Part II: Mobility of arsenic in soils ammended with poultry litter. *Environmental Science and Technology* 37: 1515–1520.

Seiter, J., and Donald Sparks. 2006. Fate and transport of arsenic in poultry litter amended Delaware soils: Impacts on water quality. *18th World Congress of Soil Science, Philadelphia, Pennsylvania*.

Seiter, J., and Donald Sparks. 2005. *Fate and transport of arsenic in Delaware soils.* Soil Science Society of America Annual Meeting, Salt Lake City, Utah.

Stolz, J. F., E. Perera, et al. 2007. Biotransformation of 3-nitro-4-hydroxybenzene arsonic acid (roxarsone) and release of inorganic arsenic by Clostridium species. *Environmental Science and Technology* 41: 818-823.

Tassi, E., M. Pedron, M. Barbafieri, and G. Petruzzelli. 2004. Phosphate-assisted phytoextraction in As-contaminated soil. *Engineering and Life Science* 4: 341–346.

U.S. Environmental Protection Agency. 2000. *Drinking water priority rulemaking—Arsenic: Office of Ground Water and Drinking Water,* accessed June 11, 2008, at http://www.epa.gov/safewater/arsenic.html

U.S. Food and Drug Administration. 2000. Freedom of Information Summary NADA 140–445.

Villaescusa, I., and J. C. Bollinger. 2008. Arsenic in drinking water: sources, occurrence, and health effects (a review). *Reviews in Environmental Science and Technology* 7: 307–323.

World Health Organization. 1999. Arsenic in drinking water: Fact Sheet 210, accessed on June 15, 2008, at *http://www.who.int/mediacentre/factsheets/fs210/en/*

In: Environmental Chemistry of Animal Manure
Editor: Zhongqi He

ISBN: 978-1-62808-641-6
© 2013 Nova Science Publishers, Inc.

Chapter 18

MERCURY IN MANURES AND TOXICITY TO ENVIRONMENTAL HEALTH

Irenus A. Tazisong[1], Zachary N. Senwo[1,],
Robert W. Taylor[1] and Zhongqi He[2]*

18.1. INTRODUCTION

Mercury (Hg) has been known to society for centuries and was heavily used during the industrial revolution (Jing et al., 2008). It occurs naturally and has an average crustal abundance by mass of approximately 0.08 parts per million (Ehrlich and Newman, 2009). Its distribution in the environment is controlled by either natural processes or anthropogenic activities. The natural processes are mostly attributed to rock weathering, volcanic activities, and crustal degassing (Dommerque et al., 2002; Guedron et al., 2006); whereas anthropogenic sources include mining and smelting, coal combustion, waste incineration, chlor-alkali facilities, and other industrial processes that require Hg usage (Voegborlo et al., 2010; Harris et al., 2007). Mercury can also be commonly found in thermometers, lamps, dental amalgam fillings, cosmetics, vaccines, and eye drops (Manahan, 1992; Global Mercury Assessment). Because it is broadly spread in the environment, it occurs at various concentrations in air, soil, water, and biomass. Mercury forms strong covalent bonds in biological systems and extremely strong ionic bond with reduced sulfur (Ravichandran, 2004). Its biogeochemical cycle is quite unique among toxic metals of environmental concerns (Monteiro et al., 1996; Lindberg, 1987). In the environment Hg is either in the organic or inorganic form. However, Hg is generally released into the environment in inorganic forms and can be microbially transformed into organic (methylmercury) forms (Monteiro et al., 1996; Harris et al., 2007; Han et al., 2007; do Valle et al., 2006).

---

* Corresponding author. Tel: (256) 372-4216. Fax: (256) 372-5906. Email address: zachary.senwo@aamu.edu
[1]School of Agricultural and Environmental Sciences, Alabama A&M University, Normal, AL 35762, USA.
[2]USDA-ARS, New England Plant, Soil, and Water Laboratory, Orono, ME 04469, USA.

Total Hg in most soils is usually low, although high levels have been reported. In soil, Hg occurs mainly as Hg (0) and Hg (II) forms depending on the redox conditions (Boening, 2000). Mercury speciation and distribution in soil define its toxicity and mobility. An assessment of its chemical forms (water soluble, exchangeable, organic matter bound, oxide bound, and residual), is of significant importance to improve and maintain environmental quality and sustainability. The water soluble form has been reported (Wallschläger et al., 1998; Neculita et al., 2005; Biester and Scholz, 1997) to account for less than 1% of the total Hg in soils. Although this form is negligible, more attention is given to it due to its ease of bioavailability and mobility (Bloom et al., 2003; Neculita et al., 2005). Mercury in exchangeable form is moderately bound or weakly associated with ligands (Di Giulio and Ryan, 1987), and can be exchanged with other metal ions at the binding sites. Mercury bound to organic matter represents the organic form reported to be the dominant fraction in most soils (Miretzky et al., 2005; do Valle et al., 2006) after the residual form. Residual Hg has been reported as the dominant form in most samples (Neculita, et al., 2005; Di Giulio and Ryan, 1987; Biester and Scholz, 1997; Lechler et al., 1997) and represents Hg bound to sulfur (HgS), silicates, and oxides of iron and manganese. Organic matter content, soil pH, Cl$^-$ ions, and the presence of sulfide are very significant parameters influencing Hg speciation in soils (Lin and Pehkonen, 1999, Ravichandran, 2004). Numerous studies (Miretzky et al., 2005; do Valle et al., 2006) have revealed that Hg sorption in Amazon top soils was mainly influenced by organic matter content, and poorly correlated with clay content. Just as pH influences heavy metal availability in soils, Hg availability has also been shown to be affected by pH (Jing et al., 2007). Jing et al. (2007) showed that Hg desorption in 0.01 M KCl is significantly enhanced by pH change and the presence of organic acids. Base on these authors, Hg desorption in soils decreased at pH range 3.0 − 5.0, leveled off at pH 5.0 − 8.0 and increased at pH 7.0 − 9.0. The influence of sulfur on Hg distribution in soils is not surprising due to its affinity for sulfur and the formation of Hg sulfide, an important Hg ore (Tazisong and Senwo, 2009). Studies by Tazisong and Senwo (2009) on Hg concentration and distribution in soils impacted by long-term applied broiler litter indicated a positive and highly significant correlation between Hg and sulfur.

18.2. MERCURY TOXICITY

Mercury is a priority pollutant due to its persistence in the environment and toxicity to organisms and humans (Jiang et al., 2006; Grigal, 2002). Its contamination of soils, aquatic ecosystems and biota is of serious concerns to human health. Methylmercury (organic) is the most lethal form of Hg. Bioaccumulation and biomagnifications of both methylmercury and inorganic Hg in organisms and along food chains have led to health risk for humans who consume predatory fish and other organisms from upper trophic levels. Studies by Tessier-Lavigne et al. (1985) indicated that exposure to Hg causes neurological and vision defects in humans. Visual impairment of Hg-exposed subjects was also noted in studies by Rodrigues et al. (2007). Mercury has also been implicated to cause Alzheimer disease (Mutter et al., 2004) and autism in children (Palmer et al., 2006). Methylmercury is a serious developmental neurotoxicant in fetus as a result of pregnant women consuming Hg contaminated seafood (Trasande et al., 2005). Effect of Hg on animal health has been reported to cause neurological developmental effects in fetus of dams, and acute oral exposure of inorganic Hg in rats and mice at 2–5 mg kg^{-1} b.w. per day led to increase kidney weight, and in some cases $HgCl_2$ has induced autoimmune glomerular nephritis in genetically susceptible strain of rats and mice [European Food Safety Authority (EFSA), 2008]. Based on EFSA (2008), studies on animal confirm Hg to be carcinogenic, toxic to cardiovascular system, reproductive organs, cause irritation of the gastrointestinal mucosa, developmental problems, and damage DNA in rodents. Khan et al. (2004) reported a drop in reproductive performance in mice when orally exposed to 0.25 – 1.00 mg kg^{-1} Hg day^{-1}.

Mercury toxicity is not only limited to humans and animals. It is mainly adsorbed in its ionic form (Hg^{2+}) by plants and binds with the sulfhydryl groups of proteins, displacing essential elements, and disrupting cellular structures. It has been reported to cause plant growth retardation and reduction of enzymatic activities (Senwo and Tabatabai, 1999; Pena et al., 2008). Its binding actions reduce photosynthesis, transpiration, water uptake, and chlorophyll synthesis (Godbold and Huttermann, 1986). However, studies by Cargnelutti et al. (2006); and Cho and Park (2000) suggest such binding increase lipid peroxidation. Jamal et al. (2006) reported a significant reduction in shoots and root growth in wheat varieties treated with various Hg concentrations. Studies reported by Cargnelutti et al. (2006) have also suggested that Hg induces oxidative stress in cucumber, resulting in plant injury. Plant injury due to Hg toxicity is due to the generation of such reactive oxygen species agents as superoxide anion, single oxygen, hydrogen peroxide and hydroxyl radicals responsible for tissue injury (Cargnelutti et al., 2006). Mercury concentrations (0.046 – 0.132 mg kg^{-1}) reported in vegetables from nine vegetable fields in the Guilin area of China, were moderately high (Qian et al., 2009). These values clearly indicate that human exposure to mercury in this area is eminent. Methyl mercury levels (>100 μg kg^{-1}) in rice (*oryza sativa* L.) grown at abandoned mercury mining sites was reported and reported to be 10 – 100 times higher than other crops plants (Qui et al., 2008). Mercury in food, water, and the environment is regulated by multiple federal agencies. For example, the United States Food and Drug Administration (FDA) limits the levels of Hg to below 1.0 mg kg^{-1} in sea food, and 0.002 mg L^{-1} in bottled water. The United States Environmental Protection Agency (USEPA) has limited the level of inorganic Hg in rivers, lakes, and streams to 0.144 μg L^{-1}.

18.3. MERCURY IN MANURE

Since Hg has no known biological functions, it is not used in animal feeds formulations. Mercury found in animal manure may likely originate from animal feeds formulated with fish and other grain components that contain significant levels of Hg. Mercury content in manure may be of concerns and limitations to the extensive use of animal manure in agriculture because of potential uptake by crops and its subsequent entry into the food chain.

Capar et al. (1978) measured the Hg levels in animal feeds, animal wastes and sewage sludge collected from the same source at Colorado State University. The Hg levels in a typical cattle feedlot diet (the dry weight composition of this ration was 70% corn, 3% hay, 5% beet pulp, 20% corn silage, and 2% mineral supplement) was <0.01 mg kg^{-1}. The metro Denver sewage sludge contained 7.8 mg kg^{-1}. The poultry wastes with or without litter contained 0.06 and <0.04 mg kg^{-1}, respectively. Whereas the sewage sludge and poultry wastes were used as feedlot diet ingredients, the cattle manure contained a higher Hg concentration than the typical feedlot diet without the addition of poultry waste products. Specifically, the Hg levels were 0.05, <0.03, and <0.09 mg kg^{-1} for cattle manure with low fiber diet, high fiber diet, and processed cattle waste pellets, respectively. Cappon (1984) reported 0.041 mg kg^{-1} in composted and 0.065 mg kg^{-1} in dehydrated cow manure from New York, USA sources. Raven and Loeppert (1997) reported that the Hg content was <0.4 mg kg^{-1} in cow manure and two cattle manure composts similar to the levels in several synthetic fertilizers. The study indicated a higher Hg content (1.1 and 1.5 mg kg^{-1}, respectively) in two sewage sludge samples that were used as organic fertilizers. Similarly, McBride and Spiers (2001) reported an average of 0.020 mg kg^{-1} for 20 dairy manure samples collected from farms in New York, ranging from 0.010 to 0.050 mg kg^{-1} (Table 18.1). Studies by Tazisong and Senwo (2009) indicated Hg levels ranging from 0.001 to 0.041 mg kg^{-1} with an average of 0.019 mg kg^{-1} (Table 18.1) in 23 poultry litter samples collected from Alabama, USA. The Hg content in dairy manure samples collected from conventional and organic dairy manure farms in Maine, USA were 0.057 and 0.050 mg kg^{-1}, respectively (Table 1). The studies indicate that the Hg concentrations in these manure samples were basically at similar levels of concentrations. However, with the Hg limit at 17 mg kg^{-1} based on USEPA (Part 503) regulation for land application of animal manure in the USA (He et al., 2005), usage for agricultural purposes is not of environmental concerns under normal circumstances.

Raszyk et al. (1996) investigated the toxic levels of harmful Hg, Cd and Pb of pig houses in Hodonin District, Czech Republic. They found that the Hg content in feed mixtures decreased 102 times in 10 years, whereas Pb and Cd levels dropped 6-8 times from 0.510 mg kg^{-1} in 1984 to 0.005 mg kg^{-1} in 1994. Correspondingly, the Hg content was also reduced to 0.003 mg kg^{-1} in the pig manure. This led to reduced environmental health risk not only for the pigs but also for their tenders. Nicholson et al. (1999) collected 183 livestock feeds and 85 animal manure samples from commercial farms in England and Wales. They found that the Hg levels were <0.10 mg kg^{-1} dry matter for nearly all the analyzed livestock feeds. The concentrations of Hg in animal manure were even lower (< 0.05 mg kg^{-1} dry matter). Sager (2007) determined the trace and nutrient elements in manure, dung and compost samples in Austria. However, Hg values obtained in compost and sewage sludge samples were 0.33 and 0.58 mg kg^{-1}, respectively. Moreno-Caselles et al. (2002) reported the Hg concentrations of 48 different types of animal manure samples collected from various farms in the Southeast Spain (Table 18. 2) showing a wide variation (0.1 mg to 5.7 mg kg^{-1} dry matter) between the various types of manure. The Hg concentration in goat manure was at least twice greater than those of other sources, which the authors attributed to the xenobiotical contamination of the refuse. It is worth to note that the Hg concentrations in these Spanish manure samples are 10 to 100 times greater than those reported in USA, England and Wales. The lower Hg levels in animal manure from the US might be due to the strict regulatory limits by USEPA in the nineties on the limits of Hg in foods especially fish and grains (main ingredients for animal feedstock). However, European Union countries have different Hg limits in animal wastes (16 – 25 mg kg^{-1} dry matter) for agricultural use. Netherlands has set up its limit to 0.75 mg kg^{-1} dry matter (He et al., 2005). Such differences in regulations in Europe may probably account for the high Hg levels in animal manures from the southeast Spain.

Table 18.1. Mercury concentrations (mg kg^{-1} dry matter) in animal manures collected from USA farms.

Manure	Sample number	Average	Range	Source	Reference
Cattle	3	0.04	<0.03-0.09	Colorado	Capar et al., 1978
Poultry	2	0.05	0.04-0.06	Colorado	Capar et al., 1978
Dairy composted	NA	40.8	NA	New York	Cappon, 1984
Dairy dehydrated	NA	65.2	NA	New York	Cappon, 1984
Cattle manure & compost	3	<0.4	NA	Texas	Raven & Loeppert, 1997
Dairy	20	0.020	0.010-0.050	New York	McBride & Spiers, 2001
Mixing compost	NA[†]	0.0065	NA	Connecticut	Bash & Miller, 2007
Poultry litter	23	0.019	0.001-0.041	Alabama	Tazisong & Senwo, 2009
Organic dairy	16	0.050	0.033-0.084	Maine	He et al., unpublished
Conventional dairy	4	0.057	0.040-0.075	Maine	He et al., unpublished

[†] Information on the sample number is not available; however, the authors reported the standard deviation (SD) of 0.0029.

Table 18.2. Mercury concentrations (mg kg^{-1} dry matter) in animal manure samples collected from Southeast Spain (adapted from Moreno-Caselles et al., 2002).

Manure	Number of samples	Average	Range
Horse	4	0.7	0.2-1.1
Cow	4	2.3	0.2-5.1
Calf	3	0.9	0.5-1.3
Pig	6	2.3	0.1-4.6
Sheep	4	0.6	0.4-0.7
Goat	4	5.7	0.4-15.5
Rabbit	6	0.3	0.1-0.6
Chicken	4	3.1	0.5-6.5
Turkey	3	0.1	0.1-0.2
Ostrich	6	2.4	0.1-8.3
Earthworm	4	0.5	0.1-0.9

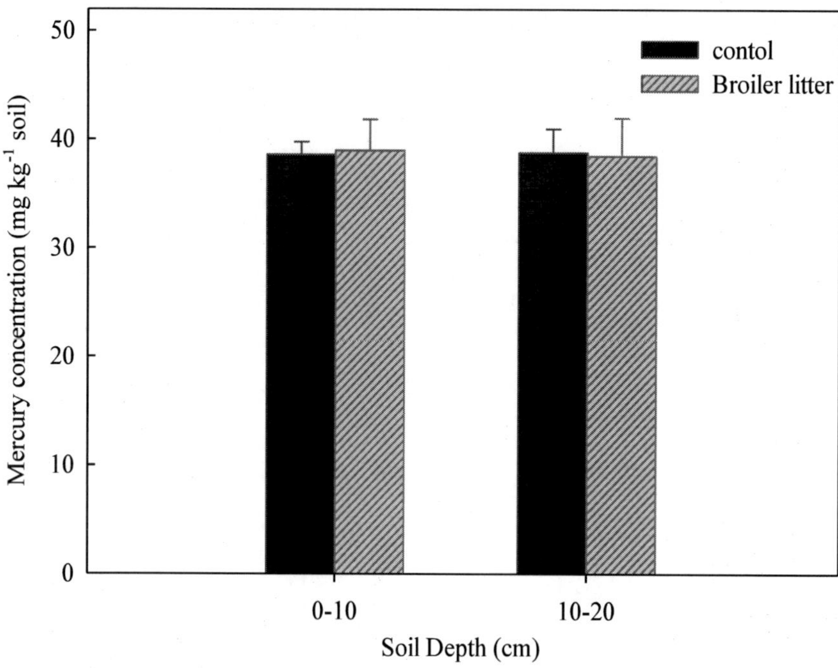

Figure 18.1. Mercury concentration in Decatur silt loam. The soil samples were collected from a 10–year cropping (cotton) system study in Alabama USA. Broiler litter applied to supply N at 100 kg N ha⁻¹. Control is NH_4NO_3 applied at 100 kg N ha⁻¹. Vertical bars indicate standard deviations (data adapted from Tazisong and Senwo, 2009).

Cang et al. (2004) reported the Hg concentrations in animal feeds and manures under intensive farming in Jiangsu Province, China (Table 18.3). The Hg levels in the animal feeds were comparable to those (<0.01 mg kg⁻¹) reported by Capar et al. (1978) and Nicholoson et al. (1999). The current (China) National Hygienical Standard for feeds is at Hg concentration ≤0.1 mg kg⁻¹. However, there has been pig feed samples with higher Hg levels of 0.290 mg kg⁻¹ (Table 18.3). Similarly, few feeds of the 183 livestock feeds examined by Nicholson et al. (1999) may contain more than 0.10 mg kg⁻¹. In comparison to the values of other feed samples, the Hg levels increased by about 5 times in chicken and pig manures (Table 18. 3). The higher Hg concentrations in manure samples indicate Hg enrichment in animal manure, which is consistent with the observations by Capar et al. (1978). Luo et al. (2009) compared Hg concentrations in four livestock manures between 2003 and the 1990's (Table 18.4). In the 1990's, the Hg concentrations of the four types of manure generated in China were at similar levels as those in USA samples (Table 18.1) and were also comparable to the levels in manures generated in Spain (Table 18.2). The Hg levels in manures generated in China have increased in an order of magnitude from tenths to hundredths mg kg⁻¹ dry matter during the past decades (Table 18.4). The increasing trend probably suggest that livestock diet changes with increasing feed additives for health and welfare reasons (Luo et al., 2009). Although, no standards for land application of organic manures containing Hg and other trace elements generated in China are currently available, it is recommended that animal manures be used with great caution for agriculture (Luo et al., 2009).

Table 18.3. Mercury concentrations (mg kg^{-1} dry matter) in animal feeds and manures collected from Jiangsu Province, China.

	Sample number	Average	Range
Chicken:			
Feed	16	0.005	<0.0002-0.033
Manure	17	0.024	0.001-0.077
Pig:			
Feed	7	0.006	<0.0002-0.290
Manure	16	0.033	<0.0002-0.118
Milch cow:			
Feed	7	0.001	<0.002-0.007
Manure	8	0.039	<0.0002-0.073
Duck:			
Feed	2	0.014	0.010-0.018
manure	2	0.029	0.008-0.051
Goose:			
Feed	2	0.0002	<0.0002-0.0003
Manure	2	0.022	<0.0002-0.045
Dove:			
Feed	3	0.001	<0.0002-0.002
Manure	3	0.052	0.019-0.118

Adapted from Cang et al., 2004.

Table 18.4. Mercury concentrations (mg kg^{-1}dry matter) in selected livestock manures during different periods in China.

Manure type	Year	Sample number	Average	Standard deviations
Chicken	2003	70	0.13	0.10
	1990's	≥22	0.03	NA[†]
Pig	2003	61	0.12	0.23
	1990's	≥33	0.07	NA
Cattle				
	2003	42	0.10	0.10
	1990's	≥66	0.04	NA
Sheep	2003	15	0.19	0.50
	1990's	≥24	0.07	NA

Adapted from Luo et al., 2009.

[†] Data not available.

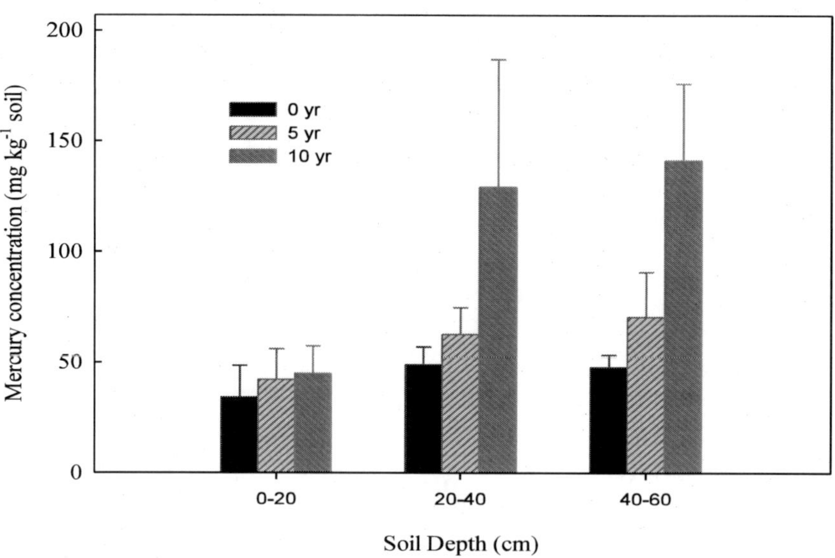

Figure 18.2. Changes of total Hg with varying year of repeated broiler litter application. Vertical bars represent standard deviation (data adapted from Tazisong and Senwo, 2009).

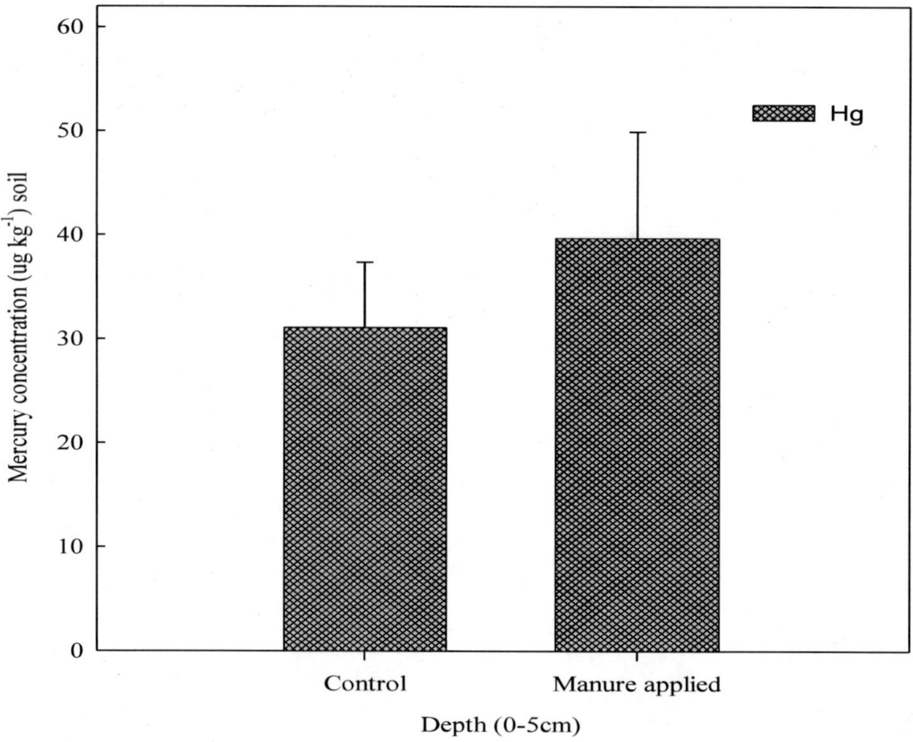

Figure 18.3. Mercury concentration in Bama/Goldsboro. The soil samples were collected in a 3–year no-till plots from Alabama Agricultural Experimentation Station E.V. Smith Research Center in central Alabama, where dairy bedding manure was annually applied at rates of ~10 Mg ha-1 yr-1 (dry matter). Control is without annual application of dairy bedding manure. Vertical bars indicate standard deviations.

Naylor et al. (1999) determined the chemical composition of feeds and biosolid fish wastes (manure) from commercial rainbow trout farms in Ontario, Canada. The Hg concentrations in the rainbow trout diets were 0.03 ± 0.01 mg kg^{-1} based on two commercially available feed samples. The average Hg content of the fresh fish manure samples collected at 12 commercial farms was 0.05 mg kg^{-1} ranging from none to 0.19 mg kg^{-1}. Zheljazkov and Warman (2002) compared the Hg concentrations in composts prepared from mixtures of racetrack manure, food waste and straw, and/or municipal solid wastes from Nova Scotia, Canada and measured using three digestion methods (nitric acid, nitric/perchloric acid and dry ashing). The average Hg concentrations of the six composts were 0.612, 0.705, and 0.702 mg kg^{-1}, respectively. These values were not statistically different at $p \leq 0.05$. Bakare et al. (2004) determined Hg background of farms in close proximity and distant to some highways in Nigeria. They reported that the Hg concentrations were 2.67 and <1.67 mg kg^{-1}, respectively, in manures samples (mainly from poultry droppings) collected from the two farms. Whereas no explicit explanation was given in the report, the location close to the highways might predispose the farm to high metal depositions.

18.4. IMPACTS OF LAND APPLIED MANURE ON SOIL MERCURY

Although there is extensive literature on the concentrations of other metals in soils amended with animal manure, little information is available on Hg concentrations in such soils. Therefore; it is imperative that we continuously monitor total and bioavailable Hg build up in soils amended with animal manure. Luo et al. (2009) compiled annual trace element inputs as well as net inputs to agricultural land in China from various sources. The data indicated that $58 - 85\%$ of the total annual inputs of As, Hg, Pb, and Ni, to agricultural land were derived from atmospheric deposition, whereas $8 - 25\%$ and $5 - 30\%$ were from livestock manures and fertilizers, respectively. Zheng et al. (2008) reported the concentration and distribution of Hg in soils from three long-term experimental sites in China that have received chemical fertilizers and manure applications for over 16 years. Mercury concentrations at the A horizon (0–20 cm) recorded for the three sites was 0.539 ± 0.055, 0.196 ± 0.036, and 0.037 ± 0.014 mg kg^{-1}. The data suggest that applying phosphate fertilizers to soils is likely to influence the soil Hg concentrations to some extent especially in soils with lower Hg background. The application of other chemical fertilizers and organic manure did not significantly change the soil Hg concentrations in the study indicating that there was little cumulative Hg risk in the soil from the long-term manure fertilization practices. Mercury content in an upland cotton production site at the Alabama Agricultural Experimentation Station in Belle Mina, Alabama amended with broiler litter showed little accumulation and downward movement in the soils (Figure 18.1). In contrast, studies conducted by Tazisong and Senwo (2009) revealed significant increases of Hg concentrations in soils after years of repeated broiler litter applications (Figure 18.2) as the Hg concentrations in the surface soil (0–20 cm) increased from 0.040 mg kg^{-1} in the control to 0.060 and 0.10 mg kg^{-1} soil respectively, in the top soil samples that received 5 and 10 year of broiler litter. The impact was more remarkable in the subsurface soils as the greater downward movement of Hg occurred with 10 year of litter application in these soils (Figure

18.2). Mercury concentrations were also tested in soils from the Alabama Agricultural Experimentation Station E.V. Smith Research Center in Central Alabama, which have received dairy manure at rates of ~10 Mg ha^{-1} yr^{-1} for two years (Terra et al., 2006). Results from this study revealed Hg accumulated in manure applied sites compared to the control (Figure 18.3). Similar results were reported by Sloan et al. (2001), in surface (0–15 cm) and subsurface (15–30 cm) soils that have received 224 Mg ha^{-1} biosolids. These conflicting observations suggest more data are needed to evaluate Hg accumulation in soils due to manure application.

The differences observed are partly due to the complicated mechanisms associated with Hg accumulation and evaporation in soils. Bash and Miller (2007) observed elevated total gaseous Hg concentrations from an agricultural production during tilling. The Hg concentrations were measured at the University of Connecticut's Hg forest flux tower during spring agricultural field operations on an adjacent corn field. Twelve tons of dry manure per acre were spread and disked into the soil. The manure was composed of 30% sawdust; 5 – 8% feed refuse, and 62 – 65% pig, sheep, cow, and horse manure. The mean Hg concentration of the manure was 0.0065 ± 0.0029 mg kg^{-1} with a mean pH of 7.6 ± 0.39. The mean plowing depth of the soil was approximately 16.5 cm. The mean Hg concentration in the soil in the plow layer was 0.573 ± 0.606 mg kg^{-1} (n = 20) with mean soil pH of 6.04 ± 0.32 (n = 20) and mean total carbon content of 3 ± 1.8% (n = 4). Mercury concentrations in the field plow layer had a much higher horizontal variability and little vertical variability when compared to samples taken in the adjacent woods which averaged 0.26 ± 0.01 mg kg^{-1} (n = 4). Analyses of the meteorological conditions and Hg content in agricultural soil, manure and the diesel consumed in the tilling operations indicate that the Hg source was from the agricultural tilling operations. Their results indicate that agricultural operations resulting in a disturbed soil surface may be a source of atmospheric Hg from Hg pool bound in the soil. Yang et al. (2008) examined the effects of dissolved organic matter obtained from humus soil (DOMH), rice straw (DOMR), and pig manure (DOMP) on the adsorption and desorption of Hg in Xanthi-Udic Ferralosols and Typic Purpli-Udic Cambosols. The presence of dissolved organic matter reduced Hg maximum adsorption capacity by up to 40% over the control in the order: DOMH (250 mg kg^{-1}) < DOMR (303 mg kg^{-1}) < DOMP (323 mg kg^{-1}) < CK (control 417 mg kg^{-1}) for the first soil and DOMH (270 mg kg^{-1}) < DOMR (313 mg kg^{-1}) < DOMP (324 mg kg^{-1}) < CK (476 mg kg^{-1}) for the second soil, respectively. Yang et al. (2008) also observed that while dissolved organic matter promoted Hg desorption from the soil, Hg adsorption rates was consistently slowed down. The studies indicate that manure application affects soil Hg status not only by direct Hg input, but also by altering the soil-Hg interactions.

18.5. MERCURY SPECIATION AND EMISSION FROM CO-COMBUSTION WITH COAL AND BROILER LITTER

Broiler litter generated from poultry industries is usually land applied, but co-combustion with coal is an increasingly important option, both as an environmentally friendly method for disposal and as an economically attractive approach to generate power or process heat (Sable et al., 2007b). Also, with the high cost of energy and fuel, broiler litter can serve as an excellent source of renewable energy. Mercury content in broiler litter is negligible, which results in low Hg emission rates during co-combustion with coal. However, broiler litter contains higher halogen content (chlorine), which may impact Hg speciation during co-firing (Cao et al., 2008). More so, the use of broiler litter especially with co-combustion in pulverized form to generate power and heat also emits NO_x and SO_x, which in effect, affect Hg interaction and behavior.

Mercury speciation and emissions due to co-combustion can be classified as gaseous, elemental, oxidized and particulate bound Hg (Li et al., 2008). Understanding the fate of Hg in co-combustion systems is difficult since Hg is extremely volatile. Mercury emissions from coal co-combustion fired plants are highly dependent upon its speciation (Sable et al., 2007b). Studies (Li et al., 2008) have indicated that in the gaseous phase, total gaseous and elemental Hg decreases with an increase in broiler litter. Cao et al. (2008) also reported a reduction of Hg emissions of over 80% when coal was co-fired with broiler litter and 50% when co-fired with wood pellets and coffee residue. Similar findings have also been reported by several investigators (Sable et al., 2007a, b; Kellie et al., 2005; Cao et al., 2005).

Mercury emissions from co-firing plants have been reported to correlate with chlorine gas concentrations, but not with chlorine concentrations in co-firing fuels (Cao et al., 2008). Researchers have explained the high reduction in emissions with broiler litter due to high chlorine content in the broiler litter while the low reduction in emissions with wood pellets has been attributed to low chlorine content. Analysis of broiler litter and coal (Li et al., 2008) revealed more chlorine content (responsible for the high ionic or oxidized Hg concentrations) in broiler litter than in coal. Liu et al. (2001) while studying Hg emissions from coal-fired combustors that used high chlorine coal to enhance conversion of elemental Hg to its oxidized form inside a fluidized bed combustion system reported that the use of high chlorine coal in a fluidized bed combustion system converted more than 99% of elemental Hg to an oxidized state, mainly ionic Hg ($HgCl_2$). Their results also indicated that with the use of high-chlorine coal in a fluidized bed combustion system the mercuric gas phase which was around 45% of the total Hg input, was primarily in the oxidized state (40% of total Hg input), whereas only a small portion (4.5% of total Hg input) still remained as elemental Hg in flue gas. It is believed that the oxidation of Hg by chlorine is a multi step reaction as illustrated by the following reactions:

$$Hg + Cl \longleftrightarrow HgCl \text{ --------------------------------1}$$

$$HgCl + Cl_2 \longleftrightarrow HgCl_2 + Cl \text{ -------------------2}$$

$$Hg + Cl_2 \longleftrightarrow HgCl_2 \text{ -------------------------3}$$

$$2Hg + 4HCl + O_2 \longleftrightarrow 2HgCl_2 + 2H_2O \text{ ----------4}$$

The initial reaction (1) is the fastest while reactions (2) and (3) are slower but dominate the entire oxidation process (Edwards et al., 2001). Kinetics models have suggested that the quantity of Cl^- ion and their lifespan in the combustor are limiting factors in oxidizing Hg (Cao et al., 2005; Kellie et al., 2005).

Coal co-combustion with broiler litter also decreases particulate bound Hg, but when olive residue and B-wood is used in co-combustion, particulate Hg increases due to unoxidized carbon, fine particles, and higher iron concentrations (Sable et al., 2007b). Coal co-combustion with broiler litter will have the following impact on Hg speciation and emissions: (i) increase Hg oxidation; (ii) reduction in elemental Hg; (iii) reduction in particulate Hg; and (iv) increase amount of Hg chemisorbed into flue ash.

The reduction in Hg emissions from coal-fired electric power plants has been emphasized recently and might further result in environmental regulations requiring the use of activated carbon as Hg sorbents (Klasson et al., 2009). The sorbents could be injected into the flue gas stream, where they could adsorb the Hg. The sorbents (now containing Hg) would be removed via filtration or other means from the flue gas. Thus, Klasson et al. (2009) evaluated broiler litter manure as raw material for Hg adsorbents in gas applications. In laboratory experiments, Klasson et al. (2009) demonstrated the use of activated carbon made from turkey cake manure to remove significant amounts of elemental Hg from a hot air stream. Other activated carbon made from chicken and turkey litter manure was also efficient. In general, unwashed activated carbon made from broiler litter manure was more efficient in removing Hg than their acid-washed counterparts. On the basis of these findings, Klasson et al. (2009) concluded that broiler litter-manure-based activated carbon may possibly be used in applications in which elemental Hg must be removed from gaseous phases.

18.6. CONCLUSION

Manure is widely used as a cheap source of fertilizer all over the world. In industrialized countries, tons of animal or organic manures per hectare each year are disposed on agricultural lands. There is lack of data to indicate that Hg is added to animal feed formulations. Published data have shown that the levels of Hg are considerably low in animal feeds than in animal wastes (manure). Chemical analysis of manures has usually indicated low Hg concentrations compared to other elements. Mercury content also differs with manure sources and regions of origins, but is mostly less than 0.1 mg kg^{-1} dry matter, much lower than the regulated concentration (e.g. < 17 mg kg^{-1}) for biosolids by USEPA. Thus, in general, land application of animal manure as a fertilizer should be free of concerns in terms of potential Hg pollution. However, some data have revealed that repeated disposal of manures to agricultural lands led to Hg buildup over time, and may be detrimental to soil, water, and environmental health. Mercury speciation and bioavailability in animal manures and wastes applied to soils should be examined periodically, to warrant the best safety use of animal manures and wastes.

REFERENCES

Bakare, S., A.A.B. Denloye, and F.O. Olaniyan. 2004. Cadmium, lead and mercury in fresh and boiled leafy vegetables grown in Lagos, Nigeria. *Environ. Technol.* 25:1367–1370.

Bash, J.O., and D.R. Miller. 2007. A note on elevated total gaseous mercury concentrations downwind from an agriculture field during tilling. *Sci. Total Environ.* 388:379–388.

Biester, H., and C. Scholz. 1997. Determination of mercury binding forms in contaminated soils: Mercury pyrolysis versus sequential extraction. *Environ. Sci. Technol.* 31:233–239.

Bloom, N.S., E. Preus, J. Katon, and M. Hiltner. 2003. Selective extractions to assess the biogeochemically relevant fractionation of inorganic mercury in sediments and soils. *Anal. Chim. Acta.* 479:233–248.

Boening, D.W. 2000. Ecological effects, transport, and fate of mercury: A general review. *Chemosphere* 40:1335–1351.

Cang, L., Y.J. Wang, D.M. Zhou, and Y.H. Dong. 2004. Heavy metals pollution in poultry and livestock feeds and manures under intensive farming in Jiangsu Province, China. *J. Environ. Sci.* 16:371–374.

Cao, Y., H. Zhou, J. Fan, H. Zhao, T. Zhou, P. Hack, C. Chan, J. Liou, and W. Pan. 2008. Mercury emissions during cofiring of sub-bituminous coal and biomass (chicken waste, wood, coffee residue, and tobacco stalk) in a laboratory-scale fluidized bed combustion. *Environ. Sci. Technol.* 42:9378–9384.

Cao, Y., Y. Duan, S. Kellie, L. Li, W. Xu, J.T. Riley, and W. Pan. 2005. Impact of coal chlorine on mercury speciation and emission from a 100-MW utility boiler with cold-side electrostatic precipitators and low NO_x burners. *Energy Fuels* 19:842–854.

Capar, S.G., J.T. Tanner, M.H. Friedman, and K.W. Boyer. 1978. Multielement analysis of animal feed, animal wastes, and sewage sludge. *Environ. Sci. Technol.* 12:785–790.

Cappon, C.J. 1984. Content and chemical form of mercury and selenium in soil, sludge, and fertilizer materials. *Water Air Soil Poll.* 22:95–104.

Cargnelutti, D., L.A. Tabaldi, R.M. Spanevello, G.d.O. Jucoski, V. Battisti, M. Redin, C.E.B. Linares, V.L. Dressler, E.M.d.M. Flores, F.T. Nicoloso, V.M. Morsch, and M.R.C. Schetinger. 2006. Mercury toxicity induces oxidativ stress in growing cucumber seedlings. *Chemosphere* 65:999–1006.

Cho, U.H., and J.O. Park. 2000. Mercury-induced oxidative stress in tomato seedlings. *Plant Sci.* 156:1–9.

Di Giulio, R.T., and E.A. Ryan. 1987. Mercury in soils, sediments, and clams from a north Carolina peatland. *Water Air Soil Pollut.* 33:205–219.

do Valle, C.M., G.P. Santana, C.C. Windmoller. 2006. Mercury conversion processes in Amazon soils evaluated by thermodesorption analysis. *Chemosphere.* 65:1966–1975.

Dommerque, A., C.P. Ferrari, F.A.M. Planchon, and C.F. Boutron. 2002. Influence of anthropogenic sources on total gaseous mercury variability in Grenoble suburban air (France). *Sci. Total Environ.* 297:203–213.

Edwards, J.R., R.K. Srivastava, and J.D. Kilgroe. 2001. A study of gas phase mercury speciation using detailed chemical kinetics. *J. Air Waste Manage. Assoc.* 5:869–877.

Ehrlich, H.L., and D.K. Newman. 2009. Geomicrobiology. pp. 265–275. In: *Geomicrobiology of mercury.* CRC press, Taylor and Francis, Boca Raton, Florida, USA.

European Food Safety Authority. 2008. Opinion of the scientific panel on contaminants in the food chain on a request from the European commission on mercury as undesirable substance in feed. *The EFSA J.* 654:1–74.

Global Mercury Assessment - *Current production and use of mercury.* http://www. chem.unep.ch/mercury/report/Final%20report/chapter7.pdf

Godbold, D.L., and A. Huttermann. 1986. The uptake and toxicity of mercury and lead to spruce seedlings. *Water Air Soil pollut.* 31:509–515.

Grigal, D.F. 2002. Inputs and outputs of mercury from terrestrial watersheds: a review. *Environ. Rev.* 10:1–39.

Guedron, S., C. Grimaldi, C. Chauvel, L. Spadini, and M. Grimaldi. 2006. Weathering versus atmospheric contributions to mercury concentrations in French Guiana soils. *Appl. Geochem.* 21:2010–2022.

Han, S., A. Obraztsova, P. Pretto, K. Choe, J. Gieskes, D.D. Deheyn, and B.M. Tebo. 2007. Biogeochemical factors affecting mercury methylation in sediments of the Venice lagoon, Italy. *Environ. Toxicol. Chem.* 26:655–663.

Harris, R., D. Krabbenhoft, R. Mason, M.W. Murray, R. Reash, and T. Saltman. 2007. Mercury emissions and deposition. pp 3-4. *In*: R. Harris et al., (eds.). *Ecosystem responses to mercury contamination.* CRC press, Taylor and Francis, New York, USA.

He, Z.L., X.E. Yang, and P.J. Stoffella. 2005. Trace elements in agroecosystems and impacts on the environment. *J. Trace Elem. Med. Biol.* 19:125–140.

Jamal, S.N., M.Z. Iqbal, and M. Athar. 2006. Evaluation of two wheat varieties for phytotoxic effect of mercury on seed germination and seedling growth. *Agric. Conspec. Sci.* 71:41–44.

Jiang, G.B., J.B. Shi, and X.B. Feng. 2006. Mercury pollution in China. *Environ. Sci. Technol.* 40:3672–3678.

Jing, Y.D., Z.L. He, and X.E. Yang. 2008. Adsorption-desorption characteristics of mercury in paddy soils of China. *J. Environ. Qual.* 37:680–688.

Jing, Y.D., Z.L. He, X.E. Yang. 2007. Effects of pH, organic acids, and competitive cations on mercury desorption i n soils. *Chemosphere* 69:1662–1669.

Kellie, S., Y. Cao, Y. Duan, L. Li, P. Chu, A. Mehta, R. Carty, J.T. Riley, and W. Pan. 2005. *Factors affecting mercury speciation in a 100-MW coal-fired boiler with low-NOx burners Energy Fuels* 19:800–806.

Khan, A.B., A. Atkinson, T.C. Graham, S.J. Thompson, S. Ali, and K.F. Shireen. 2004. Effects of inorganic mercury on reproductive performance of mice. *Food Chem. Toxicol.* 42:571–577.

Klasson, K.T., I.M. Lima, and L.L. Boihem. 2009. Poultry manure as raw material for mercury adsorbents in gas applications. *J. Appl. Poult. Res.* 18:562–569.

Lechler, P.J., J.R. Miller, L. Hsu, and M.O. Desilets. 1997. Mercury mobility at the Carson river superfund site, West-central Nevada, USA: Interpretation of mercury speciation data in mill tailings, soils, and sediments. *J. Geochem. Explor.* 58:259–267.

Li, S., S. Deng, A. Wu, and W. Pan. 2008. Impact of the addition of chicken litter on mercury speciation and emissions from coal combustion in a laboratory-scale fluidized bed combustor. *Energy and Fuels* 22:2236–2240.

Lin, C.J. and S.O. Pehkonen. 1999. Aqueous phase reactions of mercury with free radicals and chlorine: Implications for atmospheric mercury chemistry. *Chemosphere* 38:1253–1263.

Lindberg, S. 1987. Mercury. pp. 17-33. *In*: Hutchinson et al. (eds.). *Lead, mercury, cadmium and arsenic in the environment.* John Wiley & Son Ltd, Chichester.

Liu, K., Y. Gao, J.T. Riley, and W. Pan. 2001. An investigation of mercury emission from FBC systems fired with high-chlorine coals. *Energy Fuels* 15:1173–1180.

Luo, L., Y.B. Ma, S.Z. Zhang, D.P. Wei, and Y.G. Zhu. 2009. An inventory of trace element inputs to agricultural soils in China. *J. Environ. Manag.* 90:2524–2530.

Manahan, S.E. 1992. Toxic elements. pp. 249–268. *In: Toxicological chemistry.* Lewis Publisher, Inc. Chelsea, Michigan.

McBride, M.B., and G. Spiers. 2001. Trace element content of selected fertilizers and dairy manures as determined by ICP-MS. *Commun. Soil Sci. Plant. Anal.* 32:139–156.

Miretzky, P., M.C. Bisinoti, and W.F. Jardim. 2005. Sorption of mercury (II) in Amazon soils from column studies. *Chemosphere* 60:1583–1589.

Monteiro, L.R., V. Costa, R.W. Furness, and R.S. Santos. 1996. Mercury concentrations in prey fish indicate enhanced bioaccumulation in mesopelagic environments. *Mar. Ecol. Prog. Ser.* 141.21–25.

Moreno-Caselles, J., R. Moral, M. Perez-Murcia, A. Perez-Espinosa, and B. Rufete. 2002. Nutrient value of animal manures in front of environmental hazards. *Commun. Soil Sci. Plant Anal.* 33:3023–3032.

Mutter, J., J. Naumann, C. Sadaghiana, R. Schneider, and H. Walach. 2004. Alzheimer disease: Mercury as pathogenetic factor and apolipoprotein E as a moderator. *Neuroendocrinol. Lett.* 25:331–339.

Naylor, S.J., R.D. Moccia, and G.M. Durant. 1999. The chemical composition of settleable solid fish waste (manure) from commercial rainbow trout farms in Ontario, Canada. N. *Am. J. Aqualcult.* 61:21–26.

Neculita, C.M., G.J. Zagury, and L. Deschenes. 2005. Mercury speciation in highly contaminated soils from chlor-alkali plants using chemical extractions. *J. Environ. Qual.* 34:255-262.

Nicholson, F.A., B.J. Chambers, J.R. Williams, and R.J. Unwin. 1999. Heavy metal contents of livestock feeds and animal manures in England and Wales. *Biores. Technol.* 70:23–31.

Palmer, R.F., S. Blanchard, Z. Stein, D. Mandell, and C. Miller. 2006. Environmental mercury release, special education rates, and autism disorder: an eclogical study of Texas. H*ealth and Place.* 12:203–209.

Pena, L.B., M.S. Zawoznik, M.L. Tomaro, and S.M. Gallego. 2008. Hemetals effects on proteolytic system in sunflower leaves. *Chemosphere* 72:741–746.

Qian, J., L. Zhang, H. Chen, M. Hou, Y. Niu, Z. Xu, and H. Liu. 2009. Distribution of mercury pollution and its source in the soils and vegetables in Guilin area, China. *Bull. Environ. Contam. Toxicol.* 83:920–925.

Qui, G., X. Feng, P. Li, S. Wang, G. Li, L. Shang, and X. Fu. 2008. Methylmercury acumulation in rice Oryza sativa L.) grown at abandoned mercury mines in Guizhou, China. *J. Agric. Food Chem.* 56:2465–2468.

Raszyk, J., K. Nezveda, H. Docekalova, J. Salava, and J. Palac. 1996. The presence of harmful chemical elements (Hg, Cd, Pb) in the stable environment for pig herds. *Veterinarni Medicina* 41:207–211.

Raven, K.P., and R.H. Loeppert. 1997. Trace element composition of fertilizers and soil amendments. *J. Environ. Qual.* 26:551–557.

Ravichandran, M. 2004. Interactions between mercury and dissolved organic matter-a review. *Chemosphere* 55:319–331.

Rodrigues, A.R., C.R.B. Souza, A.M. Braga, P.S.S. Rodrigues, A.T. Silveira, E.B.T. Damin, M.I.T. Cortes, A.J.O. Castro, G.A. Mello, J.L.F. Viera, M.C.N. Pinheiro, D.F. Ventura, and L.C.L. Silveira. 2007. Mercury toxicity in the Amazon: Contrast sensitivity and color discrimination of subjects exposed to mercury. Brazil. *J. Med. Bio. Res.* 40:415–424.

Sable, S.P., W. de Jong, R. Meij, and H. Spliethoff. 2007a. Effect of air-staging on mercury speciation in pulverized fuel co-combustion: Part 2. *Energy Fuels* 21:1891–1894.

Sable, S.P., W. de Jong, R. Meij, and H. Spliethoff. 2007b. Effect of secondary fuels and combustor temperature on mercury speciation in pulverized fuel co-combustion: Part 1. *Energy Fuels* 21:1883–1890.

Sager, M. 2007. Trace and nutrient elements in manure, dung and compost samples in Austria. *Soil Biol. Biochem.* 39:1383–1390.

Senwo, Z.N., and M.A. Tabatabai. 1999. Aspartase activity in soils: effects of trace elements and relationships to other amidohydrolases. *Soil Bio. Biochem.* 31:213–219.

Sloan, J.J., R.H. Dowdy, S.J. Balogh, and E. Nater. 2001. Distribution of mercury in soil and its concentration in runoff from a biosolids-amended agricultural watershed. *J. Environ. Qual.* 30:2173–2179.

Tazisong, I.A., and Z.N. Senwo. 2009. Mercury concentration and distribution in soils impacted by long-term applied broiler litter. *Bull. Environ. Contam. Toxicol.* 83:291–294.

Terra, J.A., J.N. Shaw, D.W. Reeves, R.L. Raper, E. van Santen, E.B. Schwab, and P.L. Mask. 2006. Soil management and landscape variability affects field-scale cotton productivity. *Soil Sci. Soc. Am. J.* 70:98–107.

Tessier-Lavigne, M., P. Mobbs, and D. Attwell. 1985. Lead and mercury toxicity and the rod light response. Invest. *Ophthalmol. Vis. Sci.* 26:1117–1123.

Trasande, L., P.J. Landrigan, and C. Schechter. 2005. Public health and economic consequences of methyl toxicity to the developing brain. *Environ. Health Prosp.* 113:590–596.

Voegborlo, R.B., A. Matsuyama, A.A. Adimado, and H. Akagi. 2010. Head hair total mercury and methylmercury levels in some Ghanian individuals for the estimation of their exposure to mercury: Preliminary studies. *Bull. Environ. Contam. Toxicol.* 84:34–38.

Wallschläger, D., M.V.M. Desai, M. Spengler, and R.D. Wilken. 1998. Mercury speciation in floodplain soils and sediments along a contaminated river transect. *J. Environ. Qual.* 27:1034–1044.

Yang, Y.K., L. Liang, and D.Y. Wang. 2008. Effect of dissolved organic matter on adsorption and desorption of mercury by soils. *J. Environ. Sci.* 20:1097–1102.

Zheljazkov, V.D., and P.R. Warman. 2002. Comparison of three digestion methods for the recovery of 17 plant essential nutrients and trace elements from six composts. *Compost Sci. Util.* 10:197–203.

Zheng, Y.M., Y.R. Liu, H.Q. Hu, and J.Z. He. 2008. Mercury in soils of three agricultural experimental stations with long-term fertilization in China. *Chemosphere* 72:1274–1278.

INDEX

D

E

F

J

K

L

Q

T

U